Enhancing Nutrient Use Efficiency

Enhancing Nutrient Use Efficiency

Concepts, Methods and Management Interventions

Editors

Kulasekaran Ramesh
Ashis Kumar Biswas
Brij Lal Lakaria
Sanjay Srivastava
Ashok Kumar Patra

ICAR-Indian Institute of Soil Science
Nabi Bagh, Bhopal – 462038, Madhya Pradesh (India)

NEW INDIA PUBLISHING AGENCY
New Delhi – 110 034

Citation: Ramesh, K., Biswas, A.K., Lakaria, B.L., Srivastava, S. and Patra, A.K. 2017. Enhancing nutrient use efficiency: Concepts, methods and management interventions. New India Publishing Agency, New Delhi, 482 pages

NEW INDIA PUBLISHING AGENCY
101, Vikas Surya Plaza, CU Block, LSC Market
Pitam Pura, New Delhi 110 034, India
Phone: + 91 (11)27 34 17 17 Fax: + 91(11) 27 34 16 16
Email: info@nipabooks.com
Web: www.nipabooks.com

Feedback at feedbacks@nipabooks.com

ISBN: 978-93-85516-73-3

Composed, Designed and Printed in India

त्रिलोचन महापात्र, पीएच.डी.
एफ एन ए, एफ एन ए एस सी, एफ एन ए ए एस
सचिव एवं महानिदेशक
TRILOCHAN MOHAPATRA, Ph.D.
FNA, FNASc, FNAAS
SECRETARY & DIRECTOR GENERAL

भारत सरकार
कृषि अनुसंधान और शिक्षा विभाग एवं
भारतीय कृषि अनुसंधान परिषद
कृषि एवं किसान कल्याण मंत्रालय, कृषि भवन, नई दिल्ली 110 001
GOVERNMENT OF INDIA
DEPARTMENT OF AGRICULTURAL RESEARCH & EDUCATION
AND
INDIAN COUNCIL OF AGRICULTURAL RESEARCH
MINISTRY OF AGRICULTURE AND FARMERS WELFARE
KRISHI BHAVAN, NEW DELHI 110 001
Tel.: 23382629; 23386711 Fax: 91-11-23384773
E-mail: dg.icar@nic.in

Foreword

Nutrient use efficiency is a measure of how well plants use the available mineral nutrients applied to the soil system for increasing crop productivity. In spite of the best research and development efforts, use efficiency of fertilizers continues to be very low. Improvement in nutrient use efficiency of crops is needed to sustain crop productivity in the country. Though intensive cultivation with high fertilizer inputs has enhanced the production of crops, it has put enormous pressure on the soil resource, posing a threat to the sustainability of agricultural production systems. This warrants a thorough understanding of basic concepts of nutrient use efficiency, methods to be followed and practices in various crops and cropping systems.

The book entitled "Enhancing Nutrient Use Efficiency: Concepts, Methods, and Management Interventions " has 27 chapters that provide a comprehensive understanding of the subject with topics related to improvement of nutrient use efficiency of crops and cropping systems. I am sure, this book will be useful to researchers, farmers, teachers and students to understand the nutrient flow in the soil-plant system to enable agriculture to become input-use efficient and profitable particularly in the Indian soil situation.

I congratulate the editors for synthesizing this volume in academic spirit.

(T. MOHAPATRA)

Dated the 8th March, 2017
New Delhi

Preface

Nutrient use efficiency is a measure of how efficiently plants use the available mineral nutrients to produce economic produce. Improving nutrient use efficiency is very much essential under current scenario besides an understanding of the nutrient dynamics in soil-plant system."Enhancing nutrient use efficiency: Concepts, methods, and management interventions" is a comprehensive book which has a compilation of topics related to enhancement of nutrient use efficiency of various crops and cropping systems. Although it depends on the ability to plants to take up the nutrients from the soil, management practices have a key role to intervene the use efficiency. This book has 27 chapters written by eminent researchers in the field and addresses multifaceted approaches to enhance the nutrient use efficiency. Improving nutrient use efficiency is a prerequisite to reducing production costs in the wake of escalating cost of agricultural inputs in farming besides minimizing environmental contamination. Soil physical management is the foremost strategy to enhance the nutrient use efficiency. Hand held devices are popular among the farmers for nitrogen management in rice in several parts of the country, besides slow release nitrogenous fertilizers. Experiences from Long term fertilizer experiments and soil test crop response correlation are of immense use to understand the nutrient dynamics in soil and in turn designing practices for higher use efficiency. Besides describing the concepts and methods of nutrient use efficiency, management practices for dry land crops and cropping systems, cereal based cropping systems, rainfed pulses based cropping systems, soybean based cropping systems, sugarcane based cropping systems, cotton based cropping systems, tobacoo and oil seed based cropping systems, rapeseed-mustard based cropping systems and spices based cropping systems are also dealt in this book and would serve as a resource guide for enhancing nutrient use efficiency in various crops and cropping systems.

We place on record our sincere thanks to all the authors for providing updated information in the field of nutrient use efficiency in various crops and cropping systems. We wish to express our deep sense of gratitude to Dr Trilochan

Mohapatra, DG (ICAR) and Secretary (DARE) for his constant encouragement to us. We thankfully acknowledge Dr SK Chaudhari, ADG (Soil and water Management) and Dr S Baskar, ADG (Agronomy, Agroforestry and Climate change) for the continued support to us.

We do hope, this book will be of immense use to scientists, farmers, students and policymakers for enhancing nutrient use efficiency in various crops and cropping systems.

K Ramesh
AK Biswas
BL Lakaria
S Srivastava
AK Patra

List of Contributors

Adhikari, T.
Principal Scientist (Soil Science)
ICAR-Indian Institute of Soil Science
Nabi Bagh, Berasia Road
Bhopal – 462038, Madhya Pradesh

Biswas, A.K.
Principal Scientist & HOD (Soil Chemistry & Fertility)
ICAR-Indian Institute of Soil Science
Nabi Bagh, Berasia Road
Bhopal – 462038, Madhya Pradesh

Bharati, K.
Senior Scientist (Microbiology)
ICAR-Indian Institute of Soil Science
Nabi Bagh, Berasia Road
Bhopal – 462038, Madhya Pradesh

Blaise, D.
Principal Scientist (Agronomy) and Head (Crop Production)
ICAR-Central Institute for Cotton Research
Nagpur – 440010, Maharashtra

Bodake, P.S.
Professor (Agronomy)
Zonal Agricultural Research Station
Igatpuri, M.P.K.V., Rahuri
Maharashtra

Billore, S.D.
Principal Scientist (Agronomy)
ICAR- Indian Institute of Soybean Research
Indore - 452 001, Madhya Pradesh

Brahmanand, P.S.
Principal Scientist (Soil Science)
ICAR-Indian Institute of Water Management
Chandrasekharpur
Bhubaneswar - 751023, Odisha

Chaudhary, R.S.
Principal Scientist (Soil Physics) and Head
ICAR-Indian Institute of Soil Science
Nabi Bagh, Berasia Road
Bhopal – 462038, Madhya Pradesh

Dey, P.
Principal Scientist & PC (STCR)
ICAR-Indian Institute of Soil Science
Nabi Bagh, Berasia Road
Bhopal – 462038, Madhya Pradesh

Dinesh, R,
Principal Scientist (Soil Science)
ICAR-Indian Institute of Spices Research
Kozhikode - 673 012, Kerala

Elanchezhian, R.
Principal Scientist (Plant Physiology)
ICAR-Indian Institute of Soil Science
Nabi Bagh, Berasia Road
Bhopal – 462038, Madhya Pradesh

Gudadhe, N.
Assistant Professor (Agronomy)
Navsari Agricultural University
Navsari – 396445, Gujarat

Hati, K.M.
Principal Scientist (Soil Physics)
ICAR-Indian Institute of Soil Science
Nabi Bagh, Berasia Road,
Bhopal – 462038, Madhya Pradesh

Hamza, S.
Principal Scientist (Agronomy)
ICAR-Indian Institute of Spices Research
Kozhikode - 673 012
Kerala

Imade, S.R.
Assistant Professor (Agronomy)
Navsari Agricultural University
Navsari – 39645, Gujarat

Jha, P.
Senior Scientist (Soil Science)
ICAR-Indian Institute of Soil Science,
Nabi Bagh, Berasia Road
Bhopal – 462038, Madhya Pradesh

Kandpal, B.K.
Joint Director
ICAR Research Complex for North East Region
Tripura Centre, Lembucherra
Tripura West
Lembucherra, Tripura – 799210

Kandiannan, K,
Principal Scientist (Agronomy)
ICAR-Indian Institute of Spices Research
Kozhikode – 673012
Kerala

Kundu, S.
Principal Scientist (Soil Science)
ICAR-Central Research Institute for
Dryland Agriculture
Santoshnagar
Saidabad (P.O.), Hyderabad - 500 059
Andhra Pradesh

Lakaria, B.L.
Principal Scientist (Soil Science)
ICAR-Indian Institute of Soil Science
Nabi Bagh, Berasia Road
Bhopal – 462038, Madhya Pradesh

Meena, B.P.
Scientist (Agronomy)
ICAR-Indian Institute of Soil Science
Nabi Bagh, Berasia Road
Bhopal – 462038, Madhya Pradesh

Mohanty, M.
Senior Scientist (Soil Physics)
ICAR-Indian Institute of Soil Science
Nabi Bagh, Berasia Road
Bhopal – 462038, Madhya Pradesh

Mandal, K.G.
Principal Scientist (Agronomy)
ICAR-Indian Institute of Water Management
Chandrasekharpur, Bhubaneswar – 751023
Odisha

Mohanty, S.R.
Senior Scientist (Microbiology)
ICAR-Indian Institute of Soil Science
Nabi Bagh, Berasia Road
Bhopal – 462038, Madhya Pradesh

Neenu, S.
Scientist (Soil Science)
ICAR-Central Plantation Crops Research
Institute
Kasargod – 671124
Kerala

Patra, A.K.
Director
ICAR-Indian Institute of Soil Science,
Nabi Bagh, Berasia Road
Bhopal – 462038, Madhya Pradesh

Premi, O.P.
Senior Scientist (Agronomy)
ICAR-Directorate of Rapeseed-Mustard
Research
Sewar, Bharatpur
Rajasthan – 321303

Ramesh, K.
Principal Scientist (Agronomy)
ICAR-Indian Institute of Soil Science
Nabi Bagh, Berasia Road
Bhopal – 462038, Madhya Pradesh

Rajagopal, V.
Scientist (Soil Science)
ICAR-National Institute of Abiotic Stress
Management
Malegaon, Baramati – 413115, Pune
Maharashtra

Rashmi, I.
Scientist (Soil Science)
ICAR-Indian Institute of Soil Science
Nabi Bagh, Berasia Road
Bhopal – 462038, Madhya Pradesh

Ramana Rao, K.V.
Principal Scientist (Agrl. Engg)
ICAR-Central Institute of Agricultural Engineering
Nabi Bagh, Berasia Road
Bhopal 462038, Madhya Pradesh

Rathore, S.S.
Principal Scientist (Agronomy)
Division of Agronomy
ICAR-Indian Agricultural Research Institute
New Delhi – 110012

Rao, Subba, A.
Former Director
ICAR-Indian Institute of Soil Science
Nabi Bagh, Berasia Road
Bhopal – 462038, Madhya Pradesh

Reddy, K.S
Principal Scientist (Soil Science) and
Head (Division of Resource Management)
ICAR-Central Research Institute for Dryland Agriculture, Santoshnagar
Hyderabad - 500 059, Andhra Pradesh

Saha, R.
Principal Scientist (Soil Science)
ICAR- Central Research Institute for Jute and Allied Fibres
Barasat Road, Nilgunj
Barrackpore, West Bengal – 700120
Kolkata

Singh, A.B.
Principal Scientist (Biochemistry)
ICAR-Indian Institute of Soil Science
Nabi Bagh, Berasia Road
Bhopal – 462038, Madhya Pradesh

Singh, R.K.
Principal Scientist (Soil Physics)
ICAR-Indian Institute of Soil Science
Nabi Bagh, Berasia Road
Bhopal – 462038, Madhya Pradesh

Shanmugam, P.M.
Assistant Professor (Agronomy)
TNAU-Agricultural College and Research Institute, Navalur Kuttapattu
Trichirapalli – 620009, Tamil Nadu

Sangeetha, S.P.
Assistant Professor (Agronomy)
Tamil Nadu Agricultural University
Coimbatore
Tamil Nadu – 641003

Shinogi, K.C.
Scientist (Agricultural Extension)
ICAR-Indian Institute of Soil Science
Nabi Bagh, Berasia Road
Bhopal – 462038, Madhya Pradesh

Srivastava, S.
Principal Scientist (Soil Science)
ICAR-Indian Institute of Soil Science
Nabi Bagh, Berasia Road
Bhopal – 462038, Madhya Pradesh

Singh, M.
Principal Scientist & PC (LTFE)
ICAR-Indian Institute of Soil Science
Nabi Bagh, Berasia Road
Bhopal – 462038, Madhya Pradesh

Shekhawat, K
Scientist SS (Agronomy)
Division of Agronomy
ICAR-Indian Agricultural Research Institute
New Delhi – 110012

Somasundaram, J.
Senior Scientist (Soil Physics)
ICAR-Indian Institute of Soil Science
Nabi Bagh, Berasia Road
Bhopal – 462038, Madhya Pradesh

Srinivasan, V.
Principal Scientist (Soil Science)
ICAR-Indian Institute of Spices Research
Kozhikode – 673012
Kerala

Sharma, K.L.
Principal Scientist (Soil Science)
ICAR-Central Research Institute for Dryland Agriculture
Santoshnagar
Hyderabad – 500059
Andhra Pradesh

Srinivasarao, Ch
Director
ICAR-Central Research Institute for Dryland Agriculture
Santoshnagar, Hyderabad – 500059
Andhra Pradesh

Shukla, A.K.
Project Coordinator (MSPE)
ICAR-Indian Institute of Soil Science
Nabi Bagh, Berasia Road
Bhopal – 462038, Madhya Pradesh

Sinha, N.K.
Scientist (Soil Physics)
ICAR-Indian Institute of Soil Science
Nabi Bagh, Berasia Road
Bhopal – 462038, Madhya Pradesh

Thanki, J.D.
Assistant Professor (Agronomy)
Navsari Agricultural University
Navsari – 39645, Gujarat

Thakur, J.K.
Scientist (Agricultural Microbiology)
ICAR-Indian Institute of Soil Science
Nabi Bagh, Berasia Road
Bhopal – 462038, Madhya Pradesh

Thankamani, C.K.
Principal Scientist (Agronomy)
ICAR-Indian Institute of Spices Research
Kozhikode – 673012
Kerala

Upadhyay, P.K.
Scientist (Agronomy)
Division of Agronomy
ICAR-Indian Agricultural Research Institute
New Delhi – 110012

Vishwakarama, A.K.
Senior Scientist (Agronomy)
ICAR-Indian Institute of Soil Science
Nabi Bagh, Berasia Road
Bhopal – 462038, Madhya Pradesh

Wanjari, R.H
Senior Scientist (Agronomy)
ICAR-Indian Institute of Soil Science
Nabi Bagh, Berasia Road
Bhopal – 462038, Madhya Pradesh

Contents

Management Interventions and Socio-economic Issues

Enhancing Nutrient Use Efficiency in Crops and Cropping Systems

Concepts and Methods of Nutrient Use Efficiency

1

Enhancing Nutrient Use Efficiency, pp. 1-18
Editors: K. Ramesh, A.K. Biswas, B.L. Lakaria, S. Srivastava and A.K. Patra

Enhancing Nitrogen Use Efficiency - Challenges and Options

A.K. Biswas, Pramod Jha and A. Subba Rao

ICAR-Indian Institute of Soil Science, Nabi Bagh, Bhopal – 462038, India

Introduction

To meet the food needs of the burgeoning population, India will need to produce 300 million tonnes of food grains by 2020. At present more than 75% of the total food grains produced in the country are of rice and wheat. Use of nitrogenous fertilizers has contributed much to the remarkable increase in production of rice and wheat in India that has occurred during the past three decades. During the last half-decade or so while fertilizer N consumption is touching new heights, the production of both rice and wheat is showing a trend of plateauing. In fact, fertilizer N efficiency of food grain production expressed as partial factor productivity of N (PFP_N) has been decreasing exponentially since 1965. The PFP_N is an aggregate efficiency index that includes contributions to crop yield derived from uptake of indigenous soil N, fertilizer N uptake efficiency, and the efficiency with which N acquired by the plant is converted to grain yield. A decrease in PFP_N occurs as farmers move yields higher along a fixed N response function, unless other factors shift the response function up. In other words, an initial decline in PFP_N is an expected consequence of the adoption of N fertilizers by farmers and not necessarily bad within a system's context.

Nitrogen – The Most Enigmatic Element

Billions of people today owe their lives to a single discovery now century old. In 1909 German chemist Fritz Haber of the University of Karlsruhe figured out a way to transform nitrogen gas which is abundant in the atmosphere but nonreactive and thus unavailable to most living organisms into ammonia, the active ingredient in synthetic fertilizer. The world's ability to grow food exploded

20 years later, when fellow German scientist Carl Bosch developed a scheme for implementing Haber's idea on an industrial scale. Over the ensuing decades new factories transformed ton after ton of industrial ammonia into fertilizer, and today the Haber- Bosch invention commands wide respect as one of the most significant boons to public in human history. As a pillar of the green revolution, synthetic fertilizer enabled farmers to transform infertile lands into fertile fields and to grow crop after crop in the same soil without waiting for nutrients to regenerate naturally.

But this good news for humanity has come at a high price. Most of the reactive nitrogen made for the purpose of fertilizer and, to a lesser extent, as a by-product of the fossil-fuel combustion that powers our cars and industries does not end up in the food we eat. Rather it migrates into the atmosphere, rivers and oceans. Scientists have long cited reactive nitrogen for creating harmful algal blooms, coastal dead zones and ozone pollution. But recent research adds biodiversity loss and global warming to nitrogen's rap sheet, as well as indications that it may elevate the incidence of several nasty human diseases.

At the same time, fertilizer is and should be a leading tool for developing a reliable food supply in sub-Saharan Africa and other malnourished regions of the world. But the international community must come together to find ways to better manage its use and mitigate its consequences worldwide. The solutions are not always simple, but nor are they beyond our reach.

Nutrient Use Efficiency

Terminology

Efficiencies are calculated as ratios of inputs to outputs in a system. A recent scientific review identified 18 different forms of nutrient use efficiency. Four of them are very commonly used, but are often misinterpreted.

1. **Partial factor productivity, PFP** (crop yield per unit of nutrient applied) answers the question: "How productive is this cropping system in comparison to its nutrient input?"
2. **Agronomic efficiency, AE** (yield increase per unit of nutrient applied) answers a more direct question: "How much productivity improvement was gained by the use of this nutrient?"
3. **Physiological efficiency, PE** (yield increase per unit of additional nutrient uptake) answers the question: "How much productivity was gained by above-ground crop uptake?"

4. **Recovery efficiency, RE** (increase in above-ground crop uptake per unit of nutrient applied) answers the question: "How much of the nutrient applied did the plant take up?"

In the short term, all four of these ratios increase as rates of fertilizer application are decreased, even to levels well below the economic optimum. This is because, the crop response to applied nutrients typically follows a diminishing return function as yields approach the potential limit. This might cause one to falsely conclude that the lowest fertilizer rate results in the most efficient cropping system. This is untrue. Cropping systems depend on multiple inputs, including land, labor, seed, plant protection, capital, and more. At the rate where the net return to the use of a plant nutrient peaks, it is making its best contribution to increasing the efficiency of all other inputs involved. This most economic rate is also often associated with minimal nutrient loss. Best management practices ensure effective use of fertilizers in improving the efficiency of all inputs used in cropping systems. The goal of their use is to apply the most appropriate sources at the right rate, time, and place. Apparent nutrient efficiency of applied fertilizer is commonly calculated by the difference method, but tracer techniques allow us to determine recovery efficiency directly. For both techniques, NUE measurements can be based on total nutrients in the grain or on total nutrients in grain plus straw. Due to paucity of fund and analytical facilities to use tracer techniques, most Indian data reported apparent NUE. However, the apparent efficiency data assumed to become nearer to direct measurements once the system is near steady-state with respect to its nutrient content and input via external addition, wet and dry deposition and biological fixation *etc*.

Current status of nitrogen use efficiency

A recent review of worldwide data on N use efficiency for cereal crops from researcher-managed experimental plots reported that single-year fertilizer N recovery efficiencies averaged 65% for corn, 57% for wheat, and 46% for rice. However, experimental plots do not accurately reflect the efficiencies obtainable on-farm.

Differences in the scale of farming operations and management practices (i.e. tillage, seeding, weed and pest control, irrigation, harvesting) usually result in lower nutrient use efficiency. Nitrogen recovery in crops grown by farmers rarely exceeds 50% and is often much lower. A review of best available information suggests average N recovery efficiency for fields managed by farmers ranges from about 20% to 30% under rainfed conditions and 30% to 40% under irrigated conditions. N recovery averaged 31% for irrigated rice grown by Asian farmers and 40% for rice under field specific management. In India, N recovery averaged 18% for wheat grown under poor weather conditions, but 49% when grown under good weather conditions

While most of the focus on nutrient efficiency is on N, P efficiency is also of interest because it is one of the least available and least mobile mineral nutrient. First year recovery of applied fertilizer P ranges from less than 10% to as high as 30%. However, because fertilizer P is considered immobile in the soil and reaction (fixation and/or precipitation) with other soil minerals is relatively slow, long-term recovery of P by subsequent crops can be much higher. There is little information available about potassium (K) use efficiency. However, it is generally considered to have a higher use efficiency than N and P because it is immobile in most soils and is not subject to the gaseous losses that N is or the fixation reactions that affect P. First year recovery of applied K can range from 20% to 60%.

How realistic NUE values are?

Nitrogen use efficiency can mean different things to different people and is easily misunderstood. For example, a typical irrigated soil under rice-wheat cropping system in the Indo-Gangetic plains contains about 2000 kg N ha^{-1} in the top 30 cm of soil where roots derive majority of N supply. The amount of N derived from indigenous resources during a single cropping cycle typically ranges from 30-100 kg N ha^{-1} that represents only 1.5 to 5% of total soil N. Although small in size, the indigenous N supply has a very high N-fertilizer substitution value because of the relatively low RE_N from applied N fertilizer. Further, as C/N ratio of soil organic matter is relatively constant, changes in soil C balance introduced by management practices including fertilizer use affect the soil N balance. The overall fertilizer nitrogen use efficiency can thus be increased by achieving greater RE_N, by reducing the amount of N lost from soil organic and inorganic pools, or both. When soil-N content is increasing, the amount of sequestered N contributes to higher nitrogen use efficiency and the amount of sequestered N derived from applied N contributes to a higher $RE_{N.}$ Conversely, any decrease in soil N stocks will reduce overall nitrogen use efficiency and $RE_{N.}$

Reported estimates of fertilizer N-use efficiencies vary widely, as do reporting methods themselves. Foremost, the amount of fertilizer N recovered by the crop depends on the experimental approaches and equations used to calculate N recovery. Researchers defined N-use efficiency in many different ways and contexts. Since the definition can be based on economic, agronomic or physiological principles, the estimates have different expressions. Moreover, estimates of fertilizer-N recovery can be based on either the amount of N in the grain or the total N in the aboveground biomass. Fertilizer N accumulated in roots is almost never included as part of fertilizer N recovered by the crop because it remains in the soil and there are difficulties in its measurement. Additionally, the uptake of fertilizer N by the crop following the fertilized crop is rarely included in fertilizer-use efficiency studies. Both contribute to underestimation of true fertilizer N-use efficiency. Although inclusion of zero-N-fertilized treatment is required to calculate fertilizer N-use efficiency, approaches based on economic principles or regression analysis often do not include zero-N fertilizer plots. Clearly, when fertillzer N-use efficiencies are reported, it is extremely important that the method used to calculate NUE is not misunderstood.

Fertilizer N recovery efficiency (RE_N) across regions and three cereal crops varied over a large range: 0.1 to 0.7 kg N taken up per kg N taken up per kg N applied (20-90%) based on grain plus straw N (RE_{NT}; estimates falling between 10th and 90th percentiles). As expected, many of the studies that obtained large values of RE_N used lower N rates. The average estimates based on grain N (RE_{NG}) were similar among the three crops and were the highest (45%) in maize and lowest (34%) in wheat. The grain plus straw-based estimates (RE_{NT}) were greater by 20, 8, and 23% in maize, rice and wheat, respectively. A similar trend was observed in estimates based on $RE15_N$ for grain, except that the overall efficiency was lower by 11% compared with the N-difference method. The overall average RE_{NT} among regions and crops was 55% based on the N-difference method and 44% based on the ^{15}N-dilution method.

It is important to note that almost all data available in the literature come exclusively from experiments conducted at research stations. Rarely has NUE been measured in fields that are fully managed by farmers themselves. The average RE_{NT} estimates (0.31 kg N taken up per kg N applied) from on-farm assessments were smaller by 25% than the average reported researcher-managed plots in farmers' fields. Similar on-farm studies in maize and wheat seem to be nonexistent. Based on the best available information, it is likely that across the three major cereals, RE_{NT} ranging from 20 to 30% to 30 to 40% are occurring in rainfed and irrigated conditions, respectively. An RE_{NT} exceeding 40% is expected to occur in response to improved N management practiced. The bulk of N-recovery data collected in most studies globally are based on the N-difference method. More precise estimates of RE can be derived with isotope dilution methods. Ladha *et al.* (2005) compiled data on ^{15}N recovery by cereal crops and found that average RE ^{15}N was 44% in the first growing season and total recovery of ^{15}N fertilizer in the first and five subsequent crops was only around 50%. Assuming that amount of ^{15}N in the roots becomes negligible in the sixth growing season, the remaining 50% of the ^{15}N fertilizer would have either become part of the soil organic matter pool or was lost from the cropping system.

Factors affecting N use efficiency

Fertilizer N use efficiency is controlled by crop demand of N, supply of N from the soil and fertilizers and N losses from soil-plant system. Nitrogen needs of crop plants are met by applying fertilizer N and net N mineralization from soil organic matter. Fertilizer N is applied in forms readily available to plants but mineralization of N is controlled by water, temperature and aeration. Once these factors become optimum, amount of N that is mineralized depends upon the quality and quantity of organic matter in the soil. Also a strong interaction between C and N dynamics is obvious because N transformations are driven largely by biological activity. High rates of net N mineralization can result in dilution of fertilizer N. But if crop demand for N and the amount of fertilizer N remain constant, an increase in net N mineralization will lead to a decrease in observed $RE_{N.}$ Relative magnitude of different N loss mechanisms will depend upon soil, weather, fertilizer and crop management. On an overall basis, climate exerts the strongest effect on the amount and pathways of N losses.

Challenges

Decisions regarding improvements in fertilizer N use efficiency will begin at the field scale where farmers need to deal with the variability in soils, climates, and cropping patterns. As there exists a large fertilizer-N substitution value of soil N, it is important to know the amount and temporal variations of the indigenous N supply during crop growth for determining the optimal timing and amount of fertilizer N applications.

A great deal of research has been carried out during the past 50 years to improve fertilizer N use efficiency by trying to develop better fertilizers or improved N management practices based mainly on a better synchronization between N uptake by crops and supply through fertilizers. Strategies based on applying N at the right rate, right time, and in the right place have already been developed and are in use. Recent literature on improving RE_N has emphasized on achieving greater synchrony between crop N demand and the N supply from all sources throughout the growing season. This approach explicitly recognizes the need to efficiently utilize both indigenous and applied N, because losses of N via different mechanisms increase in proportion to the amount of available N present in the soil at any given time.

Options

Most of the fertilizer-N is lost during the year of application. Consequently, N and crop management must be fine-tuned in the cropping season in which N is applied. Two broad categories of concepts and tools have been developed to increase nitrogen use efficiency. Those in the first category include genetic improvement and management factors that remove restrictions on crop growth and enhance crop N demand and uptake. Management options that influence the availability of soil and fertilizer-N for plant uptake come in the second category. These include site-specific N application rates to account for differences in within-field variation in soil N supply capacity (in large fields), field specific N application rates in small scale production fields, remote sensing or canopy N status sensors to quantify real-time crop N status, better capabilities to predict soil N supply capacity, controlled release fertilizers and fertigation.

Optimizing Nitrogen Use Efficiency

Strategies to improve nitrogen use efficiency can be grouped into two:

Product strategy: Coated fertilizers, slow-release fertilizers, urease and nitrification inhibitors.

Fertilizer management strategy: Split application, N rate based on plant analysis and variable rate (time and space, precision farming) *etc.*

Product strategy

Numerous products supplying plant nutrients have emerged, offering a variety of nutrient contents, physical forms, and other properties to meet individual needs. Finding new sources of fertilizers and modifiers to minimize losses, thereby achieve higher use efficiency is a challenging task. Much of the work has been done on N with major source as urea containing high N (46%) and its proneness to losses. The complexities of N management occur due to its solubility, modality and vulnerability to denitrification. The approaches have been to manipulate the granule size variations, coatings with neem or coal tar modifiers or additives to control the nutrient release rate. Table 1 provides the details on slow-release nitrogenous fertilizers (SRNFs) being manufactured, marketed and used in USA, Western Europe and Japan. Farmers' use of SRNFs has almost doubled over the past ten years across the globe, but it still accounts for only 0.15% of the total fertilizer N used. The main reason for its limited use is its high cost, which could be almost 3 to 10 times the cost of conventional fertilizers.

Table 1: Consumption ('000 tonnes) of slow/controlled release fertilizers in 1995-96

Slow-release Fertilizer	USA	Western Europe	Japan	Total	% Total SRNF
Urea-form (UF)	190	. 30	5	225	40
Sulphur coated urea (SCU) /SCU+P	100	2	6	108	19
Polymer coated urea (PCU) /PCU NPK'$_S$	45	20	72	137	24
Isobutylidene-diurea (IBDU)/Crotylidene-diurea (CDU)	14	35	33	82	15
Others	7	-	3	10	2
Total	356	87	119	562	100

Application of controlled release fertilizers (CRFs) is an approach for minimizing non-point contamination in agriculture (Table 2). The CRFs have higher N-use efficiency, thus reducing N loss through leaching and volatilization while keeping higher yields. New gel-based CRFs were developed by mixing and processing N, P and K fertilizers with natural and semi natural organic materials and inorganic materials. These gel-based fertilizers were earlier produced, had lower price than coated CRFs, and pressed better physical and chemical properties. The gel-based CRFs showed significant positive influence on agronomical, physiological and biochemical characteristics of maize plants. These increased dry biological yield of maize by 26.8-42.3%, and improved N-use efficiency by 17.0-31.7%, P-use efficiency by 8.0-16.0% and K-use efficiency by 4.6-18.3%. Besides, the nutrients (N, P and K) in gel-based fertilizers were leached more slowly into the soil than common fertilizers.

Table 2: Slow release forms of urea to improve N use efficiency

Fertilizer forms	Example
Coated with inert material	Urea coated with polymer, lac, gypsum, neem cake, sulphur, and rock phosphate
Enlargement of the granule	Urea supergranule, granular urea
Limited solubility forms of urea	Urea form, Oxamide, Urea-Z
Coated with urease inhibitors	Hydroquinone, phenyl phosphorodiamidate (PPD)
Coated with nitrification inhibitors	Nitrapyrin. AM (acetylene. 2-amino-4-chloro-6-methyl-peyrimidine), DCD (dicyandiamide), ATC (4-amino 1,2,4-triazole).encapsulated Ca-carbide, neem cake, karanj cake, DMPP (3-4-dimethylpyrazole phosphate)

IARI's urea coating technology employing neem oil emulsion needing 0.5-1.0 kg neem oil per tonne of urea was found superior to prilled urea. The use of nitrification inhibitors might contribute to increased NUE or apparent N recovery efficiency. Maintenance of more NH_4^+available in the soil might also increase P absorption, and therefore increase P-use efficiency (PUE). The use of ammonium N source with the nitrification inhibitor 3, 4-dimetyipyrazole phosphate (DMPP) shows promise to increase NUE and PUE. Large granular forms of N and P fertilizer proved better than powdered and prilled forms for increasing nutrient uptake and grain yield of rice in irrigated lowland. Compacting phospho-gypsum (PG), diammonium phosphate (DAP), $ZnSO_4$ and KCl separately with urea slowed down urea hydrolysis and reduced NH_3 volatilization loss. The new product increased the rice yield and NUE as compared to that obtained with prilled urea. Coating P fertilizer could limit the contact of applied P with soil resulting in early season deficiencies for crops like wheat. The development of thin polymer coating has improved the opportunity to coat fertilizer granules and increased the predictability of time of availability of nutrients from the controlled release product. At IISS, the study showed that urease inhibitor (UI) or urease inhibitor-nitrification inhibitor (NI) coated urea applied at 100% rate and nitrification coated urea applied at 80% rate improved apparent N recovery in maize (Table 3).

Table 3: Effect of UI and NI coated urea and plain urea on nutrient uptake by maize.

S. No.	Treatment	Nutrient uptake (Grain + Stover) (kg ha^{-1})			Apparent N Recovery (%)
		N	P	K	
Tl	Control (no urea)	70.1	13.6	82.0	-
T2	Urea at 100% rate	117.1	19.3	117.3	47
T3	Ul coated urea at 100% rate	131.7	21.0	132.1	62
T4	UI and NI coated urea at 100% rate	136.8	21.5	138.2	67
T5	NI coated urea at 80% rate	105.6	17.6	106.9	36
T6	UI and NI coated urea at 80% rate	113.5	19.2	113.5	44

Economics of SRFNs

There are two options at the moment: (1) To increase the use of conventional N fertilizers with a recovery efficiency (REn) of 30-40 per cent or (2) To produce N fertilizres with a higher REn of 50-80 per cent. As regards the second option, the available literature suggests that this can be partly achieved by blending the conventional N fertilizres with nitrification inhibitors, urease inhibitors and urea super granules/briquettes. The real solution, however, lies in developing slow/ controlled release N fertilizres (SRNFs) that can release N as per needs of the crop so as to achieve synchrony between crop demand and supply of nutrients. As regards transplanted rice the crop uptake of N is about 1.0 to 1.5 kg N $ha^{-1}day^{-1}$

Technology is now available to make such fertilizers. The only barrier is the high cost of their production. In USA the cost of 1 kg N was about US$ 0.66 for urea, US$ 1.50 to 1.58 for UF (38-40 per cent N) and US$ 2.90 to 3.55 for IBDU. PCU is the costliest SRNF and its' cost may be 4 to 8 times depending upon the polymer and process used. Some saving in cost of crop production is possible with SRNFs because they require a single application as against 2 or 3 applications with conventional N fertilizers. The increased cost of SRNFs should, however, be also examined from the viewpoint of reduction in green house gases (GHG), which are a major environmental concern. The Kyoto protocol has provided a framework for an International Carbon Market for GHG reduction activities. The fertilizer industry has already put up 63 projects for Certified Emission Reduction (CRM) for consideration to UN Framework Convention on Climate Change (UNFCC), out of which 46 projects are in Asia (23 in China and 21 in India). It is high time that SRNF manufacturing companies take some advantages of Carbon Credit Market and submit some projects to UNFCC to produce eco- friendly SRFNs, which are the needs of the time.

Scope of nanotechnology in slow-release fertilizers

Nanotech materials (scale below 100 microns) are in development for the slow release and efficient dosage of water and plant nutrients for increasing the efficiency of nutrients and water. These inexpensive nanotech applications to increase soil fertility and crop production would be a major growth in agriculture in developing countries in near future. The nanoporous zeolites are being utilized for slow release and efficient dosage of water and fertilizer for plants. In China, different nanoparticle (clay-polyester, humus-polyester and plastic-starch) have been tried for slow release of N to wheat. The increase in yield due to clay and plastic (nanomaterial coating) was around 4.5% over chemical fertilizer application apart from saving through leaching losses of N fertilizer. There is a great scope in Indian agriculture for improving fertilizer-use efficiency of major crops through nanotechnology.

Fertilizer management strategy

Applying nutrients at the right rate, right time, and in the right place as a best management practice (BMP) is the best bait for achieving optimum nutrient efficiency.

Right rate: Most crops are location and season specific depending on cultivar, management practices and climate *etc*., and so it is critical that realistic yield goals are established and that nutrients are applied to meet the target yield. Over- or under-application will result in reduced nutrient use efficiency or losses in yield and crop quality. Soil testing remains one of the most powerful tools available for determining the nutrient supplying capacity of the soil, but to be useful for making appropriate fertilizer recommendations good calibration data is also necessary.

Other techniques, such as omission plots, are proving useful in determining the amount of fertilizer required for attaining a yield target. In this method, N, P, and K are applied at sufficiently high rates to ensure that yield is not limited by an insufficient supply of the added nutrients. Target yield can be determined from plots with unlimited NPK. One nutrient is omitted from the plots to determine a nutrient-limited yield. For example, an N omission plot receives no N, but sufficient P and K fertilizer to ensure that those nutrients are not limiting yield. The difference in grain yield between a fully fertilized plot and an N omission plot is the deficit between the crop demand for N and indigenous supply of N, which must be met by fertilizers. Nutrients removed in crops are also an important consideration. Unless nutrients removed in harvested grain and crop residues are replaced, soil fertility will be depleted. In a just concluded experiment at IISS, Bhopal nutrient omission trials conducted on farmers' fields deficient in N, P, S and Zn revealed that the balanced application of all these deficient nutrients at recommended rates helped in higher apparent recovery of applied P, K and S by soybean as compared to nutrient missing plots (Table 4).

Table 4: Apparent recovery (AR) of nutrients under balanced fertilization by soybean

Nutrient	Apparent Recovery (%)
P	22
K	68
S	31

Right time: Greater synchrony between crop demand and nutrient supply is necessary to improve nutrient use efficiency, especially for N. Split applications of N during the growing season, rather than a single, large application prior to planting, are known to be effective in increasing N use efficiency. Tissue testing is a well-known method used to assess N status of growing crops, but other diagnostic tools are also available. Chlorophyll meters have proven useful in fine-tuning in-season N management and leaf color charts have been highly successful in guiding split N application in rice (Table 5) and now maize production in Asia. Precision farming technologies have introduced, and now commercialized, on-the-go N sensors that can be coupled with variable rate fertilizer applicators to automatically correct crop N deficiencies on a site-specific basis.

Table 5: Comparison of chlorophyll-meter-based N management with fixed schedule application either by broadcasting of pilled urea or deep placement of urea tablets in transplanted rice

Treatment	Rate (kg N ha[1])	Yield (t ha[1])	AE_N(kg-')
	1995-96 wet seasons		
Control	0	5.27 b[a]	-
Prilled urea	55	6.23 a	17b
Prilled urea	110	6.61 a	12b
Urea tablet	55	6.51a	22 ab
Urea tablet	110	6.44 a	11 b
SPAD[h]	30	6.19 a	30 a
	1996 dry season; rice cv. IR64		
Control	0	4.77 c	-
Prilled urea	90	6.60 ab	20 b
Urea tablet	55	6.42 b	30 a
Urea tablet	110	7.06 a	21 b
SPAD	60	6.54 ab	30 a
	1996 dry season; rice cv. IR64		
Control	0	4.40 d	-
Prilled urea	90	6.31b	21 be
Urea tablet	55	6.29 b	34 a
Urea tablet	110	6.99 a	24 b
SPAD	60	5.28 c	15c

Another approach to synchronize release of N from fertilizers with crop need is the use of N stabilizers and controlled release fertilizers. Nitrogen stabilizers (e.g. nitrapyrin, DCD [dicyandiamide], NBPT [n-butylthiophosphoric triamide]) inhibit nitrification or urease activity, thereby slowing the conversion of the fertilizer to nitrate. When soil and environmental conditions are favorable for nitrate losses, treatment with a stabilizer will often increase fertilizer N efficiency. Controlled-release fertilizers can be grouped into compounds of low solubility and coated water-soluble fertilizers.

Most slow-release fertilizers are more expensive than water-soluble N fertilizers and have traditionally been used for high-value horticulture crops and turf grass. However, technology improvements have reduced manufacturing costs where controlled-release fertilizers are available for use in corn, wheat, and other commodity grains. The most promising for widespread agricultural use are polymer-coated products, which can be designed to release nutrients in a controlled manner. Nutrient release rates are controlled by manipulating the properties of the polymer coating and are generally predictable when average temperature and moisture conditions can be estimated.

Right place: Application method has always been critical in ensuring fertilizer nutrients are used efficiently. Determining the right placement is as important as determining the right application rate. Numerous placements are available, but most generally involve surface or sub-surface applications before or after planting. Prior to planting, nutrients can be broadcast (i.e. applied uniformly on the soil surface and may or may not be incorporated), applied as a band on the surface, or applied as a subsurface band, usually 5 to 20 cm deep. Applied at planting, nutrients can be banded with the seed, below the seed, or below and to the side of the seed. After planting, application is usually restricted to N and placement can be as a top-dressing. In general, nutrient recovery efficiency tends to be higher with banded applications because less contact with the soil lessens the opportunity for nutrient loss due to leaching or fixation reactions. Placement decisions depend on the crop and soil conditions, which interact to influence nutrient uptake and availability.

Role of organic matter and balanced fertilization

Controlling N release from organic sources depends on their nutrient content and quality, soil properties, and the environmental and management factors. The build-up of soil organic matter is required to increase the potential for N mineralization. The challenge in optimizing crop N uptake in organic and cover crop-based systems does not entirely rely on developing organic matter pools but is more important to influence the rate and timing of N mineralization. It was found that interactions among inputs (manure, cover crops and fertilizer) and soil organic matter influence the rate of soil N mineralization. It is also well known that N from many organic fertilizers often shows little effect on crop growth in the year of application because of the slow release characteristics of organically bound N. Nitrogen immobilization after application can occur, leading to enrichment of the soil N pool. This process increases the long-term efficiency of organic fertilizers.

Plant nutrients rarely work in isolation. Interactions among nutrients are important because a deficiency of one restricts the uptake and use of another. Numerous studies have demonstrated that interaction between N and other nutrients, primarily P and K, impact crop yields and N efficiency. For example, data from a large number of multi-location on-farm field experiments conducted in India show the importance of balanced fertilization in increasing crop yield and improving N efficiency (Table 6).

Table 6: Effect of balanced fertilization on agronomic efficiency of N (AE_N)

Crop	Control yield (t ha^{-1})	N applied (kg ha^{-1})	AE_N (kg grain kg^{-1}N)		Increase in AE_N (%)
			N alone	+PK	
Rice (wet season)	2.7	40	14	27	100
Rice (summer)	3.0	40	11	81	671
Wheat	1.5	40	11	20	85
Pearl millet	1.1	40	5	15	219
Maize	1.7	40	20	39	100
Sorghum	1.3	40	5	12	126
Sugarance	47	150	79	228	189

Adequate and balanced application of fertilizer nutrients is one of the most common practices for improving the efficiency of N fertilizer and is equally effective in both developing and developed countries. In a recent review based on 241 site-years of experiments in China, India, and North America, balanced fertilization with N, P and K increased first-year recoveries on an average of 54% compared to recoveries of only 21% where N was applied alone.

Tillage and weed management

Reducing tillage and optimizing N fertilization are important strategies for soil and water conservation and N-use efficiency for sustainable agriculture. Conservation tillage increased N, P and K uptake compared to minimum tillage. Soil, water and nutrients play an important role in increasing yield of crops in black soil of semi-arid tropics during post-rainy season. Tillage practices along with organic material further affected the moisture and nutrient availability to crops.

Precision Agriculture

With the introduction of geographical information systems (GIS) and global positioning systems (GPS), farmers can now refine nutrient recommendation domain to the site-specific conditions of each field. Managing a field for inherent soil variability can be done with GIS and GPS technologies without expensive and time-consuming grid-based techniques. Global positioning systems linked to yield monitors also provide field maps that are used to control variable-rate seeders and variable-rate chemical applicators. By managing a field using productivity zones and N treatments that account for spatial soil variability, N-use efficiency has been shown to increase when compared with conventional N application treatments. Although all farmers' fields can be considered to be heterogeneous for available N and yield, the degree of heterogeneity has to be large enough to recover the cost associated with site-specific application of N fertilizer. Precision farming technologies have now been developed to spatially vary N prescriptions within a field, based on various information sources (maps of soil properties, terrain attributes, remote sensing, yield maps). These practices include the timely and precise application of N fertilizer to meet plant needs varying across the landscape. Significant increases in N-use efficiency have taken place over the past 10 years using precision agriculture management practices. Achieving synchrony between N supply and crop demand without excess deficiency is the key to optimizing trade-offs among yield, profit, and environmental protection in both large scale systems in developing countries. The determination of rates and dates for N application must be more precise in this context. Models and diagnosis indicators have been developed to meet these requirements. A much better understanding of crop-soil-microclimate interactions on crop growth and nutrient demand, combined with better weather prediction, will be needed before site-specific farming management practices will be used widely.

Comparative Evaluation

Ladha *et al.* (2005) compared different strategies to improve nitrogen use efficiency on the basis of benefit cost ratio and limitations (Table 7). If a new technology leads to at least a small and consistent increase in crop yield with the same amount or less N applied, the resulting increase in profit is usually attractive enough for a farmer. With very high benefit -cost ratio and with no limitation, use of simple and inexpensive leaf colour chart assists farmers in applying N when the plant needs it. As use of leaf colour chart can adequately take care of N supply from all indigenous sources, it ensures significant increase in RE_N and reduced fertilizer N use. This tool is particularly useful for small to medium size farms in developing countries. Similarly, precision farming technologies based on gadgets like optical sensors have demonstrated that variable rate N-fertilizer application has the potential to significantly enhance nitrogen use efficiency by crops like rice and wheat.

Modern N management concepts usually involve a combination of anticipatory (before planting) and responsive (during the growing season) decisions. Improved synchrony can be achieved by more accurate N prescriptions based on the projected crop N demand and the levels of mineral and organic soil N, but also through improved rules for splitting of N applications according to phenological stages, by using decision aids to diagnose soil and plant N status during the growing season (models, sensors), or by using controlled-release fertilizers or inhibitors. Important prerequisites for the adoption of advanced N management technologies are that they must be simple, provide consistent and large enough gains in fertilizer N use efficiency, involve little extra time and be cost-effective.

Table 7: Comparative evaluation of tools and strategies for enhancing fertilizer N use Efficiency

Tools/Strategies	Benefit-Cost	Limitations
Site-specific N management	High	Has to be developed for every site infrastructure required.
Chlorophyll metre	High	Initial high cost
Leaf colour chart	Very high	None
Plant analysis	High	Facilities need to be developed
Controlled release fertilizers	Low	Lack of interest by industry
Nitrification inhibitors	Low	Lack of interest by industry
Fertilizer placement	High	Lack of equipment
Foliar N application	High	Lack of equipment, risk
Models and decision support systems	Medium	Tools are yet to be perfected
Remote sensing tools	Low	Needs fine-tuning
Precision farming technology	High	Needs fine-tuning
Breeding strategies	Medium	Limited research effort

Conclusion

Fertilizer N-use efficiency is a complex term with many components. To quantify NUE, the term most widely used is a ratio with output as the numerator and input as the denominator. Based on a large number of studies conducted in research fields in diverse agro-ecologies from across the globe, 44 to 55% (irrespective of ^{15}N or difference methods) of the applied fertilizer N was recovered in the first crop. When fertilizer-N recovery in soil and plant during succeeding seasons were considered, an additional recovery of 7% was estimated. Based on the best available information, it is likely that on-farm N recovery efficiency values ranging from 20 to 30% and 30 to 40% are occurring under rainfed and irrigated conditions, respectively. However, N recovery efficiency of 40% or more has been obtained with improved N management practices.

What is the maximum N recovery efficiency value attainable in cereal production? To achieve a substantial increase in N use efficiency will remain a huge challenge, albeit not beyond our reach. Fertilizer NUE is governed by three major factors: N uptake by the crop, N supply from soil and fertilizer, and N losses from soil-plant systems. The crop N requirement is the most important factor influencing NUE. Much research has been conducted during the post 50 years to improve NUE by trying to develop better fertilizers or improved N management practices, based mainly on a better synchronization between the supply and the uptake of N by the crop. There is significant potential to increase NUE in cereals. Many of the strategies needed to achieve such increases have already been developed. The use of an integrated crop management strategy comprising optimal soil, water, and crop management under good climatic conditions could attain a large NUE value. In addition, cereal genotypes with a large harvest index must be used to obtain high RE_N and PE_N because the harvest index is tightly linked to NUE.

Many approaches have been suggested for increasing NUE, for example, optimal time, rate, and methods of application for matching N supply with crop demand; the use of specially formulated forms of fertilizer, including those with urease and nitrification inhibitors; the integrated use of fertilizer, manures, and/or crop residues; and optimizing irrigation management. In addition, some modern tools such as precision farming technologies, simulation modeling, decision support systems, and resource-conserving technologies also help to improve NUE.

Nitrogenous fertilizers and their management will remain at forefront of measures to improve the global reactive N balance on both the short and long-term basis. Fertilizer N use efficiency is governed by N uptake by crops, N supply from soil and fertilizer and N losses from soil-plant system. Innovative fertilizer management has to integrate both preventive and field specific corrective N management strategies to increase the profitability and to ensure that there exists synchrony between crop N demand and supply of mineral N from soil reserves and fertilizer inputs. It will lead to maintenance of plant available N pool at the minimum size required to meet crop N requirements at each growth stage with little vulnerability to loss of N to environment.

Enhancement of NUE can be achieved by following the best crop management practices to achieve good yields, adopting right method and timing of fertilizer application, and practicing balanced NPK application, site specific integrated nutrient management *etc*. From the nutrient management point of view, the means and ways to enhanced NUE are: 1. Recycling of the plant nutrients by using more and more organic manures and crop residues, 2. Putting more reliance on biological nitrogen fixation and use of phosphate solubilising organisms 3. Reducing the rate of fertilizer application, 4. Using the right source of nutrients 5. Applying the nutrients at right time, 6. Placing the nutrients at right place, and 7. Balanced site-specific nutrient management (SSNM).

Selected References

Cassman KG, Dobermann A, Walters D. 2002. Agroecosystems, nitrogen-use efficiency, and nitrogen management. *Ambio* 31: 132-140.

FAO.2005. Fertilizer Use by Crop in India. Food and Agriculture Organization of the United Nations, Rome.

Galloway JN, Schlesinger WH, Levy H, Michaels A, Schnoor JL.1995. Nitrogen fixation atmospheric enhancement - environmental response. *Global Biogeochemical Cycles* 9: 235-252.

Jansson SL, Persson J.1982. Mineralization and immobilization of soil nitrogen. In: Nitrogen in Agricultural Soils (F.J Stevenson, Ed.), *Agronomy Monograph* 22:229-252 ASA, CSSA, and SSSA, Madison, WI, USA.

Ladha JK, Pathak H, Krupnik TJ, Six J, van Kessel C. 2005. Efficiency of fertilizer nitrogen in cereal production: Retrospect and Prospects. *Advances in Agronomy* 87: 85-156.

Mosier AR, Syers JK, Freney JR. 2004. Nitrogen fertilizer: An essential component of increases" food, feed, and fiber production In Agriculture and the Nitrogen Cycle Assessing the Impacts of Fertilizer Use on Food Production and the Environment (A R Mosier, J K Syers, and J R Freney, Eds), pp 3-15. SCOPE 65, Pans, France.

Townsend AR, Howarth RW.2010. Fixing the Global Nitrogen Problem. *Scientific American*. Pp 50-57.

2

Enhancing Nutrient Use Efficiency, pp. 19-29
Editors: K. Ramesh, A K. Biswas, B.L. Lakaria, S. Srivastava and A.K. Patra

Soil Testing: Basic Tool for Enhancing the Nutrient Use Efficiency

Sanjay Srivastava, P. Dey and Shinogi, K. C.

ICAR-Indian Institute of Soil Science, Nabi Bagh, Bhopal – 462 038, India

Introduction

It is well-known that soil testing is agronomically sound, beneficial and environmentally responsive tool for monitoring the nutrient as well as pollutant status of soils, and also for making precise fertilizer recommendations for various crops and cropping sequences. Soil testing with associated plant and water analysis is the only tool known which helps to control soil fertility. Nutrient supplying power of soils, crop responses to added nutrients and fertilizer and amendment needs can safely be assessed through sound soil testing programme. Monitoring of soil fertility, against depletion and accumulation of certain elements in toxic proportions over time, is also possible through appropriate soil tests. This also helps to economize on cost of fertilizers and also in increasing fertilizer use efficiency.

Traditionally, in India by balanced fertilization one is made to believe and understand use of N, P_2O_5 and K_2O in a certain ratio, ideally 4:2:1, on a gross basis both in respect to areas and crops. This ratio is not scientifically designed but is perhaps advocated as a somewhat safe and general prescription from the practical angle with a view to maintain the overall fertility of Indian soils. Blanket recommendations are static and cannot commensurate with variability and changes in soil nutrient status, crop demand and crop management. Therefore, one alternative approach could be that fertilizer is applied based on recommendations emanating from soil test crop response data or in its absence from soil test reports. Data of the on farm trials comparing the responses to applied fertilizer as per yield target *vis-à-vis* state recommendation and farmers' practice (Table 1) indicate that state recommendation gives much lower yield as compared to the fertilizer application based on actually testing the soil and yield targeting underpins the importance of soil testing in improving nutrient use efficiency.

Table 1: Results of on-farm trials with wheat, pearlmillet and mustard at Delhi

Crops	Treatment	Nutrient dose (kg ha^{-1})			Yield (kg ha^{-1})
		N	K_2O	P_2O_5	
Wheat	STCR Target 5 t ha^{-1}	126	49	41	4887
	SR	120	60	40	4567
	FP	80	57	0	3662
Pearlmillet	STCR Target 5 t ha^{-1}	100	42	43.5	2540
	SR	80	40	40	2020
	FP	46	23	0	1360
Mustard	STCR Target 2.5 t ha^{-1}	97	75.5	35	2281
	SR	100	40	40	1890
	FP	60	57	0	1312

STCR=Soil test crop response, SR=State recommendation, FP=Farmers practice
Source: Sharma *et al.* (1999)

Soil testing approach followed in India

Liebig's law of minimum states that the growth of plants is limited by the plant nutrient element present in the smallest amount, all others being in adequate quantities. From this, it follows that a given amount of a soil nutrient is sufficient for any one yield of a given percentage nutrient composition. This concept is followed in "targeted yield approach", which is being advocated in India since late 1960s. Ramamoorthy *et al.* (1967) established the theoretical basis and experimental proof for the fact that Liebig's law of the minimum operates equally well for N, P and K. This forms the basis for fertilizer application for targeted yields, first advocated by Truog (1960). The approach is unique in the sense that it not only prescribes the optimum dose of nutrient based on soil fertility status but also predicts the level of yield that a farmer can expect. The targets can be chosen based on farmers' resources. During the last four decades, the co-operating centres of Soil Test Crop Response Correlation (STCR) project of the Indian Council of Agricultural Research (ICAR), have generated numerous fertilizer adjustment equations for prescribing rates of fertilizer application for obtaining targeted yields of crops on a variety of Indian soils. The approach has been test verified in several follow-up experiments and demonstrated in a large number of farmers' fields. Recommended agronomic practices are to be followed along-with the fertilizer doses. The recommendations for different crops for specified yield targets are available and published (Subba Rao and Srivastava, 2001; Muralidharudu *et al.*, 2012). Lately, the calibrations are being developed under integrated supply of organics and fertilizers keeping into account the nutrient contribution of organics, soil and fertilizers. The technology of fertilizing the crops based on initial soil test values for the whole cropping system is also being generated.

Calibration to identify the high response plots

A separate calibration procedure has been devised by Tamil Nadu Agricultural University (TNAU), Coimbatore to select the balanced nutrition plots. In this procedure only balanced nutrition plots are used to develop the calibrations.

- The yields of grain and straw for the subplots in each of the four large calibration plots are arranged in groups. Within each group the level of a given nutrient is constant.
- The total uptake of nitrogen from the treatment producing the highest yield in the group is computed.
- The value obtained is divided by the corresponding yield of grain.
- The quotients from all nutrient groups in all four large plots are averaged to obtain Nutrient requirement (NR)
- The value of CS are obtained by analyzing the grain and straw from four control subplots in each of the four large calibration plots.
- The four subplots represent the highest, the lowest and two intermediate yields.
- The total yields of nutrients are then divided by corresponding soil test values and the quotients are averaged to obtain CS.

Saving in chemical fertilizers under IPNS using STCR approach

A calibration chart showing the savings in fertilizers under IPNS system compared to fertilizer alone system for targeted yield of some crops at different regions in India is presented in table 2. It is observed that more than 60 kg of N + P_2O_5 + K_2O can be saved for a yield target of 40 to 50 q ha^{-1} of rice and maize under IPNS system. Different types of organic manures are tested and used at different centers depending upon their local availability and accessibility to farmers (Subba Rao and Srivastava, 1998).

Table 2: Saving in fertilizers under IPNS as against alone system for targeted yield of some crops at different places in India

Center,soil	Crop	Target (q/ha)	Soil Test Value			Fertilizer dose			Organic Matter applied under IPNS(t/ha)	Saving in fertilizer under IPNS		
			N (kg/ha)	P (kg/ha)	K	N (kg/ha)	P_2O_5	K_2O		N	P_2O_5	K_2O
Jabalpur, Typic haplustert	Sunflower	10	250	15	300	82	60	58	3.5 FYM	25	29	13
Pusa, calcareous	Rice	40	250	15	186	152	72	55	5.0 Compost	28	20	16
Pusa, calcareous	Maize	50	250	15	186	79	40	22	5.0 Compost	27	22	22
Coimbatore (Molapalayam), red calcareous	Rice	60	250	15	200	155	71	96	6.25 GM + 2 kg PB	25	12	20

(Subba Rao and Srivastava, 1998)

Future Prospects in Soil Testing

Mapping of soil fertility

An attempt was made with joint venture of IISS, Bhopal and NBSSLUP, Nagpur to create spatial fertilizer recommendation maps using available validated fertilizer adjustment equations (STCR's generated) and Geographic Information System (GIS). The district level soil fertility index of 10 districts of India were prepared, which can be used to generate balanced fertilizer recommendation for specified crops for entire district based on the average soil fertility status of that district. District wise soil fertility georeferenced maps were prepared using index values for nitrogen (N), phosphorus (P) and potassium (K) for ten states. Corresponding equivalent soil nutrient values in respect of N, P and K were calculated from the index values. Reasonable limits for targeted yields were defined. The recommendations in the form equations for targeted yields developed by Subba Rao and Srivastava (2001) have been interlinked with the fertility maps. The use of this recommendation system suggested for varied applications for targeted yields in different districts of states. This can be used up to field level also, if the farmer has the knowledge of his fertility status and the yield target. The maps can also be updated from time to time based on the soil test result data base. It can be further narrowed down to block/village level depending on the availability of information. These fertility maps can also be used to study the changing

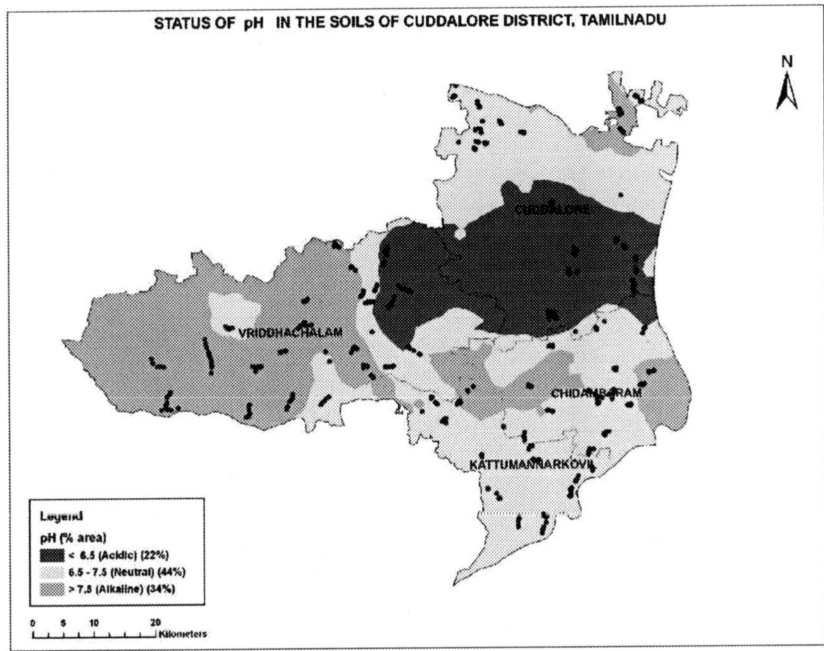

Fig 1: GIS/GPS based soil pH map of Cuddalore district, Tamil Nadu

(*See colour version on page 449*)

trends in the fertility of nutrients and can be correlated with fertilization practices of farmers of a particular region. These maps, however, only indicate the general fertility level in a district since the data collected and were not georeferenced. Recently, soil samples are being collected with georeferenced information and updated geo-referenced soil fertility maps are being prepared (Fig. 1).

Computer aided on-line software for fertilizer recommendation

India has contributed significantly in the field of information technology (IT) and has created a unique position in international arena. The spread of IT to rural India has opened new vistas in technology transfer. Not only large volume of information can be made available through these media but interactive interface can also be provided between the farmers and an expert sitting at distant place. This will help in increasing reach of expert in distant and otherwise inaccessible areas. Use of GIS for supporting farmers in decision making for management of natural resources and improvement in crop productivity can be cost effective as well. IT provides opportunity for blending advantages of inter personal and mass communication in to one media through interactive mode and wider reach.

As an attempt in this direction, an on-line application software was developed to calculate the fertilizer doses for the targeted yields at the district level. This software programs read data, perform calculations and generate graphical and

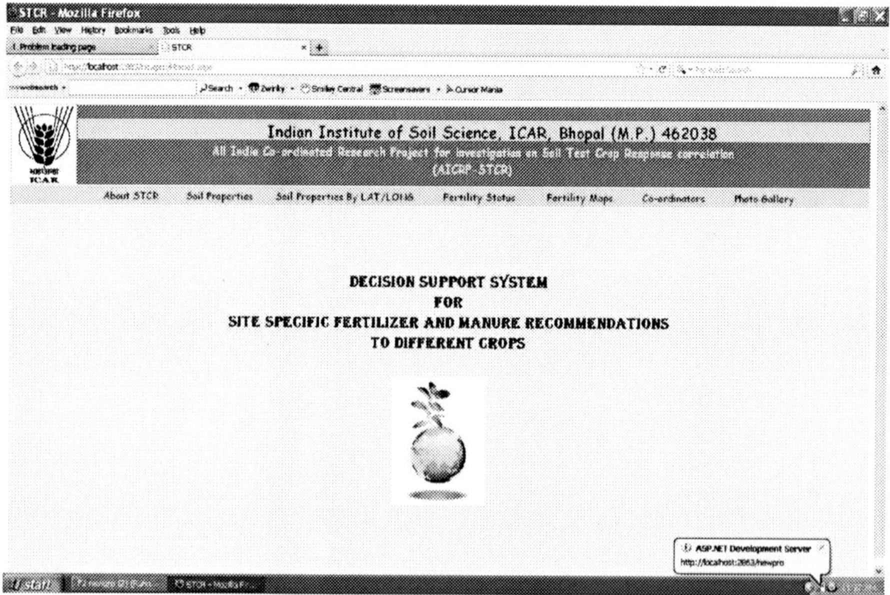

Fig 2. On-line soil test based fertilizer application software

tabular output as well as test reports. This system has the ability to input actual soil test values of the farmers' fields to obtain optimum dose of nutrients. The application is a user friendly tool. It will aid the farmer in arriving at an appropriate dose of fertilizer nutrient for specific crop yield for the given soil test values (Fig. 2).

Rapid soil testing

Soil testing has failed to gain wide acceptance in India, with only a small increase in sample numbers since inception. Reasons for the low utilisation of soil testing services include: farmers not having the time nor the expertise to collect soil samples; delays in receiving results from laboratories; lack of faith in results produced by laboratories linked to chemical companies; and cost of soil testing. The availability of simple, low cost field-based soil tests, could provide a solution to these problems.

Soil test kits for use in the field are available from various sources in India. The rapid soil test methods have been correlated with standard methods used in soil laboratories in India and can be used for the determination of pH, lime requirement, phosphate, potassium, nitrate, magnesium and calcium. However, it is not guaranted that acceptable results will be obtained for all soils and for all nutrients. The lack of calibration of these methods for Indian soils is a potential limitation of tests developed overseas. It is strongly recommended to pay a more serious attention to these rapid soil analyses methods and to train the farmers in the use of these procedures. The rapid soil testing can also be used by consultants and agronomists to analyse soil samples and make "on the spot" fertilizer recommendations.

New dimensions in soil testing

There is an interest currently to use soil testing to determine the pollution of surface and groundwater from fertilizer/waste material application. However, to identify the potential impact of pollutants, a comprehensive approach should be taken on the environment and soil testing. The farmers should be educated about the factors that need to be considered for soil testing so as to protect the environment while producing economical food. "Soil tests for environmental interpretations are more complicated than for agronomic purposes. From an environmental standpoint, one has to consider the kind of fertilizer or animal waste material being applied, the type of soil, proximity to water bodies, hydrology, and erosion hazard." This calls for a comprehensive approach to evaluating issues and identifies areas where additional information is needed. One important issue involves the use of soil tests to determine application rates of waste, such as livestock and poultry manures, so that the environment is not harmed.

Although research has demonstrated numerous benefits of using these products to enhance soil fertility, there is concern that the traditional practice of applying animal waste based on the nitrogen requirement alone may result in the buildup of certain nutrients, such as phosphorus, copper and zinc. Environmental interpretations should take into account the loss of nutrients from soils. For nutrient loss to occur from agricultural soils, both a source of the nutrient and a mechanism for transporting it to surface or ground water are needed. This indicates that a key to effective management of nutrient pollution is to focus on the critical areas where these two requirements are met. The soil testing has to become the foundation of any successful attempt to understand and control nutrient losses from the soil.

Role of fertilizer industry

The fertilizer promotional strategy in a free market economy has to be oriented to achieving complete customer satisfaction. The farmer customer has to be provided with not only the fertilizer material but associated materials and services related to its use. A fertilizer sales point therefore has to be upgraded as an agro-input sales point cum service centre. Diagnostic services for soil and crop disorders alongwith the remedial measures in terms of both knowledge input as well as material input should be available to a farmer. Services like rapid soil and tissue testing, agro-input or fertilizer quality testing, availability of inputs, application equipments, enhancing availability of amendments, providing technical know-how, linkages with credit *etc.* will go a long way in winning the customer and promoting fertilizer use. Some of the services can even be made, paid service, however, quality and reliability of the services will be of utmost importance.

Soil testing and agriclinics

The Indian government has launched an ambitious initiative to encourage private extension with the Ministry of Agriculture and the National Bank for Rural Developments (NABARD). Graduates are being trained to become 'agripreneurs' and on completion of their courses they receive a loan to establish an 'agriclinic' or 'agribusiness centre'. Farmers are expected to pay a fee for their services and the agripreneurs are expected to identify the demand for a broad range of services from soil testing to advice on organic production and food processing. So far 112 businesses have been set up in 10 states and it is intended that the new services will provide specialist advice that may be beyond the scope of the service presently offered to farmers through public extension.

Remote sensing and GIS for soil fertility management

Remote sensing is defined as a science and an art of acquiring data about material objects from measurements made at a distance without actually coming into physical contact with the object. Satellite imagery provides an opportunity to identify problems developing in the field, and especially monitor changes in the area affected. High spatial resolution satellite data is useful in site-specific nutrient management/precision agriculture. It enables converting point samples to field maps, mapping crop yield, mapping soil variability, monitoring seasonally variable soil and crop characteristics, moisture content, crop growth and phenology, crop evapo-transpiration rate, crop nutrient deficiency, crop disease, weed and insect infestation *etc.* GIS is the latest and most sophisticated technology proved to be extremely useful for natural resources management and thus for sustainable development. GIS technology helps in finding the solutions for agro forestry/reforestation/forest development and hence in resources for development and management of forest. GIS is "a Computer-assisted System for capture, storage, retrieval, analysis and display of spatial data, within a particular organisation". Information derived through remote sensing and non-remote sensing methods are integrated within GIS and specific management and developmental plans are arrived finally. The information in a GIS is presented in two basic forms; as maps and as tables to produce information that is needed by a user. Use of suitable GIS techniques helps integration of the vast data base covering a wide variety of relevant parameters in a more efficient manner.

Soil fertility assessment: Most of the studies, hitherto, have been carried out on N. Leaf N concentration is an important indicator for diagnosing plant N status. Nitrogen deficiency causes a decrease in leaf chlorophyll concentration, leading to an increase in leaf reflectance in the visible spectral region. This measurement, however, is subject to errors due to other stresses that cause yellowing of leaves. NIR reflectance has been found promising in quantification of soil moisture, organic carbon and total nitrogen. A marked difference in reflectance throughout the 0.5 to 1.1 μm wavelength region has been observed due to presence of organic matter in a soil (Swain and Davis, 1978). Spectral reflectance, generally, decreases over the entire short-wave region as organic matter content increases. Remotely sensed imagery of bare soil field could be quantified to describe the spatial variation in organic carbon. The soil organic matter distribution as estimated from Landsat-TM images was strongly correlated with the spatial distribution determined by grid soil sampling (Bhatti *et al.*, 1991)

Soil nutrient management: Soil nutrient management at field level involves delineation of homogeneous management zones (HMZs) based on physico-chemical characteristics of soils and crop yield needs to be done. Remote sensing can provide valuable information about soils and crops/vegetation conditions over large areas for making decisions related to site specific management. Research efforts are currently directed to develop and validate the remotely sensed data derived inputs to support site specific crop management.

Application of information and communication technologies (ICT)

Agricultural development and sustainability crucially depend upon relevant information access at opportune time. There is a vast scope of extending ICTs through public, private and non-governmental organizations with respect to extension, marketing and community services. The ideal delivery model for these ICTs is envisioned to be a multi-pronged strategy involving institutions under National Agricultural Research System (NARS) to go online, share their contents, and rural information kiosks providing information access to the farmers. India has 37% of world ICT enabled projects in rural areas. Agricultural resource information using GIS models, expert systems, databases on successful technologies have critical role in governance and decision making by farmers.

Farmers' acceptance of soil testing

The awareness of the farmers about the benefits of soil testing is too inadequate. Very often the farmers' acceptance of soil testing is linked with benefits in terms of crop yields. It is, however, important to know that soil testing is not an end itself. It is a means to an end. Good crop yields are a result of several management practices. Special efforts are required to bring about farmers' awareness and make the programme a farmer oriented activity. Laying out large scale field demonstrations to show the benefits of soil test based recommendations over farmers' practice are required.

Selected References

Bhatti, AU, Mulla DJ, Frazier BE. 1991. Estimation of soil properties and wheat yields on complex eroded hills using geostatistics and thematic mapper images. *Remote Sensing of Environment* 37: 181-191.

Muralidharudu Y, Sammi Reddy K, Mandal BN, Subba Rao A, Singh KN, Sonekar S. 2011. GIS based soil fertility maps of different states of India. All India Cordinated Project on Soil Test Crop Response Correlation, Indian Institute of Soil Science, Bhopal, pp:1-224.

Ramamoorthy B, Narsimham RL, Dinesh RS. 1967. Fertilizer prescription for specific yield target of Sonara-64. *Indian Farming* 1: 43-45.

Sharma BN, Singh RV, Singh KD. 1999. Balanced fertilizer use based on soil test for wheat, bajra and mustard. *Fertilizer News* 44 (3): 55-58.

Subba Rao A, Srivastava S. 1998. Annual Report 1993-98 AICRP on Soil Test Crop Response Correlation. Indian Institute of Soil Science, Bhopal pp. 1-87.

Subba Rao A, Srivastava S. (Eds.) 2001. Soil test based fertilizer recommendations for targeted yield of crops. Proceedings of the National Seminar on Soil Testing for Balanced and Integrated Use of Fertilizers and Manures, Indian Institute of Soil Science, Bhopal, India, pp 1-326.

Swain PH, Davis SM.1978. Remote sensing: The qualitative approach, McGraw-Hill, New York, 396p.

3

Enhancing Nutrient Use Efficiency, pp. 31-40
Editors: K. Ramesh, A.K. Biswas, B.L. Lakaria, S. Srivastava and A.K. Patra

Management of Soil Physical Environment for Higher Nutrient Use Efficiency

Ritesh Saha

ICAR- Central Research Institute for Jute and Allied Fibres Barrackpore, West Bengal – 700 120, India

Introduction

The Indian agriculture is very complex and carrying out multi-functionalities of providing food, nutrition and ecological security besides employment and livelihood for over 700 million people. Indeed, India has made a marvelous achievement in attaining self-sufficiency in food grain production after the introduction of Green Revolution which eventually resulted in maintaining all-time high buffer stock in warehouses of our country. Such rosy picture in production trends turned to be bleak in the past few years. There are various reasons behind this. Shrinking resources of prime lands, deforestation and accelerated erosion, deterioration of soil physical environment, increasing waterlogging and salinity in canal irrigated areas, declining water table in well-irrigated areas, poor management of rainwater, lower efficiency of inputs such as water, fertilizers and agrochemicals, rapid industrialization coupled with pollution and environmental degradation and hazards have aggravated the problem. Poor soil fertility and inappropriate nutrient management strategies are the important major constraints contributing to food insecurity and eco-system degradation. A major consequence of decline in soil fertility in agricultural land is the prevalence of sustained periods of moisture deficiency resulting from poorer water storage throughout the soil profile. The fact that at least 60% of the cultivated soils have plant growth-limiting problems associated with nutrient deficiencies makes soil fertility management and nutrient use efficiency a major promising research area. Under such circumstance, there is an imperative need to produce more from less arable land and water through meticulous management of basic agricultural resources such as soil, water and biological inputs.

Soil health and nutrient use efficiency

Soil health is a state of dynamic equilibrium between flora and fauna and their surrounding soil environment in which all the metabolic activities of the former proceed optimally without any hindrance, stress or impedance from the latter. As per the ICAR annual report, soil fertility maps of India showed that about 59% of soils are low, 36% of soils medium and only 5% are high in available nitrogen (N). With respect to Potassium (K), about 52% soils are high, 39% of soils are medium and only 9% are low. Fertilizer consumption in India increased 322 times during 1950-51 to 2007-08 period, about 72% of the total N,P and K was consumed in food grains production (Chanda, 2008). However, nutrient use efficiency has been very low in Indian agriculture. Due to imbalanced use of plant nutrients, mining of nutrients is considered as the main cause for decline in crop yield and crop response ratio. In such a precarious situation it is high time that nutrient use efficiency is optimized with integrated agronomical practices.

An efficient nutrient management plan is based on nutrient budget. Soil tests, environmental and climatic data, and production goals determine the use of from nutrients at the right time, right rate and in the right place is the best management practice for achieving optimum nutrient efficiency. Thus integrated plant nutrient supply (IPNS) system demands a holistic approach to optimize the nutrient efficiency from crop production and it involves judicious combined use of fertilizers, biofertilizers, organic manures (FYM, compost, vermicompost, green manures, crop residues etc.) and growing of legumes in the cropping systems (Prasad, 2008).

Crop response/ nutrient use efficiency to soil physical environment

Soil is the major source of crop production system, supplying essential nutrients for crop growth. Soil physical environment acts as critical sink for soil physico-chemical sustenance. It refers to the interaction effect of various factors such as soil moisture, soil aeration, temperature or thermal regimes and mechanical impedance influenced by nature of the parent materials, physiographic and climatic condition of the specific location. The soils vary greatly in supporting crop production system because of the wide variation in soil physical environment.

Many factors influence the complex chemical, physical and biological processes which govern soil quality and crop production. Crop production is affected by various soil physical properties, which directly or indirectly affect plant growth. Physical edaphic factors which are important for directly influencing plant growth are; soil moisture, soil aeration, thermal regimes and mechanical impedance. Good physical management of soils is one of the key factors to maintain or improve agricultural productivity, in tropical and subtropical regions, and to reduce soil and environmental degradation (Lal, 2000).

Interrelationships of soil physical properties

Water is the dominant controlling factor among all the four soil physical properties directly related to plant growth. The other three are affected by water content as schematically illustrated in Fig. 1. The relationship between water and aeration is opposite to that between water and mechanical resistance. Increasing water content decreases aeration which is undesirable but decreases mechanical resistance which is desirable. The effect of water on both of these parameters is intensified by an increase in bulk density and/or presence of small pores. The optimum range of water content for plant growth has generally been assessed on the basis of plant water availability. The upper limit is associated with field capacity and the lower limit is associated with permanent wilting point. However the *non-limiting water range* (NLWR) may be affected by aeration and/or mechanical resistance, particularly in poorly structured soils with high bulk density. In other words, the NLWR can be reduced by poor aeration and/or high mechanical resistance in some soils, as illustrated in Fig. 2.

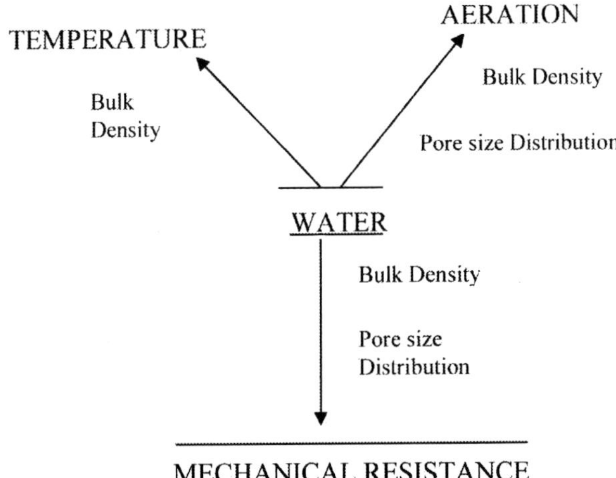

Fig. 1. Schematic representation of the relationship between water and other soil physical properties, directly affecting plant growth

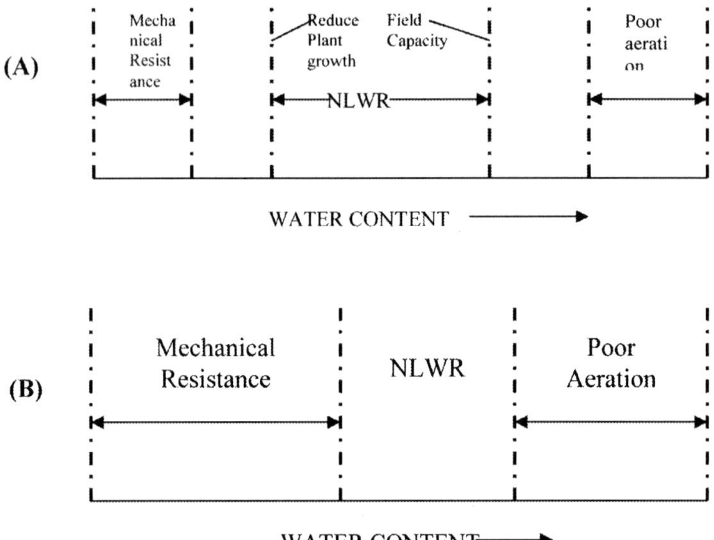

Fig. 2. Generalized relationships between soil water content and restricting factors for plant growth for soils with increasing bulk density and decreasing soil structure.

Management of soil physical environment for better nutrient use efficiency

Poor soil physical environment requires careful management to maintain conditions favorable for plant growth. The key point is that traditional soil physical properties like soil texture, structure, bulk density *etc.* affect the management practices required to maintain water potential, oxygen diffusion rate, temperature and mechanical resistance in a range suitable for good crop production. The suitable management practices such as tillage, residue management, soil amendments, drainage and irrigation are, therefore, recommended to maintain soil physical health so as to achieve the higher crop production. Some of these management practices are discussed here under:

Tillage Practices

Tillage is one of the important pillars of conservation agriculture which disrupts inter dependent natural cycles of water, carbon and nitrogen. Conservation tillage is a generic term encompassing many different soil management practices. It is generally defined as 'any tillage system that reduces loss of soil or water relative to conventional tillage; mostly a form of non-inversion tillage allows protective amount of residue mulch on the surface (Mannering and Fenster, 1983). Tillage practice directly influence bulk density of the soil. In general, it decreases bulk density of soil and the effect may last for variable lengths of time depending on the soil properties, crop effects and nature and type of wetting. zero or no tillage, on the other hand, generally causes greater soil bulk density, increase soil strength and decreased soil porosity as compared to conventional tillage practices. Tillage practice either loosens or compact the soil, means changes particle to particle contact and porosity of soil. Loosening of soil results in an increase in total porosity and large pores whereas compaction has the reverse effect. Change in porosity directly affects water and heat transmission characteristics of soil. As soil strength is a function of bulk density and water content, all the physical properties affecting seed germination, seedling emergence and root growth are influenced by tillage practices.

Tillage affects the soil water status and also the capacity of the crop to utilize water. It alters surface and subsurface soil conditions that govern infiltration, runoff and evaporation of water, weed growth, crop establishment and root growth of the crop. It helps carry-over moisture in the seed zone over longer periods compared with untilled bare soils. When water is in short supply, a tillage practice should aim at maximizing the amount of plant available water. Conservation tillage is well suited for moisture conservation on well drained and moderately well drained soils. With excess water, there is a need to minimize the risk of insufficient aeration. On poorly drained soils, conservation tillage may fail by causing aeration problem (Lal, 1976).

Effect of tillage on aeration of soil is directly linked with its effect on infiltration rates. The faster the rate, the shorter is the time the surface remains sealed to air exchange. Tillage practice increases the relative proportion of larger pores in the tilled layer which drain out rapidly and restore adequate water free porosity soon after heavy rain. Air-filled porosity at field capacity or at higher soil matric water potentials is much lower in untilled soil and therefore, O_2 diffusion is 2-6 times less than the tilled soils.

Mishra *et al.* (2005) reported that zero tillage in transplanted rice saves the operation energy without jeopardizing the grain yield (36.68 q/ha) and also conserves 25% profile soil moisture for subsequent *rabi* crops, thus, improving the soil hydro-physical environment under hilly eco-system of Meghalaya.

Residue Management

Crop residue management is well known for directly or indirectly affecting physical, chemical, biological and biochemical soil properties. It is perceived that soil quality is improved by the adoption of crop residue management practices (Lal, 1991). Proper management of residue can provide farmers with some measures to mitigate changes due to implements and traffic loads imposed during cropping cycle. As residue management increases soil organic matter, it helps in maintaining soil aggregate stability. Incorporation of organic matter either in the form of crop residues or farmyard manures has been shown to improve soil structure and water retention capacity (Bhagat and Verma, 1991), increase infiltration rates (Acharya *et al.*, 1988) and decrease bulk density (Khaleel *et al.*, 1981). Singh *et al.* (1994) reported greater amounts of water stable aggregates in no till + straw treatments. The proportion of wind-erodible (< 1 mm) and water stable micro-aggregates (< 0.25 mm) was also lower and mean weight diameter (MWD) and geometric mean diameter (GMD) were greater as compared to tillage practices. However, surface and subsurface soil density and penetration resistance may increase naturally when using a no-tillage system resulting from the raindrop effect and the structural failure of soil having low stable aggregates.

Crop residues increase soil hydraulic conductivity and infiltration by modifying mainly soil structure, proportion of macro pores and aggregate stability. Up to eight folds increase in hydraulic conductivity in zero-tillage stubble retained have been reported over treatments where stubble was removed by burning (Bissett and O'Leary, 1996). Baumhardt and Lascano (1996) observed increase in cumulative infiltration from 29 mm for bare soil to as high as 49 mm under different residue management practices.

Crop residues left on the soil surface as a mulch as compared to incorporation, removal or burning in arid and semiarid region in summer season are known for beneficial crop production. This reduces the soil temperature, thus influencing the biological processes and enhances soil N mineralization (Tian *et al.*, 1993). It is well established that increasing amounts of crop residues on the soil surface reduce the evaporation rate. Thus, soils covered with residue tend to have greater moisture content than bare soils.

Residues as a mulch on the soil surface act as a barrier restricting soil particle emission from the soil surface and also increasing the threshold wind speeds for detaching these particles. It is reported that standing residues are more effective than flat residues in reducing erosion by reducing the soil surface friction velocity of wind and intercepting the soil particles.

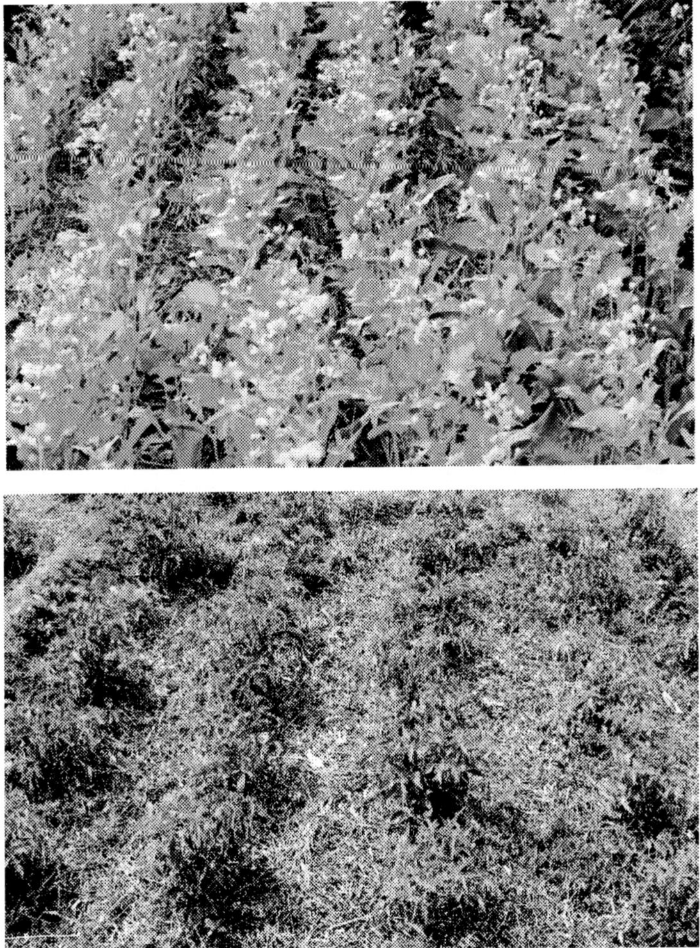

Fig. 3. Paddy straw mulching in Mustard crop and Grass mulching in Tomato crop for higher nutrient use efficiency (*See colour version on page 450*)

Land Configuration

The measures for providing surface drainage, conserving *in-situ* rain water and its safe disposal on cost effective basis include ridge and furrow system, broad bed furrow and raised-sunken bed system (Fig. 4).

Ridge and Furrows

The sowing of upland crops under ridge and furrow system was most common during rainy season on such soils having slope less than 1%. Furrows serve as an effective means of surface drainage and carry excess water into cut-off drains dug across the slope. The cut-off drains may be constructed at an interval of about 20 m depending on the slope and rainfall criteria. This system has proved highly effective in medium and high rainfall area (rainfall ranging from 700 to 1200 mm) and also helps in improving the productivity of crops.

Fig. 4. Land configuration in the form of RSB for increasing cropping intensity, nutrient and water use efficiency (*See colour version on page 451*)

Broad-bed and Furrows

This system consists of a series of broad-beds and furrows accommodated in 90-150 cm wide parallel running strips. In this system, beds and furrows are created on a grade of 0.5%. Furrows drain into grassed waterways. The system permits safe disposal of runoff, tends to conserve soil and water in-situ, reduces soil and nutrient losses and enhances crop productivity and sustainability. This system reduced runoff, soil loss, N and P loss by 13.5, 30.1, 16.1 and 13.5%, respectively over flat beds on conventional system of cultivation and enhanced soybean yield (Sharma and Gupta, 1996).

Raised and Sunken Bed system (RSB)

Raised and sunken bed system (RSB) is mostly practiced in those areas having very high rainfall where upland *kharif* crops suffer due to poor drainage during the periods of continuous and intense rainfall. The system consists of an array of alternating raised (6-8 m wide) and sunken (3-4 m wide) beds with an elevation difference of 20-30 cm. the system is created by mechanically shifting soil from demarcated 3-4 m wide strips, designated as sunken beds to adjoining 6-8 m wide strips called raised beds. Sunken beds are tied with small cross section earthen bunds of about 10 cm height at 20 m distance interval to ensure uniformity in runoff retention. The runoff from raised beds, planted to any upland crop like French bean (*Phaseolous vulgaris*) *etc.* is arrested in the adjacent sunken beds supporting a relatively water tolerant crop such as rice (*Oryza sativa*) *etc.*

Mishra and Saha (2007) reported that the raised and sunken bed of 1 m: 3 m width ratio and 30 cm elevation was effective for inter-plot water harvesting (4.9 cm. submergence) and rice–french bean cropping system. Three years experimentation revealed that it was possible to take two crops in a year in lowland situation by configuration of the land, which facilitates the soil moisture availability under raised and sunken beds. However, the best realised yield of rice in sunken bed was obtained (37.2 q/ha) in ratio of 1:3, while the maximum (105.3 q/ha) green pod yield of french bean was recorded in 1:3.5 m ratio of raised and sunken bed.

Conclusion

Efficient use of nutrients is absolutely essential for attaining the goal of sustainability in agriculture. The relevance of soil physical environment to plant growth has been recognized as very important for higher nutrient use efficiency harnessing the potential of costly inputs like energy, fertilizer, crop varieties and irrigation *etc.* and sustainable agriculture. Soils with so-called "good soil physical conditions" are easier to manage than "poor soil physical condition" soils. Improved technologies chosen judiciously would go a long way in ameliorating soil physical conditions and enhancing crop production without harming the environment. There is an urgent need to assess the possibilities of nutrient particularly N management for climate abatement and in the same time ensuring food security, while minimizing environmental and human health impacts.

Selected References

Acharya CL, Bisnoi SK, Yaduvanshi HS. 1988. Effect of long-term application of fertilizers and organic and inorganic amendments under continuous cropping on soil physical and chemical properties in an Alfisol. *Indian Journal of Agricultural Science.* 58 : 509-516.

Baumhardt RL, Lascano RJ.1996. Rain infiltration as affected by wheat residue amount and distribution in ridged tillage. *Soil Science Society of America Journal.* 60: 1908-1913.

Bissett MJ, O'Leary GJ. 1996. effects of conservation tillage on water infiltration in two soils in south-eastern Australia. *Australian Journal of Soil Research.* 34: 299-308.

Bhagat RM, Verma TS. 1991. Impact of rice straw management on soil physical properties and wheat yield. *Soil Science.* 152: 108-115.

Chanda TK. 2008. Analysis of fertilizers use by crops. *Indian Journal of Fertilizers.* 4: 11-16.

Khaleel R, Reddy KR, Overcash MR. 1981. Changes in soil physical properties due to organic waste application: a review. *Journal of Environmental Quality* 10 : 133-141.

Lal R. 1976. No-tillage effects on soil properties under different crops in Western Nigeria. *Soil Science Society of America Journal.* 40: 762-768.

Lal R. 1991. Soil structure and sustainability. *Journal of Sustainable Agriculture.* 1: 67-92.

Lal R. 2000. Physical management of the soils of the tropics: priorities for the 21st century. *Soil Science.* 165 (3): 191–207.

Mannering JV, Fenster CR.1983. What is conservation tillage? *Journal of Soil Water Conservation.* 38: 141-143.

Mishra VK, Saha R. 2007. Effect of raised and sunken bed system on productivity of rice and French bean in high rainfall areas of Meghalaya. *Indian Journal of Agricultural Sciences.* 77 (2): 73-78.

Mishra VK, Saha R, Bujarbaruah KM. 2005. Zero tillage techniques for transplanted rice in high rainfall eco-system. ICAR Research Complex for NEH Region, Umiam, Meghalaya.

Prasad R. 2008. Integrated plant nutrient supply system (IPNS) for sustainable agriculture. *Indian Journal of Fertilizers.* 4(12): 71-90.

Sharma RA, Gupta RK.1996. Integrated Research project for Soil Conservation and Water Management. Annual report, Indo-US project, AICRPDA, College of Agriculture, Indore, M.P. India.

Singh B, Chanasyk DS, McGill WB, Nyborg MPK. 1994. Residue tillage management effects on soil properties of a Typic Cryoboroll under continuous barley. *Soil and Tillage Research.* 32: 117-133.

Tian G, Brussaaard L, Kang BT. 1996. Biological effects of plant residues with contrasting chemical compositions under humid tropical conditions on soil fauna. *Soil Biology and Biochemistry.* 25: 731-737.

4

Enhancing Nutrient Use Efficiency, pp. 41-49
Editors: K. Ramesh, A.K. Biswas, B.L. Lakaria, S. Srivastava and A.K. Patra

Non-Monetary and Low-Cost Agronomic Measures for Enhancing Nutrient Use Efficiency

K. Ramesh[1], Pravin K. Upadhyay[2] and KG Mandal[3]

[1]ICAR-Indian Institute of Soil Science, Nabi Bagh, Bhopal – 462038, India
[2]ICAR-Indian Agricultural Research Institute, New Delhi– 110012, India
[3]ICAR-Indian Institute of Water Management, Bhubaneswar – 751023, India

Introduction

"Non-monetary inputs are defined as those cultural operations which help to achieve high yield at no extra cost and whose cost does not change with the level of output".

After the advent of green revolution, nutrient management has received the significant attention by farmer as well as researchers; as a result of which and coupled with the requirement for high yielding varieties, consumption of plant nutrient (N, P and K) increased tremendously. Variability was observed in fertilizer consumption between the states. However, the use efficiency of applied fertilizer nutrients from fertilizer sources is not quite encouraging, for example, N use efficiency hovers around 20-50 % only while P is 15-20% of applied P. The use efficiency of applied K is also low due to various reasons. The use efficiency of micronutrients is still lesser than the major nutrients. These nutrient related problems could be addressed, at least partially, through the low-cost agronomic measures.

Efficiency of any system is the obtainable output per unit amount of input; and fertilizer use efficiency (FUE) or nutrient use efficiency (NUE) is perceived and determined differently by soil scientists, agronomists and physiologists. It can be expressed as apparent recovery fraction (ARF), physiological efficiency (PE) and / agronomic efficiency (AE). Actually these nutrient use efficiency parameters are interrelated. AE is the product of ARF and PE. However, the common numerator in different approaches is crop yield; the denominator is amount of nutrient applied. Any measure, which increase crop yield, will lead to an increase in nutrient use efficiency within a specified level of nutrient. Thus nutrient use efficiency is an index to which nutrient and management practices have worked together with the soil-climate complex in producing greater yield.

Simple agronomic practices increase crop yield and enhance nutrient use efficiency (NUE). These practices, particularly the non-monetary and low-cost, are briefly described in the following sections.

1. **Cultivar selection:** Number of cultivars, high yielders of different crops, resistant/tolerate to pest and disease have been developed.
2. **Time of sowing:** In time seed bed preparation and timely sowing as per the season. (*Kharif/ Rabi/* summer) for any crop, results in even and better germination, well establishment and thereby minimize the crop loss.
3. **Plant population:** Keeping the total plant population constant. Inter and intra row plant population can be adjusted to minimize the humidity buildup within the crop canopy.
4. **Timely intercultivation:** Manual or- mechanical inter- cultivation suppresses the pest, diseases and weeds similar to that of preparatory tillage. Weeds that could be serving as alternate host to insect and pathogen can be effectively controlled by inter- cultivation.
5. **Crop rotation:** Helps as complementary effect, that legumes inclusion in system add the nitrogen and residue, thereby increases the water holding capacity of soil as well as reduction in infestation of weeds which are associated with crop and minimize the build-up of pest and diseases.
6. **Strip cropping:** inclusion of erosion permitting and erosion resisting crop help in conservation of moisture and increases in water intake.
7. **Tillage:** In sequential cropping, when cropping intensity is increased, there is less time between two crops and this affects the intensity of tillage and its cost.

Cultivar selection

The choice of appropriate crop and efficient variety is one of the prime non-monetary and low-cost technique. Crop diversification is also essential for risk aversion, system sustainability and improving NUE of crop (s). Number of varieties (high yielding and hybrids) that are available for use in modern agronomic systems is large. In most of the field crops only a part of the total dry matter produced is accumulated as crop yield (*i.e.* economic yield), thus high harvest index is desirable for the chosen variety. In addition, varieties should have specific disease and insect resistance besides, herbicide resistance, quality characteristics for specific markets, and above all the nutrient efficient. Thus for higher yield and optimum use of resource we should go for a proven variety released at regional level and which suits to location specific problems.

Experiments conducted at Tamil Nadu Agricultural University (TNAU) for comparing the ability of hybrid and non-hybrid to use fertilizer P and K indicated that non-hybrids and hybrids of rice and cotton had large responses to external P and K application, but the degree of agronomic response is greater in hybrid crops. For instance, the cotton hybrid 'TCHB 213' produced more yield per unit of K fertilizer (agronomic efficiency of K or AE_k) than the non-hybrid 'MCU 5' (MYR Annual Report, 1997-98). Similarly, the rice hybrid 'CORH-1' is more effective in utilizing applied P compared to the non-hybrid 'ASD 18'.

Thus, the availability of fertilizer responsive high yielding cultivars has been driving force for the increased response to fertilizer application, which encourages the farmers to apply desired rate of fertilizers. Moreover, the genetic potential of any cultivar cannot be fully exploited without adopting proper agronomic management practices. The cultivar having high yield potential needs more quantity of nutrients to realize high economic yield.

Optimum time of planting

Planting time is the most vital non-monetary input affecting crop yield and nutrient use efficiency (NUE). Even in photo and thermo-insensitive crops, it is a critical input for higher yield. Decline in yield due to delay in sowing cannot be compensated by excess use of other inputs. Of course, time of sowing vary with the variety, agro-climatic condition and crop season. Actually, production efficiency of different crop genotypes greatly differs under different planting dates depending upon their foliage characteristics and canopy structure. A genotype, which quickly attains optimum leaf area index and retains it for a longer period is more efficient as it can take better advantage of solar energy available during the growing season. Sowing time is the most vital non-monetary input to achieve target yields in mustard. Production efficiency of different genotypes greatly differs under different planting dates. Soil temperature and moisture influence the sowing time of rapeseed-mustard in various zones of the country. Sowing time influences phenological development of crop plants through temperature and heat unit. Sowing at optimum time gives higher yields due to suitable environment that prevails at all the growth stages.

The optimum sowing time for soybean in the soybean-growing regions is June-July, depending on the onset of the monsoons. Any delay in the onset of monsoon should be judiciously analyzed and the suitable variety should be selected (Ramesh *et al.,* 2017).

Under field conditions, temperature is most universal in influencing growth. The yield gets reduced when the development coincides with the period of relatively high or low temperature. The yield potential of the genotype or the benefit of applying nutrients can be fully exploited through providing appropriate microclimate- temperature at different growth and development phases. Planting time influences phenological development of crop plants through temperature and heat units. Thus, an agronomist take a lead by providing the planting time of various field crops after studying the relationship between temperature and plant's growth and development. Early planting in temperate regions is usually associated with climate frequently is determined by the onset of rains. In addition, soil temperature more than 35°C as well as low moisture availability may restrict germination of cool season species. The third week of November in the north-plain region of India was earlier found to be the most suitable time for sowing dwarf wheat in variance to the second half of October for the tall wheat. Interestingly, this practice has reversed in the context that the production by planting them during the second fortnight of October, thereby demonstrating their variable response with the change in the environment.

Field experiment with split plot design conducted to study the effect of monetary (three N levels (125, 150 and 175 kg/ha) and non-monetary inputs (two varieties (BSR 1 and BSR 2), three planting time (15 May, 15 June and 15 July); and three spacing (30 x 15 cm, 45 x 15 cm, 60 x 15 cm) on yield and economics of turmeric showed that turmeric variety BSR 2 out yielded BSR 1 in terms yield. Planting the turmeric during middle of May (15 May) was superior compared to 15 June and 15 July plantings. Among spacing, 30 x 15 cm recorded significantly higher growth, nutrient uptake and yield than 45 x 15 cm and 60 x 15 cm. The crop response was better for higher rate of nitrogen (175 kg/ha) than other levels. Economic evaluation indicated that combination of non-monetary inputs *viz.*, planting BSR 2 at 15 May with monetary inputs *viz.*, 30 x 15 cm spacing with 175 kg/ha N would increase the turmeric production and income of the farmers (Kandiannan and Chandragiri, 2008).

Field experiment conducted during rabi season of 2006-07 and 2007-08 to study the effect of sowing dates and integrated nutrient management on growth, yield and quality of winter maize (three sowing dates :15 October, 25 October and 5 November in main plots and three levels of urea: 50, 100 and 150 N20 kg ha^{-1} and two organic fertilizer (FYM andAzospirillum) indicated that crop sown on 25 October significantly enhanced the growth and grain yield than either early or late sowing while, 150 kg of N ha^{-1} application significantly increased over 100 and 50 kg N ha^{-1}. However, N application through FYM was found statistically at par with Azospirillum. But, application of 100 kg ha^{-1} with 7.50 t ha^{-1} FYM at the sowing of 25 Oct significantly influenced the growth, yield and quality of maize and has recorded 9.35 and 23.07% more grain yield (Verma, 2011).

Though different varieties have a differential response to date of sowing, mustard sown on 14 and 21 October took significantly more days to 50% flowering (55 and 57) and maturity (154 and 156) as compared to October 7 planting. Delayed sowing resulted in poor growth, low yield, and oil content. The reduction in yield was maximum in "RH-30" and minimum in "Rajat".Date of sowing influence the incidence of insect-pest and disease also. Sowing on October 21 resulted in least *Sclerotinia* incidence. The maximum (20.5–25.4°C) and minimum (3.9–10.7°C) temperatures at the flowering stage of crops established through sowing on October 21 were negatively correlated with the development of Sclerotina stem rot. Mustard aphid (*Lipaphis erysimi*) has been reported as one of the most devastating pests in realizing the potential productivity of Indian mustard. Normal sowing (1st week of November) also helps in reducing the risk of mustard aphid incidence (Shekhawat *et al.,* 2012).

Precision planting & optimum plant population

Precision planting encompasses the following: optimum distribution of seeds/ planting material, crop geometry (i.e. row width and plant-to-plant distance) and planting depth. Modern planters can be equipped to vary seeding rates, and thus plant population changes between soils and fields are more easily achieved than in past history. The optimum spacing between rows and between plants may vary with the variety, season, soil type and its nutrient status and availability of soil moisture. The optimum plant geometry or the arrangement of plants in the field has a direct impact on the interception of solar radiation by the crop canopy. Thus row-to-row and plant-to-plant distance should be optimum. It helps to derive the benefit of applied nutrients and enhance NUE. Based on past experiences, the preparation of a database of site specific and crop specific information that are available on plant population, row widths, depths and planting dates *etc.* is the way to make decisions that optimize performance of the crops without additional input costs.

The stand of crops in farmer's field is generally low and uneven. The establishment of optimum plant population per unit area geometry of the crop is one of the most important factors to ensure efficient utilization of available resources, especially nutrient and a good crop yield. Sub-optimal or superabundant plant population is one of the major factors responsible for low yield in most of the crops. Lower seed rate, low field germination per cent of seeds, abundance of weeds having a smothering effect on plants, incidence of diseases and insect pests, etc., are the general causes of superabundant plant population that favors inter-plant competition and parasitism among plants reduces crop yield.

Under normal sowing conditions the optimum plant density recommended for soybean is 2.25 lakh plants ha^{-1}; spacing of 45 × 15 cm is followed. Under delayed sowing conditions, the row to row spacing needs to be narrowed down to 30 cm to accommodate more plants m^{-2}. However, under field conditions, farmers use excess seed rate, follow row to row spacing of 20.0 - 22.5 cm with the same seed drill which is used for wheat. It adversely affects the productivity of the soybean crop through crowding of branchesreducing sunlight to the ground and lower branches, leading to excess vegetative growth, weak flowering and poor pod set *etc*. This hampers water and nutrient management too (Ramesh *et al.,* 2017).

At the high rate of fertilization, the hybrids effectively use applied P and produce higher yield per unit quantity of P, which is improved even further when closer (more plant population density) per unit area is adopted. Contrary to this, conventional varieties are shown to be less efficient in using higher rates of P at close spacing. Although hybrids are efficient P user, low plant populations fail to show their fullest potential. While targeting for high yields, it is not only important to select the right plant type, but also to optimize population to improve NUE, especially at higher rates of nutrients.

Nutrient management

Required plant nutrients can be most successfully applied following the 4R Nutrient Stewardship Principles either it be chemical fertilizers or organic fertilizers. This applies to all nutrients that are required above what the soil can supply, and for all crops, cropping situations, and nutrient sources. Having an adequate and balanced supply of plant nutrients is essential for optimum yielding crops, especially as cropping practices have improved using higher yielding varieties, moisture conserving conservation tillage, and more effective pest control options. As crop yields increase there is greater demand on our soils to supply both a sufficient and balanced supply of nutrients.

4R nutrient stewardship provides a framework to achieve cropping system goals, such as increased production, increased farmer profitability, enhanced environmental protection and improved sustainability.

To achieve those goals, the 4R concept incorporates the following

- Right fertilizer source
- Right rate

- Right time
- Right place

Although Soybean needs a starter dose of nitrogen alongwith adequate P and K, in reality, most of the farmers don't apply fertilizers due to short sowing window and dependence on fertilizer-cum-seed drill for sowing where diammonium phosphate (DAP) is the only fertilizer source for soybean (Ramesh *et al.*, 2017) and in few cases only urea is applied as fertilizer dose for Soybean in Madhya Pradesh. The simple agronomic practice of split-applying nutrients does not increase production cost by much, but it significantly improves NUE. It is now well established that for most crops, N must be applied in 2-3 or more spilt does coinciding with the crop growth stages when N requirement is high. Appropriate rate, source and application method increase nutrient utilization by crop plants. In case of mobile nutrients, such as N, application of higher doses at one time may result in nutrient loss due to leaching in humid climates, and N efficiency is greatly reduced. The source and method of applying urea-containing fertilizers affect N-use efficiency. Reduced N efficiency from surface application of urea-containing N sources is primarily due to losses through NH_3volatilization as urea is hydrolyzed. In case of P, under conditions of low P availability, banded P is usually more effective than broadcast application. Seed or row placement of P has been the most effective method of P application on low P soils. A common belief is that row application is twice as effective as broadcast application.

Crops do vary greatly in their ability to utilize P from fertilizer. This differential behavior may be related to their demand pattern for P. A quantum of research evidence also indicates that P utilization by crops vary with source of P fertilizers.

Potassium is normally applied at the time of planting, owing to its large availability by diffusion. But in coarse textured soils, leaching of K can occur, and split application may increase K use efficiency plants in such soils.

Increasing NUE is an offspring of balanced fertilization and sound management practices and decisions. Balanced fertilizer use is not only the first requirement but it is a pre-requisite also because agronomic manipulation cannot produce high efficiency out of imbalanced nutrient application. When balanced fertilization is practiced, one nutrient increases the efficiency of others through a synergistic effect. Traditionally, in India, balanced fertilization indicates the use of N, P and K in a certain ratio (ideally 4:2:1) on gross basis both in respect of areas and crops.

Weed Management

Weeds compete with crop plants for nutrient, soil moisture, sunlight and space. Of course, the intensity of weed competition depends upon the type of weed species, severity of infestation, duration of weed infestation etc. Consequently, crop yield is reduced. Reduction in crop yield has a direct correlation with competition. Of the total annual loss of agricultural produce from carious pests in India, weeds cause highest loss, about 33%. The most important point is that weeds require the same or greater amount of nutrients and at same time weeds remove plant nutrients more efficiently than crop plants. Therefore, timely and appropriate weed control greatly increases the crop yield and NUE. In absence of an effective weed control measure, they remove considerable quantity of applied nutrients, resulting in loss of yield.

Conclusion

Thus, while summing up, if our goal is to increase the nutrient use efficiency, it is important to know the role of each factor in the uptake and utilization processes of nutrients as well as agronomic management factors. Simple non-monetary measures could enhance the use efficiency of nutrient to a great extent. The best management practices, either individually or collectively, has direct bearing on the yield and NUE of crops. Furthermore, NUE greatly depends upon how efficiently nutrient and crop management practices work together in a soil-climate complex in producing higher yield.

Selected References

IPNI. 2013. Plant Nutrition Today. Spring 2013, No. 2.

Kandiannan K, Chandaragiri KK. 2008. Monetary and non-monetary inputs on turmeric growth, nutrient uptake, yield and economics under irrigated condition. *Indian Journal of Horticulture* 65(2): 209-13.

Mandal KG. 2005. Non-monetary and low-cost agronomic measures for enhancing nutrient use efficiency. Lecture delivered in the summer school on "Advances in frontier approaches to increase nutrient use efficiency in crop production" held during 22 Jul-11 Aug 2005, IISS, Bhopal.

Prasad R, Rai RK, Sharma SN, Singh P, Prasad R, Ahlawat IP (Eds.) 1998. Fifty Years of Agronomic Research in India, Indian Society of Agronomy, New Delhi, pp. 51-86.

Prasad P. 1996. Management of fertilizer nitrogen for higher efficiency. In: Tandon, H.L.S. (Ed.) Nitrogen Research and Crop Production Fertilizer Development and Consolation Organization, New Delhi, India, pp. 104-115.

Ramesh K, Patra AK, Biswas AK. 2017. Best Management Practices for Soybean under Soybean-Wheat System to Minimize the Impact of Climate Change. *Indian Journal of Fertilisers* 13 (2): 42-55.

Shekhawat K, Rathore SS, Premi OP, Kandpal, BK, Chauhan, JS. 2012. Advances in Agronomic Management of Indian Mustard (*Brassica juncea* (L.) *Czernj & Cosson*): An Overview. *International Journal of Agronomy* 408284: 1-14.

Singh G. 1993. Integrated weed management in pulses. In: Proc. of International Symposium on Integrated Weed Management for Sustainable Agriculture, Vol. I, Indian Society of Weed Science, Hisar, India, pp. 335-342.

Verma NK. 2011. Integrated nutrient management in winter maize (*Zea mays* L.) sown at different dates. *Journal of Plant Breeding and Crop Science* 3(8):161-167.

5

Enhancing Nutrient Use Efficiency, pp. 51-58
Editors: K. Ramesh, A.K. Biswas, B.L. Lakaria, S. Srivastava and A.K. Patra

Plant Mechanisms for Nutrient Uptake and its Efficient Use

R. Elanchezhian[1], V. Rajagopal[2] and K. Ramesh[1]

[1]ICAR-Indian Institute of Soil Science, Nabi Bagh, Bhopal – 462038, India
[2]ICAR-National Institute of Abiotic Stress Management, Baramati – 413 115, India

Plant mineral nutrition is a sub branch of plant physiology which addresses issues on uptake of elements that plant needs to absorb from air or water or soil to grow and to live. Besides function of nutrient elements in plants, it also includes deficiency symptoms and how to address deficiency symptoms in a plant. Plant mineral nutrition and mechanism of uptake has two major aspects:

a) Acquisition of mineral nutrients.

b) Functions of those minerals in the plant.

a) Acquisition of mineral nutrients

Plants, just like any other living organism require certain elements for normal growth, reproduction and completion of their lifecycle. The ultimate reservoirs of these elements on earth are the rocks, oceans and the atmosphere. Rocks are weathered into soil over a period of geological time scale. Oceans and the water that is released into lakes, streams and soils as well as the atmosphere itself make up the compounds, aggregates, solution and gases which living organisms mine and exploit for the essential elements, which plants absorb to form their bodies. Only green plants and certain microorganisms are capable of extracting simple ions and inorganic compounds from the environment without having to rely on complex energy rich compounds previously synthesized by other living organisms. These organisms are autotrophic. The photosynthetic plants, algae in oceans, streams and lakes and the green plants on land are by far the most important agents in the primary acquisition of elements from the external environment. Once these elements are acquired, they are translocated to different parts for further use in metabolism by using the elaborate structures and mechanisms, which effect a long distance translocation of water and solute within the plant bodies.

b) Functions of nutrients

If one of the elements essential to the plant is present in the environment in insufficient amount/unavailable forms or in difficultly available forms, the deficiencies of these elements in the cells of plant will bring about derangement or destruction in metabolism *e.g.* P- fruiting, rooting, flowering, N – vegetative growth (yellowing). These metabolic disturbances will manifest in visible symptoms such as stunted growth, yellowish/purpling of leaves and other abnormalities. These symptoms of deficiency are characteristic for a given nutrient element. However, the symptoms depend also on the severity of deficiency, the particular species of plant and many other environmental factors and state of the plant (healthy or not). All these reasons put together will make us focus attention on the functional role of elements in the metabolism of plants. The elements also constitute structural components of compounds and metabolites in the plant.

Historical perspective of plant nutrition

Early theories on plant nutrition were of philosophical in nature in the absence of the knowledge of chemistry and experimental techniques. The advancement of science was hindered because of this situation, particularly by the general acceptance of a speculative scheme that was advanced by the Greek philosopher Aristotle (384-322 B.C). He held the view that all matter consisted of four elements viz. i) Earth, ii) Water, iii) Air and iv) Fire.

In the modern era of science, it was Nicholas de Cusa (1450 AD) who recognized that plants take up ash constituents in small amount from the soil and these are transported through water, which forms the bulk of the plant. Thereafter, Job van Helmont (1577-1644 AD) a Belgian physician, who was honoured as being the first man to conduct qualitative experiment in plant nutrition, investigated the source of materials that plants are composed of.

Before 18th century, scientists believed that plant obtained all their nutrition through soil. However, Steven Hales, considered as the father of plant physiology, suggested in 1727 that plant obtain part of their nourishment through atmosphere paving the way for involvement of gaseous non-mineral elements in nutrition. In 1771, Joseph Priestly, an English chemist implicated the role of O_2 in green plants for purifying the air. The Dutch physician, Jan IngenHousz, demonstrated that light was essential for this purification and his work also showed that leaves are the primary organs of food production and carbon from carbohydrate is the primary source of carbon for plants in spite of its low concentration in the air. Jean Senebier in 1782 repeated several experiments of IngenHousz and demonstrated that only the green portion of leaves were photosynthetically active. He was the first investigator to give a reasonable insight into photosynthesis. He found out that the amount of O_2 given out by green plants kept in water was proportional to the concentration of CO_2 dissolved in water. He also showed that it was light and not the heat from the sun, which induces giving off O_2. The work of Lavoisier and others also made it clear that CO_2 and O_2 are essential for photosynthesis of plants. The role of water was implicated by NT de Saussure in 1804 when he made the quantitative estimation of H_2O during photosynthesis.

For an element to be regarded as an essential nutrient, it must satisfy the following criteria, as propounded by Arnon and Stout (1939):

- A deficiency of an essential nutrient element makes it impossible for the plant to complete the vegetative or reproductive stage of its life cycle.
- The deficiency is specific to the element and can be prevented or corrected only by supplying that element.
- The element is involved directly in the nutrition of the plant as for example as a constituent of an essential metabolic or required for the action of an enzyme system.

All mineral nutrients together make up less than 4% of plant mass, yet plant growth is very sensitive to nutrient deficiency. The various forms of macro and micro nutrient used by plants (Marschner, 1985) are given in Table 1. The concentration of macro and micronutrients in plant dry biomass and their abundance in plant parts with respect to Molybdenum is given in Table 2.

Table 1: Forms of macro and micro nutrient used by plants

Macronutrient	Form Used	Micronutrient	Form Used
Carbon	CO_2	Iron	Fe^{2+}
Oxygen	H_2O, O_2	Manganese	Mn^{2+}
Hydrogen	H_2O	Boron	$H_2BO_3^-$
Nitrogen	NO_3^-, NH_4^+	Molybdenum	MoO_4^{2-}
Phosphorus	H_2PO_4& HPO_4^{2-}	Copper	Cu^{2+}
Potassium	K^+	Zinc	Zn^{2+}
Calcium	Ca_2^+	Nickel	Ni^{2+}
Magnesium	Mg^{2+}	Chlorine	Cl^-
Sulfur	SO_4^-		

Table 2: Tissue concentration and relative abundance of atoms of macronutrients in plant biomass

Element	Chemical symbol	Concentration in dry matter (%)	Relative number of atoms with respect to molybdenum
Non-mineral elements			
Hydrogen	H	6	60,000,000
Carban	C	45	40,000,000
Oxygen	O	45	30,000,000
Macronutrients			
Nitrogen	N	1.5	1,000,000
Potassium	P	1.0	2,50,000
Calcium	Ca	0.5	1,25,000
Magnesium	Mg	0.2	80,000
Phosphorus	P	0.2	60,000
Sulfur	S	0.1	30,000
Silicon	Si	0.1	30,000
Micronutrients		**Concentration in dry matter (ppm)**	
Chlorine	Cl	100	3,000
Iron	Fe	100	2,000
Boron	B	20	2,000
Manganese	Mn	50	1,000
Sodium		10	400
Zinc	Zn	20	300
Copper	Cu	6	100
Nickel	Ni	0.1	2
Molybdenum	Mo	0.1	1

Source: Epstein 1972; 1999.

The values for the non-mineral elements (H, C, O) and the macronutrients are percentages. The values for micronutrients are expressed in parts per million.

In general, soil particles are negatively charged and so bind positively charged ions (cations). The cation exchange capacity (CEC) refers to soil's ability to bind cations. Roots are the primary entry point for the nutrients in to the plants. Root hairs provide large surface area for water and nutrient uptake. Water and mineral salts from the soil enter the plant through the epidermis of roots and ultimately flow to the shoot system.

- The endodermis is the innermost layer of cells in the root cortex
- It surrounds the vascular cylinder and is the last checkpoint for selective passage of minerals from the cortex into the vascular tissue
- Water can cross the cortex via the symplast or apoplast

- The waxy Casparian strip of the endodermal wall blocks apoplastic transfer of minerals from the cortex to the vascular cylinder

Mineral uptake mechanisms: There are several mechanisms by which nutrients move into plant systems, which are as follows.

- Mineral movement to root by diffusion or bulk flow
- Uptake controlled at root endodermis
- Uptake by either simple diffusion (no protein), facilitated diffusion (protein channel), or active uptake (requires energy and a protein carrier)
- Plants concentrate minerals and most other substances
- Usually biggest energy expenditure of roots

Nutrient acquisition efficiency and transporter systems

The spatial configuration and distribution of these roots determine root system architecture in the soil, which in turn primarily regulates the acquisition of soil resources like nutrients and water (Wang *et al.,* 2006). A key step in mineral nutrient acquisition is the initial trans-membrane transport step. In many cases, for any individual nutrient, there are gene families encoding multiple homologs. In *Arabidopsis*, for example, there are two gene families for nitrate transporters, *NRT1* and *NRT2*, with 53 and 7 members, respectively, a gene family of 14 sulfate transporters (Hawkesford, 2003; Kataoka *et al.,* 2004) and 9 members of the phosphate transporter family pht1 (Smith *et al.*, 2003; Buchner *et al.,* 2004). While in most cases, there are families specific for a single nutrient, there are instances of non-specificity: Sulfate transporters effectively transport selenate and molybdate (Shinmachi *et al.*, 2010).

While there is some potential redundancy of function with these large gene families, it has become apparent that there is tissue, developmental, and even membrane specificity with regard to expression patterns. Functionally, there are usually both high and low affinities for the substrate ions, depending on functional requirements: In relation to primary uptake into root cells, the most common functionality is for high-affinity uptake, as required for effective acquisition from soil solutions with low concentrations of ions. Patterns of expression within the root are often complex to effectively transfer the respective ions from the soil solution to the vasculature for transfer to the shoot material. In some instances, vacuolar storage may also play an important part (Kataoka *et al.*, 2004). Many studies have focused on the impacts of nutrient limitation on patterns of transporter expression and the contribution to overall nutrient use efficiency strategies of plants in limiting nutrient availability (Buchner *et al.*,

2010). For phosphate and sulfate, there is an apparent de-repression system controlling gene expression, facilitating increased expression when nutrient demand exceeds availability (Hawkesford and De Kok, 2006). For nitrate, the pattern is more complex, with some transporters induced and others repressed, depending on the presence of nitrate and the nutritional status of the plant.

The transporters play essential roles, contributing to nutrient use efficiency, for the most part extremely effectively scavenging nutrients from the soil (potentially present at low concentrations), and particularly in conjunction with effective root proliferation. As targets for improvement of NUE, sophisticated strategies are likely to be important. Modifications to the selectivity (Rogers *et al.*, 2000) may enhance preferential uptake of beneficial ions and exclude toxic ions. Overriding negative feedback mechanisms may facilitate luxury uptake, but appropriate sinks or temporary storage would also be required. One approach that apparently overrides limits on nitrogen uptake is the overexpression of alanine amino transferase in root exodermal tissues, thus channeling nitrogen away from metabolites involved in negative feedback. In some instances, enhancing remobilization and optimizing partitioning to harvested organs may require optimization of transporter expression.

Metabolic responses to nutrient availability

Plant responses to nutrient availability are complex and involve changes in pathway fluxes, in activity of pathway enzymes mediated by post-translation modifications and/or changes in substrate/inhibitor ratios (allosteric effects), as well as changes in expression of genes encoding the pathway enzymes and many additional proteins. The challenge for the plant is to optimize growth and development given the available nutrient inputs. Matching availability to demand may entail many regulatory steps and sensory mechanisms. It is essential to understand these networks before intervention through trans-genesis or molecular breeding. For the most part, our knowledge of these regulatory loops is restricted in plants (Gojon *et al.*, 2009).

Nutrient use efficiency, although simply divided into uptake and utilization, encompasses all processes of plant growth and development, and all aspects of metabolism. Potential targets for nutrient use efficiency improvement are therefore diverse. Obvious targets in, for example, nitrogen metabolism include genes of the assimilatory pathway. Glutamine synthetase has been a specific target for transgenic approaches, as it is not only involved in primary assimilation but also has a role in efficient recycling of ammonia during senescence processes (Kichey *et al.*, 2006). Generally, results of single-gene manipulation have been disappointing, in part because metabolic pathways form networks that have a great plasticity in responding to perturbations, whether due to gene targeting or environmental fluctuations, for example, in nutrient supply. Typically, nutrient uptake is balanced by nutritional requirement for growth, and a coordination of pathway expression and activity is seen (Hawkesford and De Kok, 2006; Gojon *et al.*, 2009) and excess uptake of nutrient is avoided. Excess accumulation of some ions does occur but only to the point at which available storage pools are saturated (for example, nitrate accumulation in vacuoles); this is a strategy to aid with fluctuating supplies of nutrients but is not helpful when one nutrient becomes permanently limiting.

Future perspectives for integrated approach

- The existence of genotypic variation in nutrient uptake and variable performance of crop genotypes in limited environment provides an opportunity for saving nutrient resources through selection and breeding of efficient cultivars
- The genetic variability should be used in targeted plant breeding to tailor new crop genotypes for fitting them in nutrient limiting soils
- Agronomic management also be included for effective NUE

Nutrient use efficiency in its broadest sense indicates how effectively a plant is able to capture and utilize nutrients to produce biomass. It is most usually specified for nitrogen as this is a main driver for production. However, healthy and productive crop growth requires a balanced nutrition including several macronutrients and many micronutrients. Irrespective of the quantity needed, all are essential and any limitation will impact on plant growth and crop yields.

In almost all cases, the nutrient in question must be obtained from the pedosphere and therefore uptake processes dependent on architecture and functioning of the roots are critical. Subsequent to this, partitioning within the plant is a vital prerequisite to efficient utilization of the element as part of the plant's growth and developmental cycle. Independent but simultaneous selection for both of these traits must be performed. A radical and alternative solution to providing nitrogen fertilizer would be the transfer of nitrogen fixation capacity, or the ability to form the required symbioses, to non-legume crops.

NUE is an essential component of crop production, and irrespective of the agronomic system, low-input or intense, efficient utilization of valuable resources will be essential for future sustainable food production. NUE is a complex trait that can be broken down into subtraits, all of which are also complex in nature. Few instances can be expected where single genes or a single locus will have a huge benefit; dwarfing genes were an exception. Modern tools and resources available to plant scientists and the agronomy and breeding communities should aid further improvements in NUE and hence crop production. Great variability exists in the extent to which individual crops have been optimized in relation to NUE, and while large improvements may be anticipated for some crops, for the major world grain crops such as wheat maize and rice, smaller incremental improvements are likely. The prospect of step changes in primary production by engineering the photosynthetic process itself will require additional concomitant improvements in nutrient acquisition efficiency.

Selected References

Buchner P, Parmar S, Kriegel A, Carpentier M, Hawkesford MJ. 2010. The Sulfate Transporter Family in Wheat: Tissue-Specific Gene Expression in Relation to Nutrition. *Molecular Plant* 3 (2): 374–389.

Buchner P, Takahashi H, Hawkesford MJ. 2004. Plant sulphate transporters: co-ordination of uptake, intracellular and long-distance transport. *Journal of Experimental Botany* 55: 1765–1774.

Epstein, E.1972. Mineral Nutrition of Plants: Principles and Perspectives, New York: John Wiley & Sons.

Gahoonia, TS, Nielsen NE.1998. Direct evidence on the participation of root hairs in uptake of phosphorus from soil. *Plant and Soil* 198: 147–152.

Gahoonia, TS, Nielsen, NE. 2003. Phosphorus (P) uptake and growth of a root hairless barley mutant (bald root barley, *brb*) and wild type in low- and high-P soils. *Plant, Cell and Environment* 26: 1759–1766.

Gojon A, Nacry P, Davidian JC. 2009. Root uptake regulation: a central process for NPS homeostasis in plants. *Current Opinion in Plant Biology* 12: 328 – 338.

Hawkesford MJ, De Kok. 2006. Managing sulphur metabolism in plants. *Plant Cell and Environment* 29 (3): 382–395.

Hawkesford MJ.2003. Transporter gene families in plants: The sulphate transporter gene family—redundancy or specialization? *Physiologia Plantarum* 117: 155–163.

Kataoka T, Watanabe-Takahashi A, Hayashi N, Ohnishi M, Mimura T, Buchner P, Hawkesford MJ, Yamaya T, Takahashi H. 2004. Vacuolarsulfate transporters are essential determinants controlling internal distribution of sulfate in Arabidopsis. *Plant Cell* 16: 2693–2704.

Kichey T, Heumez E, Pocholle D, Pageau K, Vanacker H, Dubois F, Le Gouis J, Hirel B. 2006. Combined agronomic and physiological aspects of nitrogen management in wheat highlight a central role for glutamine synthetase. *New Phytologist* 169(2): 265.

Marschner H. 1985. Mineral Nutrition of higher plants. Academic Press London.

Rogers EE, Eide DJ, Guerinot ML. 2000. Altered selectivity in an Arabidopsis metal transporter. *Proceedings of National Academy of Science. U.S.A.* 97: 12356–12360.

Shinmachi F, Buchner P, Stroud JL, Parmar S, Zhao FJ, McGrath SP, Hawkesford MJ. 2010. Influence of sulfur deficiency on the Expression of specific sulfate transporters and the distribution of sulfur, selenium, and molybdenum in wheat. *Plant Physiology* 153: 327–336.

Smith, FW, Mudge, SR, Rae AL,Glassop, D. 2003. Phosphate transport in plants. *Plant and Soil* 248: 71–83.

Wang H, Inukai Y, Yamauchi A. 2006. Root development and nutrient uptake. *Critical Reviews in Plant Science* 25(3): 279-301.

6

Enhancing Nutrient Use Efficiency, pp. 59-79
Editors: K. Ramesh, A.K. Biswas, B.L. Lakaria, S. Srivastava and A.K. Patra

Controlled Release Fertilizers for Enhancing Nitrogen Use Efficiency

B.P. Meena, K. Ramesh, Neenu S., Pramod Jha and I. Rashmi

ICAR-Indian Institute of Soil Science, Nabi Bagh, Bhopal – 462038, India

Introduction

Nitrogen (N) fertilization assured centre stage for enhancing food production in developing countries especially after the introduction of high yielding and fertilizer responsive crop varieties. Almost half of the human population relies on N fertilizer for food production (Ladha *et al*., 2005), and about 56 percent of the N fertilizer is used for producing rice, maize, and wheat (IFA, 2002). It has contributed an estimated 40% to the increase in per capita food production over the past 50 years (Smil, 2002). However, excess use of chemical N fertilizer may negatively affect surface water, as well as groundwater, and the atmosphere (through leaching and runoff) through volatilization of N (Galloway *et al*., 2008). The average nitrogen-use efficiency (NUE), especially through chemical fertilizers such as urea, in India ranges from 20% to 50% for rice (Prasad *et al*., 1998). Raun and Johnson (1999) have estimated NUE below 33% for cereal production at the global scale because of 50% of the N applied is not assimilated by the plant (Tilman *et al*., 2002). Although the remaining N may remains in the soil, its recovery to succeeding crops is very limited (<7% of applied N up to six consecutive crops), it will be lost from the soil–plant system via volatilization, denitrification, leaching and stabilization into SOM causing serious disruptions in ecosystem function (Ladha *et al*., 2005). Thus quite a high proportion of the applied N is lost in one way or through the others resulting in discouragingly low N use efficiency. Under these situations, increasing crop yield per unit area through the use of efficient N management strategies has become an essential component of modern crop production (Fageria and Barbosa Filho, 2001). When N fertilizers like urea are applied to soils, there is inevitable hydrolysis fairly rapidly to ammonium carbonate over a maximum period of 3-7 days (Mohanty *et al.*, 1999). The ammoniacal form of N is subsequently converted to nitrite and then nitrate by the action of nitrifying bacteria viz., *Nitrosmonas* spp. and *Nitrobacter* spp., respectively. Therefore, nitrate formed as a result of relatively rapid hydrolysis, being highly soluble is liable to be leached down the soil profile, beyond the active root zone of the crops. Moreover, under waterlogged conditions, nitrates are reduced to nitrous oxide and elemental N by the action of denitrifying bacteria to meet their oxygen demand. This is led to the development of N deficiency and extremely low N uptake even under best management condition. The complexities of N management occur due to its solubility, mobility and vulnerability to denitrification (Hegde *et al.*, 2007). Reducing water solubility of the nitrogen fertilizer sources would eliminate N losses so that crop uptake may be more or if the crop does utilize more N could be retained in the soil for use of subsequent crop. Therefore, approaches have been to manipulate the granule size variations, different coatings materials, modifiers or additives to slowing the nutrient release rate.

Controlled/slow-release fertilizers

A fertilizer containing a plant nutrient in a form which extends its availability to the plant significantly for longer period than quickly available nutrient fertilizers like urea, ammonium nitrate, ammonium phosphate or potassium chloride. The delay of initial availability or extended time of continued availability may occur by a variety of mechanisms. These include reduced water solubility of the material by semi-permeable coatings, occlusion, protein materials, or other chemical forms, by slow hydrolysis of water-soluble low molecular weight compounds, or by other materials. The water solubility of nitrogenous compound can be reduced by the physical and chemical methods. Physical method is comprised of coated or encapsulated fertilizers and chemical methods include the conversion of N to polymeric form that has reduced water solubility. Controlled/slow-release fertilizers (SRFs) is a possible alternative to common compound fertilizer (CCF) to increase N uptake efficiency and minimise N losses to the environment. Conceptually, controlled/slow-release fertilizers (SRFs)/ nitrification inhibitors (NI)/ urease inhibitors (UI) provide the only effective solution to the problems created by the ever increasing demand and use of water soluble nitrogenous fertilizers. Slow release N fertilizers have been largely used to improve fertilizer use efficiency, reduce fertilizer input and improve crop yield and quality (Shoji *et al.*, 2001). Thus, the controlled/slow release fertilizers (CRFs) may have enhanced N-use efficiency, through reduction in N loss through leaching and volatilization while keeping higher yields (Mao *et al.*, 2005). The CRFs showed advantages in slowing down the N release pattern, promoting N uptake by the crops, increasing grain yield and reducing N application rate. Reports suggested that single basal application of rice-specific CRFs could increase the nitrogen use efficiency by 12.2–22.7% and 17.1% in pot and filed condition, respectively (Chen *et al.*, 2005). Extensive research have been conducted over the past decades on the evaluation of modified urea forms (Katyal *et al.*, 1985; Prasad *et al* , 1996 and Trenkel 1997) for improving NUE. The condensation product of urea and urea aldehydes, of which are the most significant type on the market are urea formaldehyde (UF), isobutylidene diurea (IBDU), and crotonylidene diurea (CDU). Simultaneously, synthetic chemical nitrification inhibitors such as PPDA (Phenyl phosphorodiamidate), Nitrapyrin or N-Serve, AM (2-amino-4-chloro, 6-methyl pyridine), DCD (diacynamide), ATC, thiourea, hydroquinone, calcium carbide and several natural products like neem cake, karanj cake, neem oil, nimin etc evolved through research. Some researchers have suggested the use of urease inhibitors for reducing NH_3 volatilization losses and the most widely tested urease inhibitor is NBTPT or NBPT (Hendrickson, 1992). In India, neem cake coated urea (NCCU), developed by Prasad (1980) was shown to have nitrification inhibiting properties (Thomas and Prasad, 1982) and increased yield in rice (Sudhakara and Prasad, 1986) and rice-wheat cropping systems (Prasad *et al.*, 1981). Application of CRFs are for regulating the N supply to crop by slowing down the rate of hydrolysis or nitrification or both to ensure continuous and need based supply of N to match the demand of crops at different growth stages. Nutrient release from coated fertilizer may depend on coating thickness (Shaviv *et al.*, 2003). Different longevity ratings of slow release fertilizers could well be synchronized with nutrient demand at various growth stages with nutrient release pattern, with uniform efficiency on both coarse as well as fine textured soils (Alila and Srivastava, 2008) (Table 1).

Table 1: Characteristic features of various types coating materials and their effectiveness.

Coating material	Nutrient supplied	Effectiveness
SCU	N, S	Variable approximately per cent/day
Osmocote	N,P,K	3 to 4 months, 8 to 9 months
IBDU	N	58% in 21 weeks for 1.0 to 1.2 mm particles
CDU	N	31% in 10 weeks
Triazimes	N	10% for 15 weeks
MagAmp	N, P, Mg	100 days for incorporated coarse granules
Gypsum	N, Ca, S	3 to 4 months
Ureaform	N	60% in 6 months for 75% insoluble material
Activated sludge (Milor ganite)	N,P,K	2-4 material
Ammelide	N,P,K	35 per cent in 7 days at 31°C ± 1.5°C

(SCU, IBU and CDU stand for sulfur coated urea, isobutylidene urea, and crotonylidene diurea, respectively)

Condensation products of urea

The condensation of urea with aldehydes (and particularly with formaldehyde) is one of the most common methods for preparing SRFs. Urea formaldehyde/ Urea forms (UF), urea-isobutyraldehyde/isobutylidene diurea (IBDU), and urea-acetaldehyde/cyclo diurea (CDU) are the most popular condensation products of urea used for the slow release of N fertilizers. In the past, urea-aldehyde reaction products had the largest share of the slow-release fertilizers, but now a days, the polymer-coated / polymer-sulphur-coated products have taken the lead; mainly due to increased production capacity by Kingenta (China) and Agrium/Hanfeng (Canada/China).

Urea formaldehyde or Urea forms

UF is the most popular organic-N compound used for the slow release of N, and the most widely used as slow release fertilizers (Trenkel, 1997). UFs are white, odourless solids containing about 38% N which are made by reacting urea with formaldehyde in the presence of catalyst and excess urea under controlled conditions of pH, temperature, mole proportion, reaction time, etc., resulting in a mixture of methylene ureas with different long-chain polymers. UF fertilizers show a significant slow release of N with a good compatibility with most crops. Because of its low solubility it will not scorch vegetation or impare germination. Because it is more effective at higher temperatures and it is widely used in USA for fertilizing lawns.

Isobutylidene diurea

Isobutylidene diurea (IBDU) is formed as a condensation product by a reaction of isobutyraldehyde with urea. In contrast to the condensation of urea with formaldehyde which results in a number of different polymer chain lengths, the reaction of urea with isobutyraldehyde results in a single oligomer. Theoretically, it contains 32.2% N, but fertilizer grades contain about 30% N. The solubility of IBDU in water is initially very low at room temperature but hydrolysis occurs rapidly once dissolution begins. The rate of hydrolysis is not stable but is greater under acid conditions and at higher temperature (Hauck, 1972). Based on agronomic response and safety margin, IBDU is good for turf, but phytotoxicity has sometimes been observed with greenhouse crops. IBDU gave 20% more rice grain yield than ammonium sulphate at equivalent rate of N application (Rajale and Prasad 1975). Release of N from IBDU depends upon particle size (the finer the particle size the more rapid the release), moisture, temperature and pH and it is preferably applied at lower temperatures. Hughes (1976) reported that N release from IBDU was much more rapid in acid than in alkaline soils.

Crotonylidene diurea

Crotonylidene diurea (CDU) is another such type of slow release N fertilizer is formed by the reaction of ureas with crotonaldehyde or acetaldehyde. Powdered CDU containing 31% N has been directly used as a fertilizer (Prasad *et al.*, 1971). CDU decomposes by both hydrolysis and microbial processes in the soil, and temperature, soil moisture and biological activity affect the rate of release. The degradation is slower than that of IBDU, even in acid soils (Trenkel, 2010).

Coated or encapsulated fertilizers

These are conventional soluble N fertilizer materials with rapidly available plant nutrients, which after granulation, prilling or crystallization are given a protective, water-insoluble coating to control water penetration and thus dissolution rate, nutrient release and duration of release. Association of American Plant Food Control Officials (AAPFCO) (1995) defined them as products containing sources of water soluble nutrients, the release of which in the soil is controlled by a coating applied to the fertilizer. Coated/encapsulated fertilizers offer flexibility in determining the nutrient release pattern (Shoji and Takahashi, 1999). Nyborg *et al.* (1999) reported slowing the release pattern of fertilizer in soil and improved yield by coating fertilizer granules (polymer coating) in greenhouse and field condition. Coated/encapsulated fertilizers are classified based on the following coating materials such as sulphur, sulphur plus polymers, including wax polymeric materials (*e.g.*, polyvinylidene chloride (PVDC)-based copolymers, gel-forming polymers, polyolefine, polyethylene, ethylene-vinyl-acetate, polyesters, urea formaldehyde resin, alkyd-type resins, polyurethane-like resins, etc), fatty acid salts (*e.g.* calcium-stereate); latex, rubber, guar gum, petroleum derived anti-caking agents, wax; calcium and magnesium phosphates, magnesium oxide, magnesium ammonium phosphate and magnesium potassium phosphate; phosphogypsum, phosphate rock, attapulgite clay; peat (encapsulating within peat pellets: organo-mineral fertilizers); neem cake/nimin extract.

Sulphur coated urea

Sulphur-coated urea (SCU) has been developed at the TVA (Tennessee Valley Authority) in 1961. It is prepared by spraying molten sulphur over preheated urea granule. Elemental sulphur, a low-cost secondary plant nutrient, was found to be convenient for coating due to its ability to melt at about 156°C, thus enabling spraying molten S over granular urea and possibly on other fertilizers as well. The release of N from SCU depends on coating quality (Oertli, 1980). The product contains between 31 to 38% N. After coating the urea with sulphur, a wax sealant is sprayed in order to seal cracks in the coating and reduce the microbial degradation of the sulphur coating (Allen *et al.*, 1971). This is followed by a layer of conditioner such as attapulgite or talc. The sealant may be replaced by an organic polymer layer (thermoplastic or resin) to produce a polymer coated SCU called PSCU (Polymer coating of sulphur coated urea). The additional polymer layer was also intended to improve the attrition resistance of the coated granules and showed a much better performance than the SCU (Goertz, 1995). The polymer sulphur-based coated fertilizers and SCU form the majority of coated products are used in non-agricultural markets, such as turfs, landscaping, and horticulture. The release of N from SCU may be controlled by the thickness of the coating, placement, microbicides, temperature, and time of the contact with the soil and release of N from the SCU was about 1% per day (Lunt, 1971).

Similarly, Sharma (1994) also reported the relative performance of different modified urea fertilizer and found that sulphur coated urea (SCU) has shown better performance in low land rice than other modified urea fertilizers due to extended duration of N release, synchronize with important stages of crop growth (Table 2). Eman and El-Ashry (2009) reported that application of SCU was more effective in facilitating N uptake and resulted in increased apparent N recovery, N agronomic efficiency and physiological efficiency as compared to other fertilization treatments of lemon balm (Table 3). SCU was significantly superior to other N treatments including split applied prilled urea. This effect might be due to higher utilization of N by the plants as results of reduction of losses of fertilizer by the regulation of urea hydrolysis and nitrification (Kiran and Patra, 2003).

Table 2: Relative performance of modified urea fertilizers applied (basally@40 kg N/ha) in flood prone low land rice at Cuttack

N fertilizers	Relative grain yield	Agronomic efficiency (kg grain/kg/kg N)	Apparent N Recovery (%)
No nitrogen	100	-	-
Prilled urea	140	26	18
Urea super granules	158	39	40
Sulphur coated urea	164	45	45
Lac coated urea	140	29	26
Neem cake coated urea	135	22	16
Coaltar coated urea	122	18	15
FYM enriched urea	147	30	29

Sharma (1994)

Table 3: Influences of slow release urea fertilizer on N uptake and N use efficiency of lemon balm plant (N rate 340 kg N/ha).

N fertilizers	N uptake (kg/ha)	Apparent N recovery (%)	Agronomic efficiency (kg grain/kg/kg N)	N physiological efficiency (kg grain/kg/kg N)
No nitrogen	77.5	-	-	-
Prilled urea (full dose)	151.8	20.8	5.5	26.7
Prilled urea (2 split)	174.2	26.9	7.7	28.6
Prilled urea(4 split)	199.3	33.9	10.4	30.7
Sulphur coated urea	227.8	41.8	13.0	31.1
Neem coated urea	195.8	32.7	9.9	29.8
LSD (P=0.05)	3.45	0.96	0.40	0.63

Eman and El-Ashry (2009)

Polymer coated urea

Polymer-coated urea (PCU) is an important slow release fertilizer for improving the N use efficiency (NUE). PCU as controlled- release N sources such as POLYON-coated urea by Pursell, 'ESN' by Agrium, 'Osmocote' by Scotts, Meister by Chisso-Asahi, and many others (Trenkel, 1997). The coatings are usually resins or thermoplastic materials and their weight can be as low as <1% of the granule mass without significantly reducing the N content. Unlike SCU which releases urea through small pinholes that can result in a more difficult controlled-N release pattern, PCU releases N by diffusion of urea through the swelling polymer membrane. The release pattern is related to the coating composition and usually depends on soil moisture and temperature (Christianson,1988). It requires in a smaller quantity (3-6%) unlike sulphur or neem cake, which are needed in larger quantities. Salman (1998) reported that the PCU has slower dissolution rate than urea and reduce ammonia volatilization and leaching losses of N. In a field experiment although PCU did not give a significant increase in yield, recorded higher agronomic efficiency and apparent recovery of PCU was used and performs better than urea (Fashola *et al.*, 2002). In field trials, Singh *et al.* (1995) reported that grain yield of lowland rice from a single application of PCU was equivalent to or better than 3-4 well-timed split urea application and fertilizer recovery with PCU was 70-75% compared with 50% with prilled urea. Hutchinson *et al.* (2002) found that PCU significantly improved potato yield tuber yield as compared with ammonium nitrate. Shoji *et al.* (2001) reported that N recovery was almost twice with PCU as compared to urea in flood irrigated barley, irrigated potato and maize, due to greater NUE and reduced N fertilization rates. The nutrient release rate of PCU fertilizers better matches the nutrient demands of crop plants, and thus significantly enhances NUE and crop yields (Nelson *et al.*, 2009). However,

PCU slow release fertilizers have a complicated manufactory process with higher price.

Resin-coated fertilizers

A resin-coated fertilizer is prepared by in-situ polymerization resulting in the formation of a cross-linked, hydrophobic polymer (thermosettic). It was first produced in California in 1967. The resins in practical use are the alkyd-type resins (Osmocote) and polyurethane-like coatings (Polyon, Plantacote and Multicote) (Trenkel, 1997). The osmocote market has mainly been limited to high value plants such as commercial ornamental nurseries, greenhouses, citrus and strawberry production. Other technology for coating granular fertilizers is through utilization of thermoplastic resins as coating substances. The coatings are dissolved in fast-drying chlorinated hydrocarbon solvent. Because the thermoplastic polymers used are highly impermeable in water, to obtain the desired diffusion characteristics ethylene-vinyl acetate and surfactants must be added as a release controlling agents. The release pattern is controlled by the level of release-controlling agents. Some commonly marketed products are Meister products. Many of the experiments using coated fertilizers compared similar levels of nutrients supplied by ordinary fertilizers, ignoring the potential to reduce application levels, and still maintain or even increase yield levels. Improved coated fertilizers products offering "tailor-made" release characteristics and an increased awareness of the importance of matching release rate and pattern with plant demand, have started to motivate agronomists and consumers to adapt a management oriented approach.

Pine oleoresin coated urea

Pine oleoresin is a gum-like substance extracted from pine trees. The crude resin is composed of levopimaric, palustric, l-abietic and neoabietic acids in different proportions. All these acids have antibacterial and antifungal properties. Pine oleoresin acts as a physical barrier around the urea granules thereby reducing the release of N from coated urea; it inhibits urease activity through antibacterial properties, and being acidic in nature, it inhibits volatilization loss by reducing alkaline microsites. A protocol has been developed to coat urea with oleoresin and the coated urea contained 3.82–4.36% pine oleoresin and 44.07– 44.31% N (Kundu *et al.*, 2013). Irrespective of the soil type, urease activity decreased considerably in the pine oleoresin-treated soils compared to control due to acidic and antimicrobial properties of pine oleoresin. The amount of urea extracted at different times from Vertisols fitted to a first-order kinetic equation showed that time required for hydrolysis of 90% of the applied urea markedly increased from 88.56 to 328.94 h in the presence of pine oleoresin

(Kundu *et al.*, 2013). This indicates that pine oleoresin is a potential urease inhibitor. The volatilization loss of pine oleoresin-coated urea from a Vertisol decreased from 16.99% to 10.12% after 240 h. Thus, pine oleoresin- coated urea can be a better substitute for neem coated urea.

The amount of N volatilized from urea and POR coated urea treated Vertisol at different time intervals and the percentage of loss are summarized in Table 4. The total amount of NH_4–N volatilized from normal urea and POR-coated urea after 240 h was 31.278 and 18.079 mg N pot^{-1} respectively, and the observed reduction in volatilization loss was 40.43% under POR-coated urea pre-treated soil (Table 4). The acidic nature of POR might have reduced the microsite pH around the urea granules and thereby NH_3 loss through volatilization was reduced. Also, the antimicrobial properties of POR might have reduced the rate of hydrolysis of urea to $(NH_4)_2CO_3$ and subsequently to NH_4OH and CO_2.

Table 4: Volatilization loss of N from normal urea and pine oleoresin (POR.)-coated urea

	NH_4 volatilized (mg N/pot) after				
Treatment	48 h	96 h	168 h	240 h	Volatilization loss (%)
Normal urea	6.436	17.634	26.148	31.278	16.99
POR-coated urea	3.108	10.626	14.462	18.079	10.12

Neem coated urea

Nitrification inhibiting (NI) properties of neem and its role in increasing NUE in rice was first reported by Bains *et al.* (1971). They treated the urea with ethanol extract of neem seeds. This was followed by a series of studies at the Indian Agricultural Research Institute (IARI), New Delhi and farmers fields have indicated an increase in yields and apparent N recovery of rice by using neem cake (Sharma and Prasad, 1980), neem-oil emulsion coated urea (Shivay *et al.*, 2001), pusa neem golden urea (Prasad *et al.*, 1999) and other modified fertilizer materials. Besides, increased apparent recoveries of applied N and yields, NIs have been found to reduce the emissions of green house gases. Nimin, a commercially available extract prepared from neem cake, and neem-oil have been reported to reduce N_2O-N emission from urea N by 9% and 63% respectively (Malla *et al.*, 2005) at IARI, New Delhi, India. Beside rice, the application of N through neem coated urea (NCU) or ordinary urea up to 120 kg N/ha significantly increased grain yield of wheat at Gurdaspur (Table 5) but the increase was significant only up to 96 kg N/ha at Ludhiana for both the sources of N (Thind *et al.*, 2010). When N was applied in 2 equal splits at sowing and first irrigation, the NCU did not out perform urea in increasing grain yield of wheat at any level of N application either at Ludhiana or Gurdaspur.

Similar trend followed for straw yield also (Table 5). However, it was interesting to note that NCU gave 9.4, 5.6 and 2.5% higher grain yield over urea with the application of 48, 96 and 120 kg N/ha respectively, at Ludhiana. While at a level magnitude of 3.2, 4.5 and 1.6% at Gurdaspur.

Table 5: Grain and straw yield and nitrogen uptake by wheat at Ludhiana and Gurdaspur under different sources and rates of N (Pooled data over 3 years)

Treatment	Ludhiana			Gurdaspur		
	Grain yield (tones/ha)	Straw yield (tones/ha)	Total N uptake (kg/ha)	Grain yield (tones/ha)	Straw yield (tones/ha)	Total N uptake (kg/ha)
Control	1.67	2.56	36	2.34	3.35	41
Urea (48 kg N/ha)	2.66	4.60	59	3.78	5.50	76
Urea (96 kg N/ha)	3.78	5.50	86	4.19	6.13	114
Urea (120 kg N/ha)	3.97	6.20	100	4.88	6.98	138
NCU (48 kg N/ha)	2.91	5.04	70	3.90	5.72	85
NCU (86 kg N/ha)	3.09	5.05	100	4.60	6.81	125
NCU (120 kg N/ha)	4.07	6.70	108	4.96	7.06	141
Urea (120 kg N/ha applied in 3 doses; 48, 48 and 24 kg N/ha)	4.54	6.40	116	4.66	6.38	119
NCU (120 kg N/ha applied in 3 does: 48: 48 and 24 kg N/ha)	4.57	6.72	127	4.72	6.67	121
Urea (96 kg N/ha drilled at sowing)	4.23	6.38	110	4.41	7.89	109
NCU (96 kg N/ha drilled at sowing)	4.49	6.94	112	4.45	6.23	106
LSD (*P*=0.05)	0.349	0.582	10.7	0.255	0.434	8.8

NCU, Neem coated urea

The NCU was at par with SCU giving 13.4% increase in grain yield (Table 6). However since 15-20% neem cake was needed to make NCU, this could not be manufactured at the factory level. Again since nitrification inhibiting chemicals in neem are lipid associate, a neem oil micro-emulsion was developed for coating urea (Suri *et al.*, 2004). Only 0.5 kg neem oil per ton of urea was needed in the technique. This technique or its modifications are currently used in India by several fertilizer manufactures. The cost of coating urea with neem oil emulsion is only US$ 2-3 t^{-1} urea. Increase in rice yield with NCU so made varies from 6.1 to 11.9% over urea (Table 7). Singh and Singh (1986) observed a significant reduction in NO_3-N leaching when urea was blended with neem cake. Neem seed cake blending with urea reduced ammonia volatilization by 31.4% (Reddy and Misra, 1983) and an increase of 22.3% in rice grain yield was observed with the application of neem seed extract with treated urea (Bains *et al.*, 1971).

Superiority of NCU over prilled urea was also reported from Punjab (Thind *et al.*, 2009) and therefore, field experiment showed that neem coated urea (NCU) increased its efficiency, resulting in increased yield of crops like rice, wheat, potato, and sugarcane under irrigated conditions (Sharma *et al.*, 2004). Recently, Laijawala (2010) has pinched the attention to neem as a possible nitrification inhibitor, showing particularly that neem oil-coated urea significantly reduces ammonia volatilization.

Table 6: Relative efficiency of urea, sulphur coated over (SCO) and neem cake coated urea (NCCU) in the rice-wheat cropping system.

Source of N	Grain yield (t ha^{-1})			Increase over urea (%)
	Rice	Wheat	Rice + Wheat	
Urea	4.13	3.03	7.16	–
SCU	4.61	3.67	8.28	15.6
NCCU	4.56	3.56	8.12	13.4

Prasad *et al.* (1981)

Table 7: Relative efficiency of neem coated urea (NCU) and urea for rice in on-farm trials

Location	No. of trials	Grain (t ha^{-1})		PFPn	
		Urea	NCU	Urea	NCU
Delhi	9	4.5	4.9	37.5	40.8
Punjab	45	4.9	5.2	40.8	43.8
Haryana	88	4.3	4.9	35.8	40.8

Prasad (2007)

Zeolite coated urea

In the past, organic materials such as manure have been mixed with sandy soil to improve water and nutrient retention. One disadvantage of organic amendments is that they decompose over time, reducing their beneficial effects (Bigelow *et al.*, 2004). The ideal amendment should be relatively stable to provide water and nutrient retention and release that are comparable with organic amendments. Zeolites are natural minerals, first discovered in 1756 by a Swedish mineralogist, who named the porous minerals from the Greek words meaning 'boiling stone' (Mumpton, 1999). In recent years, more common products like porous ceramics, diatomaceous earth and zeolites have been increasingly used as inorganic soil amendments. Some of the characteristics of these products that potentially make them desirable for improving the properties of soft soils are a large internal porosity that results in water retention, uniform particle size distribution that allows them to be easily incorporated and high cation-exchange capacity (CEC) that retains nutrients (Bigelow *et al.*, 2004). The unique physical and chemical properties of natural zeolites, in combination with their abundance in sedimentary deposits and rocks derived from volcanic parent materials, have made them useful in many industrial applications. These properties have also increased their use in agronomic and horticultural applications (Dwairi, 1998). Eberl (2002) described the production of slow release fertilizers using zeolites. Zeolites are heated to 400°C to drive out pore water which is then replaced by molten urea and the product contains about 17%N. The sequestering action of zeolite reduces leaching and slows down nitrification. Zeolite based fertilizers have been studied by NASA as means to sustain plant growth in space environments. These are currently being sold under the brand name Zeoponix for use in golf course, sports playing fields and for green house horticultural use as a proven combination fertilizer and soil amendment. Clinoptilolites are one type of zeolite, and although not the most well known, is one of the most useful. Extensive deposits of clinoptilolites are found in western USA, Bulgaria, Hungary, Japan, Australia and Iran (Mumpton, 1999).

Nitrification inhibitors

Another approach has been to use nitrification inhibitors to retard nitrification of applied NH_3 or urea-N and to reduce leaching and denitrification losses and might contribute to increased NUE or apparent N recovery efficiency (Ortega, 2006). The use of ammonium N sources with the nitrification inhibitor 3, 4- dimethylpyrazole phosphate (DMPP) shows promise to increase NUE and P-use efficiency (PUE). A large number of chemicals are known to have nitrification inhibiting properties. These include, N-serve or nitrapyrin (2-chloro-6-(trichlormethy1) pyridine), DCD (dicyndiamide), AM (2-amino-4-chloro-6-methyl pyrimidine), CMP (1-carbamoyle-3-methylpyrazole), terrazole (etridiazole), CP (2-cyanimino-4-hydroxy-6-methy1 pyrimidine), AT/ATc (4 aminotriazole), ST (sulphatiazole or sodium thiosulphate), ATS (ammonium thiosulphate), ZPTA (thiosulphory1 triamide) (Prasad and Power, 1995; Trankel, 1997), neem extractives (containing epinimbin, deacetylnimbin, azadirachtin and pyrites. Of this long list of NIs, only nitrapyrin, AM and DCD have been widely tested for rice and wheat. Nitrapyrin, discovered in USA, was reported to inhibit nitrification and increase corn yield (Goring, 1962). The AM was discovered in Japan (Toyo Koatsu Industries, 1965) and DCD in Germany.

Table 8: Some of the patented nitrification inhibitors

Chemical name	Common name	Developer	Inhibition by day 14 (%)
2-chloro-6-(trichloromethyl-pyridine)	Nitrapyrin	Dow Chemical	82
4-amino-1,2,4-6-triazole-HCl	ATC	Ishihada Industries	78
2,4-diamino-6-trichloro-methyltriazine	CI-1580	American Cyanamid	65
Dicyandiamide	DCD	Showa Denko	53
Thiourea	TU	Nitto Ryuso	41
1-mercapto-1,2,4-triazole	MT	Nippon	32
2-amino-4-chloro-6-methyl-pyramidine	AM	Mitsui Toatsu	31

Frye (2005)

Table 9: Commonly used nitrification inhibitors in agriculture

Name (chemical, trademark)	Solubility in water (g/l)	Relative volatility	Mode of application
2-chloro-6-(trichloromethyl) pyridine (Nitrapyrin; N-serve) ammo-nia for soil injection	0.04 (at 20°C)	High	Suitable with anhydrous
2-amino-4-chloro methyl pyrimidine	1.25 (at 20°C)	High	Coatings on solid N fertilizers
Dicyandiamide (DCD), cyanoguanidine	23.0 (at 13°C)	Low	Blend with urea or other solid N fertilizers
DMPP		Low	Blend with urea or other solid N fertilizers

Subba Rao *et al.* (2006)

Also application of dicyandiamide (DCD) and use of DCD containing fertilizers such as Agrotain Plus delayed nitrate formation and hence could improve NUE by reducing nitrate leaching and denitrification losses (Singh *et al.*, 2010). Patrick *et al.* (1968) reported no advantage with nitrapyrin in rice, while Nishihara and Tsunyoshi (1969) reported that ammonium sulphate treated with nitrapyrin increased rice yield by 15-20%. In India, neem cake coated urea (NCCU), which was shown to have nitrification inhibiting properties (Thomas and Prasad, 1982). It increased yield in rice and rice-wheat cropping systems (Prasad *et al.*, 1981). In rice-wheat cropping system, NCCU was as good as sulphur coated urea; the major factor responsible for N regulation was nitrification inhibition by the triterpenes in neem (Devakumar and Mukherjee, 1985). Since coating of urea with neem cake was industrially not feasible due to the large volumes involved (*e.g.*, 20% w/w of urea), a neem oil microemulsion technique was developed (Prasad *et al.*, 1999). This technique or its modification is currently being used by the National Fertilizes Ltd., Indo-Gulf Fertilizers, and

Shriram Fertilizers and Chemical Ltd. and about 0.4 Tg of neem coated urea (NCU) are being manufactured in India. Furthermore, on-farm trials in Delhi, Punjab, Haryana, and Uttar Pradesh have shown that NCU results in 6 to 11% increase in rice yield. PFPn for NCU ranged from 41 to 43% compared to 36 to 41% for prilled urea (Prasad, 2007).

Urease inhibitors

Urease inhibitors prevent or suppress over a certain period of time the transformation of amide-N in urea to ammonium hydroxide and ammonium through the hydrolytic action of the enzyme urease. By slowing down the rate at which urea is hydrolyzed in the soil, volatilization losses of ammonia to the air (as well as further leaching losses of nitrate) is either reduced or avoided. Thus, the efficiency of urea and of N fertilizers containing urea (e.g. urea ammonium nitrate solution), is increased and any adverse environmental impact from their use is decreased. Urease inhibitors reduce the rate of hydrolysis of urea to ammonium and can reduce loss of N due to ammonia volatilization when urea is surface applied (Sarker *et al.*, 1991). A number of urease inhibitors have been reported which include PPD/PPDA (pheny1 phosphorodiamide), NBPT (N-n-butyl) thiophosphorice triamide), TPT (thiophosphory1 triamide), CHTPT (cyclohexy1 phosphoric triamide), CNPT (cyclohexy1 phosphoric triamide), PT (phosphoric triamide), HQ (hydrquinone), ATS (amino thiosulphate), and HACTP (hexaamidocyclo triphosphazine) (Trenkel, 1997; Kiss and Simihaian, 2002). Boric acid was also reported to have an inhibitory effect on urease and it appears that metals react with sulfhydryl groups of the urease enzyme rendering it inactive, whereas boric acid appears to show competitive inhibition with urea (Benini *et al.,* 2004). However, the effectiveness of these inorganic products is somewhat low (Reddy and Sharma, 2000) and some of them are heavy metals which have restrictions for soil application. Byrnes and Freney (1995) reported that application of PPDA significantly increased rice grain yields in only two out of eight flooded-field rice trials. One of the reasons they attributed was a rapid degradation of PPDA due to high pH or temperature of flooded water. Broadbent *et al.* (1985) found that adding PPDA to a urea solution applied to corn did not affect the rate of urea hydrolysis, N uptake of corn or corn yield. However, Joo and Christians (1986) found that 2% PPDA added to liquid urea increased the fresh weight of Kentucky bluegrass turf by 20-31%. Lately attention has been focused on the most widely tested urease inhibitor, N-(n-butyl) thiophosphoric triamide (NBTPT), trade named "Agrotain." In a greenhouse study, Byrnes (1988) reported that NBTPT was more effective than PPDA at retarding urea hydrolysis and reducing the ammonium-N concentration in flooded water. Similar results that showed NBTPT was more effective than PPDA in retarding urea hydrolysis were also reported by Buresh *et al.* (1988). Addition of NBTPT increased flooded rice grain yield by an average of 40% over unamended urea (Byrnes, 1988). Results of over 400 field trials with NBTPT and showed that on average, treating urea or UAN with NBTPT brought about maize grain yield increases of 0.89 and 0.56 t ha^{-1} compared to yields obtained with untreated fertilizers (Trenkel, 1997). Volatilization process with untreated urea took place within 2–3 days after N fertilizer application. The addition of NBTPT postponed volatilization by 2-8 days, depending on the soil moisture and temperature conditions, and slowed down the rate of ammonia loss. Centarella *et al.* (2005) also reported that reduction of NH_3 volatilization from surface-applied urea due to NBTPT compared to no inhibitor.

Urea super granules (USG)

USG was tested at research centres as well as on farmers' fields during the 1980's and the results were very encouraging on most soils except the highly porous coarse textured soil. Thomas and Prasad (1982) have reported that in addition to the advantage of N placement, which reduces volatilization losses, placement of such a high amount of urea at a micro-locus produced very high concentrations of NH_3, inhibiting nitrification. Increased rice yield and NUE with USG has been reported from Indonesia (Pasandaran *et al.*, 1999) and Bangladesh (Balasubramanian *et al.,* 2004). The USG was developed taking a lead from the mud-ball technique of application of fertilizer N. In this technique a small amount of fertilizer N is placed in the centre of a mud ball and the balls are sealed. Mud balls are then dried in the shade are deep-placed in rice field. The USG in rice on medium to heavy soils can increase the NUE by 10-20%. Savant and Stangel (1998) have shown that N loss is significantly reduced (NH_3 volatili-zation and denitrification losses), which results in a significant increase in rice grain yield under flooded conditions compared with split- applied prilled urea (PU). In addition, ammonia concentration in the vicinity of USG is high enough to inhibit nitrification for some time (Thomas and Prasad, 1982). Tandon (1989) also reported that same yields of rice can be obtained with 26-59% less N with USG. The reason for producing USG is that it makes it easier for farmers to apply USG by hand. Use of USG has one great advantage in that it requires only one-time application after rice transplanting, whereas surface application of PU requires two to three split applications that can still result in significant N loss through NH_3 volatiliza-tion (Chien *et al.*, 2009). The negative aspect of applying USG is that it is a labour-intensive practice that some rice farmers in developing countries are not willing to adopt, for example, China. Also, it is not an alternative to commercial rice farms in the USA, Europe, and Latin America due to high labour costs. However, the use of USG has been successfully promoted in several Asian countries, notably in Vietnam and

Bangladesh. During 2008 USG were used in 500,000 ha, which increased rice productivity by 268,000 million tonnes in Bangladesh and reduced country's urea import by 50,000 million tonnes saving US$ 22 million in fertilizer imports (IFA, 2009).

Conclusion

N derived from fertilizers and not taken up by plants may be immobilized in soil organic matter or may be lost to the environment. Thus the economic losses both due to low nutrient use efficiency and nutrient wastages into environment are the serious problems. Therefore, increasing the NUE is crucial to improvement in yield and quality of crops, reduce nutrient input cost and improve soil, water, and air quality. An improved NUE in plants can be achieved by careful manipulation of plant, soil, fertilizer, biological and environmental factors and losses reduced by the matching supply with demand and optimizing slow release fertilizers and inhibitors. Slow release N fertilizer (SRFs), nitrification (NIs) and urease inhibitors (UIs) offers a good options to reduce the N losses from the soil-plant system because their delayed N release pattern may improve the match crop demand. Most of SRFs give higher yields as compared to conventional fertilizers, higher N use efficiency in different crops. However, the high cost of these materials is twice or thrice or some times more than conventional fertilizers, making them beyond the reach of common farmers especially Indian conditions.

Future strategies

- Slow release N fertilizers (SRFs) deserve more attention to better NUE and a more environmentally friendly. But improved utilization of advanced technologies and development of new concepts for preparing more cost-effective SRFs.
- Development of standard methods or strategies for the evaluation of nutrient release rates from controlled-release fertilizers.
- Assessment of expected benefits to the environment by using SRFs. This should also include estimates of the economic significance of reducing pollution of ecosystems (air, water and soil) and sustaining soil productivity.
- Improved quantification of the economic advantages resulting from reduced losses of nutrients and savings in labour costs.
- Development of new coating and encapsulating materials, specifically of more rapidly degradable synthetic materials/polymers.
- Development of new, improved, lower-cost and environment-friendly technologies in coating/encapsulating processes.

- Also very important are the physical characteristics of these substrate on which the coating is applied. These include particle size, shape and surface irregularity, prills, which often have holes on their surface, or granules which can be smooth or rough and irregular depending on whether the granulation process was agglomeration or compaction.

Selected References

Alila P, Srivastava AK. 2008. Slow release fertilizers and citrus: Emerging facts. *Agricultural Reviews* 29 (2): 99–107.

Allen SE, Hunt CM, Terman GL. 1971. Nitrogen release from sulfur coated urea, as affected by coating weigh, placement and temperature. *Agronomy Journal* 63:529–533

Association of American Plant Food Control Officials (AAPFCO) 1995. Official Publication No. 48. Association of American plant food control officials, Inc., West Lafayette, Indiana, USA.

Balasubramanian V, Alves B, Aulak MS, Bekunda M, Cai Zc, Drinkwater ., Mugend, D, Van Kassel C, Oenema O. 2004. Crop environment and management factors affecting N use efficiency. In Agriculture and the Nitrogen Cycle: Assessment the Impact of Fertilizer Use on Food Production and the Environment eds. A.R. Mosier, J.K. Syers J.R. Freney, PP 19-33, SCOPE 65, Paris, France.

Benini S, Rypniewski WR, Wilson KS, Mangani S, Ciurli S. 2004. Molecular detail so furease inhibition by boric acid: Insights into the catalytic mechanism. *Journal of the American Chemical Society* 126: 3714–3715.

Bigelow CA, Bowman DC, Cassel DK. 2004. Physical properties of three sand size classes amended with inorganic materials of sphagnum peat moss for putting green root zones. *Crop Science* 44: 900–907.

Broadbent FE, Nakeshima T, Chang GY. 1985. Performance of some urease inhibitors in field trials with corn. *Soil Science Society of American Journal* 49:348–351.

Buresh, R.J., S.K. De Datta, J.L. Padilla, and T.T. Chua. 1988. Potential of inhibitors for increasing response of lowland rice to urea fertilization. *Agronomy Journal* 80: 947-952.

Byrnes BH, Freney JR. 1995. Recent developments on the use of urease inhibitors in the tropics. *Fertilizer Research*. 42: 251–259.

Byrnes BH. 1988. The degradation of the urease inhibitor phenyl phosphorodiamidate in soil systems and the performance of N-(n-butyl) thiophosphoric triamide in flooded rice cultural. Ph.D. Thesis, Technical University of Munich, Weihenstephan, Germany.

Cantarella H, Quaggio JA, Gallo PB, Bolonhezi D, Rossetto R, Martins JLM, Paulino VJ, Alcantara PB. (2005). Ammonia losses of NBPT-treated urea under Brazilian soil conditions. In "IFA International Workshop on Enhanced-Efficiency Fertilizers", Frankfurt, Germany, 28–30 June 2005 (CD-ROM).

Chen J, Xu P, Tang S, Zhang F, Xie C. 2005. Nutrient-use efficiency and yield-increasing effect of single basal application of rice specific controlled release fertilizer. *Ying Yong Sheng Tai Xue Bao* 16(10): 1868–1871.

Chien SH, Prochnow LI, Cantarella H. 2009. Recent developments of fertilizer production and use to improve nutrient efficiency and minimize environmental impacts. *Advances in Agronomy* 102: 267-322.

Christianson CB.1988. Factors affecting N release of urea from reactive layer coated urea. *Fertilizer Research* 16: 273–284.

Devakumar C, Mukherjee SK. 1985. Nitrification retardation by neem products. *Neem News* 1: 2-11

Devakumar C, Mukherjee SK. 1998. Evaluation of Jordanian zeolite tuff as a controlled slow-release fertilizer for NH_4. *Environmental Geology* 34:1–3.

Eberl DD. 2002. Controlled release fertilizer using zeolites. US Geological survey, 19 august 2002, Fact sheet http://www.usgs.gov/tech-transfer/factsheet/94-066b.html 02 December 2002.

Eman EA, El-Ashry SM. 2009. Efficiency of slow release urea fertilizer on herb yield and essential oil production of lemon balm (*Melissa officinals* L.) plant. American- *Eurasian Journal of Agriculture and Environment Science* 5(2):141-147.

Fageria NK, Barbosa FMP. 2001. Nitrogen-use efficiency in lowland rice genotypes. *Communications in Soil Science and Plant Analysis* 32: 2079–2089.

Fashola OO, Hayashi K, Wakatsuki T. 2002. Effect of water management and polyolefin-coated urea on growth and nitrogen uptake of indicia rice. *Journal of Plant Nutrition* 25:2173–2190.

Frye WW. 2005. Nitrification inhibition for nitrogen efficiency and environment protection. IFA International Workshop on Enhanced-Efficiency Fertilizers, Frankfurt. International Fertilizer Industry Association, Paris, France.

Galloway JN, Townsend AR, Erisman JW, Bekunda MC, Freney JR, Martinelli, LA, Seitzinger SP, Sutton MA. 2008. Transformation of the nitrogen cycle: Recent trends, questions, and potential solutions. *Science* (Washington, DC). 320: 889–892.

Goertz HM. 1995. Technology development in coated fertilizers. In: "Proc. Dahlia Greidinger Memorial Int. Workshop on Controlled/Slow Release Fertilizers", (Y. Hagin *et al*. Eds.) Mar. 1993, Technion, Haifa. Israel.

Goring CAI. 1962. Control of nitrification by 2-chloro-6-trichloromethyl.-pyridine. *Soil Science* 93:431–439.

Hauck RD. 1972. Synthetic slow-release fertilizers and fertilizer amendments. pp. 633-690. In C.A.I. Goring and J.W. Hamaker (eds.) Organical chemicals in the soil environment. Vol. 2. Marcel Dekker, Inc., New York.

Hegde DM, Sudhakara Babu SN, Qureshi AA, Murthy IYLN. 2007. Enhancing nutrient-use efficiency in crop production- A review. *Indian Journal of Agronomy* 52 (4) : 261-274

Hendrickson LL. 1992. Corn yield response to the urease inhibitor NBPT: five years summary. *Journal of Production Agriculture* 5, 131-137.

Hughes TP. 1976. Nitrogen release from isobutylidene diurea: soil pH and fertilizer particle size effects. *Agronomy Journal* 68:103-106.

Hutchinson CM, Tilton WA, Livingston-Way PK, Hochmuth GJ. 2002. Best management practices for potato production in Northeast Florida. EDIS, Florida Cooperative Extension Service Publication HS877. Last accessed 11/05/04. http://edis.ifas.ufl.edu/cv279.

IFA, 2002. Fertilizer use by crop. International Fertilizer Industry Association (IFA), International Fertilizer Development Centre (IFDC), International Potash Institute (IPI), Potash and Phosphate Institute (PPI), Food and Agriculture Organization (FAO), Rome.

IFA, 2009. The global "4R" nutrient stewardship framework: developing fertilizer best management practices for delivering economic, social and environmental benefits. International Fertilizer Industry Association, Paris, France.

Joo YK, Christians NE. 1986. The response of Kentucky blue grass turf to phenyl phosphorodiamidate (PPD) and magnesium (Mg^{2+}) applied in combination with urea. *Journal of Fertilizers* 3: 30–33.

Katyal JC, Singh B, Vlek PLG, Buresh RJ. 1985. Fate and fertility of nitrogen fertilizer applied to wet land rice. II. Punjab, *Indian Fertilizer Research* 6: 279-290.

Kiran U, Patra DD. 2003. Medical and aromatic plant materials as nitrification inhibitors for augmenting yield and nitrogen uptake of Japanese mint (*Mentha arvensis* L. Var *Piperascens*). *Bioresource Technology* 86:267-276.

Kiss S, Simihaian M. 2002. Improving efficiency of urea fertilizers by inhibition of soil urease activity. Kluwer Academic Publishers, Dordrecht, The Netherlands.

Kundu S, Tapan Adhikari, Vassanda Coumar M., Rajendiran S, Bhattacharyya R, Saha JK, Biswas AK, Subba Rao A. 2013. Pine oleoresin: a potential urease inhibitor and coating material for slow-release urea. *Current Science*, 104: 8, 25.

Ladha JK, Pathak H, Krupnik TJ, Six J, Kessel CV. 2005. Efficiency of fertilizer nitrogen in cereal production: Retrospect's and prospects. *Advances in Agronomy* 87: 85–156.

Laijawala K. 2010. Neem as a nitrification inhibitor. IFA International Conference on Enhanced-Efficiency Fertilizers, Miami. International Fertilizer Industry Association, Paris, France.

Lunt OR. 1971. Controlled-release fertilizers: achievements and potential. *Journal of Agriculture Food Chemistry* 19: 797-800.

Malla G, Bhatia A, Pathak H, Prasad S, Jain N, Singh J. 2005. Mitigating nitrous oxide andmethane emissions from soil under rice–wheat system with nitrification and urease inhibitors. *Chemosphere* 58: 141–147.

Mao XY, Sun KJ, Wang DH, Liao ZW. 2005. Controlled release fertilizer (CRF): A green fertilizer for controlling non-point contamination in agriculture. *Journal of Environmental Science* 17: 181–184

Mohanty SK, Singh U, Balasubramanian V, Jha KP. 1999. Nitrogen deep placement technologies for productivity, profitability and environmental quality of rainfed lowland rice system. *Nutrient Cycling* and *Agro-ecosystem* 53: 43-57

Mumpton FA. 1999. La roca magica: uses of natural zeolite in agriculture and industry. *Proceeding of National Academy Science* USA. 96:3463–3470.

Nelson KA, Paniagua SM, Motavalli PP. 2009. Effect of polymer coated urea, irrigation and drainage on nitrogen utilization and yield of corn in a clay pan soil. *Agronomy Journal* 101: 681-687.

Nishihara T, Tsunyoshi Y. 1969. The effect of some nitrification inhibitors on the availability of basic fertilizer nitrogen by the rice plants on dry paddy field. Bull Faculty Agriculture Kagoshima University Japan PP. 133-141.

Nyborg M, Malhi S, Solberg E, Zhang M. 1999. Influence of polymer-coated urea on mineral nitrogen release, nitrification, and barley yield and nitrogen uptake. *Communications in Soil Science and Plant Analysis* 30:1963–1974.

Oeritli JJ. 1980. Controlled –release fertilizers. *Fertilizer Research* 1:103–123

Ortega R. 2006.Increasing nitrogen and phosphorus fertilizer use efficiency by using the nitrification inhibitor 3, 4- dimethylpyrazole phosphate (DMPP) in Chile. 18th World Congress of Soil Science, held during 9-15 July, Philadelphia, Pennsylvania, the USA.

Pasandran E, Gultomm B, Sri Adisingih JH, Rochayati. 1999. Government policy support for technology promotion and adoption: A case study of urea tablet technology in Indonesia, in Balasubramanian, V., Ladha, J.K., Gupta, and GL Deung (Eds.) Resources management in rice system: nutrient pp 180-190 Kluwer Academic Netherlands & IRRI, Los Banos Philippines.

Patrick WH Jr,Peterson FJ, Tumer FT. 1968. Nitrification inhibitors for lowland rice. *Soil Science*. 105:103-105.

Prasad M, Prasad R. 1980. Yield and nitrogen uptake by rice as affected by variety, method of planting and new nitrogen fertilizers. *Fertilizer Research* 1:207–213.

Prasad R, Power J. 1995. Nitrification inhibitors for agriculture, health and the environment. *Advances in Agronomy* 54:233-281.

Prasad R, Rajale GB, Lakhdive BA. 1971. Nitrification retarders and slow-release nitrogen fertilizers. *Advances in Agronomy* 23: 337–383.

Prasad R, Saxena VC, Devakumar C. 1998. Pusa neem golden urea for increasing nitrogen-use efficiency in rice. *Current Science* 75: 15.

Prasad R, Sharma SN, Prasad M, Reddy RNS. 1981. Efficient utilization of nitrogen in rice–wheat rotation. In: *National Symposium on Crop Management to Meet the New Challenges*, pp. 37–43. New Delhi: Indian Society of Agronomy.

Prasad R, Singh S, Saxena VS, Devakumar C. 1999. Coating of prilled urea with neem (*Azadirachta indica* A. Juss) oil for efficient use in rice. *Naturwissenschaften* 86: 538-539.

Prasad R. 2005. Rice-wheat cropping systems. *Advances in Agronomy* 86:255–339.

Prasad R. 2007. Nitrogen in Indian Agriculture. In Agricultural Nitrogen Use and its Environmental Implications eds., Y.P. Abrol, N. Raghuram, M.S. Sachde pp 29-54 IK International Publishing, New Delhi.

Prasad R.1996. Management of fertilizer N for higher efficiency. P. 104-115. In nitrogen research and crop production (HLS Tandon ed.), FDCO, New Delhi.

Rajale GB, Prasad R. 1975. Nitrogen and water management for rice. *Il Riso* 24:117-125.

Raun WR, Johnson GV. 1999. Improving nitrogen-use efficiency for cereal production. *Agronomy Journal* 91: 357–363.

Reddy DD, Sharma KL. 2000. Effect of amending urea fertilizer with chemical additives on ammonia volatilization loss and nitrogen-use efficiency. *Biology and Fertility of Soils* 32:24–27.

Reddy VRM, Mishra B. 1983. Effect of altered soil urease activity and temperature on ammonia volatilization from surface applied urea. *Journal Indian Society of Sol Science* 31:143–5.

Salman OA.1988. Polymer coating on urea prills to reduce dissolution rate. *Journal Agricultural Food Chemistry* 36: 616–621.

Sarkar MC, Banerjee N., Rane DS, Uppal KS. 1991. Field management of ammonia volatilization loss of nitrogen for urea applied to wheat. *Fertilizer News* 36 (11): 25-28.

Savant NK, Stangel PJ. 1998. Urea briquettes containing diammonium phosphate: A potential NP fertilizer for transplanted rice. *Fertilizers Research* 51: 85–94.

Sharma AR. 1994. Fertilizer management in rainfed lowland rice under excess water condition. *Fertilizer News* 39(5):35-44.

Sharma SN, Prasad R, Gautam RC. 2004. Neem for higher nitrogen use efficiency P 20 TB-ICN 19/2004, IARI, New Delhi.

Sharma SN, Prasad R. 1980. Effect of rates of N and relative efficiency of sulfur-coated urea and nitrapyrin-treated urea in dry matter production and nitrogen uptake by rice. *Plant and Soil* 55: 389–396.

Shaviv A, Smadar R, Zaidel E. 2003: Model of diffusion release from polymer coated granular fertilizers. *Environmental Science and Technology* 37:2251-2256.

Shivay YS, Prasad R, Singh S, Sharma SN. 2001. Coating of prilled urea with neem (*Azadirachta Indicia*) for efficient nitrogen use in lowland transplanted rice (*Oryza sativa*). *Indian Journal of Agronomy* 46: 453–457.

Shoji S, Delegado J, Mosier AR, Minura Y. 2001. Controlled release fertilizers and nitrification inhibitors to increase nitrogen–use efficiency and to conserve air and water quality. *Communications in Soil Science and Plant Analysis* 32: 1051–1071.

Shoji S, Higashi T. 1999. Enhancing quality and safety of farm products – tea and spinach. In: Meister controlled release fertilizer – Properties and utilization. Shoji, S. (ed). Konno Printing Company Ltd. Sendai, Japan.

Singh M, Singh TA. 1986. Leaching losses of nitrogen from urea as affected by application of neem-cake. *Journal of Indian Society of Soil Science* 34: 766–73.

Singh U, Cassman KG, Ladha JK, Bronson KF. 1995. Innovative nitrogen management strategies for lowland rice systems. In "Fragile Lives in Fragile Ecosystems"pp. 229-254. International Rice Research Institute, Manila, Philippines.

Smil V. 2002. Nitrogen and food production: Proteins for human's diets. *Ambio* 31: 126–131.

Subbarao GV, Sahrawat KL, Berry WL, Nakahara K, Ishikawa T, Watanabe T, Suenaga K, Rondon M, Rao IM. 2006. Scope and strategies for regulation of nitrification in agricultural systems – challenges and opportunities. *Critical Reviews in Plant Sciences* 25: 302-335.

Sudhakara K, Prasad R. 1986. Relative efficiency of prilled urea, urea super granules (USG) and USG coated with neem cake or DCD for direct seeded rice. *Journal Agricultural Science Cambridge* 106:185-190.

Suri IK, Prasad R, Devakumar C. 2004. Neem coating of urea – present status and future trends. *Indian Journal of Fertilizers* 49: 21–4.

Tandon HLS. 1989. Urea super granules for increasing nitrogen use efficiency in rice An overview. Pp.14-22. In. Urea super granules for increasing nitrogen use efficiency (Soil fertility and fertilizers use Vol.III). Eds. V Kumar, GC Shrotriyia, V S Kaore), Indian farmers fertilizers cooperative Ltd., New Delhi.

Thind HS, Singh B, Pannu RPS, Singh Y, Singh V, Gupta RK, Vashistha M, Singh J, Kumar A. 2009. Relative performance of neem (*Azadirachta indica*) coated urea vis-à-vis ordinary urea applied to rice on the basis of soil test or following need based nitrogen management using leaf colour chart. *Nutrient Cycling in Agro ecosystems* 87: 1–8.

Thind HS, Singh B, Pannu RPS, Singh Y, Singh V, Gupta RK,Singh G, Kumar A, Vashistha M. 2010. Managing neem (*Azadirachta Indica*)-coated urea and ordinary urea in wheat (*Triticum aestivum*) for improving nitrogen-use efficiency and high yields. *Indian Journal of Agricultural Sciences* 80 (11): 960–964.

Thomas J, Prasad R. 1982. Mineralization of urea, coated urea and nitrification inhibitors treated urea in different rice growing soils. *ZPfanzen, Boden*. 146:341-347.

Tilman D, Cassman KG, Matsan PA, Naylor RL, Polaskey S. 2002. Agricultural sustainability and intensive production practices. *Nature* 418: 671-677.

Trenkel ME. 1997. Improving fertilizer use efficiency. Controlled-release and stabilized fertilizers in agriculture. The International Fertilizer Industry Association, Paris.

Trenkel ME. 2010. Slow- and controlled-release and stabilized fertilizers: an option for enhancing nutrient use efficiency in agriculture. The International Fertilizer Industry Association, Paris.

Upendra S, Wilkens P, Jahan I, Sanabria J, Kovach S. 2010. Enhanced efficiency fertilizers. 19th World Congress of Soil Science, Soil Solutions for a Changing World 1– 6 August 2010, Brisbane, Australia. pp 9-12.

7

Enhancing Nutrient Use Efficiency, pp. 81-92
Editors: K. Ramesh, A.K. Biswas, B.L. Lakaria, S. Srivastava and A.K. Patra

Handheld Devices for Judging Nutrients in Crops and Their Role in Enhancing Nutrient Use Efficiency

K. M. Hati, R. K. Singh and M. Mohanty

ICAR-Indian Institute of Soil Science, Nabi Bagh, Bhopal – 462 038, India

Water and nutrients are the most important, and readily manageable, variables for producing a profitable crop. Efficient management of water and essential plant nutrients is imperative for achieving sustainable agriculture and maintaining necessary increase in food production to feed the burgeoning population while minimizing economic losses and environmental impacts. Technology here can play a catalytic role in striking a common ground between these environmental and economic goals. Recent advances in sensor technologies indicate that efficient nutrient management in crop fields can be attained through the application of Precision Agriculture (PA)-based geo-spatial technologies such as global positioning system, geographical information system, remote sensing, geostatistics and variable rate application. Variable-rate fertilizer application for site specific nutrient management, one of the basic tenets of PA, has been found to optimize fertilizer use efficiency by overcoming the problem of over- and under-fertilization. Development of cost-effective site specific nutrient management strategies is envisaged to increase crop yields and quality, reduce resource wastage and control environmental pollution. Spatial and temporal variability in crop and/or soil productivity are influenced by both intrinsic (e.g., soil forming factors such as parent material, climate, topography, fauna/flora and time) and extrinsic factors (e.g., farm management practices and maintenance operations). Quantifying the spatial and temporal variability of soil properties and responding to such variability via carefully designed site- and time-specific input application are believed to enhance nutrient assimilation in crops. Besides these, estimation of nutrient particularly nitrogen (N) status in crops at an early growth stage can also help in optimizing the dose of fertilizers to be applied to the crops through foliar spray or band placements. N stress in plants can be estimated from the spectral reflectance characteristics of the crops in the visible and near infra-red region. Thus the nutrient use efficiency in arable crops could be increased either through rapid estimation of soil nutrient status or through estimation of nutrient status in crop canopy using the optical remote sensing techniques. Considerable progress has been made for estimation of plant growth and N status in crop canopy using commercially available optical and radiometric sensors which use different vegetative indices for estimation of crop nutrient status. However, sensors developed for rapid and non-destructive measurement of soil nutrient status using diffused reflectance spectroscopy is yet to come in commercial stage, though many promising technologies are being tested for development of on the go variable nutrient applicator for site specific nutrient management. This chapter will deal with both the aspects for improvement of nutrient use efficiency.

N, because of its high demand in the plant and variability within the soil, is the most intensively managed plant nutrient in crop production and is also the largest agricultural input in most of the cropping systems. Applying right amount of N in the right place at the right physiological stage is a significant challenge for crop growers, which could improve the use efficiency of this valuable input which was otherwise quite low (average 33% on a global scale). However, matching placement of N-supply to demand of the crop requires an adequate assessment of N and water status in agricultural landscapes, especially early in the season when management decisions can impact the yield. The main goal of farmers is to detect crop N status at an early growth stage, and apply an appropriate amount of N fertilizer for optimal yield, high N-use efficiency, and minimal N losses to the environment. Several studies have documented that N status of field crops can be assessed using leaf or canopy spectral reflectance data. Remote sensing techniques using both spectral and thermal approaches can provide field scale diagnostic methods for rapid identification of crop N status and water stress, permitting precise N inputs to areas with sufficient plant available water.

Use of spectral radiance as a tool to determine the nutrient status in plants has several advantages compared to other non-destructive methods. Spectral radiance measurements can be obtained without attaching the meter to a specific leaf and many readings can be acquired in a short time thus reducing variability. In addition, sensors combined with other technologies such as GIS and GPS could be able to make location specific fertilizer recommendations (Schepers, 1994). Therefore, undesired environmental effects associated with excess fertilization might be reduced and crop yield and use efficiency of nutrients could be improved. Besides this, spectral measurements from crop have potential use in monitoring crop condition and predicting yields. Vegetation indices computed from different wave bands have been related to various crop biophysical properties like, green leaf area index, canopy biomass, absorbed photo-synthetically active radiation and yield. Numerous spectral vegetation indices have been developed to identify and characterize vegetation canopies. The most common of these indices, which utilize red and near infrared (NIR) wavelengths, are the simple ratio of infrared to red, or normalized difference vegetation index (NDVI = NIR-Red/NIR+Red), or its linear combination like, perpendicular vegetation index (PVI). These indices have been found to correlate well with various vegetation parameters including green leaf area, biomass, percent green cover, productivity, and photosynthetic activity. Gitelson *et al.* (1996) proposed a green normalized difference vegetation index (GNDVI), where the green band is used in the equation for NDVI instead of the red band, and showed that the green band, in combination with the NIR band, is more closely associated with the variability in leaf chlorophyll, N content, and grain yield than the red band particularly at a later growth stage. Some indices have been in use to predict biological growth (biomass) and nutrient status in plant tissue. For example, the normalized pigment chlorophyll index (NPCI = Reflectance at 430 nm [(R430) - reflectance at 680 nm (R680)/(R680+R430)] proposed by Penuelas *et al.* (1993) was used to predict carotenoid/ Chl A ratio. Some of the recently developed commercially available hand-held devices like Green seeker, weed seeker, Crop Spec, Crop Circle, SPAD etc. use spectral reflectance from the crop canopy, calculate vegetation indices like NDVI, normalised difference red edge (NDRE) etc. from the reflectance values and determine N doses for crops using calibration equations or algorithms developed for specific crops. Working principles and use of some of important optical sensor based nutrient estimation devices available for commercial use are discussed below:

Optical crop sensor

Optical crop sensors evaluate crop conditions by shining light of specific wavelengths at crop leaves, and measuring the type and intensity of the light wavelengths reflected back to the sensors. Not all optical sensors use the same light wavelengths. Different color light waves can be used to measure different plant properties. Commercially available crop sensors use two or more of red, green, blue or near infrared (NIR) color light waves. The basic principles of optical sensing are similar to visual observations. Energy in the form of light waves travels from the sensor to the plant leaves. Green plants absorb much of the visible light wavelengths, particularly the blue and red light waves, and reflect much of the green light waves. This is why plant leaves appear green to us. Sensing the reflectance of green light wavelengths from plants can provide a relative measure of chlorophyll in the leaves. Green reflectance can be used to evaluate crop N status, the degree of iron deficiency chlorosis, sulphur deficiency, or any other condition, causing reduction in green color. Plants absorb much less NIR light than red light. Darker green leaves reflect more NIR light and absorb more red light than lighter green leaves. These reflectance characteristics for visible and NIR light of plants are used to develop vegetative indices to compare the relative health of crops. The Normalized Difference Vegetation Index (NDVI) is calculated using the reflectance of red and NIR wavelengths. The formula for NDVI gives values from -1.0 to +1.0. Typical plant sensing operations give values from 0.1 to 0.9, with values ranging from 0.1 to 0.2 for soil surfaces and 0.2 to 1.0 for crop canopies. NDVI values increase when the crop cover increases, and to a lesser degree, the crop greenness increase.

Most of the commercially available crop sensors are called active sensors because they emit their own source of light onto the crop canopy and then measure the percent of light reflected from the canopy back to the sensor. The commercial handheld devices like Green Seeker TM, Opt Rx TM and the Holland Scientific Crop Circle TM sensors use light emitting diodes, and the Crop Spec TM uses laser diodes for emitting light. The light reflectance is recorded by all sensors using photodiodes, which convert light waves into electrical charges, which are then digitally quantified. Since sensors use a modified light source, the reflectance can be distinguished from natural sunlight, allowing the sensors to function in any daylight or darkness, with minimal disturbance by passing cloud cover. The crop sensors can be used for in-season N fertilizer, increasing nitrogen use efficiency and in-season crop yield estimate.

How crop sensors determine in-season nitrogen fertilizer rates

Crop sensors can be mounted on N fertilizer applicators equipped with computer processing and variable rate controllers to apply fertilizer during the growing season. Each commercial sensor manufacturer recommends specific operating procedures to calculate a fertilizer rate based on a vegetative index. The sensor software uses a step-by-step mathematical procedure, called an algorithm, to estimate a fertilizer rate.

Typical procedures for using optical crop sensors

Step 1: Nitrogen-rich Strips: Plant a N-rich strip in each field within each variety with an abundance of nitrogen fertilizer applied prior to planting the crop. This strip must have enough N to make sure the crop plants will not show any N deficiency symptoms. The sensor is operated over the N-rich strip to establish a N-sufficient reference area in the field. The remaining areas of the field are compared to the N-rich strip to make N rate recommendations.

Step 2: Scan N reference Strip with Sensors: The user takes sensor readings from the N-rich reference strip. The Opt Rx sensor uses a "virtual reference strip", which is the healthiest area of the field. Reference values are stored in the computer for on-the-go calculations.

Step 3: Select the Appropriate Crop Algorithm: The sensor manufacturer incorporates algorithms for each crop. The algorithms are generally based on in season estimated yield (INSEY). INSEY is determined by dividing the sensor NDVI readings by the growing degree days. Generally, the sensor operator must enter the growing degree days for the field area into the sensor computer.

Step 4: Sensing the Field: The operator then drives over the rest of the field, automatically sensing, calculating and applying a variable rate of N fertilizer to the crop. Sensor values may be different if the crop leaves are wet or dry. Temperature changes throughout the day may also cause changes in sensor values. The N-rich or virtual strip should be re-sensed if the field conditions change during fertilizer application.

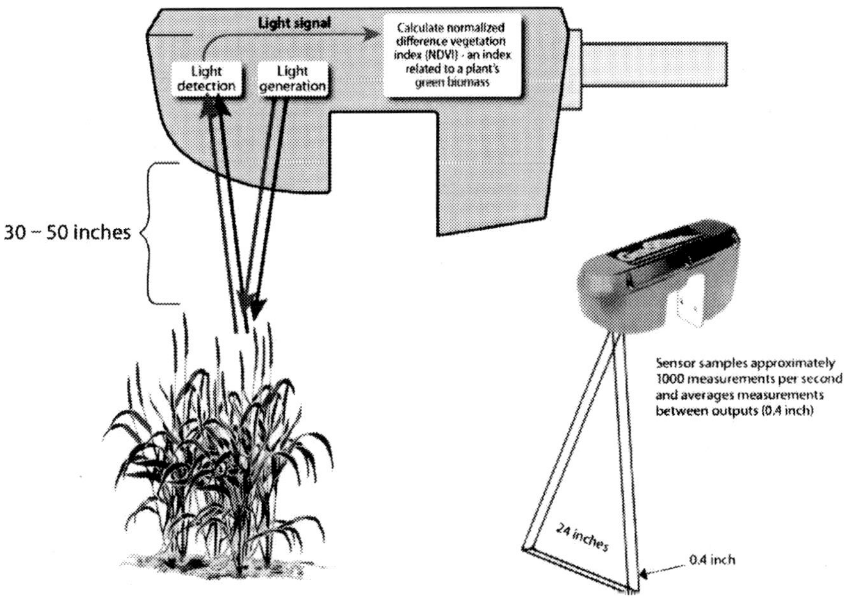

Fig. 1. A diagram representation of a Green Seeker optical sensor system and outputs measurement. (*See colour version on page 452*)

How the green seeker works?

Fig. 2. Handheld Greenseeker an active remote sensing instrument for NDVI calculation

- The sensor emits brief bursts of red and infrared light, and then measures the amount of each type of light that is reflected back from the plant
- The sensor continues to sample the scanned area as long as the trigger remains engaged
- The sensor displays the measured value in terms of an NDVI reading (ranging from 0.00 to 0.99) on its LCD display screen
- The strength of the detected light is a direct indicator of the health of the crop; the higher the reading, the healthier the plant

Chlorophyll meter or SPAD (Soil plant analysis development)

The chlorophyll content meter (or SPAD meter) is a commercially available portable piece of equipment that is used to measure greenness based on optical responses when a leaf is exposed to light that in turn is used to estimate foliar chlorophyll concentrations. The meter makes instantaneous and nondestructive readings on a plant based on the quantification of light intensity (peak wavelength: approxi-mately 650 nm: red light-emitting diode [LED]) absorbed by the tissue sample. A second peak (peak wavelength: approximately 940 nm: infrared LED) is emitted simultaneous with red LED to compensate for thickness of the leaf. The SPAD meter enables quick, easy measurement of the chlorophyll content of plant leaves without damaging the leaf. Chlorophyll content is one indicator of plant health, and can be used to optimize the timing and quantity of applying additional fertilizer to provide larger crop yields of higher quality with lower environmental load. The SPAD offers a new strategy for synchronizing N application with actual crop demand in crops. The chlorophyll meter indicates the need of a N top dressing that would result greater agronomic efficiency of N fertilizer than commonly pre-application of N. The SPAD based N management needs considerably lower amount of N than the standard N

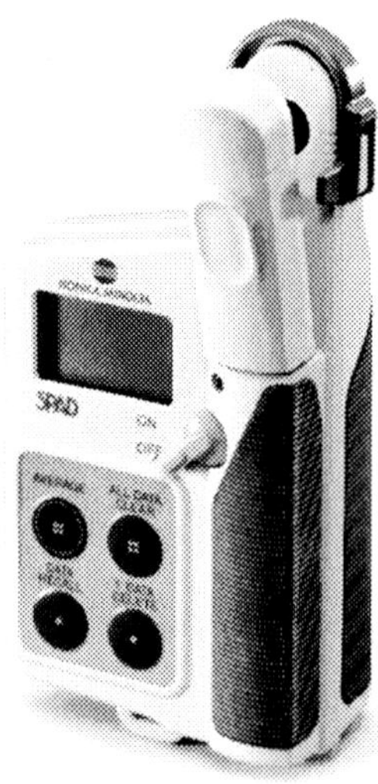

Fig. 3. SPAD meter

management practices without any yield losses resulting in improved N use efficacy of crops.

Leaf colour chart

The leaf color chart (LCC) is an innovative cost effective tool for real-time or crop-need-based N management in Rice, Maize and Wheat. LCC is a visual and subjective indicator of plant N deficiency and is an inexpensive, easy to use and simple alternative to chlorophyll meter /SPAD meter. It measures leaf color intensity that is related to leaf N status. LCC is an ideal tool to optimize N use in Rice/Maize at high yield levels, irrespective of the source of N applied, *viz.*, organic manure, biologically fixed N, or chemical fertilizers. Thus, it is an eco-friendly tool in the hands of farmers. Now, it is manufactured with 4 colors called Four Panel LCC & 6 colors called Six Panel LCC. Moreover, LCC is provided with water-proof laminated instruction sticker in the required regional language. LCC has been jointly developed by International Rice Research Institute (IRRI) and Philippines Rice Research Institute (PhilRice) from a Japanese prototype, for the purpose of measuring the required quantity of N to be applied in Rice field and thereby to get a maximum productivity. The LCC is also suitable for maize & wheat providing farmers with a good diagnostic tool for detecting N deficiency. The LCCs relevant to use for Sugarcane, Potato, Cotton, Cassava, *etc.* are under development in order to maximize the yield of these crops.

Use of LCC helps to determine N demand of the crop and guide right time of fertilizer N application so as to prevent unwanted N losses and their serious impacts on the ecosystem. The excessive dose of urea otherwise increases insect-pest attack and thus leads to consumption of high doses of insecticides and pesticides, thereby, resulting in increased cost of production, environmental pollution and deterioration of the quality of produce. Also nitrates trickle down the earth with use of excessive urea causing nitrate contamination to the ground water which if reaches 10 milligram or more in per liter water then water becomes unfit for human consumption.

Procedure

1. Randomly select 10 healthy plants in field where plant distribution is uniform.
2. Select the top most, fully expanded, and healthy leaf of each of the 10 plants. Take LCC readings by placing the middle part of the leaf on top of the LCC's color strips for comparison. Do not detach the leaf and do not

expose the LCC to direct sunlight during readings. The same person should take the LCC readings at same time of the day between 8:00 and 10:00 a.m. from first up to the last reading.

3. If more than 5 out of 10 leaves have readings below 4 in transplanted rice and below 3 in direct wet-seeded rice, apply 30 kg N/ha during dry season (DS) or 23 kg N/ha during wet season (WS). Use 1.5 bags of urea per hectare during DS or 1 bag urea during WS.
4. Repeat LCC readings every seven days until the first flowering. Different sets of 10 leaves can be used for each weekly reading.

Use of LCC in maize

1. Apply 25 kg Urea per acre at the time of sowing of maize.
2. Take readings of ten randomly selected maize plants, by matching the color of the first fully exposed leaf from the top with the LCC starting from 21 days after sowing of maize till initiation of silking at 10 days interval.
3. Match the color of maize leaves with LCC shade 5. If 6 or more leaves out of 10 leaves are lighter than the specified threshold, apply 25 kg urea per acre. When color of 5 or more leaves is equal to or darker than the specified threshold no urea need not be applied.
4. Always match the color of leaves with LCC in the shade of your body to avoid direct sunlight.
5. Use of LCC should be discontinued after silking in maize and no urea should be applied afterwards.

Use of LCC in wheat

1. Apply a basal dose of 55 kg di-ammonium phosphate (DAP) per acre.
2. Apply 40 kg urea per acre for timely sown and 25 kg urea per acre for late sown wheat with first irrigation.
3. Match leaf colour of youngest fully exposed leaf from the 10 randomly selected insect/disease free plants from each field with LCC before 2nd irrigation.
4. If the greenness of 6 or more out of 10 leaves is less than shade 4 on LCC, broadcast 40 kg urea per acre for timely sown wheat and 25 kg urea per acre for late sown wheat with second irrigation.

5. If greenness is equal to or more than shade 4 on LCC apply only 25 kg urea per acre for timely sown wheat and 15 kg urea per acre for late sown wheat with second irrigation.
6. Always compare colour of the wheat leaf with LCC under shade of your body.

Fig. 4. Use of LCC for N estimation in rice

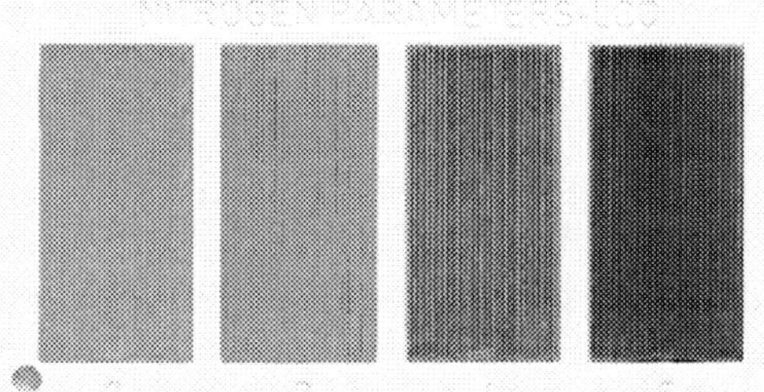

Four panel

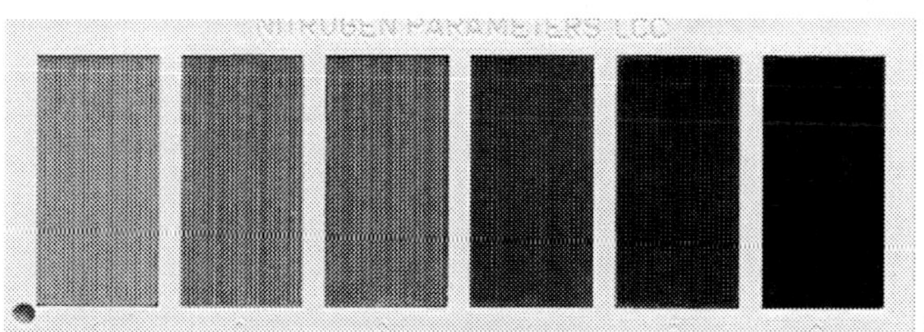

Six panel

Fig. 5. Four panel and six panel leaf color chart

(*See colour version on page 452*)

Spectroradiometer

Spectroradiometer is an instrument for measuring the spectral concentration of radiant energy. It combines the functions of a spectrometer with those of a radiometer. Spectroradiometers are designed to measure the spectral power distributions of illuminants. Spectroradiometers are used in the laboratory, field, aircraft, and on satellites. It measures the spectral radiance of an object in the electromagnetic radiation wavelength range from 300 to 2500 nm. It also measures the irradiance of the incident solar radiation. The ratio of the incident-to-reflected radiant flux measured from an object or area over specified wavelengths is known as spectral reflectance. Unlike radiance and irradiance values, reflectance is an inherent property of an object and is independent of time, location, illumination intensity, atmospheric conditions and weather.

Principles of measurement

Spectroradiometer unit consists of a main spectrometer integrated with computer based processing unit with built-in storage device, user interface, display unit and light source, fiber-optic cable, a pistol grip, and different field of view (FOV) lens. Inside the spectrometer instrument light is projected from the fibre optics onto a holographic diffraction grating where wavelength components are separated and reflected for independent collection by the detector(s). The detector is a fixed 256-element photodiode array. Each photodiode or pixel is a capacitive device that discharges while it is exposed to light. During a scan, each pixel is exposed for a pre-defined time (the integration time). Each detector converts incident photons into electrons that are stored, or integrated, until the detector is read out. At the read out time, the photoelectric current for each detector is converted to a voltage and is digitized by a 16-bit analog to digital (A/D) converter. This data is directly transferred to the computer main memory which in turn available for further processing by the controlling software. Gathering spectra at any given location involved optimizing the integration time (typically set at 10 milliseconds), providing foreoptic information, and recording dark current, collecting white reference reflectance and then obtaining target reflectance. The target reflectance is the ratio of energy reflected off the target (*e.g.*, crops) to energy incident on the target (measured using a $BaSO_4$ white reference). Since the dark current varies with time and temperature it was gathered for each integration time (virtually for each new reading).

Measurement on spectral reflectance from individual leaves using a UniSpec-SC

Spectroradiometer

For many "Plant physiological" applications, measurements for single leaf need to be non-destructive, *in-situ,* and measured in the VIS/NIR regions (310-1100 nm). In addition to the main unit (UniSpec-SC) which is portable, the system provides a light source suitable for reflectance measurements in the VIS/NIR, a bifurcated fiber optic for delivery of light to the sample and for directing reflected light back to the detector, a sample holder (leaf clip) for sealing and holding the leaf in place at a fixed angle (60^0) and a reference standard for calibration of the light source and detector.

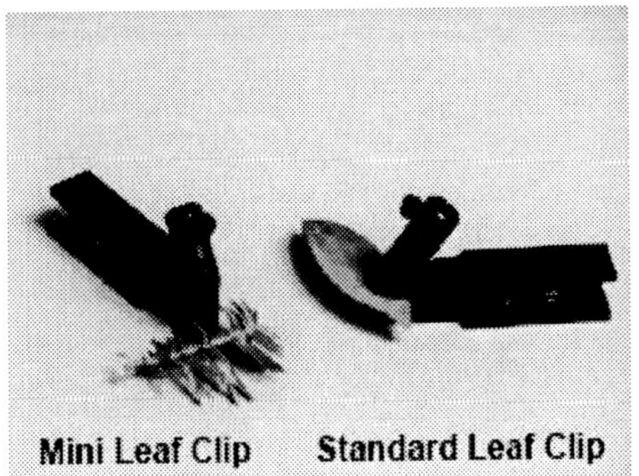

Fig. 6. Measurement of reflectance from individual leaves using leaf clip
(*See colour version on page 453*)

Nutrient sensing technologies

Intensive grid sampling is generally regarded as one of the most accurate methods of mapping the variability of crop and soil attributes in precision agriculture. However, intensive grid sampling is laborious, time consuming and expensive and thus impractical for implementation in large scale. It is, therefore, desirable to develop a more rapid means of obtaining spatial and temporal data for detailed variability mapping. The efficiency of site and time-specific crop-soil management and monitoring strategies can be improved by using low-cost sensors to estimate soil properties that impact crop yields. On-the-go soil sensor technologies that can serve as a rapid method for measuring soil mechanical, physical and chemical properties are steadily developing (Christy, 2008). Soil sensors can be used to generate real-time soil data, such as pH, electrical conductivity, salinity, dissolved oxygen and nutrient concentration, which are subsequently turned into geo-referenced maps to facilitate site-specific nutrient application. Few on-the-go sensors have been manufactured to measure mechanical, physical and chemical soil properties and most of them have been based on electrical and electromagnetic, optical and radiometric, mechanical, acoustic, pneumatic and electrochemical measurement concepts.

Optical and radiometric sensors: Optical sensing technology uses visible and near-infrared wavelength ranges to rapidly quantify soil properties. The principle of this approach is the interaction between incident light and soil surface properties, such that the reflected lights vary as a function of soil physical and chemical properties. Optical nutrient sensing techniques are non-destructive and are often more favoured in comparison to electrochemical sensing. Optical soil sensors have a high potential for estimation of soil organic matter content based on soil color. In optical sensing of soil, the visual and near-infrared spectral reflectance can potentially estimate texture, moisture, CEC and other soil parameters if proper data analysis techniques are applied.

Conclusion

Optical remote sensing devices have the potential to estimate nutrient content in plants and thus can be used to adjust the dose and time of nutrient application in crops which in turn improve the productivity and reduce the loss of nutrient from the rhizosphere. Handheld devices are becoming cost effective and their ease of use makes them suitable for use in commercial cultivation. Development of sensors for soil nutrient estimation has a great potential for cost effective soil nutrient analysis and use in site specific nutrient management.These nutrient sensors will help in improving the use efficiency of the costly fertilizer nutrients in future.Increasing population growth coupled with the increasing risks associated with climate change inevitably requires a commensurate increase in agricultural productivity. Key to this challenging task is to ensure sustainable soil productivity while maintaining high crop yields and reducing environmental pollution. To this end, the implementation of sensor technologies for soil nutrient management and monitoring is a step in the right direction.

Selected References

Buscaglia HJ, Varco JJ. 2002. Early detection of cotton leaf nitrogen status using leaf reflectance. *Journal of Plant Nutrition* 25: 2067-2080.

Christy CD. 2008. Real-time measurement of soil attributesusing on-the-go near infrared reflectance spectroscopy. *Computers and Electronics in Agriculture*, 61: 10-19.

Cozzolino D , Moron, A. 2006.The potential of near-infraredreflectance spectroscopy to analyze soil chemical and physicalcharacteristics. *Journal of Agricultural Science*, 140: 65-71.

Gitelson, AA, Kaufman, YJ, Merzlyak MN. 1996. Use of a green channel in remote sensing of global vegetation from EOSMODIS. *Remote Sensing Environment*. 58: 289-298.

Peñuelas J, Gamon JA, Griffin KL, Field CB. 1993. Assessing community type, plant biomass, pigment composition, and photosynthetic efficiency of aquatic vegetation from spectral reflectance. *Remote Sensing Environment*. 46:110-118.

Schepers JS. 1994. New diagnostic tools for tissue testing. *Communications in Soil Science and Plant Analysis* 25: 817-826.

8

Enhancing Nutrient Use Efficiency, pp. 93-106
Editors: K. Ramesh, A.K. Biswas, B.L. Lakaria, S. Srivastava and A.K. Patra

Nutrient Management for Enhancing Productivity and Nutrient Use Efficiencies in Long Term Fertilizer Experiments

Muneshwar Singh and R.H. Wanjari

ICAR-Indian Institute of Soil Science, Nabi Bagh, Bhopal – 462 038, India

With the introduction of high yielding varieties and use of high analysis fertilizer in intensified agriculture under irrigation condition resulted in green revolution in our country. But continuous use of chemical fertilizer in indiscriminate manner without assessing soil also had adverse effect on productivity and environment. Though use of high analysis fertilizer ensured the high production but also resulted in acceleration of mining of native sources of nutrient. This has led to appearance of hidden hunger of several other essential nutrients and sustainability of our agriculture has become venerable. High analysis fertilizer (NPK) resulted in deficiency of Zn and S, in Indo-Gangetic Plain (IGP) and caused decline in productivity. The ever increasing population and shrinking natural resources (land and water) are building pressure on us to produce more and more from a unit quantity in a unit time. The pressure on natural resources in Indian context could be assessed from the fact that it supports 18 percent of human and over 15 percent of livestock of the world from 2.3 percent of world's geographical area. The over stressing on natural resources by disproportionate human and live stock populations led to non sustainability of many agro-systems. Therefore, it has become essential not only to sustain our agro-system but also to keep the pace of acceleration with increasing demand of food and fiber.

Under present scenario it is beyond imagination to sustain the productivity without fertilizer input. But due to over mining of the nutrient, Indian soils are working with negative nutrient balance to the tune of 12-14 m tons yr^{-1} and the negative balance is likely to increase in future even after using the full potential of fertilizer industry. The increase in nutrient efficiency not only would help in reducing the negative nutrient balance but also curtail the input cost and acts as key to safe guard the environment. The nutrients use efficiency (NUE) and productivity goes hand in hand and compliment each other. Under this situation integrated nutrient management not only reduce the nutrient gap between addition and removal but also ensures the higher nutrient use efficiency, sustainability of the system and minimize the environmental pollution. In this paper, attempts is made to have an over view of nutrient management on productivity and nutrient use efficiency in long term fertilizer experiments.

Integrated Plant Nutrient Supply System (IPNS)

The survey of literature revealed that to convey the meanings of integrated plant nutrient different terms are beings used. Integrated plants nutrient supply system (IPNS), integrated nutrient management (INM), balance nutrient *etc.* The basic principle of IPNS is based on nutrient supply through different sources FYM, crop residue, organic manure and fertilizer to plant for sustaining the productivity. Number of workers attempted to define IPNS taking into account, productivity, maintaining living organism and social economic environment.

1. Integrated Plant Nutrition System (IPNS) is defined as maintenance or adjustment of soil fertility and of plant nutrient supply to an optimum level for sustaining the desired crop productivity through optimization of benefit from all possible resources of plant nutrient in an integrated manner (Roy and Ange, 1991).

2. Integrated Plant Nutrition System (IPNS) is used to maintain or adjust soil fertility and plant nutrient supply to achieve a given level of crop production. This is done by optimizing the benefits from all possible sources of plant nutrient (FAO, 1998).

3. Integrated Nutrient Management (INM) it actually the technical and managerial component of achieving the objective of IPNS under farm situations. It takes into account of all factors of soil and crop management including management of all other inputs such as water, agrochemicals, amendment *etc.*, besides nutrients (Goswami, 1998).

I. Potential source of organic material

The annual potential of organic resources in the country available for use through excreta of live stock and human, crop residue, and sewage sludge, municipal waste *etc* is estimated to be around 20 m tons currently expected to be around 28 m tons together N, P_2O_5 and K_2O exclusive of micronutrient (Bhardwaj and Gaur, 1985). In addition to this following source are also available.

a. **Organic Manure:** Organic manures like FYM and composts are the important component of IPNS and should be assessed for the quantity available. According to one estimate nearly 800 m tons of organic manure available which comes out to be 2.5 t ha^{-1} yr^{-1} on an average if spread over 329 mha^{-1} geographical area of the country. The waste from household and cattle sheds are major components of the FYM and compost. But a major portion of cattle dung is used as domestic fuel and crop residue for cattle feed. Survey conducted by IISS indicated more than half (52%) of cow dung is used for making dung cake for fuel purpose and the remaining is used for making compost (Reddy *et al.*, 2015).

b. **Press Mud:** Sugarcane industry is the main source of press mud which contains 1.0-2.5% N, 0.25 to 0.65% P and 0.4 to 0.85% K but being acidic in nature and can be applied to alkaline soils. However, other type of press mud which is obtained from carbonation process contains lime and can be used in acidic soils. Nearly 2.7 m tons press mud is produced every year in our country.

c. **Green Manure and BGA:** Green manure is a good source of biologically fixed N and organic carbon. Several green manure crops are grown in our country such as *Sesbania*, *Crotalaria juncea*, *Tephrosia purpurea* and sunhemp *etc.* A 40-45 days old crop can supply 100-125 kg N which is almost equal to average N applied in Indian agriculture. However, green manure is suited for rice based system. But practicing of green manure is restricted to a limited area where water is available during summer season like Kerala and North East part of the country. In North though water is available but farmer prefer to grow summer mung rather than growing green manure crop. In addition to green manure crop, growing of legume trees on bund or outside the field for the purpose of green manure is also practiced in pocket. Blue green algae (BGA) are also component of INM which is practiced in rice based cropping system. BGA is being practiced in Eastern and North Eastern part of the country.

d. **Legume Residue:** As discussed in previous section in North Western part of the country short duration summer mung is grown in between rice-wheat during fallow period. After harvesting pod, the C residue is incorporated which save 60 kg N (Rekhi and Meelu, 1983 and Singh and Dwivedi, 1996).

II. Integrated Nutrient Management and Productivity of the System

Since ages Indian farmers are firm believers of INM and were using all possible sources of nutrients including pond silt and earth scraping of fig tree in addition to FYM. The introduction of chemical fertilizer had brake on use of organic manure for a short period but increasing cost of fertilizer farmer has again started using organic manure to supplement the nutrient and to reduce dependence on chemical fertilizer. A large volume of work scanned and package of INM made available for different agro-system. (Rao,1998). Enhancing the productivity is prime objective and priority of Indian agriculture to feed the ever growing population from shrinking natural resources. Inclusion of various components to supply the nutrient under INM depends on cropping system and availability of resources in a particular location. To study the impact of INM on productivity and NUE requires long term fertilizer experiment. Productivity and NUE are directly related to each other, therefore impact of INM on productivity is essential before assessing NUE. Long-term studies give an opportunity to evaluate the impact of INM on productivity of the system.

a. Rice-Wheat: Rice-wheat is most predominant cropping system of the country and contributes major portion to food basket. The yield data (Table 1) of Pantnagar (Mollisols), Barrackpore (Inceptisols) and Raipur (Vertisols) indicated that integration of nutrients resulted increase in productivity of the system at all the three places and incorporation of FYM further enhanced the productivity. At Pantnagar, application of 100% N alone for 34 years, average yields of rice and wheat recorded were 4780 and 3115 kg ha^{-1}. Integration of P with N and K with NP resulted in increase in yield of rice to 4865 and 5095 kg ha^{-1} and of wheat to 3779 and 3794 kg ha^{-1} respectively. Incorporation of FYM with NPK further increased productivity of rice and wheat to 5788 and 4497 kg ha^{-1} respectively. The similar trend in yield on integration of nutrient was also noted at other two locations differing level of magnitude in productivity because of climatic condition. Increase in yield on incorporation of FYM is not only due to additional supply of major nutrients but also ruled out the hidden hunger of Zn and secondary nutrient at many places. Moreover, organic manure also helped in turnover process of nutrients in soil which is responsible for nutrient transformation in soil.

Table 1: Effect of integrated nutrient management on average productivity of rice-wheat (kg ha^{-1}) cropping system under LTFE at different locations

Treatments	Pantnagar		Barrackpore*		Raipur	
	1972 to 2007		1972 to 2007		1999 to 2007	
Crop	Rice	Wheat	Rice	Wheat	Rice	Wheat
Control	2930	1450	1507	747	2382	1057
100 % N	4780	3115	3194	1962	3678	1544
100 % NP	4860	3779	3573	2206	5095	2210
100 % NPK	5090	3794	3719	2321	5158	2239
NPK + FYM	5780	4497	3515	3620	5541	2697
100% NPK + GM	-	-	-	-	4666	1744
50% NPK + BGA	-	-	-	-	4225	1624
CD 5%	611	448	265	316	328	346

*Saha *et al.* (2008)

To evaluate the INM impact on productivity of rice and wheat, number of trials were conducted through out the country under AICRP Agronomy. The data summarized in Table 2 indicated that combined use of fertilizer and organic manure (FYM) result in larger productivity of the system with residual effect on subsequent wheat crop (Singh and Singh, 1998). Eight years study on integrated nutrient management in rice-wheat in Vertisols (Singh *et al.*, 2001) at Jabalpur revealed that conjunctive use of 5 t FYM and 6 t green manure (Parthenium) with 90 kg N not only sustained the productivity but also saved nearly 90-100 kg ha^{-1} fertilizer N annually. In additional to saving of nitrogen, INM practices also improved the SOC and nutrient status of soil (Table 3).

Table 2: Effect of combined use of organic and inorganic fertilizer average crop yields (t ha^{1}) in rice-wheat at different location

Treatment		Location					
		Bhagalpur (87 trials)		Manipur (96 trials)		Ludhiana (5 years)	
Rice	Wheat	Rice	Wheat	Rice	Wheat	Rice	Wheat
FoN_{60}	FoN_{60}	4.21	2.75	4.41	1.04	5.6	2.8
$^{a}F_{12}N_{60}$	$F_{12}N_{60}$	4.14	2.95	5.44	1.33	5.7	3.9

At Ludhiana a= N_{80} to rice and N_{90} to wheat
Singh and Singh (1998)

Table 3: Integrated Nutrient management in rice wheat for 7 years and average productivity of rice-wheat (t ha^{-1}) and nutrient status in Vertisol at Jabalpur

Treatment	Rice	Wheat	Nutrient Status after 7 years (mg kg^{-1})		
			OC%	P**	K*
N_{90}	4.42	4.19	0.58	21.1	138
N_{180}	5.08	4.70	0.71	18.7	125
N_{90} + FYM	4.95	4.49	0.74	40.1	230
N_{90} + GM	4.58	5.07	0.72	39.1	240
Initial	-		0.60	19.5	195

Singh *et al.* (2001), *Singh *et al.* (2002) and **Singh *et al.* (2007).

b. Soybean-Wheat: Soybean-wheat is predominant cropping system of Madhya Pradesh and becoming popular in Rajasthan and Maharashtra. Thirty five years data summarized in table 4 clearly indicated that integration of N, P and K sustained the yield at a higher level compared to application of N and NP. However, the maximum average productivity is recorded with conjunctive use of NPK and FYM. The impact of INM on productivity scenario recorded at Ranchi (Alfisols) is different. Continuous application of N alone compared to control indicates that N has adverse effect on productivity. Integration of P and PK with N resulted increase in yield of soybean from 296 kg ha^{-1} (100% N alone) to 861 and 1496 kg ha^{-1} respectively and the corresponding yields of wheat recorded were 1049 (100% N) 2432 kg ha^{-1} and 2789 kg ha^{-1}. (Table 4) It is interesting to note that integration of NPK with FYM further increased the yield of both soybean and wheat. At Ranchi use of lime as a component of INM also improved the productivity significantly. Thus big jump in yield denotes that integrated nutrient management becomes more important in soils which have poor buffering capacity.

Seven years study on soybean-wheat at IISS farm also demonstrated that integration of nutrient (FYM and fertilizer) sustained the yield at larger level compared to sole application of either organic or fertilizer (Rao *et al.*, 1998). Similarly, trials conducted at farmers' field of Sehore and Bhopal districts for three years (Table 5) further confirmed that integrated nutrient management is best option as far as productivity and profit of the farmers are concerned (Singh *et al.*, 2008).

c. Maize-Wheat: Maize-wheat is another important cropping system of North-West India. The yield data of 36 years old maize-wheat cropping system at Ludhiana and 10 years old experiment at Udaipur and Delhi indicated that integration of P and K with N enhanced the productivity of the cropping system. Integration of P with N resulted increase in yield of maize from 1771 kg ha^{-1} (N alone) to 2209 and wheat from 2932 kg ha^{-1} to 4108 kg ha^{-1} At Ludhiana. Addition of K and incorporation of FYM further enhanced the productivity (Table 6). At Palampur (Alfisols) continuous application N resulted sharp decline in yield with time and at present there was zero yields in this particular plot. However, integration of PK along with FYM/lime sustained the productivity of the system over last 36 years (Table 4).

Table 4: Effect of integrated nutrient management on average productivity of soybean/ maize-wheat system under LTFE at different locations

Treatments	Ranchi[a] 1972 to 2007		Jabalpur[b] 1972 to 2007		Palampur[c] 1999-2007	
Crop	Soybean	Wheat	Soybean	Wheat	Maize	Wheat
Control	617	705	838	1206	261	365
100 % N	296	375	1049	1651	449$	425$
100 % NP	861	2432	1728	4034	1953	1642
100 % NPK	1496	2789	1888	4353	3175	2289
NPK + FYM	1832	3333	2069	4770	4588	3029
NPK + Lime	1771	3218	-	-	5580	2480
CD 5%	212	390	282	441	706	502

$ at present yields are zero. [a]Dwivedi *et al.* (2007), [b]Mahapatra *et al.* (2007), [c]Sharma *et al.* (2005).

Table 5: Effect of INM on productivity of soybean wheat (t ha^{-1}) and economics

Treatment	Soybean	Wheat	Total output cost*
FYM	1.81	3.87	45631
PM	1.90	3.69	46390
IPNS	1.97	4.32	50392
Conventional	1.74	3.98	45617
CD (5%)	NS	0.39	-

*The cost of output was calculated on the basis of price prevailed during that period.

Table 6: Effect of integrated nutrient management on average productivity (kg/ha^{-1}) maize-wheat system under LTFE

Treatments	Ludhiana 1972 to 2007		New Delhi 1995-96 to 2007		Udaipur 1997-98 to 2007	
Crop	Maize	Wheat	Maize	Wheat	Maize	Wheat
Control	783	1141	1182	2403	2000	1941
100 % N	1771	2932	1577	3564	2550	3206
100 % NP	2209	4108	1793	3785	2897	3683
100 % NPK	2927	4739	2151	4514	3064	3909
NPK + FYM	3601	4949	2551	4959	3390	4357
CD 5%	244	266	192	286	223	242

d. Finger Millet-Maize: Finger millet-maize covers large area in southern India. The yield data of both finger millet and maize indicate that integration of all the three major nutrients (NPK) always had the larger productivity compared to application of N and NP. Integration of FYM with NPK further boosted the productivity of the system (Table 7). Rice-Rice is quite common in some parts of southern and eastern states of the country. The 10 year average yield data (Table 7) indicated that integrated nutrient management is the only option to sustain productivity of rice-rice Alfisols of Pattambi (Kerala). At Pattambi green manure and liming are essential components of INM for sustaining the productivity at higher level.

Table 7: Effect of integrated nutrient management on average productivity (kg ha^{-1}) of finger-millet and rice-rice system under LTFE

Treatments	Bangalore* 1987-2007		Coimbatore 1971-2007		Pattambi 1998-2007	
Crop	F. Millet	Maize	F. Millet	Maize	Rice	Rice
Control	582	284	1094	917	1532	2120
100 % N	740	387	1346	1179	2120	3053
100 % NP	948	644	2976	3079	1997	2902
100 % NPK	4241	2172	3063	3211	2210	3020
NPK + GM	-	-	-	-	3162	3629
NPK + Lime	4762	2597	3537	3767	2634	3536
CD 5%	126	566	400	316	447	351

*Sudhir *et al.* (2004); Singh and Wanjari (2008)

Thus the data generated under long-term experiment and in other studies clearly demonstrated that integrated nutrient management is the only option to sustain the productivity at higher level. Application of FYM ensures the sustainability at higher level by taking care of the hidden hunger of micro and secondary nutrients which other wise could have been limiting factor in sustaining the productivity and NUE. Applications of FYM in addition to supply of nutrient also act as conditioner for physical condition of soil like infiltration and bulk density which improves aeration and water movement in soil required for better root and plant growth. In Alfisols of Ranchi, Bangalore and Palampur application of N alone had adverse effect on productivity. The reason for decline in productivity in Alfisols is due to non availability of P and K due reduction in soil pH on application of N alone. Whereas at Jabalpur P and K were not limiting nutrient for initial few years which made possible to sustain the yield. However, application of P along with N and K along with NP had negative effect on productivity of rice-wheat, soybean-wheat and maize-wheat system.

III. Integrated Nutrient Management and Use Efficiency

Nutrient use efficiency has been all time prime concerned for agriculture scientist because of less nutrient need to obtain a targeted production, more return per unit amount invested and reduced the risk of environmental pollution. The increasing cost of fertilizer further attracted to develop technique to improve nutrient use efficiency. On the basis of present fertilizer consumption (12.3 m tons N and 5 m tons P), one percent increase in the efficiency of N and P would lead to saving of 3.0 lakhs tones of N and 2.2 lakhs tones of P annually which cost together in saving of Rs. 7000 millions. In addition to economic benefit it would reduce the risk of pollution. Nutrient use efficiency can be improved by taking care of following issues in our management system.

1. Reducing the loss of nutrients
2. Use of nutrient on soil test basis
3. Integrated/balanced use of nutrients
4. Adopting nutrient efficient variety
5. Modified form of fertilizers
6. Synchronizing the demand and supply of nutrient

While scanning the literature, it was found that a lot of work has been done on this aspect but most of work is based on two to four years experiment. To get the maximum benefit of INM on nutrient use efficiency you need some long-term study. To illustrate the facts, here are few examples from long-term fertilizer experiment. Nitrogen is most mobile element and applied in larger quantity. The residual effect of nitrogen is very low as it does not remain stable in the system. The efficiency of a nutrient depends on the adequacy of other essential nutrient. Perusal of data on nutrient use efficiency calculated on the basis of 32 years study indicated that when the nitrogen is integrated with other nutrient (P and K) and organic manure, increase in use efficiency of N irrespective of crop and soil. For instances, in Inceptisols at Ludhiana, the N use efficiency in maize recorded in 100% N (alone) was 16.7% which increased to 23.5, 36.4, 40.2 percent on integration with P, PK, and FYM, respectively (Table 8). Similar trend was noted in Mollisols of Pantnagar in rice and wheat crop sequence. The N use efficiency recorded in 100% N treatment was 37.5 percent which increased to 40.7%, 44.4% and 61.2 percent in the plots received 100% NP, 100% NPK and 100% NPK+FYM, respectively. Similar effect of nutrient management was also observed in subsequent wheat crop.

Table 8: Nitrogen use efficiency in different crops as affected by balanced and imbalanced use of nutrient in long-term fertilizer experiment

Soil Type	Location	Crop	Mean nitrogen use efficiency, %			
			100% N	100% NP	100% NPK	100% NPK+FYM
Inceptisol	Ludhiana	Maize	16.7	23.5	36.4	40.2
Alfisol	Palampur	Maize	6.4	34.7	52.6	63.7
Mollisol	Pantnagar	Rice	37.5	40.7	44.4	61.7
Inceptisol	Ludhiana	Wheat	32.0	50.6	63.1	67.8
Alfisol	Palampur	Wheat	1.9	35.6	50.6	72.6
Mollisol	Pantnagar	Wheat	42.4	46.1	48.4	47.9

Integrated nutrient management also led to increase P and K use efficiency. The data presented in Table 9 indicated that integrated nutrient management resulted in increase in use efficiency of P and K at all the locations and in all the cropping systems. For instance, the P use efficiency recorded in maize at Ludhiana (Inceptisols) and Palampur (Alfisols) in NP treatment was 10.3 and 21.8 percent, respectively. On addition of K and organic manure the corresponding values of P use efficiency observed were 21.4 and 35.6 percent, respectively. The greater P use efficiency at Palampur is due to jump in yield on application of K and FYM and little decline in yield over the years in control treatment. Moreover, these soils are low in K and application of K and FYM has additive effect on crop productivity.

Though in the beginning at many places crop did not respond to applied K because of high status. The data indicate that application of organic manure irrespective of soil and cropping system induced K use efficiency. The K use efficiency in rice at Pantnagar in NPK treatment recorded was 34.5% which increased to 108.3% on incorporation of FYM, whereas K use efficiency in wheat increased from 13.7 to 35.8 percent. Recording of more than 100% increase in K use efficiency is due to application of less amount of K and proportionately more mining of K by crop on application of FYM. Plant also takes more K for luxury consumption. In problematic soils (an acidic) for maximising the nutrient use efficiency, there is need to get rid of the problem. To sustain the productivity in acid soil (Alfisols) there is need to raise the pH of the soil by liming. The results of long term fertilizer experiment clearly indicated that application lime has positive effect on productivity and nutrient use efficiency as well. The data generated (Table 7) from 18 years old experiment clearly demonstrated that application of lime significantly improved the productivity which would result in increase in efficiency of nutrient.

Table 9: Phosphorus and K use efficiency as affected by integrated use of nutrient

Soil Type	Location	Crop	P use Efficiency (%)			K use Efficiency (%)	
			100% NP	100% NPK	100% NPK+FYM	100% NPK	100% NPK+FYM
Ludhiana	Inceptisol	Maize	10.3	21.4	26.3	43.8	58.2
Palampur	Alfisol	Maize	21.8	35.6	51.1	23.0	38.9
Pantnagar	Mollisol	Rice	18.2	23.3	53.0	34.5	108.3
Ludhiana	Inceptisol	Wheat	20.6	30.7	34.8	88.1	112.8
Palampur	Alfisol	Wheat	10.7	15.2	24.6	22.6	66.8
Pantnagar	Mollisol	Wheat	11.2	10.4	23.3	13.7	35.8

Nutrient use and soil health

Soil Organic C

Data depicted in Fig. 1 on soil organic carbon (SOC) revealed that continuous balanced nutrients application maintained SOC whereas incorporation of FYM resulted in build-up in SOC in Alfisols of Bangalore and Palampur. At Ranchi, only NPK+FYM could maintain SOC whereas in all other treatments decline in SOC is recorded. Thus, results indicated that under high rainfall incorporation of FYM is essential to sustain soil productivity.

Organic carbon data in Inceptisol of Ludhiana and Vertisol of Jabalpur and Alfisol of Bangalore revealed that balanced application of fertilizer has maintained the soil organic carbon (Table 10 and 11) and the yields are also sustained.

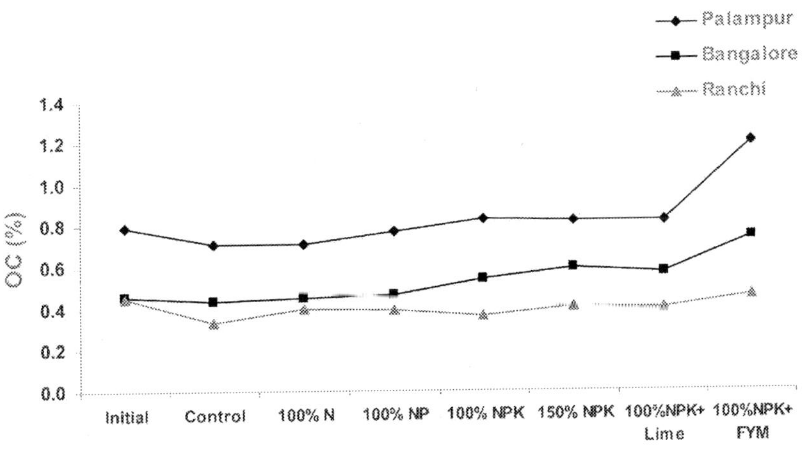

Fig. 1. Organic C status after 34 years in different treatments under long-term fertilizer in Alfisols

Table 10: Organic carbon (g/kg soil) in different soils as affected by continuous application of manure and fertilizer

Treatment	Inceptisol (Udaipur)	Inceptisol (Ludhiana)	Vertisol (Jabalpur)
Control	6.83	2.6	5.3
N	7.23	3.2	5.9
NP	7.13	3.2	6.9
NPKS	8.10	3.7	8.0
NPK+FYM	9.92	5.2	9.4
CD (5%)	0.5	0.4	
Initial	8.0	3.1	5.4
Year after	9.5	34	34

Table 11: Organic carbon (g/kg soil) content in Alfisol

Treatment	Vertisol (Bangalore) Finger millet-maize	Alfisol (Ranchi) Soybean-wheat
Control	4.06	2.9
N	4.39	3.6
NP	4.83	4.0
NPKS	5.01	4.3
NPK+FYM	5.97	4.6
Initial	4 60	4.5
Year after	17	34

From the above results and discussion it is clear that to sustain the yield, maintenance of soil organic carbon is essential. Therefore, an attempt was made to quantify the soil organic carbon in an 7 years old experiment on soybean-wheat system in Vertisols by quantifying C input and output and their relationship.

Nutrient status

Perusal of data given in table 12 revealed that the available status of N in all the soils except at Palampur remained more or less same. It is because of application of all the three major nutrients which are responsible for good yield and thereby incorporating good amount of organic material to soil through root biomass. The increase in Availble N is because of soybean which adds nealy 2000 kg C and nealy 30 kg N to soil. Continuous application of P resulted increase in P status at all place though differ in magnitude, may because of inherent P fixation capacity of soil and crop requirement. The build up P in soil is expected as the the amount of P applied is larger than the uptake. On contrary to P, significant decline in available K was observed in all the soils. Obviously it is due to larger uptake of K from than applied.

Table 12: Available N, P and K status in 100% NPK treatments at various LTFEs centers after 2005-06

Center	Cropping sequence	Initial status (kg/ha)			Available status (kg/ha)		
		N	P	K	N	P	K
Bangalore	F.millet-Maize	257	34	123	258	86	148
Coimbtore	Fmillet-Maize	178	11	810	190	29	584
Ludhiana	Maize-Wheat	100	9	100	107	82	91
Delhi	Maize-Wheat	210	16	155	225	25	291
Palampur	Maize-Wheat	729	12	194	328	126	164
Udaipur	Maize-Wheat	245	22	671	255	23	647
Ranchi	Soyb.-Wheat	236	12	158	351	77	137
Jabalpur	Soyb.-Wheat	226	8	370	240	29	278
Pantnagar	Rice-Wheat	392	18	125	229	18	129

Conclusion

The research carried out in the country proved that with out using the chemical fertilizer we cant keep the pace in productivity to meet the demand of the people of this country. However, increasing use of fertilizer especially nitrogenous is becoming important to consider the fertilizer induced groundwater pollution in heavily fertilized area. To reduce the risk of pollution, increase in the nutrient efficiency (NUE) and the productivity is most suited alternatives. Integrated nutrient management is one of the most appropriate options for sustaining the productivity and reducing the nutrient losses by increasing the nutrient use efficiency. The reports indicated that integrated nutrient management (INM) improved the nutrient use efficiency in addition to productivity irrespective of cropping system and the soil. The INM practice becomes more important in the soil having low buffering capacity like Afisols. In Alfisols, balanced use of NPK along with lime increased the nutrient use efficiency to great extent and minimized the possibility of losses of nitrogen from the system which other wise would have been lost from the system either through leaching or surface runoff and would have polluted the groundwater and ssurface water bodies surface water bodies.

Selected References

Bhardwaj KR, Gaur AC. 1985. Recycling of Organic Waste, ICAR New Delhi, pp 104.

FAO. 1998 Guide to Efficient Plant Nutrition Management Land and Water Development Division, FAO Rome 1998, pp 1-19.

Goswami NN. 1998. Some Thoughts on Concept, Relevance and Feasibility of IPNS under Indian Condition. In: Integrated Plant Nutrient Supply System for Sustainable Productivity. Bulletin No. 2 IISS, Bhopal Pp 3-9.

Rao AS, Singh M, Reddy DD, Saha JK, Manna MC, Singh MV. 1998. Integrated Plant Nutrient Supply System to Improve and Sustained Productivity of Soybean-Wheat System on a

Typic Haplusterts. In: *Integrated Plant Nutrient Supply System for Sustainable Productivity. Bull.* NO. 2 IISS, pp 78-91.

Reddy KS, Mohanty M, Rao DLN, Singh M, Subba, Rao A, Pandey M, Blamey F. Pax C, Dalal RC, Dixit SK, Menzies NW. 2015. Nutrient mass balances and leaching losses from a farmyard manure pit in Madhya Pradesh. *Journal of the Indian Society of Soil Science* 63 (1): 64-68.

Rekhi RS, Meelu. 1983. Effect of complementary use of mung straw and inorganic fertilizer nitrogen on nitrogen availability and yield of rice. *Oryza* (India) 20: 125-129.

Roy RN, Ange AL. 1991. In: Integrated Plant Nutrition System (IPNS) and Sustainable Agriculture. Proc. FAI Annual Seminar, FAI, New Delhi. Pp SV/1-1-SV/1-12.

Saha MN, Singh M, Wanjari RH, Majumder A, Gorai D, Saha AR, Majumder B. 2008. Soil quality crop productivity and sustainability of jute –rice –wheat cropping system after 36 years of long term fertilizer experiment in Inceptisols, IISS, Bhopal, p. 65.

Singh B, Singh Y. 1998. Integrated Nutrient Management for Sustainable Rice-Wheat System. In: *Integrated Plant Nutrient Supply for Sustainable Productivity Bulletin* No. 2. IISS: 46-58.

Singh GB, Dwivedi BS. 1996. Integrated Nutrient Management for Sustainability. *Indian Farming* 46(8): 9-15.

Singh M, Singh M, Kumrawat B. 2008. Influence of Nutrients Supply Systems on Productivity of Soybean-Wheat and Soil Fertility of Vertisol of Madhya Pradesh. *Journal of the Indian Society of Soil Science* 56: 436-441.

Singh M, Singh VP, Reddy DD. 2001. Potassium Balance and Release Kinetics Under Continuous Rice-Wheat Cropping System in Vertisol. *Field Crop Research* 77:81-91.

Singh M, Tripathi AK, Reddy DD. 2002. Potassium balance and release of kinetics of non-exchangeable K in Typic Haplustert as influenced by cattle manure application under soybean-wheat system. *Australian Journal of Soil Research* 40: 533-541.

Singh, M, Wanjari RH. 2007. Research Bulletin on Lessons Learnt from Long-Term Fertilizer Experiments and Measures to Sustain Productivity in Alfisols. AICRP on Long-Term Fertilizer Experiments (LTFE) to Study Changes in Soil Quality, Crop Productivity and Sustainability, Indian Institute of Soil Science (ICAR), Bhopal.

Sudhir K, Singh MV, Jayaprakash SM. 2004. Soil Quality, Crop Productivity and Sustainability: Experiences under Long-Term Finger Millet-Maize Cropping in Alfisol. AICRP-LTFE, Indian Institute of Soil Science (ICAR) P. 1-130.

9

Enhancing Nutrient Use Efficiency, pp. 107-121
Editors: K. Ramesh, A.K. Biswas, B.L. Lakaria, S. Srivastava and A.K. Patra

STCR-Based Fertilizer Recommendation and Improving Nutrient Response Ratio for Major Crops in India

P. Dey

ICAR-Indian Institute of Soil Science, Nabi Bagh, Bhopal – 462 038, India

Introduction

The major challenges in 21st century are food security, environmental quality and soil health. Besides, shrinking land holdings and increasing cost of inputs in India merit adoption of scientific use of plant nutrient for higher crop productivity. The soil fertility and fertilizer use project initiated in 1953 following a study by Stewart in 1947 which was the first systematic attempt in India to relate the knowledge of the soils to the judicious use of chemical fertilizers. The soil testing programme was started in India during the year 1955-56 with the setting up of 16 soil testing laboratories under the Indo-US Operational Agreement for "Determination of Soil Fertility and Fertilizer Use". In 1965, five of the existing laboratories were strengthened and nine new laboratories were established to serve the Intensive Agricultural District Programme (IADP) in selected districts. Chemical indices of nutrient availability chosen for use in the soil testing laboratories consisted of organic carbon or alkaline permanganate oxidizable nitrogen, as a measure of available nitrogen, sodium bicarbonate (Olsen's extractant) extractable phosphorus, as a measure of available phosphorus and neutral normal ammonium acetate extractable potassium, as a measure of available potassium. Muhr and co-workers described sets of critical values that characterized the estimates as low, medium or high in a monograph on soil testing in India in 1965. Background research for the choice of critical values consisted of a few pot culture and field experiments with paddy and wheat, carried out in the Division of Soil Science and Agricultural Chemistry at Indian Agricultural Research Institute, New Delhi. Taking a simplistic view of the situation, the differences among soil groups in the range of properties, which influence the susceptibility to absorption by plants of native and applied nutrients, were ignored. The generalized recommendations of fertilizer use developed for the soil testing laboratory were thought applicable to the medium category of soil testing estimates with an arbitrary adjustment (decrease or increase by 25-50 per cent) for high and low categories of soil test estimates. The ICAR project on soil test crop response AICRP (STCR) has used the multiple regression approach to develop relationship between crop yield on the one hand, and soil test estimates and fertilizer inputs, on the other. Nutrient supplying power of soils, crop responses to added nutrients and amendment needs can safely be assessed through sound soil testing programme. Soil test calibration that is intended to establish a relationship between the levels of soil nutrients determined in the laboratory and crop response to fertilizers in the field permits balanced fertilization through right kind and amount of fertilizers.

Liebig's law of minimum states that the growth of plants is limited by the plant nutrient element present in the smallest amount, all others being in adequate quantities. From this, it follows that a given amount of a soil nutrient is sufficient for any one yield of a given percentage nutrient composition. Ramamoorthy and his co-workers in the year 1967 established the theoretical basis and experimental proof for the fact that Liebig's law of the minimum operates equally well for N, P and K. This forms the basis for fertilizer application for targeted yields, first advocated by Truog in the year 1960. Among the various methods of fertilizer recommendation, the one based on yield targeting is unique in the sense that this method not only indicates soil test based fertilizer dose but also the level of yield the farmer can hope to achieve if good agronomic practices are followed in raising the crop.

Basic data requirement

The essential basic data required for formulating fertilizer recommendation for targeted yield are

- Nutrient requirement in kg/q of produce, grain or other economic produce
- The per cent contribution from the soil available nutrients
- The per cent contribution from the applied fertilizer nutrients

The above mentioned three parameters are calculated as follows.

Nutrient requirement of (N, P and K) for grain production

$$\text{kg of nutrient/q of grain} = \frac{\text{Total uptake of nutrient (kg)}}{\text{Grain yield (q)}}$$

Contribution of nutrient from soil

$$\text{\%Contribution from soil (CS)} = \frac{\text{Total uptake in control plots (kg ha}^{-1}\text{)} \times 100}{\text{Soil test values of nutrient in control plots (kg ha}^{-1}\text{)}}$$

% Contribution of nutrient from fertilizer

Contribution from fertilizer (CF) = Total uptake of nutrients - (Soil test values of nutrients in fertilizer treated plots x CS)

$$\text{\% Contribution from fertilizer} = \frac{\text{CF}}{\text{Fertilizer dose (kg ha}^{-1}\text{)}} \times 100 \text{ fertilizer}$$

Calculation of fertilizer dose

The above basic data are transformed into workable adjustment equation as follows :

$$\text{Fertilizer dose} = \frac{\text{NR}}{\text{\% CF}} \times 100\ \text{T-} \frac{\text{\% CS}}{\text{\% CF}} \times \text{soil test value}$$

= a constant × yield target (q ha^{-1}) - b constant × soil test value (kg ha^{-1})

NR= Nutrient requirement in kg/q of grain

Similarly the contribution of nutrients from organic can also be determined.

The differentiation of significant multiple regression equations provides a basis for soil test-fertilizer requirement calibration for maximum yield per hectare, maximum profit per hectare and maximum profit per rupee investment on fertilizer. The resultant fertilizer adjustment equations have been tested in follow up and frontline demonstrations conducted in different parts of the country. In these trials soil test based rates of fertilizer application helps to obtain higher response ratios and benefit: cost ratios over a wide range of agro-ecological regions (Dey and Srivastava, 2013).

Use of targeted yield equation and development of prediction equation for cropping sequence

Nutrient availability in the soil after the harvest of a crop is influenced by the initial soil nutrient status, the amount of fertilizer nutrients added and the nature of the crop raised. But recently, the monoculture is replaced by cropping sequence approach. To apply soil test based fertilizer recommendations, the soils are to be tested after each crop, which is not practicable. Hence it has become necessary to predict the soil test values after the harvest of the crop. It is done by developing post-harvest soil test value prediction equations, making use of the initial soil test values, applied fertilizer doses and the yields obtained or uptake of nutrients following the methodology outlined by Ramamoorthy and coworkers in 1971. The post-harvest soil test values were taken as dependent variable and a function of the pre-sowing soil test values and the related parameters as yield/uptake and fertilizer nutrient doses.

Prediction equation for cropping sequence

The method of calculation for prediction of post harvest soil test values for cropping sequences is given below for use.

YP/H = f (F, IS, yield/nutrient uptake)

Where, YP/H is the post harvest soil test value, F is the applied fertilizer nutrient and IS is the initial soil test value. The mathematical form is

YP/H = a + b1F + b2 IS + b3 yield/uptake

Where, a is the absolute constant and b1, b2 and b3 are the respective regression coefficients. Prediction equations for post-harvest soil test values were developed from initial soil test values, fertilizer doses applied and yield of crops/uptake of nutrients to obtain a basis for prescribing the fertilizer amounts for the crops succeeding the first crop in the cropping sequence.

During the past one and half decades, the various centres of AICRP on STCR developed prediction equation by using the targeted yield equation for different cropping sequence like rice-rice, rice-maize, rice-wheat, maize-tomato, maize-wheat, potato-yellow sarson, paddy-ragi, maize-cotton, wheat-groundnut, okra-wheat, paddy-chick pea, soybean-wheat, rice-pumpkin, bajra-wheat, cotton-maize and soybean-onion. The predicted values can be utilized for recommending the fertilizer doses for succeeding crop thus eliminating the need of soil test after each crop. This provides the way for giving the fertilizer recommendations for whole cropping sequence based on initial soil test values. For example, in potato– yellow sarson cropping sequence:

Potato (Kufri Jyoti)

PHN = 104.94 + 0.28 FN – 0.041 SN – 0.11 Y (R^2 = 0.35**)

PHP = -2.74 + 0.091 FP + 0.84 SP + 0.013 Y (R^2 = 0.78**)

PHK = 31.28 + 0.71 FK + 0.45 SK – 0.17 Y (R^2 = 0.70**)

Yellow Sarson (PYS-I)

PHN = 107.91 + 0.36 FN – 0.08 SN – 0.79 Y (R^2 = 0.72**)

PHP = 23.19 + 0.26 FP + 0.011 SP + 0.24 Y (R^2 = 0.70**)

PHK = 153.25 + 0.42 FK + 0.02 SK – 0.54 Y (R^2 = 0.56**)

Benefits of soil test based targeted yield approach

During the last more than four decades the STCR project has generated numerous fertilizer adjustment equations for achieving targeted yields of important crops on different soils in different agro ecological regions of the country. In these trials soil test based rates of fertilizer application helped to obtain higher response ratios and benefit: cost ratios (Table 1 and 2) over a wide range of agro-ecological regions (Dey, 2012). It is evident from above tables that STCR based approach of nutrient application has definite advantage in terms of increasing nutrient response ratio over general recommended dose of nutrient application. Yields and response ratios can be increased if the fertilizer prescriptions are made as per the table 1 for specified crops and locations.

Table 1. Response ratios in existing and improved practice for different crops at different sites in India: Results from AICRP on STCR

Crop	Location/ AER	Soil type	Fertilizer Response Ratio (kg grain/kg nutrient)				
			Present practice		Improved practice		
			Fertilizer dose	RR	Fertilizer dose	RR	Type of intervention
Rice	Coimbatore/8.1, Hot dry semi-arid	Alfisol	GRD: 120-38-38	16.5	STCR: 7 t/ha 185-51-19 STCR: 7 t/ha under IPNS (GM @ 6.25 t/ha and Azospirillum @ 2 kg/ha 150-67-10	17.0 19.7	Soil test based fertilizer recommendation under IPNS
Rice	Coimbatore/8.1, Hot dry semi-arid	Alfisol	GRD: 120-38-38	15.4	STCR: 7 t/ha 179-71-19	16.1	Soil test based balanced fertilization
Rice	Hisar, Haryana/ 2.3 Hot typic arid	Podzolic	Farmers' Practice 75-30-0	18.31[2]	STCR: 7 t/ha 139-63	23.49[2]	Soil test based balanced fertilization
Rice	Jabalpur/ 10 Hot sub-humid	Medium black	GRD: 80-70-40	8.47	STCR: 3.5 t/ha 76-66-0	11.13	Soil test based balanced fertilization
Rice	Kalyani, WB/ 15.1 Hot moist sub-humid	Deep loamy to clayey alluvial	80-40-40	8.02	Soil test based 62.5-28-62 + 7.5 t/ha FYM	13.19	Soil test based fertilizer recommendation under IPNS
Rice	Narsinghpur, MP		GRD: 80-70-40	11.45**	STCR: 4 t/ha 91-74-0	19.07**	Soil test based balanced fertilization
Rice	Pantnagar, Uttaranchal/14.5 Warm humid/perhumid	Medium to deep loamy tarai	Farmers' Practice 120-0-0 GRD: 120-40-40	12.5 8.5	STCR: 4.0 t/ha 94-36-0	16.15	Soil test based balanced fertilization
Wheat	Jabalpur, MP/10 Hot sub-humid	Medium black	GRD: 100-60-30	14.77**	STCR: 4 t/ha 59-57-28	41.01**	Soil test based balanced fertilization
Wheat	Palampur, HP***/14.3 Warm humid to per humid transitional	Podzolic	Farmers' Practice 30-0-0 GRD: 120-60-30	14.83 3.52	STCR: 4.0 t/ha 176-187-75	6.95	Soil test based balanced fertilization
Wheat	Pantnagar, Uttaranchal/14.5 Warm humid/perhumid	Medium to deep loamy tarai	Farmers' Practice 115-20-0 GRD: 120-60-40	6.67 10.68	STCR: 4.0 t/ha 104-60-57	11.31	Soil test based balanced fertilization
Finger millet	Kolhapur, Maharashtra Rainfed submontain zone	Black soil	GRD: 60-30-0	10.1	STCR: 1.6 t/ha 45-34-17	10.9	Soil test based balanced fertilization
Maize	Palampur, HP***/14.3 Warm humid to per humid transitional	Podzolic	Farmers' Practice 40-0-0 GRD: 120-60-40	13.1 7.14	STCR: 4.0 t/ha 189-0-73	8.91	Soil test based balanced fertilization

Chickpea	Durg, Chattisgarh/11 Hot/moist/dry sub humid transitional	Vertisol	Farmers' Practice 10-30-0 GRD: 20-50-20	2.78 2.76	STCR: 1.2 t/ha 20-0-0	7.90	Soil test based balanced fertilization
Chickpea	Jabalpur/ 10 Hot sub-humid	Medium black	GRD: 20-60-20	9.00	STCR: 1.5 t/ha 22-36-0	12.76	Soil test based balanced fertilization
Urid	Jabalpur/ 10 Hot sub-humid	Medium black	GRD: 20-50-20	0.361** (Mean of three trials)	STCR: 1.2 t/ha 25-35-0	0.464**	Soil test based balanced fertilization
Groundnut	Coimbatore, north western zone of TN/8.1, Hot dry semi-arid	Red soil, Irugur series	GRD: 18-36-54	4.62	STCR: 2.5 t/ha 55-55-71 STCR: 2.5 t/ha with 12.5 t/ha FYM 17-37-31	5.5 5.92	Soil test based fertilizer recommendation under IPNS
Groundnut	Kakapalayam, TN/8.1, Hot dry semi-arid	Red soil, Irugur series	GRD: 18-36-54	6.7	STCR: 2.5 t/ha 50-43-72 STCR: 2.5 t/ha with 12.5 t/ha FYM 15-25-32	6.9 7.4	Soil test based fertilizer recommendation under IPNS
Groundnut	Kolhapur, Maharashtra	Typic haplustert	GRD: 25-50-0	20.9	STCR: 2.5 t/ha 55-62-24	13.8	Soil test based balanced fertilization
Groundnut	Tumkur, Karnataka		GRD: 25-75-38	5.50	STCR: 2.0 t/ha 16-144-53	6.20	Soil test based balanced fertilization
Linseed	Jabalpur, MP/10 Hot sub-humid	Medium black	GRD: 60-40-20	5.21	STCR: 2.0 t/ha 89-51-19	8.29	Soil test based balanced fertilization
Mustard	Durg, Chhattisgarh	Vertisol	Farmers' Practice 60-40-0 GRD: 120-80-40	2.71 6.53	STCR: 1.3 t/ha 103-83-0	6	Soil test based balanced fertilization
Mustard	Jabalpur, MP/10 Hot sub-humid	Medium black	GRD50-30-20	4.38	STCR: 1.6 t/ha 68-42-16	5.44	Soil test based balanced fertilization
Mustard	Jabalpur/ 10 Hot sub-humid	Medium black	GRD: 50-30-20	2.29	STCR: 2 t/ha 88-46-35	2.34	Soil test based balanced fertilization
Mustard	New Delhi/ 4.1 Hot semi-arid	Alluvial soils	Farmers' Practice 60-57-0 GRD: 80-40-40	6.4 [1] 7.8 [1]	STCR: 2.5 t/ha 90-43-48	8.6 [1]	Soil test based balanced fertilization
Onion	Coimbatore Tamil Nadu/ 8.1, Hot dry semi-arid	Red Inceptisols	FP: 80-80-60 GRD: 60-60-30	41.7 61.8	STCR: 20 t/ha: 118 to123-32 to 43 - 15 to 78	62.1	Soil test based balanced fertilization
Safflower	Bangalore, Karnataka/8.2 Hot moist semi arid	Black soil, sandy clay loam	GRD: 38-50-25	5.78	STCR: 1.5 t/ha 54-0-13	10.9	Soil test based balanced fertilization

Soybean	Durg, Chhattisgarh/11 Hot/moist/dry sub humid transitional	Vertisol	Farmers' Practice 12-30-0 GRD: 20-50-20	20.2 15.0	STCR: 2.0 t/ha 20-35-0	20.1	Soil test based balanced fertilization
Soybean	Jabalpur, MP/10 Hot sub-humid	Medium black	GRD: 20-80-20	8.28	STCR: 2.5 t/ha 15-52-0	13.77	Soil test based balanced fertilization
Sunflower	Coimbatore, north western zone of TN/8.1, Hot dry semi-arid	Red soil, Irugur series	GRD: 60-40-40	4.34	STCR: 2.0 t/ha 87-63-13 STCR: 2.5 t/ha with 12.5 t/ha FYM 52-45-0	7.05 7.57	Soil test based fertilizer recommendatio n under IPNS
Sunflower	Jabalpur/ 10 Hot sub-humid	Medium black	GRD: 80-40-25	4.31	STCR: 2 t/ha 197-27.4-0	5.10	Soil test based balanced fertilization
Sunflower	Kalipalayam, TN/8.1, Hot dry semi-arid	Mixed black (Inceptisol)	FP: 50-40-50 GRD: 40-20-20	4.76 6.26	STCR: 2 t/ha 92-28-10 STCR: 2 t/ha with 12.5 t/ha FYM 62-13-5	6.86 7.33	Soil test based balanced fertilization
Bhendi	Kalipalayam, TN/8.1, Hot dry semi-arid	Mixed black (Inceptisol)	FP: 100-60-60 GRD: 40-50-30	30.7 25.4	STCR: 1.5 t/ha 72-21-15	77.9	Soil test based balanced fertilization
Bhendi	Suradevanapura, Bangalore/ 8.2 Hot moist semi-arid	Medium to deep red laom	125-62.5-62.5	17.88	STCR: 8 t/ha 91-74-56	24.25	Soil test based balanced fertilization
Brinjal	Rahuri, Maharashtra/ 6.1 Hot dry semi-arid	Typic ustorthent	GRD: 150-75-75	73.3	STCR: 5 t/ha 140-20-110	124.9	Soil test based balanced fertilization
Cabbage	Rahuri, Maharashtra/6.1 Hot dry semi-arid	Ustorthent	GRD: 180-80-60	6.88	STCR: 3.5 t/ha 256-129-193	5.33	Soil test based balanced fertilization
Chilli	Thirumalayampal ayam, Madukarai Block. Coimbatore, TN/8.1, Hot dry semi-arid	Red Inceptisol	GRD: 75-35-35	3.7	STCR: 2 t/ha 108-62-68	4.1	Soil test based balanced fertilization

IPNS = Integrated Plant Nutrient Supply; STCR = Soil Test Crop Response; * Higher yield obtained with lesser fertilizer dose than farmers' practice; ** Response ratio calculated over farmers' practice; 1 Average of two demonstrations; 2 Average of four demonstrations; *** In case of wheat and maize at Palampur the high response ratio in farmers' practice is due to very low rates of fertilizer application. Even though the response ratio is high the level of yields the farmers are getting is very poor. In STCR technology the response ratio is not as high as in farmers' practice but the yields are very good.

Table 2: Average response ratios (kg grain/kg nutrients)

Crop	No. of trials	Farmer's practice	STCR- IPNS recommended practice
Rice	120	11.4	16.8
Wheat	150	10.3	14.2
Maize	35	12.7	17.7
Mustard	45	8.0	8.2
Raya	25	4.8	7.6
Groundnut	50	5.1	6.8
Soybean	17	9.6	12.2
Chickpea	35	6.1	9.4

Economic analysis of fertilizer doses associated with different yield targets

An appraisal of the effect of nutrients (NPK) applied on crop yield and benefit: cost ratios (BCR), both under (NPK) alone and under IPNS for 15 agricultural and horticultural crops (Dey and Santhi, 2014) is furnished in Table 3. The input output prices used in these analyses were: Produce prices:— Paddy (rough rice) and wheat grain Rs 12,000/t, rice straw Rs. 1,200/t wheat straw Rs.500/t, maize grain Rs. 8,000/t maize straw Rs. 500/t, cotton Rs. 25,000/t onion Rs. 9,000t, okra (Bhendi) Rs. 10.000/t cabbage Rs. 3500/t, potato Rs. 7000/t carrot Rs. 5,000/t, beetroot Rs. 3300/t. radish Rs. 3,000/t, tomato Rs. 3,300/t and *Ashwagandha* Rs. 82,000/t Input prices –Rs. 11.76/kg N through urea, Rs 47.63/kg P_2O_5 through SSP and Rs. 28.00/kg K_2O through MOP, 250/T for FYM and Rs 750/t for vermicompost. Economic analysis of the data showed that out of 66 crop x target combinations, the BCR was between 1and 2 in 35 % cases and between 2.1 and 3.0 in 62% cases. In 3% cases BCR was above 3.Irrespective of the crops, higher yield has been recorded at higher yield targets over lower target coupled with higher net return and BCR. As in the case of yield, wherever three targets (low, medium and high) were tried, the BCR was relatively higher between low and medium target levels then between medium and high target levels both under NPK alone and IPNS. Again, irrespective of the crops and yield targets, yield increase was higher with IPNS then under NPK applied through fertilizers alone. In the regard, farmers can choose the desired yield targets according to their investment capabilities and availability of organic manures but would generally benefit form adopting an appropriate IPNS package as apart form contributing nutrients, organic manures also improve soil physical conditions. At present, the soil test based recommendations are relatively on a stronger footing when these involve only fertilizers as compared to IPNS. This is because there are several issues concerning the nutrient which need to be sorted out as illustrated using STCR information form Andhra Pradesh. One of the outstanding problems is that while the composition of fertilizers is fairly standard, that of organic manures can vary several fold even within the same location or from lot to lot.

Table 3. Economic analysis and benefit: cost ratios of fertilizer doses for different yield targets

Crop	Yield target and Treatments	Fertilizer Doses (kg/ha) N	P_2O_5	K_2O	Yield (t/ha)	Fertilizer Cost Rs./ha	BCR
1. Rice – Flooded (TNAU Farm, Coimbatore)	6 t ha^{-1}NPK fertilizer	137	56	23	6.01	4922	1.52
	7 t ha^{-1}NPK fertilizer	172	56	23	6.94	5286	1.74
	7 t ha^{-1}IPNS package*	118	36	23	7.11	3746	1.75
2. Rice – SRI (TNAU Farm, Coimbatore & Farmer's fields, Coimbatore Dt.)	7 t ha^{-1}NPK from fertilizer	173	62	74	6.68	7059	1.92
	8 t ha^{-1}NPK from fertilizer	222	83	99	7.65	9336	2.10
	9 t ha^{-1}NPK from fertilizer	271	100	100	8.18	10775	2.18
	7 t ha^{-1}IPNS package*	133	39	39	6.94	4514	1.97
	8 t ha^{-1}IPNS package*	182	60	67	7.94	6874	2.15
	9 t ha^{-1}IPNS package*	232	80	95	8.34	9199	2.16
3. Wheat – Hill Farmer's fields, Kalrayan foothills, Salem Dt. and Kolli foothills, Namakal Dt.).	3.5 t ha^{-1}NPK from fertilizer	156	111	41	3.55	8269	1.54
	4.0 t ha^{-1}NPK from fertilizer	194	129	59	4.07	10077	1.66
	3.5 t ha^{-1}IPNS package*	109	100	28	3.63	6828	1.49
	4.0 t ha^{-1}IPNS package*	147	117	37	4.16	8337	1.62
4. Wheat – Plains (Farmer's fields, Coimbatore Dt.)	3 t ha^{-1}NPK from fertilizer	119	91	51	2.90	7161	1.33
	4 t ha^{-1}NPK from fertilizer	196	110	72	3.93	9560	1.66
	3 t ha^{-1}IPNS package*	65	74	37	2.98	5325	1.31
	4 t ha^{-1}IPNS package*	152	105	55	4.05	8328	1.61
5. Maize – Hybrid (Farmer's fields, Coimbatore and Dindigul Dts.)	9 t ha^{-1}NPK from fertilizer	244	98	82	8.97	9833	2.04
	10 t ha^{-1}NPK from fertilizer	290	124	97	10.09	12032	2.17
	11 t ha^{-1}NPK from fertilizer	329	139	113	10.33	13653	2.13
	9 t ha^{-1}IPNS package*	202	78	52	9.23	7546	2.05
	10 t ha^{-1}IPNS package*	249	104	69	10.5	9813	2.21
	11 t ha^{-1}IPNS package*	289	119	86	10.59	11474	2.14
6. Rain fed Bt Cotton (Farmer's fields, Perambalur Dt.)	2.8 t ha^{-1}NPK from fertilizer	103	71	74	2.67	6665	1.46
	3.2 t ha^{-1}NPK from fertilizer	123	81	89	3.07	7796	1.64
	2.8 t ha^{-1}IPNS package*	78	51	48	2.84	4690	1.51
	3.2 t ha^{-1}IPNS package*	99	67	64	3.24	6147	1.68
7. Onion (Farmer's fields, Coimbatore Dt.)	17 t ha^{-1}NPK from fertilizer	89	18	26	17.24	2632	2.34
	20 t ha^{-1}NPK from fertilizer	117	33	31	18.75	3816	2.50
	17 t ha^{-1}IPNS package*	37	14	14	17.72	1494	2.33
	20 t ha^{-1}IPNS package*	71	34	32	19.48	3350	2.50
8. Okra (Bhendi) (Farmer's fields, Coimbatore Dt.)	15 t ha^{-1}NPK from fertilizer	78	45	15	15.5	3480	2.01
	17 t ha^{-1}NPK from fertilizer	97	54	15	16.59	4133	2.14
	15 t ha^{-1}IPNS package*	51	30	8	15.95	2253	2.10
	17 t ha^{-1}IPNS package*	70	39	8	17.36	2905	2.18
9. Cabbage (Farmer's fields, Coimbatore Dt.)	60 t ha^{-1}NPK from fertilizer	139	64	23	59.90	5327	2.43
	70 t ha^{-1}NPK from fertilizer	194	93	41	70.50	7859	2.78
	60 t ha^{-1}IPNS package*	109	49	15	61.00	4036	2.44
	70 t ha^{-1}IPNS package*	164	78	29	71.80	6456	2.79
10. Potato (Farmer's fields, Nilgiris Dt.)	30 t ha^{-1}NPK from fertilizer	105	306	59	27.60	17460	1.85
	40 t ha^{-1}NPK from fertilizer	175	446	131	38.80	26967	2.38
	30 t ha^{-1}IPNS package*	98	297	55	29.20	16837	1.94
	40 t ha^{-1}IPNS package*	168	437	127	40.40	26344	2.46
11. Carrot (Farmer's fields, Nilgiris Dt.)	40 t ha^{-1}NPK from fertilizer	125	145	39	40.90	9468	2.45
	50 t ha^{-1}NPK from fertilizer	173	256	66	52.50	16075	2.92
	60 t ha^{-1}NPK from fertilizer	221	367	127	59.90	23634	3.07
	40 t ha^{-1}IPNS package*	117	139	36	43.30	9004	2.54
	50 t ha^{-1}IPNS package*	165	250	64	53.70	15639	2.92
	60 t ha^{-1}IPNS package*	213	361	120	62.60	23058	3.15

(*Contd.*)

12. Beetroot (Farmer's fields, Coimbatore and Dindigul Dts.)	40 t ha^{-1}NPK from fertilizer	99	134	80	39.50	9786	2.37
	50 t ha^{-1}NPK from fertilizer	163	186	131	49.98	14443	2.77
	40 t ha^{-1}IPNS package*	59	113	54	40.48	7588	2.37
	50 t ha^{-1}IPNS package*	123	165	99	51.37	12077	2.78
13. Radish (Farmer's fields, Coimbatore and Dindigul Dts.)	40 t ha^{-1}NPK from fertilizer	112	50	71	37.20	5686	2.23
	50 t ha^{-1}NPK from fertilizer	181	78	114	48.80	9035	2.74
	40 t ha^{-1}IPNS package*	67	24	52	38.40	3387	2.24
	50 t ha^{-1}IPNS package*	136	52	87	50.60	6512	2.78
14. Tomato (Farmer's fields, Coimbatore and Dindigul Dts.)	70 t ha^{-1}NPK from fertilizer	164	135	148	70.90	12502	2.47
	80 t ha^{-1}NPK from fertilizer	209	177	188	80.80	16152	2.71
	90 t ha^{-1}NPK from fertilizer	254	219	228	88.30	19801	2.86
	70 t ha^{-1}IPNS package*	120	113	113	72.20	9957	2.50
	80 t ha^{-1}IPNS package*	165	155	153	81.30	13606	2.71
	90 t ha^{-1}IPNS package*	210	197	193	89.50	17256	2.88
15. Ashwagandha (Farmer's fields, Salem Dt.)	0.7 t ha^{-1}NPK from fertilizer	49	74	66	0.671	5949	1.31
	0.9 t ha^{-1}NPK from fertilizer	79	109	77	0.871	8276	1.24
	0.7 t ha^{-1}IPNS package*	20	51	40	0.696	3784	1.28
	0.9 t ha^{-1}IPNS package*	59	87	68	0.905	6741	1.19

Linking soil fertility maps with STCR parameters for spatial fertilizer recommendation

An attempt was made with joint venture of IISS, Bhopal and NBSSLUP, Nagpur to create spatial fertilizer recommendation maps using available validated fertilizer adjustment equations (STCR's generated) and Geographic Information System (GIS). The district level soil fertility index of 10 districts of India were prepared, which can be used to generate balanced fertilizer recommendation for specified crops for entire district based on the average soil fertility status of that district. District wise soil fertility georeferenced maps were prepared using index values for nitrogen (N), phosphorus (P) and potassium (K) for ten states. Corresponding equivalent soil nutrient values in respect of N, P & K were calculated from the index values. Reasonable limits for targeted yields were defined. The recommendations in the form of equations for targeted yields developed by Subba Rao and Srivastava (2001) have been interlinked with the fertility maps. The use of this recommendation system suggested for varied applications for targeted yields in different districts of states. This can be used up to field level also, if the farmer has the knowledge of his fertility status and the yield target. The maps can also be updated from time to time based on the soil test result data base. It can be further narrowed down to block/village level depending on the availability of information. As an example, N soil fertility map in the year 2001 is depicted in Fig. 1. These fertility maps can also be used to study the changing trends in the fertility of nutrients and can be correlated with fertilization practices of farmers of a particular region. These maps, however, only indicate the general fertility level in a district since the data collected and used was not georeferenced.

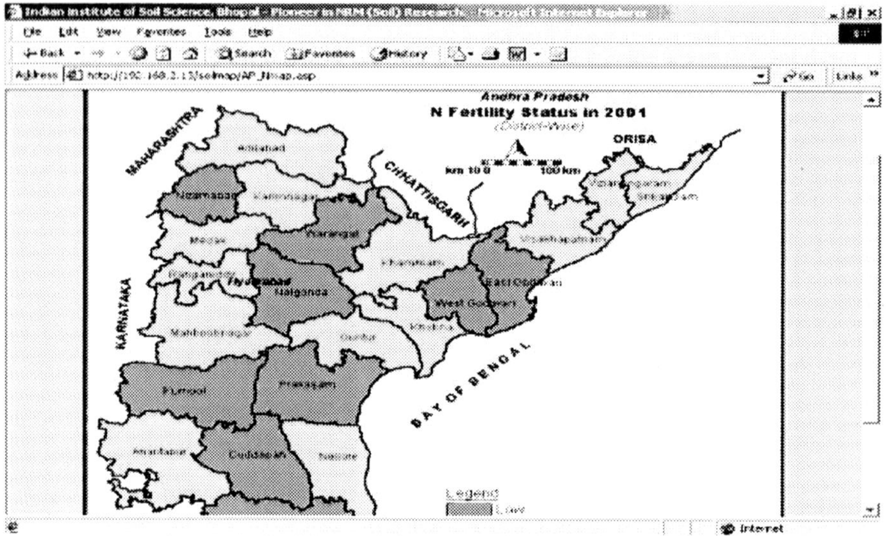

Fig. 1: Nitrogen soil fertility map

Subsequently, IISS prepared a web based soil fertility map for two districts i.e., Hoshangabad and Guna of Madhya Pradesh by following multistage random sampling for collection of soil samples where tehsils were considered as strata in each district. The GIS based soil fertility map for N, P, K was prepared for Hoshangabad district and the soil test values were revalidated. The method has been found to be better and soil fertility can be monitored over a period of time.

On similar lines, soil fertility maps of a total of 175 districts have been completed. Fertilizer recommendation equations derived from STCR studies i.e. soil nutrient efficiency (Es), fertilizer nutrient efficiency (Ef) and nutrient requirement (NR) of a particular crop are now being linked with the soil fertility values on the map which makes possible to provide spatial fertilizer recommendation in the form of maps. The recommendations can be obtained by an extension agent/ farmer simply by locating his area on the map.

Application of information and communication technologies (ICT)

Agricultural development and sustainability crucially depend upon relevant information access at opportune time. There is a vast scope of extending ICTs through public, private and non governmental organizations with respect to extension, marketing and community services. The ideal delivery model for these ICTs is envisioned to be a multi-pronged strategy involving institutions under National Agricultural Research System (NARS) to go online, share their contents, and rural information kiosks providing information access to the farmers.

India has 37% of world ICT enabled projects in rural areas. Agricultural resource information using GIS models, expert systems, databases on successful technologies have critical role in governance and decision making by farmers.

On-line fertilizer recommendation systems developed by AICRP (STCR)

http://www.stcr.gov.in

All India Coordinated Research Project on Soil Test Crop Response (AICRP-STCR) based at Indian Institute of Soil Science has developed a computer aided model that calculates the amount of nutrients required for specific yield targets of crops based on farmers' soil fertility. (http://www.stcr.gov.in). This software program reads data, performs calculations and generates graphical and tabular outputs as well as test reports. This system has the ability to input actual soil test values of the farmers' fields to obtain optimum dose of nutrients. The application is a user friendly tool. It will aid the farmer in arriving at an appropriate dose of fertilizer nutrient for specific crop yield for given soil test values (Fig. 2). Efforts are under way in developing bioinformatics, E-choupals, digital libraries and e Governance that can benefit agriculture immensely by way of providing information and assisting the users in adopting the newer technologies.

Computer software, including spreadsheets,GIS, and other types of application software are readily available. The global positioning system (GPS) has given the farmer the means to locate position in the field to within a few feet. By tying position data in with the other field data mentioned earlier, the farmer can use the GIS capability to create maps of fields or farms. Sensors are under development that can monitor soil properties, crop condition, harvesting, or post harvest processing and give instant results or feedback which can be used to adjust or control the operation.

Fig 2. Internet enabled soil test based fertilizer application software
(*See colour version on page 453*)

DSSIFER

Decision Support System for Integrated Fertilizer Recommendation (DSSIFER) is an user friendly software and the updated version (DSSIFER 2010) encompasses soil test and target based fertilizer recommendations through Integrated Plant Nutrition System developed by the AICRP-STCR, Department of Soil Science and Agricultural Chemistry, TNAU, and the recommendations developed by the State Department of Agriculture, Tamil Nadu. If both recommendations are not available for a particular soil – crop situation, the software can generate prescriptions using blanket recommendations but based on soil test values. Using this software, fertilizer doses can be prescribed for about 1645 situations and for 190 agricultural and horticultural crops along with fertilisation schedule. If site specific soil test values are not available, data base included in the software on village fertility indices of all the districts of Tamil Nadu will generate soil test based fertilizer recommendation. Besides, farmers' resource based fertilizer prescriptions can also be computed. Therefore, adoption of this technology will not only ensure site specific balanced fertilisation to achieve targeted yield of crops but also result in higher response ratio besides sustaining soil fertility. In addition, the software also provides technology for problem soil management and irrigation water quality appraisal. Moreover, soil testing laboratories of all the organisations can generate and issue the analytical report and recommendations in the form of Soil Health Card (both in English &Tamil) which can be maintained by the farmers over long run.

Computer software, including spreadsheets, databases, geographic information systems (GIS), and other types of application software are readily available. The global positioning system (GPS) has given the farmer the means to locate position in the field to within a few feet. By tying position data in with the other field data mentioned earlier, the farmer can use the GIS capability to create maps of fields or farms.

Epilogue

Among the various methods of formulating fertilizer recommendation, the one based on yield targeting has found popularity. This method not only indicates soil test based fertilizer dose but also the level of yield the farmer can hope to achieve, if good agronomy is followed in raising the crop. It provides the scientific basis for balanced fertilization not only between the fertilizer nutrients themselves but also that with the soil available nutrients.

Selected References

Dey P, Santhi R. 2014. Soil test based fertilizer recommendations for different investment capabilities. In *Soil Testing for Balanced Fertilisation – Technology, Application, Problem Solutions* (H.L.S. Tandon ed.), pp. 49-67

Dey P. 2012. Soil-Test-Based Site-Specific Nutrient Management for Realizing Sustainable Agricultural Productivity. In: *Book-International Symposium on "Food Security Dilemma: Plant Health and Climate Change Issues* (Eds. Khan *et al.*), held at FTC, Kalyani on December 7-9, 2012, pp. 141-142.

Dey P, Srivastava S. 2013. Site specific nutrient management with STCR approach. In Kundu *et al.* (Eds): IISS Contribution in Frontier Areas of Soil Research, Indian Institute of Soil Science, Bhopal, 259-270.

Subba Rao A, Srivastava S. (Eds.) 2001. Soil test based fertilizer recommendations for targeted yield of crops. *Proceedings of the National Seminar on Soil Testing for Balanced and Integrated Use of Fertilizers and Manures*, Indian Institute of Soil Science, Bhopal, India, pp 1-326.

10

Enhancing Nutrient Use Efficiency, pp. 123-140
Editors: K. Ramesh, A.K. Biswas, B.L. Lakaria, S. Srivastava and A.K. Patra

Improving Phosphorus Use Efficiency

I. Rashmi, A.K. Biswas, Neenu S., B.P. Meena and Shinogi, K.C.

ICAR-Indian Institute of Soil Science, Nabi Bagh, Bhopal – 462 038, India

Introduction

Phosphorus (P) is a major nutrient in crop production, but is also a major constraint due to its low bioavailability in soils. It is 11th most abundant element on earth's crust and a vital component of DNA, RNA, ATP and photosynthetic system and catalyses a number of biochemical reactions from the beginning of seedling growth through to the formation of grain and maturity. Many factors influence soil P availability like type of parent material from which the soil is derived, degree of weathering and climatic conditions. The only economic source of P in the world is rock phosphate (RP) whose availability is estimated to peak by 2030 and thereafter a steep decline is envisaged. However, others, by calculating RP reserve longevity using current reserves and production, have indicated that the world has over 300 years of reserves (Steven *et al.*, 2013). About more than 21.1 mt reserve of rock phosphate (RP) reserve belong to high grade (30% P_2O_5) and most other reserve belong to low grade (10-20% P_2O_5). Fertilizer Statistics of India (2012-2013) showed that between 1980-81 and 2012-2013, P_2O_5 consumption almost increased six times (increased from 1.1 to 6.7 mt). In India, the phosphate resources are estimated to be 260 mt with major proportion being low grade and the recoverable phosphate are approximately 142 mt (FAI, 2012). In India 60-65% of P fertilizer is used in rice and wheat cropping system (Raju *et al.*, 2005). Grain production consume nearly 44% of world phosphatic fertilizers, nearly 18% is consumed for production of fruits and vegetables. The phosphate consumption is recorded highest in Asian countries accounting for 60% of which China consumes the highest approximately 30% of total P fertilizers (Fig: 1). The low recovery of P by crops, high retention by soil and residual fertilizer P in different P pools necessitates understanding P dynamics in soil and crop management for efficient use of P resources in Indian agriculture.

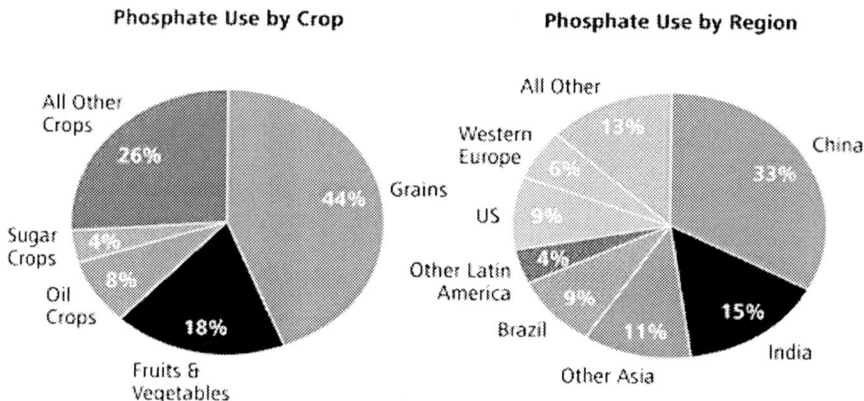

Fig. 1: Phosphate fertilizer use by crop and countries (CRU, Fertecon, IFA, 2013)

Phosphorus pools across soil types can be (i) soluble P in soil solution; (ii) labile P in the solid phase that is easily exchangeable from the mineral surface and (iii) non-labile P that is slowly exchangeable or non-exchangeable from the mineral surface. The concentration of phosphate ions in soil solution can range from very high 10^{-4} M to low 10^{-6} M, to as very low as 10^{-8}M in some very poor tropical soils (Johnston *et al.,* 2014). Phosphorus in most Indian soils is tightly held to soil colloids and does not leach down easily unlike nitrogen. The phosphate adsorbed on active soil sites that are insoluble and organic compounds resistant to mineralization are the fixed P pool. Some slow conversion between fixed P pool and active P pool always occur in soils. In tropics 1018 million hectares of area is occupied by high P fixing soils (Sanchez and Logan, 1992). Inorganic P is chemically occluded by hydrous oxides of Al, Fe, Mn at a pH < 5, while at pH of 6-7 P is considered more available to plants, but at a higher pH > 7.5 P is again sorbed by silicate and calcium minerals. Phosphorus in soil reacts with Al, Fe and Ca gets transformed into sparingly soluble compounds which become unavailable for crop uptake.The reaction between soil colloids and phosphorus is slow which could be due to diffusive penetration of adsorbed P ions into soil components and resulting in less extractability, isotopic exchangeability and plant availability with time. The soil available P is considered as 'working capital' for maintaining P in soil solution and determines the productivity of crops. Soil phosphorus is influenced by many factors like type of soil, clay content, soil reaction, organic matter, temperature, liming and management practices. All these factors influence to establish equilibrium between soil P pools that vary in plant availability. The growing crops utilize the immediately available soil solution P which is very small, and as the concentration decreases some P is released from active P pools. Fertilizers are the key inputs for increasing food production in India. In terms of consumption, India is second to China in nitrogen and phosphorus. Some estimates of N, P and K removed by the major cereal crops are depicted in the table 1 (Prasad, 2012). This table indicates less removal of P by cereal crops revealing more amount of P gets fixed in soil once applied. The GIS based fertility mapping revealed that 49% of Indian soils are under low category, 45% under medium and 6% under high category of soil P (Muralidharudu *et al.,* 2011). The data indicated that 94% of Indian soils were classified under low to medium soil P fertility. The consumption of P fertilizer was the highest in Punjab followed by Andhra Pradesh and Tamil Nadu, whereas Rajasthan and some states in North East were in the low consumption category. Overuse of P fertilizers in excess of crop demand can result in buildup of P in pockets that lead to its inefficient use and consequent environmental pollution. A recent study reported a positive balance of 1.02 t P in Indian soils indicating only 20 per cent of applied P is recovered by crops and the rest is assumed to be locked up in soil (Pathak *et al.*, 2010). However, with excessive application of manures and fertilizers, improper management can lead to P accumulation in soil resulting downward movement within soil profile as indicated by long term fertility experiments in Punjab (Aulakh *et al.*, 2007). Soil P management should aim to maintain a sufficient available P level in soil solution at the appropriate time at a reasonable cost, thus increasing P use efficiency through use of organic and inorganic P sources, lime application, rock phosphate utilization, waste recycling, new generation P fertilizers, nanotechnology, improving biological system and phosphate solubilising microorganisms.

Table 1: Removal of NPK (kg t^{-1} grain) of major food grain crops

Crop	N	P	K
Rice	20.4	3.6	20.4
Wheat	22.4	3.8	28.2
Maize	24.3	6.4	18.3
Sorghum	26.1	4.5	21.5
Pearl millet	27.1	8.2	39.7
Chickpea	50.6	8.6	29.7
Pigeonpea	92.1	8.2	30.7

Strategies to improve P use efficiency in Indian soils

The approach "P bank in soil' is more applicable for developed countries with enriched soil P whereas for developing countries like India with low to medium soil P status measure should be taken for judicious use of P fertilizers in conjunction with other resources like organic residues, industrial and municipal solid waste, RP with improved technologies.

(1) Phosphorus management in cropping system: Unlike N, P application is known to benefit crop growth and productivity of more than one crop in rotation. The residual effect of P can form soluble fertilizer reaction products and crops ability to utilize such P. Phosphorus applied to one crop during the season showed residual effects in two succeeding crops. In a study P supplied to soybean was more efficiently utilized compared to that applied to wheat crop. The recoveries of added P were greater in smaller rates and higher in first two residual crops (Subba Rao *et al.*, 1996). Better utilization P is also illustrated in intercropping systems. Phosphatic fertilizers applied as band placement in legume crops showed better PUE. Some studies also indicated application of P to legume crops can increase yield of succeeding crop. Inclusion of green manure in cropping system is also a good option for improving P utilization. The use of green manures like *dhaincha, sunhemp* and *azolla* on crops like rice improved P use efficiency and P recycling in waterlogged condition. The green manures like *sesbania, sunhemp* and *clusterbean* absorb heavy amount of P from 20-25cm soil layer due to their deep root system. Thus they can tap subsoil P and make it available to shallow rooted crops. The deep penetration of legume roots allow exudates in greater soil depth and upon decomposition, microbial metabolites of these exudates may serve as solubilising or chelating agents for fixed forms of P. Green manure response was found to be more on acid and alkaline than neutral soils.

(2) Agronomic and soil management techniques: Method and timing of P fertilizer application can enhance P use efficiency. The best results are obtained with basal P dressing to crops. In upland crops placement of P gave overwhelming results compared to broadcasting. Correct placement of P fertilizer in crop root zone will improve fertilizer use efficiency, seedling vigor and reduce P loss from soil. Phosphorus fertilizer should not be applied when crop growth is minimal and runoff potential is high. As compared to band placement broadcasting application can lead to more wastage and accumulation of P on surface soil which remains mostly unutilized by crops. Band placement is good alternative option on many crops. Similarly sub surface placement is important under reduced tillage systems to achieve maximum yields.

The major limitation for root growth and development is the presence of dense sub surface layer (e.g. plough pans) and surface soil compaction. Therefore soil management is perquisite for managing P loss. This may be avoided by timely cultivation, or zero tillage, incorporation of crop residues to soil surface and by minimizing encroachment by livestock. Soil erosion from agricultural fields is a significant contributor of soil P loss. Management practices like conservation tillage, crop cover, conservation cropping system, delayed seed bed preparation, grass filter strip, contour farming and strip farming can control runoff and erosion loss. Application of P fertilizers excess to crop requirement can adversely affect environment and economic efficiency. Presently, there is no agronomic justification for building soil P test levels higher than crop sufficiency levels. Therefore, once crop sufficiency levels are reached, P should be applied based upon soil test values.

(3) Different Sources of P

(a) **Inorganic sources of P:** Phosphorus fertilizers are the best inorganic sources of P applied to soil and are divided into three classes based upon their solubility: water soluble mono ammonium phosphate and di ammonium phosphates, superphosphate, citrate soluble (Dicalcium phosphate, Thomas slag, basic slag, deflourinated phosphate and fused magnesium phosphate) and acid soluble (phosphate rock and bone meal). The MAP, DAP have water solubility > 90% and are the best P sources for crops. The rock phosphate and bone meal are applied in large quantities on acid soils which are sparingly soluble and converted to usable form for plant uptake over a period of time. The water soluble P fertilizer is readily available for crop uptake, but the major problem is that these fertilizers quickly get fixed and become unavailable for crops. The polyphosphate fertilizer produced by dehydration of phosphoric acid. Liquid ammonia polyphosphate can be used as N and P source. One unique and advantageous characteristics of APP is its chelating and sequestering ability. Nowadays P fertilizers coated with polymer substance are available for improving PUE.

(b) **Organic sources of P:** Organic-matter content of soils on an average comprises less than 1% of soils but has a great impact on soil fertility Organic matter is a huge reservoir of immobilized P accounting for 20-80% of P in soil.The organic manure during decomposition forms organic acids, humic acids and chelating substances which help in liberation of insoluble P into soil solution. Organic acids are released by root exudates and micro organisms which are by- products of degradation of complex organic molecules that decreases phosphate fixation for easy P uptake by crops.The organic sources can even replace 20 to 40% of recommended dose of P fertilizer. The availability of organic resources are shown in Table 2 indicate a huge amount of organic sources of P can be tapped for agricultural purpose. Studies reported that long-term use of fertilizers and FYM decreased P adsorption even more than a super-optimal application of P fertilizers. Composting of biodegradable organic sources with microbial inoculums and chemical amendments can convert bulky organic waste into more cost effective and environmentally viable option of nutrient resources. For efficient recycling of organic waste improved technologies like vermicomposting, phospho-sulfo-nitro compost and P enriched compost from poultry manure, pressmud, distillery effluents by incorporating rock phosphate, pyrites and other minerals can improve the potential of nutrient supply from locally available resources.

Table 2: Availability of P_2O_5 from organic resources

Resource	2010	2025
Nutrients (theoretical potential)		
Human excreta (million t P_2O_5)	0.64	0.76
Livestock dung (million t P_2O_5)	1.31	1.45
Crop residues (million t P_2O_5)	0.21	0.36
Nutrients (considered tappable)		
Human excreta (million t P_2O_5)	0.51	0.62
Livestock dung (million t P_2O_5)	0.66	0.73
Crop residues (million t P_2O_5)	0.07	0.12
Total P_2O_5 (million t)	1.24	1.47

(Tandon, 1997)

(4) Liming: Liming improves the soil pH and enhances P availability. Addition of lime increases the soil pH of acid soils upto 6-7 and increase P availability. Liming is done to avoid Al toxicity and to improve soil pH in tropical soil upto 5.5 thus increasing base saturation by 50% to improve nutrient uptake by plants (Fig: 3). Addition of lime also increases P availability by improving mineralization of soil organic P. Initially in soils with high exchangeable Al^{3+}, lime applied reacts with Al^{3+} to form high P adsorbing surface and precipitates as insoluble polymeric hydroxyl Al cation species. But if an air dried soil reacts with lime and phosphate will reduce P adsorption (Hayens, 1982). High Al in soil at pH less than 5.4 is chemically active and reacts with P fertilizer and makes insoluble Al-P. Under some instances fertilizer P serves inadvertently as liming material and reduce Al solubility. The different types of liming materials used are calcium carbonate ($CaCO_3$), calcium oxide (CaO), calcium hydroxide (Ca $(OH)_2$), calcium magnesium carbonate (dolomite) and calcium silicate (slag) etc.

(5) Opportunities of rock phosphate: Most of the rock phosphates are reasonably suitable for direct use in acid soils, but have not given satisfactory results in neutral to alkaline soils. Indian RP is low grade having 25-30% P_2O_5.The efficiency of the RP applied to the soil depends upon the chemical and mineralogical composition, physical nature of the material, soil properties and crop management practices. In Asian countries RP is major P source in rice based cropping system.Use of low grade RP with pyrites and elemental Sulphur (S) leads to considerable solubilisation of RP. The efficiency of this material is facilitated by bacterial species *Thiobacillus oxidans* which oxidizes pyrite and release S to crops. Partial acidulated RP (PARP)prepared from low grade RP treated with minimal amount of mineral acid to produce super phosphate is a cheaper source and easy to handle as fertilizer material. Alterative approach of using low grade RP is by dry granulation (compaction) with water soluble P fertilizers. This material is equivalent to PARP and can be more cost effective. This can improve P use efficiency by water soluble P taken up by crops initially as starter dose resulting better root development which in turn can utilize RP in later stages. Another approach is fusion of low grade RP with olivine or serpentine (magnesium silicate) at high temperature is also a useful strategy for utilizing Indian low grade RP. This product contains about 20% P_2O_5 and 15% MgO and nearly 90% of P is citric acid soluble (Ray *et al.*, 2013). Zeolite saturated with cations such as NH_4 and K ions increase solubility of RP. Crops take NH_4 and K ions and free exchange sites occupied by Ca, thus lowering soil solution Ca and further dissolution of RP with the release of phosphate into soil solution as the reaction given below:

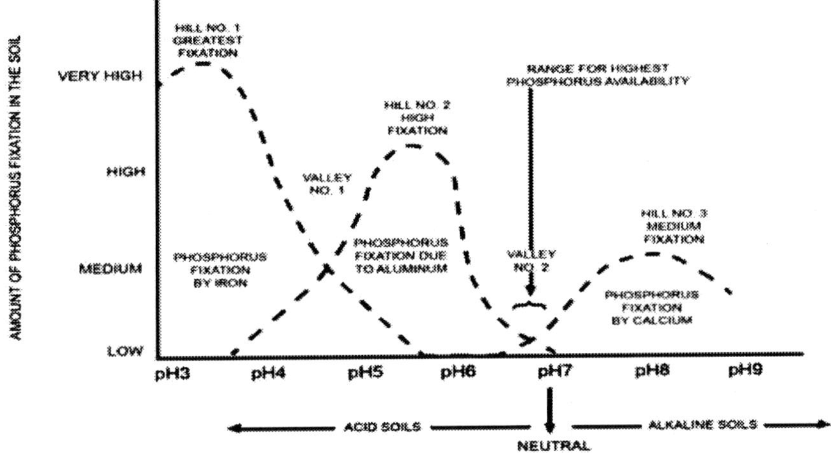

Fig 2: Availability of P at different pH levels

$$Ca_{10}(PO_4)_6F_2 + NH_4^+\text{-Zeolite} \rightarrow Ca^{2+}\text{-zeolite} + NH_4^+ + PO_4^{3-}$$

Preparation of RP enriched compost using crop residue holds a lot of promise in developing countries like India. The organic acid produced during composting result in release of insoluble P into water and citrate soluble P is utilized by crops. Enriched compost prepared by mixing Udaipur, Purulia, Mussorie and Jhabua RP with crop residue or biodegradable wastes. For preparing 1000 kg of enriched compost, 200 kg of low grade RP is mixed with 1000 kg of crop residues and 100 kg of fresh cattle dung. Initially the Olsen P was less, but with time it increased significantly as illustrated in Fig: 3 (Biswas and Narayanasamy

2006). The use of soil micro organisms for effective utilization of RP is a good alternative for improving P use efficiency. The micro organism secretes organic acids like citric, oxalic, tartaric, acetic, malic, lactic, gluconic, α ketogluconic and other enzymes like acid and alkaline phosphatase which leads to transformation of sparingly soluble to labile P forms. The P content of manures vary from 0.04 % in cattle dung to 1.3% in poultry waste. These organic waste can be mixed and composted with other farm waste like bedding material, food waste and human excreta which has good soil conditioning properties that does not occur with only application of organic waste (Cordell, 2013).

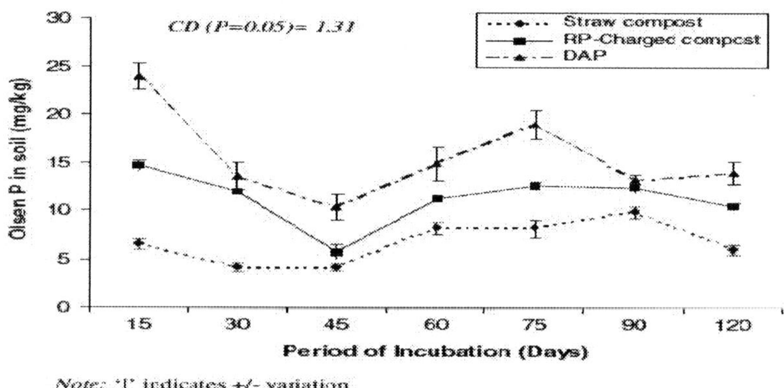

Fig 3: Change in 0.5 M $NaHCO_3$ P as affected by RP enriched compost

(6) New generation P fertilizers: The fertilizers with high P recovery efficiency like polymer coated water soluble fertilizers, rhizosphere controlled fertilizer (RCF fertilizer), organic complexed superphosphate, slow release P fertilizers with superabsorbent property and nanophosphate fertilizers are gaining importance (Ray *et al.*, 2013). In soil, rhizosphere is the major active zone of interaction between soil, plants and micro organisms. The bio chemical processes in rhizosphere are known to influence soil micro organism on one hand and will improve nutrient use efficiency on the other (Zhang *et al.*, 2010). Therefore rhizosphere controlled fertilizer usage had gained momentum in recent years. In rhizosphere controlled fertilizers water soluble fraction acts as starter dose and insoluble P fraction becomes soluble by crop and micro organisms rhizospheric activity (Erro *et al.*, 2011a) and are proved to be more efficient than superphosphate in both alkaline and acid soils. Organic complexed superphosphate is another option for enhancing agronomic efficiency of superphosphate by reducing P fixation by introducing organic chelating agent during superphosphate production (Erro *et al.*, 2011b). Slow release polymer coated water soluble P fertilizer is a major intervention for improving PUE. Some of these fertilizers are DAP-Star (Australia based), Avail-MAP (USA based). Such fertilizers have very high surface charge density (1800 cmol (+) kg^{-1}) which can inhibit P precipitation by Al, Fe, Ca and Mg in soils (IFDC, 1986). A new type of superphosphate by fusing it with organic chelating agents like humic acids had been developed to reduce P fixation and enhance PUE is known as organic super phosphate. This product was evaluated in wheat crops on both acid and alkaline soils under glass house conditions and was found to perform better than SSP (Table: 3) (Erro *et al.*, 2012).

Table 3: Effect of complexed superphosphate and SSP on shoot P concentration of wheat crop (at harvest) cultivated in different soils

Treatment	P concentration in wheat shoots (µg P g^{-1} dry shoot)		
	Alkaline soil	Acid soil with low organic matter	Acid soil with high organic matter
Control	1155b*	852 b	1203 b
SSP	1279 b	1004 b	1249 b
Complexed super phosphate 1	1423 a	1433 a	1474 a
Complexed super phosphate 2	1341 ab	1389 a	1389 a
Complexed super phosphate 3	1422 a	1383 a	1480 a
Complexed super phosphate 4	1240 b	1480 a	1337 ab

*Different letters indicate significant difference for $p<0.05$

(*Source*: Erro *et al.*, 2012)

(7) Nanotechnology: Nano science is being visualized as a rapidly evolving field that has potential to revolutionize agriculture. Presently, the application of nanotechnology in soil science research is concentrated on formulation of nano fertilizers, smart delivery systems for nanoscale fertilizers, nanoforms zeolites for slow release and efficient dosage of water and fertilizers for plants, nanosensors for soil quality and plant health monitoring, nano induced polysaccharide powder for moisture retention or soil aggregation carbon build up and nanomagnets for removal of contamination from soil and water. The cutting edge research areas are expected to emerge in the coming years. In a study carried out at IISS, Bhopal maize crops were treated with Udaipur nano RP (34% P_2O_5) recorded the highest grain and stover yield (5.44 and 7.13 t ha^{-1}) as compared to control, which was 44.68 and 13.17% more than control. It was equal to SSP followed by Udaipur nano RP (31%P_2O_5) treated plants. The P content and its uptake was more in SSP treated plants followed Udaipur nano RP (34% P_2O_5) which was comparable, where as less P content and uptake was observed in control. The total P uptake was 40.29, 38.29, 34.42 kg ha^{-1} under SSP, Udaipur nano RP (34% P_2O_5) and Udaipur nano RP (31%P_2O_5)

treated plants which was 44.82, 37.63 and 21.24% more respectively over control (Table 4) whereas the lowest was recorded by no fertilizer treated plants (17.27 kg ha^{-1}) (Adhikari *et al.*, 2013).

(8) Crop root morphology: Crop P nutrition can be improved by altering root architecture, morphology, length and density etc. Understanding the dynamics of P in rhizosphere- crop continuum provide effective strategies for P management in improving P use efficiency for crop production (Fig. 4). Different crops have different root pattern and distribution for nutrient absorption. Recovery of P from legumes residues in microbial biomass was found to be 15% to 28%, in contrast to 5% recovery from mineral fertilizer (Bunemann *et al.*, 2004). The root growth angle in crops like maize (seminal and crown roots) and soybean (basal roots) result in P acquisition which is related with genotype of the crop. Phosphorus efficient cultivars of groundnut like M522 due to higher root growth absorbed more P in shoot similarly to maize. In moderate to high P fixing soils 'root foraging strategies' can be used and thus will improve yield and will reduce the accumulation of P (Simpson *et al.*, 2011). Introduction of P efficient genotypes and P mining crops can improve P uptake in low P input systems. Phosphorus efficient cultivars of groundnut like M522 with higher root growth absorbed more P in shoot whereas maize cv Paras proved more efficient with more roots and low internal P requirement (Gill and Bhadoria, 2011). Among the efficient legumes P use efficiency are found in the order of Pea> lentil>Bengalgram. Cereals are known to more efficiently utilize P from soil due to their fine root length. In rhizosphere protons are released to acidify the environment, with carboxylate exudations to mobilize sparingly soluble P by chelation or ligand exchange or solubilisation of organic P by secretion of phosphatase enzymes (Zhang *et al.*, 2010).

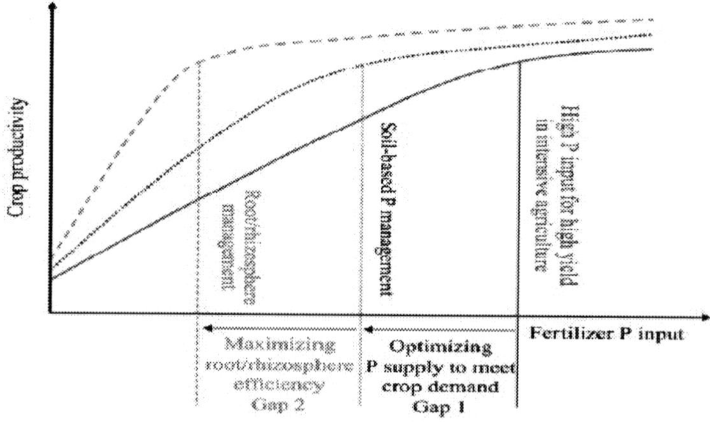

Fig 4: Conceptual model of rhizosphere and soil based nutrient management for improving PUE (Shen *et al.*, 2011)

Table 4: Agronomic parameter, biomass yield, P content and P uptake of maize as affected by different rock phosphate fertilizer application

Treatments	Biomass Yield (t ha^{-1})		Harvest index (%)	Shelling (%)	1000 grain weight (g)	P content (%)		P uptake (kg ha^{-1})	
	Grain	Stover				Grain	Stover	Grain	Stover
Control	2.38	4.11	36.67	47.07	179.3	0.332	0.228	7.90	9.37
NP 100%	3.76	6.30	37.38	59.93	214.8	0.416	0.193	15.64	12.18
NPK 100%	5.50	7.41	42.60	62.58	235.8	0.426	0.227	23.43	16.86
NK (100%)+60 kg P_2O_5 as nano RP (Udaipur 31% P_2O_5)	4.95	7.04	41.38	60.91	225.0	0.377	0.224	18.66	15.76
NK (100%)+60 kg P_2O_5 as nano RP (Udaipur 34% P_2O_5)	5.44	7.13	43.28	61.47	230.3	0.408	0.226	22.19	16.11
CD (p=0.05)	0.57	0.96	4.21	5.28	19.58	0.04	N.S.	1.92	1.71

The formation of root cluster is a specialized mechanism by which plant take up P from soils with low or sparingly soluble sources. These type of plants are mostly found in highly weathered and high P fixing soils where total P is high and not available for plant uptake that does not have this special mechanism to mobilize P. The root clusters release organic acids, enzymes, phenolic acids and protons, organic anion exudation to mobilize sparingly soluble P in soil (Lamberts *et al.,* 2006).

(9) Role of soil microorganism: Soil P management mediated by micro organism is an eco friendly and cost effective approach for sustainable crop production and improving P efficiency. Plant root infection by AM fungi is a rule rather than exception for almost all the agricultural crops, although levels of colonization varies depending upon crop management practices, soil type, crop species and environmental conditions.Arbuscular mychrozzial fungi (AMF) colonise almost all the crop species in agricultural crops and exploit larger volume of soil for P uptake in P deficient soil. Positive response with application of phosphatic biofertilizers like phosphate solubilising microorganism (PSM) and VAM increases the solubility of native P and applied P. In general PSM constitutes 0.5 to 1.0% of soil microbial population with bacteria outnumbering fungi (2-150 fold). The crop species, in extremely low available soil P, develops root clusters effective in capturing P by releasing root exudates like organic anions, enzymes, phenolic acids and protons. Co-inoculation of VAM and phosphate solubilising bacteria result in high root colonization, high VAM spore density and viable counts of *Pseudomonas striata* significantly improve P uptake and yield under temperate conditions.

Among the different P solubilizers fungi like *Aspergillus awamori*, *A. niger, A. candidus, A. fumigatus*, *Pencillium digitatum*, bacteria like *Bacillus polymyxa, B. circulans, B. subtilis, Pseudomonas striata etc.,* yeasts like *Schwanniomyces occidentalis* and actinomycetes like *Streptomyces* sp. are found to be more efficient in P utilization in most soils. In India for the first time Microbiology division of IARI, New Delhi produced biofertilizer named '*Microphos*' which contains *Bacillus megatherium* var. *phosphobacterium.* These biofertilizers are carrier based preparation containing viable cells of micro organism which improves P solubilisation and increase crop production. Inoculants, plant growth promoting rhizobacteria and AMF are the potential sources for solubilization of inorganic and organic phosphate. Thus a large consortium of micro organisms is involved in P solubilisation which includes bacteria, fungi, actinomycetes and even algae. A list of PSM involved in solubilization P sources is given in table 5. The long term application of P fertilizers can inhibit substrate induced respiration by streptomycin sulphate (Fungal activity) and actidione (bacterial activity) and microbial biomass C (Bolan *et al.*, 1996).

Therefore for improving PUE a proper integration of chemical fertilizers and biofertilizer should be planned before application.

Table 5: Biodiversity of phosphate solubilising microorganism

Bacteria	*Alcaligenes* sp., *Aerobactor aerogenes, Achromobacter* sp., *Actinomadura oligospora, Agrobacterium* sp., *Azospirillum brasilense, Bacillus* sp., *Bacillus circulans, B.cereus, B. fusiformis, B. pumils, B. megaterium, B. mycoides, B. polymyxa, B. coagulans B,.chitinolyticus, B. subtilis, Bradyrhizobium* sp., *Brevibacterium* sp., *Citrobacter* sp., *Pseudomonas* sp., *P putida, P. striata, P. fluorescens, P. calcis, Flavobacterium* sp., *Nitrosomonas* sp., *Erwinia* sp., *Micrococcus* sp., *Escherichia intermedia, Enterobacter asburiae, Serratia phosphoticum, Nitrobacter* sp., *Thiobacillus ferroxidans, T. thioxidans, Rhizobium meliloti, Xanthomonas* sp.
Fungi	*Aspergillus awamori, A. niger, A. tereus, A. flavus, A. nidulans, A. foetidus, A. wentii. Fusarium oxysporum, Alternaria teneius, Achrothcium* sp. *Penicillium digitatum, P lilacinium, P balaji, P. funicolosum, Cephalosporium* sp. *Cladosprium* sp., *Curvularia lunata, Cunnighamella, Candida* sp., *Chaetomiumglobosum, Humicolainslens, Humicolalanuginosa, Helminthosporium* sp., *Paecilomyces fusisporous, Pythium* sp., *Phoma* sp., *Populosporamytilina, Myrothecium roridum, Morteirella* sp., *Micromonospora* sp., *Oideodendron* sp., *Rhizoctonia solani, Rhizopus* sp., *Mucor* sp., *Trichodermaviridae, Torulathermophila, Schwanniomy cesoccidentalis, Sclerotium rolfsii.*
Actinomycetes	*Actinomyces, Streptomyces*
Cyanobacteria	*Anabena* sp., *Calothrixbraunii, Nostoc* sp., *Scytonema* sp
VAM	*Glomus fasciculatum*

(10) Waste recycling: Though substantial amount of P is generated through city wastes, agricultural waste, aquaculture and poultry in the form of solid and liquid waste, not efficiently utilized in agriculture. In India the annual production of crop residues or byproducts from different sources are estimated to be 1566.37 mt and 48.56 million pieces of hide or skin from animals (NAAS, 2012). The recycling of P from municipal and industrial waste is cheap source of P, but due to lack of processing technologies using them is restricted for agriculture. There are 2 lakh units for agro industrial units operating in the country generating huge amount of organic wastes which can be converted into enriched compost of about 121 mt annually. Human waste like urine is a major single source of phosphate (0.07% P) emerging from cities which has higher population. Human excreta produces approximately 1-1.5 g P per person per day which means globally, the human population produces nearly 3mt of P in excreta alone (Cordell, 2013). Thus water and sanitation providers should treat and manage such resources carefully for high nutrient recovery and reuse of nutrient through energy and cost effective means. It is estimated that urine from one person alone can provide more than half per capita P for cereal crops (Drangert, 1998).

In many developing countries like Zimbabwe, Burkina Faso, Sweden, Vietnam, Germany etc had developed small scale decentralized P recovery system (Cordell, 2013). In many developing countries, as much as 200 poor farmers use by directly untreated mixed waste water in their fields as nutrient source. Around the world nearly two third of fishes are fertilized by waste water. The reuse of waste water by treating before diverting it for agricultural purpose must essentially take minimum precautionary measures to avert any health risk on human and environment health. World health organization had also set risk based guidelines for safe reuse of human excreta in agriculture and aquaculture (WHO, 2006).

Similarly effluents from waste water treatment plants are also recycled as P source worldwide. Many commercial plants like those of industries (detergents) contain P in addition to human excreta in mixed waste water. Biosludge a byproduct from waste water treatment plant can be used, however heavy metal content in such waste water should be minimized prior to soil application. In India there is large untapped potential of urban waste (Table 6) which can be utilized with mineral fertilizer as a valuable P source for sustaining crop productivity. The sludge generated is more than 1.1 MT and an average P content of sludge water usually ranges between 0.6 to 0.9 % P, it means 0.06 mt P_2O_5 is generated annually (Vision, 2030). The industrial resources like those from sugar industry, oil cakes, sewage water, organic residues, poultry manure and urban solid waste compost have a potential of 2.18 Mt of total P_2O_5 which can be used for alternative P sources in future. The aquatic weeds like water hyacinth and swamp morning glory can harvest high amount of P in their tissues which can be decomposed for agricultural use.

Byproducts from agro processing industries like oil mills also provide P sources. Sugarcane filter cake contains about 1-2% N, 0.5-2.4% P, 0.4-0.5% K and 2-3% Ca, pH of 7.5-9.0 can be effective P source for wetland rice and sugarcane. It was also proved useful in rice grown in acid sulphate soils of Thailand (Hedley *et al.,* 1990). Rubber and oil palm factory effluents contains about 20 and 120 ppm P which are in Malayasia to vegetables can significantly increase yield by 10% (Chan *et al.,* 1980). The weed water hyacinth is reported to accumulate about 400-1600 kg P ha^{-1} from Sinkawa estuary of Japan.

Table 6: Phosphorus content in some wastes generated from different sources

Waste generated	Source	Content of P_2O_5 %
Animal waste	Cattle dung	0.1-0.15
	Cattle urine	0.01-0.02
	Sheep and goat dung (mixed)	0.5
	Night soil	0.8
	Hair and wool waste	0.1
FYM compost	Farmyard manure	0.15-0.2
	Poultry manure	2.90
	Town compost	1
	Rural compost	0.2
	Water hyacinth compost	1
Oil cakes	Castor	1.8
	Groundnut	1.7
	Cotton seed	1.8
	Coconut	1.8
	Neem	1
	Rapeseed	1.8
	Sesame	2.0
Animal meals	Blood and meat (mixed)	3.7
	Horn and hoof	0.3-1.5
	Fish	3-9
Industrial waste	Biogas slurry	1.10-1.72
	Sewage sludge	0.84-2.14
	Pressmud	2.12-2.43
	City refuse	0.59—0.71

(Tandon, 1992)

(11) Best management practices (BMPs): Best management practices for P should aim in ensuring P availability in soil solution at appropriate time at a reasonable cost, thus increasing P use efficiency (PUE) in sustaining crop productivity. This can be achieved by using suitable P source which minimize reaction with soil components and makes P pools available to crop, modifying soil component or application method (of P fertilizer) to reduce P fixation. For P '4R nutrient management' *i.e.,* the right fertilizer, the right amount (soil test based), the right time of application (crop growth stage) and the right application method (band placement) and precision application based on management zones is the need of the hour. The best management practices (BMPs) are based on nutrient requirement of individual crops, the extent of response to crops to P application and the capacity of crops to utilize the residual effect for succeeding crops as shown in the table 6 for some cropping systems of India.

Table 7: Some best P management practices (BMPs) for Indian soils

Management Practices	Situation/ Condition
Phosphorus broadcast	Under high speed operations and heavy P application rates
Phosphorus placement	In low soil test P where early season stress
Fertigation	Good under intensive agriculture; increase P fertilizer efficiency; protects environment; sustains irrigated agriculture
Use treated rock phosphate	Incubation with organic matter; addition of P solubilizer, *A.awamori*, during composting
Increasing the effective rooting area	Root symbiosis with arbascular mycorrhizal fungi (AMF)
Increase P availability through rhizosphere modification	Root exudates: phosphatase, oxalates (genotypic difference)
Use of earthworms	Enhance nutrient availability mainly in tropical soils through casting
Organic residue amendments	A rise in pH in acid soils accompanied by P solubilization; Production and release of organic anions; increased enzymatic activity; compleaxation of exchangeable ions such as Al^{3+}, Fe^{3+}

(*Source:* Subba Rao and Sivastava, 2012)

Conclusion

Soil P is a finite, non-renewable and limited resource. The supply of phosphate rock is expected to decrease year upon year after attaining a peak in year 2035, constrained by economic and energy costs. The approaches of P management may be different from P rich to P poor soils. Enhancing use efficiency of applied P in different crops and cropping systems would be an economically viable and environmentally benign way of efficient P management. For more efficient P utilization of P integration of fertilizer P with different organic waste, green manure, cattle manure, poultry manure, industrial and municipal solid waste, are some beneficial P sources to improve soil fertility and productivity. Sustainable phosphorus use will require an integrated approach that combines efficiency and reuse, in order to reduce current phosphorus losses, minimize environmental impacts, conserve a finite resource and ensure all the farmers to have access to phosphorus.

Selected References

Adhikari T, Kundu S, SubbaRao A. 2013. *Nanotechnology in Soil Science and Plant nutrition.* New India Publishing Agency, New Delhi.

Aulakh MS, GargAK,Kabba BS.2007. Phosphorus accumulation, leaching and residual effects on crop yields from long term applications in the subtropics. *Soil Use and Management.*23: 417-427.

Biswas D.R, Narayanasamy G. 2006. Rock phosphate enriched compost: An approach to improve low-grade Indian rock phosphate. *Bioresources Technology* 97: 2243–2251.

Bolan NS, Currie LD, Baskaran S. 1996. Assessment of the influence of phosphate fertilizers on the microbial activity of pasture soils. *Biology and Fertility of Soils* 21:284–292.

Bunemann EK, Steinebrunner F, Smithson PC, Frossard E, Oberson A. 2004.*Soil Sci. Soc. Am. J.*68:645–1655.

Chan KW, Watson I, Linn KC. 1980. Soil Science and agricultural development. (Eds Puspharajah, E. and Chin, S.C.), *Malaysian Society of Soil Science* Kaula Lampur, Malaysia

Cordell D, White S. 2013. Sustainable Phosphorus Measures: Strategies and Technologies for Achieving Phosphorus Security. *Agronomy* 3: 86-116.

Drangert JO.1998. Fighting the urine blindness to provide more sanitation options. *Water SA*, 24: 157–164.

Erro J, Urrutia O, Baigorri, R. Aparicio-Tejo, P. Irigoyen, I. Torino, F.Mandado, M. Yuvin, J.C. and Garcia-Mina, J.M. 2012 *Journal of Agriculture and Food Chemistry* 60: 2008-2017.

Erro J, Baigorri R. Garcia Mina JM, Yuvin JC. 2011a. Patent WO/ 2011/ 080496.

Erro J, Baigorri R. Garcia Mina, JM,Yuvin JC.2011b.*Journal of Agriculture and Food Chemistry* 59: 1900-1908

FAI Fertilizer Statistics 2011-12. 2012. *The Fertilizer Association of India*, New Delhi.

Gill AAS,Bhadoria PBS. 2010. Identification of P efficient cultivars of maize and groundnut. *Journal of Indian Society of Soil Science* 58(2): 245-247.

Haynes RJ. 1982. Effects of liming on phosphate availability in acid soils. *Plant and Soil* 68 (3):289-308.

Hedley MJ, Hussain A, Bolan NS.1990. Phosphorus for sustainable agriculture, IRRI, Manila, pp 125-142.

Jianbo S, Lixing Y, Zhang J, Li H, Bai Z, Chen X, Zhang W, Zhang F. 2011. Phosphorus Dynamics: From Soil to Plant. *Plant Physiology* 156: 997–1005.

Johnston AE, Poulton PR, Fixwn PE, Curtin D. 2014. Phosphorus: its efficient use in agriculture. *Advances in Agronomy* 123: 177- 228.

Lambers H, Shane MW, Cramer MD, Pearse SJ,Veneklaas EJ. 2006. Root structure and functioning for efficient acquisition of phosphorus: Matching morphological and physiological traits. *Annals of Botany* 98:693–713.

Muralidharudu Y, Reddy KS, Mandal BN, Rao AS, Singh KN,Sonekar S. 2011. *GIS based soil fertility maps of different states of India. All Inia Coordinated Research Project on Soil Test Crop Response Correlation.* IISS, Bhopal.

National Academy of Agricultural Sciences (NAAS) 2012. Agricultural waste management. Policy paper 49.

Pathak H, Mohanty S, Jain N, Bhatia A.2010. Nitrogen, phosphorus, and potassium budgets in Indian Agriculture. *Nutrient Cycling and Agro ecosystems* 86:287–299.

Prasad R.2012. Fertilizers and manures.*Current Science* 102 (6): 894-898.

Raju RA, SubbaRao A,Rupa TR. 2005. Strategies for integrated phosphorus management for sustainable crop production. *Indian Journal of Fertilizers* 1(8): 25-28 & 31-36.

Ray P, Rakshit R, Biswas DR. 2013. Developments in production of phosphatic fertilizers retrospect and prospect. *Indian Journal of Fertilizers* 9 (10): 44-53.

RichardsonAE.1994. Soil microorganisms and phosphorus availability. In: Pankhurst CE, Doubeand BM, Gupta VVSR (eds) Soil biota: management in sustainable farming systems. CSIRO, Victoria, Australia, pp 50–62.

Sanchez P, Logan T. 1992. Myths and science about the chemistry and fertility of the tropics. Soil Science Society of America, Madison, WI, pp 35–46 soils in the tropics.

Simpson RJ,Oberson A, Culvenor RA, Ryan MH, Veneklaas EJ, Lambers H. 2011. Strategies and agronomic interventions to improve the phosphorus-use efficiency of farming systems.*Plt. Soil,* 349(1–2): 89–120.

Steven J, Van Kauwenbergh, Stewart M, Mikkelsen R.2013.World reserves of phosphate rock- a dynamic and unfolding story.*Better Crops*, 97: 18-20.

Subba Rao A, Reddy KS, Takkar PN.1996. Residual effects of P applied to soybean or wheat in soybean-wheat cropping system on a TypicHaplustert. *J Agril. Sci.*127: 325-330.

Tandon HLS.1992. Fertilizers, organic manures, recyclable wastes and biofertilizers.Fertilizer Development sand Consultation Organisation. New Delhi. 37-74.

Tandon HLS. 1997. In: *Plant Nutrient Needs, Supply, Efficiency and Policy Issues* : 2000-2005. NAAS, New Delhi, pp 15-28.

Vision 2030 document of Indian Institute of Soil Science, Bhopal. 2011. Published by IISS, Bhopal.

Zhang F, Shen J, Zhang J, Zuo Y, Li L, Chen X.2010. Rhizosphere processes and management for improving nutrient use efficiency and crop productivity: implications for China. *Advances in Agronomy* 107: 1–32.

Management Interventions and Socio-economic Issues

11

Enhancing Nutrient Use Efficiency, pp. 141-152
Editors: K. Ramesh, A.K. Biswas, B.L. Lakaria, S. Srivastava and A.K. Patra

Preparation of Organic Nutrient Sources for Enhancing Nutrient Use Efficiency in Organic Farming

A. B. Singh, K. Ramesh, Brij Lal Lakaria and J. K. Thakur

ICAR-Indian Institute of Soil Science, Nabi Bagh, Bhopal – 462 038, India

Introduction

Increasing consciousness about conservation of environment as well as health hazards associated with agrochemicals and consumers' preference to safe and hazard free food are the major factors that lead to the growing interest in environmental friendly forms of farming in the world. Organic agriculture is one such broad spectrum of production methods that are supportive to the environment. The basic requirement in organic farming is to increase input use efficiency at each stage of farm operation. This is achieved partly through reducing losses and partly through adoption of new technologies for enrichment of nutrient content in manures. Technologies to enrich the nutrient supplying potential of manures including farm yard manure by three to four times are being widely used at organic farms. Recycling of crop residues and organic wastes through composting methods is the key technology for disposal and production of organic manures and minimization of environmental pollution. Recycling of organic wastes by incorporating chemical amendments and bio-inoculums are the recent developments in the composting technology to manage voluminous wastes with economic advantages as well as environmental and social perspectives. For efficient recycling of biodegradable organic wastes, composting technologies such as vermicompost, phosphocompost, and P enriched vermicompost have been developed utilizing agro-industrial wastes, municipal solid wastes, distillery effluents, press mud, poultry waste and are also co-composted with certain mining wastes materials such as rock phosphate and pyrites for increasing the nutrient supplying potential of the compost. In addition to these composting techniques, there is need to adopt biodynamic preparations in organic nutrient management practices to improve the soils' physical, chemical and biological conditions, thereby, enhancing quality of produce and helping in sustainability of agriculture.

Method of Organic Manure Preparations

i) Vermicomposting

ii) Phosphocompost

i) Vermicomposting

Vermicomposting is a method of composting with worms and differs from conventional composting in several ways. In vermicomposting, there is a saving of nearly two months in composting time compared to conventional compost. Vermicomposting has been arising as an innovative eco-technology for the conversion of various types of wastes into vermicompost. Vermicomposting is the conversion of biodegradable materials/wastes into high quality manure with the help of earthworms and growth of earthworms in organic waste is referred as vermiculture. Earthworms are important drivers of degradation process by fragmenting and conditioning the substrate for microbial degradation. In vermicomposting process the earthworms and microorganisms work symbiotically to speed up the organic matter decay. The vermicompost has more available nutrients per kg weight than the organic substrate from which it is produced (Buchaman, 1988). Vermicasts are believed to contain enzymes and hormones that stimulate plant growth and discourage pathogens. Vermicompost can be prepared in different ways:

a) Preparation of Vermicompost (Pit or Heap method)

In the pit method, open permanent pits of 10 feet length, 3 feet width, and 2 feet depth are constructed under the tree shade. The pits should be about 2 feet above ground to avoid entry of rainwater into the pits. Brick walls are constructed above the pit floor and perforated with 5-6 holes having a diameter of 10 cm in the pit wall for aeration. The holes in the wall are blocked with nylon screen (100 mesh) so that earthworms may not escape from the pits. Partially decomposed dung (dung about 2 months old) is spread on the bottom of the pits to a thickness of about 3-4 cm. This is followed by addition of layer of litter/ residue and dung in the ratio of 1:1 (w/w). A second layer of dung is then applied followed by another layer of litter/crop residue in the same ratio up to a height of 2 feet. Two species of epigeic earthworms viz., *Eisenia foetida* and *Perionyx excavatus* are introduced in the pit.Moisture content is maintained up to 60-70% throughout the decomposition period. Jute bags (gunny bags) are spread uniformly on the surface of the materials to facilitate maintenance of suitable moisture regime and temperature conditions. Watering by sprinkler is done oftenly. The materials are allowed to decompose for 15-20 days to stabilize the temperature to reach the mesophilic stage. The process has to pass the thermophilic stage, which comes in about 3 weeks. Earthworms are inoculated in the pit or heap @1 kg per quintal of waste material to each pit or heap. The entire material is allowed to decompose for 110 days. However, the forest litter decomposes much earlier (75 to 85 days) than farm residues (110 +/- 10) days, (Singh *et al.,* 2005). Earthworms have been used for vermicompost of urban, rural, industrial and agro-industrial wastes to produce biofertilizer (Suthar, 2006, Gupta and Garg, 2008). Sustainable reuse of rice residues as feedstocks in vermicomposting for organic fertilizer production has been studied by Shak *et al.* (2014). The studied showed that rice straw which was amended by two parts of cow dung not only encouraged the growth of earth worms but also provided the best quality vermicompost with the highest nutrient contents and the lowest C/N ratio after 60 days of vermicomposting.

In the heap method, the waste materials and partially decomposed dung (1:1 w/w) are made in heaps of dimension: 10 feet length, 3 feet width, and 2 feet high and during inoculation, channels are made by hand and earthworm @ 1 kg per quintal of waste are inoculated and then watering is done by sprinkler. Jute cloth pieces are used as covering material. The earthworms ensure bio-conversion of crop residues into rich organic manures. Earthworms acts in the soil as aerators, grinders, crushers, chemical degraders and biological stimulators.

P- Enriched Vermicompost

In India, about 260 million tonnes of rock phosphate deposits have been estimated at present. Low-grade rock phosphate is used as direct source of P for crop production, especially in acid soils and long-duration plantation crops. However, there is very little experimental work available on the effect of mixing rock phosphate with different qualities of organic sources and their application in neutral or alkaline soils. For production of P enriched vermicompost, earth worms species viz., *Eisenia foetida* and *Perionyx excavatus* are inoculated in the pit. Moisture content is maintained at 60-70% throughout the decomposition period. Jhabua rock phosphate (30-32% P_2O_5) is used @ 2.5 % P_2O_5 of waste material with the same dimension of pit or heap as mentioned earlier. Jute bags (gunny bags) are spread uniformly on the surface of the materials to facilitate maintenance of suitable moisture regime and temperature conditions. Watering by sprinkler is often done. The materials is allowed to decompose for 15-20 days to stabilize the temperature to reach the mesophilic stage, the process has to pass the thermophilic stage, which comes in about 3 weeks. Earthworms are inoculated in the pit or heap with 1 kg adult earthworms per quintal waste materials are added to each pit or heap. The entire material is allowed to decompose for 110 days. Bansal and Kapoor (2000) studied the vermicomposting of mustard residue and sugarcane trash mixed with cattle dung resulted in a significant reduction in C: N ratio and increase the mineral nitrogen, after 90 days of composting with *Eisenia foetida.* Garg *et al.* (2006) observed that *Eisenia foetida* lowered pH, EC, C: N ratio, potassium and enhance nitrogen and phosphorus contents of cow, goat and buffalo manures.

Vermicast

The food material after passing through the alimentary canal of the earthworm emerges as compact concentrated mass termed as "Vermicast". The earthworm casts contain more microorganisms, organic matter and inorganic minerals in a form that can be used by plants. Vermicast is rich in vitamins, antibiotics and enzymes. These enzymes continue disintegration of organic matter after excretion from the worms as casts.

Earthworm species

Depending on their ability to make burrows, earthworms in ecological termsareclassified mainly into three groups:

(i) ***Epigeic earthworms:*** Earthworms of this group cannot make burrows in the soil. They can only move through the crevices of the surface. They are found in aggregates in litter heaps or in loose soil with high level of nitrogen. They feed exclusively on decomposing organic wastes. They remain active throughout the year if conditions are favourable in the environment. *Eudrillus euginea* and *Eisenia foetida* are being used as prominent composting earthworms, two more species namely; *Perionyx excavatus* and *Perionyx sansibaricus* are also used for the purpose.

(ii) ***Endogeic earthworms:*** They are subsoil dwellers. These burrows are not permanent and thus are disturbed easily during rains. Secretions from body wall of earthworms cements and smoothen the walls of burrows and protect the walls from collapsing easily. They move below 30 cm or more in the soil.

(iii) ***Anecic earthworms:*** They are found in soil, which is not frequently disturbed. They make very complicated burrows in the soil and they firmly pack their burrow walls with their castings. When this feed material gets softened in the burrow by microbial activity, they feed on it. These are very large in size. They have very slow growth rate and long life cycles.

They come out from the burrows and drag the ground cover litter into their burrows. They move within 30 cm depth from the surface.

Vermiwash

Advances in vermiculture technology have recently led to novel products like vermiwash. This product has now not only caught the attention of commercial vermiculturists but also the farmers. Farmers in their own way have started collecting vermiwash for foliar application. For preparation of vermiwash, one-kilogram adult earthworms devoid of casts (approximately numbering 1000-1200 worms) is released into a trough containing 500 ml of lukewarm distilled water (37°C-40°C) and agitated for two minutes. Earthworms are then taken out and again washed in another 500 ml at room temperature (+ 30°C) and released back into the tanks. The agitation in lukewarm water makes the earthworms to release sufficient quantities of mucus and body fluids. Earthworms are transferred into ordinary water to wash the mucus still sticking on to their body surfaces and this also helps the earthworms to revive from the shock.

Vermicomposting in Silpaulin Portable Vermi-bed

For production of vermicompost, silpaulin portable vemi-bed of 12 feet length, 4 feet width, and 2 feet depth are kept under the tree shade, which is about ½ feet above ground to avoid entry of rainwater into the beds. In portable vermi-bed, partially decompose organic waste are added. Crop residues and dung are mixed in the ratio of 1:1 (w/w). Two species of epigeic earthworm viz., *Eisenia foetida* and *Perionyx excavatus* are inoculated in the bed. Moisture content is maintained at 60-70% throughout the decomposition period. Jute bags (gunny bags) are spread uniformly on the surface of the materials to facilitate maintenance of suitable moisture regime and temperature conditions.

ii) Phospho-compost Production Technology

Crop residues or farm wastes are valuable resources for the preparation of good quality compost having nutrient contents higher than FYM or ordinary compost. Straws usually contain high C: N ratio. This can be lowered to favourable C: N ratio either by using nitrogen fertilizers or residues from legumes crops or water hyacinth. Chopped rice straw is amended with nitrogen (1%w/w), phosphate rock (10-25% w/w); Pyrite (5-10% w/w) and mixed with soil, the whole mixed material is either heaped layer by layer over the ground or put inside a pit. Sufficient water is added to each layer to bring its moisture content between 60-75%. Mixture of decomposing microorganisms and phosphate solubilizing fungi like *Aspergillus awamori* are sprayed on the decomposing material. The material is turned after 30, 60 and 90 days intervals. Chemical analysis of the composted materials showed that it contains N 1.4-2.0%; P_2O_5 3-7%; K, 1.5-2.0% with C: N ratios ranging from 8-15 (Singh 2003).

Animal manures are the most common amendments applied to the soil. Cattle account for about 90% of the total animal dung and nutrients. The quantity of nutrients in manures varies with the type of animal, feed composition, quality and quantity of bedding material, length of storage conditions. These manures have capacity to fulfill nutrient demand of crops to a great extent and promote the activity of beneficial macro and micro flora in soil (Mohan Singh, 2003). Most of the Indian soils are deficient in Phosphorus. Also, yearly removal of P is more than its addition through P fertilizers during continuous and intensive cropping. Bio-solids produced in cities, agro-industries and at farms normally have low nutrient value, particularly of P content. Compost production from these bio-degradable wastes is presently not an economically viable. The traditional technology of composting, if improved in terms of nutrients content, may help in arresting trends of nutrient depletion to a greater extent. Further, the use of mineral additives such as rock phosphate and pyrites during composting has been found beneficial. A phosphocompost/N-enriched phosphocompost technology has thus, been developed using phosphate solubilizing microorganisms, namely, *Aspergillus awamori*, *Pseudomonas straita* and *Bacillus megaterium*; phosphate rock, pyrite and bio-solids to increase the manurial value compared to ordinary FYM and compost.

Various studies have been carried out to evolve efficient ways to ensure improvement in available nutrient status e.g. NH_4-N, NO_3-N, water soluble P (WSP), citrate soluble P (CSP) etc. in phopho-sulpho-nitro-compost (PSN compost) as compared to conventional compost. The content of NH_4-N and NO_3-N in PSN compost varied from 0.12 to 0.54 and 0.28 to 0.90 g per kg compost (Manna *et al.*, 2006).

Raw material used

- For the production of one ton of phosphocompost, materials, 1900 kg organic/ vegetable wastes/straw, 200 kg cow-dung (dry weight basis) and 250 kg phosphate rock (18% P_2O_5) are used.

Method

- Prepare a base of the heap out of hard, woody materials such as sticks, bamboo sticks etc. This base should be 15 cm thick, 3 m wide and 3 m long depending upon the quantity of materials to be composted.
- Place crop residue over the base made above. The layer should be around 30 cm + 10 cm thick.

- Sprinkle slurry prepared by mixing cow dung and rock phosphate over the crop residues to moisten the material.
- Make another layer of crop residue and moisten it with slurry.
- Continue with alternate layer of crop residue (30 cm) and slurry until the heap is 1.5 m high. Reduce the area of each layer so that the heap tapers by about 0.5 m at the top. Add water to the heap so that moisture remains about 60 to 70%.
- Cover the heap with soil or polythene and mix the material after 15 days. Give two turnings after 30 and 45 days. Add water at each turning to maintain the moisture content about 60-70%.
- The compost becomes ready for field application with in 90-100 days period.
- The phosphocompost contains 2-3.5% P and 17-18 C:N ratio.
- The cost incurred to obtain one kg P_2O_5 through phosphocompost is around Rs.9.00 as compared to Rs.16.0-17.0 supplied through single superphosphate or diammonium phosphate. Thus, supplying phosphorus @ 60 kg P_2O_5 ha^{-1} in a crop would save around Rs. 400-500.

Biodynamic and Panchagavya preparations

In bio-dynamic farming, the farm is considered as a"living whole" where all farm activities are inter-related and influence each other. The knowledge of the planetary and cosmic rhythms and its influence on plants is used to plan agricultural activities to produce truly enriched crops. Judicious management and utilization of available resources, while using bio-dynamic preparations (BD 500, BD 501, herbal extract and CPP) can improve soil fertility and enhance the health of plants and the quality of their produce finally for human use.

(i) Horn Manure (BD 500)

This unique formulation made from cow dung increases the humus content, improves the biological life of the soil, helps the plant roots to grow deeper, encourages earthworm activity and allows for better nourishment of the plants. It may be used @ 75 g BD 500 ha^{-1} twice or thrice a year in moist soil conditions and is an ideal foundation soil bio activator for a farm that is in transition from chemical to bio-dynamic farming methods.

Application

For every hectare of farm, dissolve the above mentioned (75 g/ ha) quantity of BD500 in 40 liters of warm (40°C) water. This preparation has to be potentised using homeopathic principles before application. First stir the water mixed with BD500 in the clockwise direction until a deep vortex is created; Repeat it in the anti-clockwise direction and continue this homeopathic potentising process for an hour without break. This liquid mixture is now ready and can be sprinkled as big droplets in the farm soil in the evenings (3.00-6.00 PM) during the descending period of the moon.

(ii) Horn Silica (BD 501)

This product is made from good quality quartz crystal and can be used as a fine mist spray above the foliage. The BD 501 gives the plant immunity to resist pests in general and is effective particularly against fungal infections. When used during the time of moon in opposition to Saturn it gives very good results. Regular use enhances the photosynthetic activity of the plant and improves the food quality of the cultivated produce.

Usage

Short duration crop

a) leaf stage in nursery.

b) 2 weeks before harvest.

Fruit orchards

a) at fruit development stage.

b) 2 weeks before the fruit harvest.

Application

For every hectare of farm, dissolve the 2.5 g BD 501/ha quantity in 40 liters of water, stirring in clockwise direction until a deep vortex is created. Repeat it in an anti –clockwise direction and continue this process for an hour. Within an hour of this homeopathic potentising, this mixture is to be sprayed as a fine mist in the atmosphere over the crop, early in the morning (6.00-9.00am) during the ascending phase of the moon.

BD Herbal preparations: This is unique combination of six herbs that have been potentised according to homeopathic principals indicated by Rudolph Steiner, to help heal sick soils & plants. The combined use of these six herbal homeopathic medicines, known as the BD compost set, helps to fix and mineralize the soil with all the essential micro nutrients (trace elements) needed for plants growth.

It also has the unique ability to harness the abundant, unused cosmic forces to improve crop health and productivity.

Application

Compost (1set /10m^3)

Insert 1 set (1g each) of these 6 herbs in 10m^3 of newly made compost. Make capsules of BD502 to BD506 herbal preparations each of the 5 separately with compost, and insert them along the sides of the compost heap. Dissolve 10 ml of 5% BD 507 in 5 liters of pure water, stir in clockwise direction until a deep vortex is created and anti-clockwise alternatively for 10 minutes. Pour some of this liquid into the heap of compost, and sprinkle the rest around the compost heap.

Liquid manure (1 set/200 liters)

Make capsule of each of BD 502-506 herbal preparationsseparately with compost and allow it to float on the liquid manure with partly decomposed leaves, straw or coconut fibre. Dissolve 10 ml of 5% BD507 in 2-3 litres of pure water, stir in clockwise direction until a deep vortex is created and anti-clockwise, alternatively, for 10 minutes. Pour this over the liquid manure.

CPP manure (3 sets/60 kg cow dung): Place 3 g of BD502-506 in five spots inside the cow dung CPP mixture (60 kg)in brick –lined pit. Add 30 ml of 5% BD507 in 2-3 litres of water and stir for 10 minutes, it is added to the CPP mixture.

(iii) CPP manure

This preparation is made from a unique combination of good quality cow dung, egg shell and rock dust powder, and six BD homeopathic herbal medicines. This CPP manure has been reported to play a vital role in distributing the benefit of the BD herbal preparation when used in foliar and soil application. The CPP also enhances the germination of seeds and the rooting of plants cutting, besides rectifying micro nutrient deficiency in plants and soil, it also enables the plants to resist pests and diseases and get healthy. The CPP manure can also be applied for various other farm requirements, such as seed dressing, seedling root dipping,graftwound dressing and making of tree pastes as indicated below:

Application

- Foliar nourishment (5kg/ha of foliage / 100 liters of water): 5.2 kg /ha of CPP is soaked overnight, and dissolved in 100 liters of pure water in food grade plastic drum or bucket. Stir in a clockwise direction until a deep vortex is created and repeat the same in an anti-clockwise direction. Repeat alternative for 20 minutes, filter and spray on the foliage. The dosage can be increased according to the plant canopy surface area, and should be sprayed in the ascending period.

- Soil application (1.5 kg/ha/40 ltrs of water): 1.5 kg /ha CPP is dissolved in 40 liters of pure water in a food-grade container. Stir to potentiseas above for 20 minutes and sprinkle over the soil using brushes or sprayers under low pressure to make big droplets. It can also be used along with horn manure (BD500) for soil application in the afternoon during the descending period.
- Seeding root dipping (1 kg for once acre): Make slurry with 1.0 kg of CPP using clay and water. Dip the seedlings for 5 minutes before transplanting.
- Seed treatment (200g/kg of seed): Makeslurry of 200 g CPP with clay and water. Sprinkle and mix this slurry with the seeds. Dry them in shade before sowing.
- Grafting wound dressing (As paste): Make a paste with the required amount of CPP and water. Use this paste to coat the wound of trees and plant after grafting or pruning.
- Tree Paste: Mix 1 kg CPP with 1 kg Rock Dust and 1 kg clay, and make it into well mixed slurry. Apply this paste on lower trunk of trees, and on pruned areas in horticulture tree crops.

(iv) Panchagavya Preparations

Panchagavya is prepared by taking five kg fresh cow dung, 3 lit cow urine, 500g cow ghee, 2 lit cow milk, 2 lit cow curd, 3 lit sugarcane juice or 500g jiggery dissolved in 3 lit water 3 lit tender coconut water and 12 ripe bananas. Mix the fresh cow dung and ghee thoroughly and keep it for three dayswith stirring it twice daily at least for 15 minutes. On 4th day add rest of the ingredients and mix them thoroughly. This solution must be stirred twice daily for 15 minutes. On 19th day panchagavya is ready for use, keep it always covered with a mosquito net or cotton cloth under shade. Stir twice daily to keep panchagavya for 6 months without deterioration in quality. Add cow urine if solution becomes thick. Apply panchagavya 3 per cent solution as foliar spray at pre flowering, flowering and grain filling stage. It has been experimented by various organic farmers. It has significant role in providing resistance to pests and disease and increasing the overall yield.

Conclusions

The widening gap between plant nutrient input and withdrawal from the soil, emerging deficiencies of micro-nutrients and increasing loss of soil organic matter (SOM), affect not only the crop yields but also adversely affect the micro-organism population and quality of environment. It disturbs the soil, plant, relationship and natural cycle of all the processes above ground and below ground. Organic farming is an alternative agriculture, which has been proposed as a solution to problems associated with intensive use of agro-chemicals. Recycling of organic wastes by chemical amendments and bio-inoculants are the recent developments in the technology for composting to manage voluminous wastes with economic advantages as well as environmental and social perspectives. Earth worms helps in accelerating the decomposition processes and brings down the composting period from six to eight months to 2- 3 months. Application of panchagavya, influences the crop growth and also gives resistance against pests and diseases, while using bio-dynamic preparations (BD 500, BD 501, herbal extract and CPP) can improve soil fertility and enhance the health of plants and the quality of their produce finally for human use.Thus, enriched compost helps to produce higher yields of crops and also improve soil physical, chemical and biological conditions of soil, thereby, enhancing productivity and sustainability of Agriculture.

Selected References

Abbasi SA, Ramasamy, EV 1999. Biotechnological methods of pollution control. Orient Longman, Universities Press India Ltd., Hyderabad pp 168.

Bansa IS, Kapoor KK. 2000. Vermicomposting of crop residues and cattle dung with *Eiseniafetida. Bio-resource Technology,* 73: 95-98.

Bhat, S.S, Singh J and Vig A.P. 2014. Genotoxic Assessment and Optimization of Pressmud with the help of Exotic Earthworm Eisenia Fetida. *Environmental Science* and *Pollution Research.* 21(13): 8112-23

Buchanam MA, Rusell E, Block SD. 1988. Chemical Characterization and nitrogen mineralization potentials of vermicompost derived from different organic wastes. In: C.A Edwards and E.F. Neuhauser (Eds.), Earthworms in Environmental and Waste Management. SPB Academic Publishing, the Netherlands. pp231-240.

Garg VK, Yadav Y K, Sheoran A, Chand S and Kaushik P 2006. Livestock excreta management through vermicomposting using an epigeicearthworm *Eisenia fetida. Environmentalist,* 26:149-154.

Gupta R, Garg VK. 2008. Stabilization of primary sewage sludge during vermicomposting. *Journal of Hazardous Materials.* 153:1023-1030.

Hanc A, Chadimova Z. 2014. Nutrient recovery from apple pomace waste by vermicomposting technology, *Bio-resource Technology.* 168: 240-244

Manna MC, Ganguly TK, Ghosh PK, Singh AB and Tripathi AK 2006. Phospho-Sulpho-Nitro- Compost production and Evaluation, Bulletin No. 02/2006, pp 1-11.

Singh M. 2003. Enrichment Technology for improving the compost quality. Published In: efficient use of On-farm and Off-farm resources in organic farming (Edited by P Ramesh and Subba Rao), pp 126-138.

Shak KPY, Wu TY, Lim SL and Lee CA 2014. Sustainable reuse of rice residues as feedstocks in vermicomposting for organic fertilizer production. *Environmental Science and Pollution Research*, 21:1349-1359.

Singh AB, Manna MC, Tripathi AK and Ganguly TK. 2005. Vermicomposting: A Technology for organic waste recycling. *IISS Bulletin No*. 01/2005 pp 1-11.

Suthar S. 2006. Potential utilization of Guar gum industrial waste in vermicompost production. *Bio-resource Technology*. 97:2474-2477.

12

Enhancing Nutrient Use Efficiency, pp. 153-159
Editors: K. Ramesh, A.K. Biswas, B.L. Lakaria, S. Srivastava and A.K. Patra

Organics in Meeting Nutrient Demands of Crops – Contents and Use Efficiency

Brij Lal Lakaria, K. Ramesh, Pramod Jha and A. B. Singh

ICAR-Indian Institute of Soil Science, Nabi Bagh, Bhopal – 462 038, India

Indian agriculture has grown over past 50 years from subsistence type of farming to a market surplus one with the present food grain production above 250 million tones (Fig. 1) which was only 50 million tones in 1950. At the same time the country has struggled over the years to meet the fertilizer requirement. More than 50 per cent fertilizer consumption is imported. The response ratio (kg grain/ kg nutrient) in food grain crops in irrigated areas in India (Fig. 1) substantially declined between 1960 and 2008 (Biswas and Sharma, 2008a). The need to improve NUE is therefore of paramount importance both for economical as well as environmental reasons. At the same time, overexploitation and degradation of natural resources is one of the major constraints that agriculture faces worldwide.

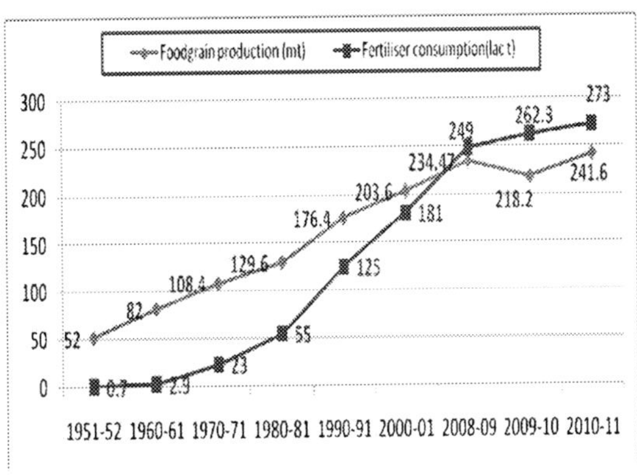

In India, currently the gap between annual output of nutrients (NPK) from soil due to crop removals and the nutrient inputs from external sources (fertilizers) is negative by about 10 million tonnes (Sundaram, 2001). Also, 15% of the animal population, which, apart from supplying milk and draught power in agricultural operations, also contribute valuable plant nutrients acting as supplement to the fertilizer nutrients. Increasing consciousness about conservation of environment as well as health hazards associated with agrochemicals and consumers' preference to safe and hazard free food are the other major factors that lead to the growing interest in environmental friendly forms of farming involving the use of organic sources of nutrient to crops. The greatest challenge in 21st century is to feed the ever increasing population. Globally, 842 million people (12% of global population) were unable to meet their dietary energy requirement during 2011-13 (FAO). It indicates still the food insecurity will remain to be an issue especially confined to the developing countries. India is endowed with a vast potential of plant nutrients locked up in organic, biological and industrial byproducts. However, their use in INM system is limited due to their alternate utilities as animal feed, fuel and building material *etc*. Major crop improvement strategies for sustainable agricultural production mainly pursue the use of natural processes such as nutrient cycles, biological nitrogen fixation and pest predator relationship in agricultural production. For sustainable agriculture, the main organic resources use for plant nutrients are FYM, compost, crop residues, green manure, bio-fertilizers, legumes, vermicompost, biogas slurry, *etc*., Apart from these organic sources of plant nutrients, other sources include soil reserves, human wastes, urban and rural wastes, sewage sludge, tree and aquatic wastes, agro-industrial wastes like press mud, coir pith from coconut, industry, distillery wastes fruit and vegetable wastes, marine wastes, sea wastes and fishmeal, etc. India has a potential of about 15 million tones of nutrients from organic sources which is quite enough to bridge the gap between consumption and supply of fertilizers. The food grain production of the country has increased five folds since independence. This achievement has been due to so many agricultural inputs among which fertilizer in one of the important one. Thus, increased attention is now needed to adopt integrated nutrient management (INM) practices that maintain or enhance soil productivity through a balanced use of mineral fertilizers combined with organic sources of plant nutrients. The INM focuses first on the annual cropping system, rather than an individual crop; secondly on the management of plant nutrients in the whole farming system; and, thirdly, on the concept of village community area rather than individual fields. In many areas, still most of the manure and nutrient management strategies and the fertilizer recommendations for crops are adopted to a very

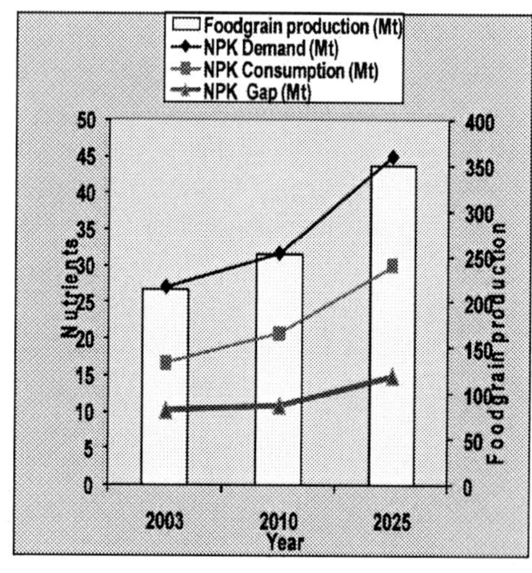

limited scale due to perceptions such as the fertilized foods are harmful to soil health. At present even the farmers are experiencing fatigue in crop yields and realizing decline in soil health due to imbalanced fertilizer application. Therefore efforts towards the use of organic sources, such as crop residues, FYM, poultry manures, cow dung manure, oil seed cakes, green manures and agro-industrial byproducts as a source of plant nutrients (Table 1&2) would be essential in improving soil fertility and productivity. In addition to these sources ample forest litter is available for communities residing adjoining to forest area which can provide about 15 million tones of compost. In addition to it there are vast scope to promote organic inputs to field such as green manuring, biofertilizer additions, Rhizobium, BGA, Azolla and PSB *etc.,* (Table 3) along with inorganic fertilizer addition to sustain the productivity.

Table 1: Nutrient potentials of organic manures and residues and their availability

Manure	Total quantity available (m t)	Total nutrients (000, t y^{-1})			
		N	P_2O_5	K_2O	Total
Cattle dung manure	279	2813	2000	2069	6882
Crop residue	273	1283	1966	3904	7153
Rural compost	285	1431	861	1423	3715
City refuse	14	98	64	112	294
Sewage sludge	0.5	5	3	3	11
Press mud	3	33	79	56	168

Table 2: Availability of various residues (t ha^{-1}) and nutrient content therein (%)

Crop Residues	Quantity	N	P	K
Mustard straw	1.8-2.2	0.45	0.53	1.48
Castor shell	0.5	1.01	0.35	1.86
Wheat straw	1.8-2.5	0.53	0.10	1.10
Cotton stalk	4.0-5.0	0.44	0.10	0.66
Paddy straw	2.20	0.36	0.08	0.71
Rice husk	0.25	0.45	0.25	0.45
Sugarcane trash	9.0	0.35	0.12	0.6
Maize cobs	0.8-1.2	0.42	1.57	1.65
Groundnut shell	0.8	1.25	0.46	1.50
Banana pseudo stem	9.0	0.61	0.12	1.00
Pigeonpea stalk	1.5-2.0	1.10	0.58	1.28
Pearlmillet straw	0.4	0.65	0.75	2.50

Table 3: Biofertilizer production in India during 2011-12

Zone	States	Tonnes
West zone	M.P., Chhatisgarh, Rajasthan, Goa, Maharashtra	11528
North Zone	H.P., Uttrakhand, Punjab, Haryana, Delhi	12183
South zone	AP, Karnataka, TN, Pondicherry, Kerala	11674
East zone	Bihar, Orissa, W.B., Jharkhand	1277
North East zone	Assam, Mizoram, Nagaland, Tripura	1624

Source: National centre of organic farming, Department of Agriculture and Co-operation

In another estimate by Tandon (1997) the availability of organic manure has been proposed to be around 32 m tones by 2025 from different sources (Table 4)

Table 4: Aavailability of nutrients ($N+P_2O_5+K_2O$) by 2025

Resources	Potential	Tapable*
Human excreta	2.60	2.60
Livestock dung	7.54	7.54
Crop residues	22.27	22.27
Total	32.41	32.41

* Tapable 30% dung, 80%human excreta, 30% residues

Advantages of organic manure additions

- Creates optimal conditions in soil for high yields and good quality produce
- Supply all the nutrients required by the plants
- Improvement in the growth and physical activity of the crop plants
- Improvement in granulation, tilth, aeration, root penetration, and water holding capacity of the soil
- Fibrous portion of the roots improve the soil aggregation and permeability of soil
- The carbon added through manures serves as a source of energy for the microbes which help in aggregation and thus increases water retention and decreases the soil erosion.

Manure Efficiency

Improving Nutrient Use Efficiency (NUE) is of paramount importance both from an economic as well as an environmental point of view. Among several strategies to improve NUE, balanced nutrition, particularly balancing N and potassium (K) nutrition and tapping into the synergistic effect of N and K, is recently gaining importance. The efficiency of fertilizer N is only 30-40% in rice and 50-60% in other cereals. The N use efficiency is low mainly because of immobilization, volatilization, denitrification and leaching losses. There are several reports indicating the positive effects of N and K interaction in terms of crop productivity and economics, but the balance of N and K application is not appropriately practiced in many parts of the world. Positive interaction of N and K has been reported in many rainfed cereals like sorghum, pearl millet, finger millet, maize, and minor millets. In a study based on 241 site-years of experiments in China, India, and North America, balanced fertilization with N, phosphorus (P), and K increased first-year recoveries by an average of 54% compared to recoveries of only 21% where N was applied alone (Brar *et al.* 2011). It has also been demonstrated that with the application of PK, the agronomic efficiency of N (AEN) improved substantially (6.7 kg grain/kg N) in sorghum and (10.3 kg grain/kg N) in pearl millet. Balanced fertilization improved both AEN as well as partial factor productivity of fertilizer N at on farm locations. Phosphorus use efficiency ranges between 15-20% on account of fixation in soils as Al-P, Fe-P and Ca-P. The use efficiency of potassium is maximum compared to other nutrients and it can be as high as even 70-80% . Its fixation in clay-lattice also prevails but it is still not permanent one.

The use efficiency of manures such as FYM may be adjudged in three ways viz., Efficiency in comparison to fertilizers, advantages of its application in addition to NPK (add-on series) and possibilities of economy of N/NPK by FYM (replacement). Numerous studies conducted have revealed improvement in soil properties with the application of manures along with significant improvements in yields of various crops. Large amounts of data are published form experiments conducted under AICRP on long term in this respect. Application of FYM along with NPK has been found to sustain the productivity in different cropping systems. In a study, application of FYM with NPK increased the recovery efficiency of N by 45.2% in maize and 68.2% in wheat. Similarly recovery efficiency of P increased by 28.4% and 34.2% and that of K by 59 and 108% by maize and wheat respectively in a long term experiment. Significant improvement in crop yield under different long term fertilizer experiments has been demonstrated while substituting about 50% of N requirement by organic sources at various centres of AICRPs (Table 5) under rice-rice cropping system. In Titabar (Assam) use of moderate level of N, P and K i.e., 40:20:20 kg ha^{-1} to each crop is recommended. Use of organic sources, such as FYM, compost, green manure, Azolla etc to meet 25-50% of N requirement in *Kharif* rice help curtailing NPK by 25-50%. In another study application of 75% of recommended NPK along with rice straw or azolla is recommended to supply 25% of fertilizer N in monsoon rice. This IPNS strategy improves the fertilizer use-efficiency and helps in saving of 25% fertilizer dose in acid soils (Entisols and Inceptisols) of Assam. When Azolla or rice is applied to monsoon rice, it is possible to curtail 25% fertilizer NPK even in subsequent summer rice without any yield reduction. In rice – wheat cropping system at Malan (H.P.), a temperate region, application of 5 t FYM+50% RDF resulted in 13.7% more rice and 32.2% more wheat yield as compared to 100% RDF. These results clearly suggest that in a two crops a year cropping system at least 25% economy of the recommended dose for the entire cropping can be achieved by the FYM.

Table 5: Effect of N substitution through organic sources on rice yields (t ha^{-1})

Treatment	Bhubaneswar	Maruteru	Karjat	Karamana
100% RDF in rabi and Kharif	9.0	10.7	8.2	7.6
50% N through FYM/compost+50% RDF in kharif 100% RDF Rabi	9.8	10.5	7.7	8.1
50% N through crop residues +50% RDF in kharif 100% RDF Rabi	9.4	10.4	7.9	7.9
50% N through green manure +50% RDF in kharif 100% RDF Rabi	9.7	10.7	8.5	8.5

Epilogue

The dependency on commercial fertilizers is mainly because of their quick effects on the crop growth and yields. From each corner, the soils are experiencing a decline in crop yields due to similar management strategies and there has been decline in crop response ratio. There seems no way out rather ensuring that plants receive an adequate and balanced supply of nutrients. The appropriate environment must exist for nutrients to be available to a particular crop in the right form, in the correct absolute and relative amounts, and at the right time for high yields to be realized in the short and long term. In this regard it is important to utilize the available and tapable organic resources to recycle plant nutrients to soil system for better soil health and crop production since there exists a huge gap between the demand and supply of plant nutrients for sustainable crop production. Research need to be further carried out for promoting biological nitrogen-fixation as a low-cost "organic" approach to increasing nitrogen availability and organic matter content in soils. The application of targeted, sufficient, and balanced quantities of organic sources along with inorganic fertilizers will be necessary to make nutrients available for high yields without polluting the environment. Also our holistic efforts are required to improve the availability and use of secondary nutrients and micronutrients, organic fertilizers, and other soil conditioners.

Selected References

Biswas PP, Sharma PD. 2008. A new Approach for Estimating Fertilizer Response Ratio: The Indian Scenario, *Indian Journal of Fertilizers* 4(7):59-62.

Brar MS, Bijay Singh, Bansal SK, Srinivasarao, Ch. 2011. Role of Potassium Nutrition in Nitrogen Use Efficiency in Cereals. Research Findings, International Potash Institute, e-ifc No. 29, December 2011.

13

Enhancing Nutrient Use Efficiency, pp. 161-178
Editors: K. Ramesh, A.K. Biswas, B.L. Lakaria, S. Srivastava and A.K. Patra

Water Management *vis-a-vis* Nutrient Use Efficiency of Crops and Cropping Systems

K.G. Mandal and P.S. Brahmanand

ICAR-Indian Institute of Water Management, Bhubaneswar – 751023, India

Introduction

Crop production system includes the components like soil, climate and plant (genetic) material. Within a particular genetic make up of the crop, productivity of the system is limited either by climatic factors, that is, radiation, temperature and rainfall (water), or by inherent edaphic environment that govern the root development and consequent uptake of water and nutrients. Where soils are cultivated over long periods, the edaphic conditions reach at its low levels. When water, nutrients and CO_2 are plentiful, yield is determined by radiation and temperature regimes during the growing season. However, under actual field situations, the yields are limited by factors other than climate, which is water and nutrients. These are not only limited in amounts but also are becoming costlier every year. It is, therefore, important that available fertilizers be used efficiently.

Water, together with seeds and fertilizers, is a major input for agriculture (Srinivasarao, 2011). By providing an optimum and timely water supply to crops, irrigation favours the plant's capacity to absorb the applied nutrients and, thus, contributes positively to fertilizer (nutrient) use efficiency. Soil water is involved in the transport and transformation of nutrients in soil, and absorption by plants. For efficient use of fertilizer nutrients, adequate quantities of water must be available throughout the crop growth period. In fact, water and nutrients have been shown to exhibit strong interactions with respect to yield and yield components.

Water favours root growth and development, soil moisture extraction by crops, better water relation in plants, crop growth (dry matter and leaf area), nutrient uptake (transport, transformation and absorption) and yield components and yield of crops. It is a concern that nutrient-use efficiency has been declining continuously. The efficiency of fertilizer N is only 30-40% in rice and 50-60% in other cereals, while the efficiency of fertilizer P is 15-20% in most crops. The efficiency of potassium (K) is 60-80%, and that of sulphur (S) is 8-12%. Regarding micronutrients, the efficiency of most of them is below 5% (NAAS, 2006). Low nutrient recovery efficiency is not only a cause for increasing cost of crop production but also a cause of environmental pollution (Srinivasarao *et al.*, 2013).

Nutrient use efficiency

Roberts (2008) stated that agronomic efficiency may be defined as the nutrients accumulated in the above-ground part of the plant or the nutrients recovered within the entire soil-crop-root system. Fertilizer nutrients applied, but not taken up by the crop, are vulnerable to losses from leaching, erosion, and denitrification or volatilization in the case of N, or they could be temporarily immobilized in soil organic matter to be released at a later time, all of which impact apparent use efficiency. Dobermann *et al.* (2005) introduced the term system level efficiency to account for contributions of added nutrients to both crop uptake and soil nutrient supply. In general, nutrient loss to the environment is only a concern when fertilizers or manures are applied at rates above agronomic need. Though perspectives vary, agronomic nutrient use efficiency is the basis for economic and environmental aspects.

Nutrient use efficiency can be expressed in several ways. Mosier *et al.* (2004) described four agronomic indices commonly used to describe nutrient use efficiency: partial factor productivity (PFP, kg crop yield per kg nutrient applied); agronomic efficiency (AE, kg crop yield increase per kg nutrient applied); apparent recovery efficiency (RE, kg nutrient taken up per kg nutrient applied); and physiological efficiency (PE, kg yield increase per kg nutrient taken up). Crop removal efficiency (removal of nutrient in harvested crop as % of nutrient applied) is also commonly used to explain nutrient efficiency. Available data and objectives determine which term best describes nutrient use efficiency. Fixen (2005) provided a good overview of these different terms with examples of how they might be applied.

Status and an account of NUE

The worldwide data on N use efficiency for cereal crops from researcher-managed experimental plots shows that single-year fertilizer N recovery efficiencies averaged as 65% for corn, 57% for wheat, and 46% for rice (Ladha *et al.*, 2005). Differences in the scale of farming operations and management practices (i.e. tillage, seeding, weed and pest control, irrigation, harvesting) usually result in lower nutrient use efficiency. Nitrogen recovery in crops grown by farmers rarely exceeds 50% and is often much lower. A review of best available information suggests average N recovery efficiency for fields managed by farmers ranges from about 20 to 30% under rainfed conditions and 30 to 40% under irrigated conditions. Cassman *et al.* (2002) looked at N fertilizer recovery under different cropping systems and reported 31 and 40% recovery for rice grown in farmer's practice and field specific management in Asia, respectively; 18 and 49% recovery for wheat with unfavourable and favourable weather in India, respectively. Fertilizer recovery is impacted by management, which can be controlled, but also by weather, which cannot be controlled (Cassman *et al.*, 2002).

Agricultural activities in India account for more than 80% of the total N_2O emissions, including 60% from the use of N-fertilizers and 12% from burning of crop residues (Ladha *et al.*, 2005; Singh and Singh, 2008). As for cereal-based agriculture, recovery of N by rice and wheat at on-farm locations in India rarely exceeds the 50% mark (Singh and Singh, 2008).

Reducing N losses in porous soils

Porous soils are characterized by high infiltration, low moisture retention and poor fertility due to limitation of organic matter and nitrogen (Aulakh and Singh, 1997). These soils can be managed by reducing losses of N via different mechanisms, thereby productivity of crops can be enhanced. Losses of N through ammonia volatilization can be reduced if fertilizer N is applied before an irrigation or rainfall event. Ammonia-N is transported to a soil depth along with percolating water, this cannot move back to soil surface where it is prone to be lost as NH_3 (Aulakh and Singh, 1997). Under upland conditions, nitrification proceeds rapidly in porous soils. Due to high water percolation rates in porous soils, continuous flooding for rice production usually cannot be maintained and alternate flood and drained conditions are created. Nitrification proceeds rapidly during drained conditions and nitrates thus produced are subsequently reduced to N_2 and N_2O through denitrification upon reflooding. Indirect N-budget estimates show that up to 50% of the applied N may be lost via nitrification-denitrification in irrigated porous soils under wetland rice (Aulakh and Singh, 1997).

Interdependence of irrigation and nutrients

The interaction effect of irrigation and nutrient on grain yield and yield attributing characters of wheat was demonstrated by Mandal *et al.* (2005) (Table 1).

Moreover, it is found that the combined effect of water and nutrients was exhibited in the root length density (Fig. 1), soil moisture extraction and Evapotranspiration (ET) (Hati *et al.*, 2001), leaf area index and dry matter yield of wheat in Vertisols of Bhopal.

Table 1. Grain yield of wheat as influenced by irrigation water and nutrient management

Irrigation	Nutrient management		
	F_0	F_1	F_2
		Grain yield (kg/ha)	
I_0	957f	1640e	2233d
I_1	1188f	2641c	3288b
I_2	1676e	3377b	4088a

I_0, without post-sowing irrigation; I_1, two irrigations; I_2, three irrigations; F_0, control; F_1, 100% NPK (100-21.5-24.9 kg ha^{-1}); F_2, 100% NPK + FYM @ 10 t ha^{-1}; Mean values followed by different letter differ significantly at $P < 0.05$ according to DMRT
(Mandal *et al.*, 2005)

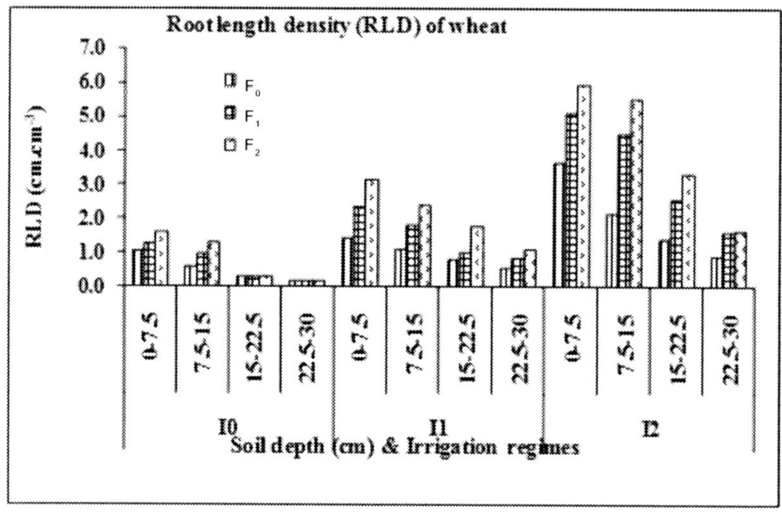

I_0, without post-sowing irrigation; I_1, two irrigations; I_2, three irrigations; F_0, control, F_1, 100% NPK (100-21.5-24.9 kg ha^{-1}); F_2, 100% NPK + FYM @ 10 t ha^{-1}

Fig. 1. Root length density (RLD) of wheat as influenced by irrigation and nutrient management (Mandal *et al.*, 2003)

Prihar *et al.* (1985) showed that an increase in water supply from 300 mm to 450 mm, without N application increased four years average yield of wheat from 2.2 to 2.7 t ha^{-1}. An addition of 80 kg N ha^{-1} at 300 mm water supply increased yield to 3 t ha^{-1}, but when water supply increased to 450 mm, with the same amount of N (i.e., 80 kg ha^{-1}), yield increased to 4.5 t ha^{-1}. They propounded that the increase in N use efficiency and apparent recovery in wheat with an increase in water supply was related to the NO_3^- -N use from the soil profile. For achieving maximum use efficiency of both inputs, application of water and fertilizer should be in concert to minimize NO_3^- -N losses from the soil profile and to maximize uptake by the crop.

Addition of fertilizer is meaningless unless sufficient water is available to support a response. Similarly, increasing plant available water is futile unless the soil fertility problems are attended with a view to ensure adequate fertility. Supply of adequate plant nutrients helps in quick establishment of crop canopy. Increased leaf area is likely to result in increased loss of soil water through transpiration, but simultaneously it can reduce the solar energy reaching the surface, thereby reducing evaporation from the soil. Further, more vigorous shoot growth with nutrient application stimulates deeper root development and greater extraction of soil water. Kanwar *et al.* (1984) showed that fertilizer use efficiency (FUE) in pearl millet under low soil moisture conditions (50 mm) decreased with applications of higher doses of fertilizer. However, FUE under higher soil moisture conditions (110 mm) increased with addition of fertilizer. Vijaya Kumar *et al.* (1998) found that a N dose of 75 kg was optimum for wheat when water supply was 458 mm.

Nitrogen use efficiency (NUE) in wheat decreased with a successive increase in N rate on sandy loam and loamy sand soils in Ludhiana (Gajri *et al.*, 1993). However, NUE was linked to seasonal water supply. It was lower on less water retentive loamy sand than on sandy loam. At 80 kg N/ha, NUE increased with an increase in irrigation up to 200 mm with 120 kg N ha^{-1}, the NUE did not increase with an increase in irrigation from 50 to 125 mm, but increased markedly when irrigation was further increased to 200 mm with 40 kg N ha^{-1}, NUE at 200 mm irrigation was lower than that at 125 mm (Table 2). This implies that the use efficiencies of the two inputs are interdependent; the efficiency of one input is influenced by the adequacy of the other.

Table 2: Irrigation and N effects on grain yield of wheat and NUE

Irrigation water (mm)	Grain yield (t ha^{-1})				N- use efficiency (kg grain kg^{-1} N)		
	N rate (kg ha^{-1})						
	0	40	80	120	40	80	120
0	0.69	1.03	1.13	0.87	8.5	5.5	1.5
50	1.28	2.09	2.75	3.42	20.2	18.4	17.8
125	1.63	2.96	3.67	3.67	33.2	25.5	17.0
200	1.80	3.01	4.22	4.64	30.2	30.3	23.7

Source: Gajri *et al.* (1993)

Therefore, it is very clear that nutrient application should be synchronized with the availability of water. This is particularly important in the context of enhancing nutrient use efficiency. Now, the question is- how to determine the required amount of nutrient as per differential water availability. This is generally done by the multiple regression technique as adopted by previous researchers (Gajri *et al.*, 1993).

Regression technique for water x nutrient interaction

The *first step* is to conduct experiments. A factorial experiment involving different water regimes and rate of fertilizer nutrients is to be planned and conducted in the location. Thus, a database is to be developed on the crop yield with respect to different combinations of water regime and nutrient rate. For example, we consider the database (Gajri *et al.*, 1993) of the wheat crop as indicated in Table 1. The *second step* is to have statistical treatment of the data. Statistical treatment of the yield data is required. The analysis of variance technique (ANOVA) is to be used as applicable to factorial randomized block design (RBD), split-plot design (SPD) or strip-plot design. The significance of the interaction effect of water and nutrient is to be tested. If the interaction effect is significant at a specified probability level (normally, $P = 0.05$), then go for further steps to be followed. It may be noted that in most of the cases, interaction effect of water and nutrient is significant, especially for irrigated crops. The *third step* is the development of the regression equation. With the help of data analysis tool pack in MS Excel, the multiple regression equation can be developed following the proper procedure. The yield (Y) will be the dependent variable, and water (W) and nutrient (say, Nitrogen, N) will be the independent variable. The standard form of the equation will be as follows:

$$Y = \alpha + \beta_1 W + \beta_2 N + \beta_3 W^2 + \beta_4 N^2 + \beta_5 WN$$

where, α is the intercept, β_1 to β_5 are the regression coefficients. Normally, of the different functional forms, perfect quadratic polynomial gives the best fit. The *fourth step* is the development of yield isoquants. The above multiple regression equation is to be used to calculate W and N combinations for different yield isoquants. Thus, the same yield could be obtained by different combinations of W and N as one substitute for the other (Fig. 2).

The *fifth step* is the computation of the marginal productivity of nutrient. Derivatives of the above multiple regression equation with respect to N at various levels of W are to be used to compute the marginal productivity of N as a function of N rate at various levels of W (Fig. 3).

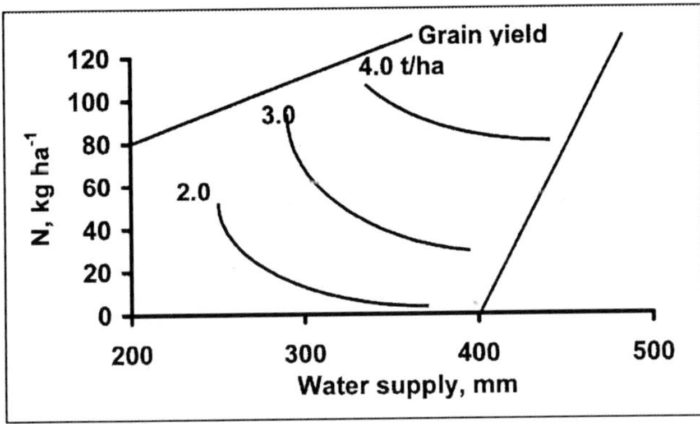

Fig. 2. Yield isoquants for different combinations of water supply and N rates (Gajri *et al.*, 1993)

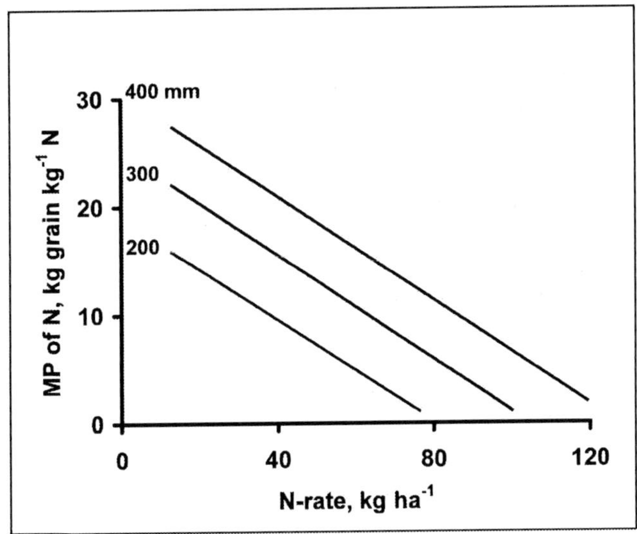

Fig. 3. Marginal productivity (MP) of N in relation to the levels of the N rates and water supply

Water management to enhance NUE

As water is becoming a limiting factor for crop production, soil and water management should be the key to the development of sustainable agriculture for both irrigated as well as rainfed areas. Water management in agriculture is defined as those methods, systems and techniques of water conservation, remediation, application, use and removal that provide a socially, and environmentally favourable level of water regime to agricultural production system at least economic cost (Singh, 1997). Because it is essential for most plant functions, the amount of water applied during irrigation, the time and method of water application, the quality of the irrigation water, and prevailing micrometeorological conditions are important in plant health and yield (Ati *et al.*, 2012). Water management influences the transformation of N sources applied to the soil and transport of the nitrate form of N in the soil. Nitrate-N is the final product of N transformations and is quite mobile in soils with the water front (Alva *et al.*, 2008). Leaching of nitrate below the root zone is an economic loss and contributes to non-point source pollution of groundwater (Alva *et al.*, 2008). An estimate shows that water management together with seeds and fertilizers (Table 3) increase crop yields by about 130% (Srinivasarao, 2011).

Table 3: Share of different agricultural inputs in increasing the yield over traditional systems in dryland agriculture in India

Practice	Increase in yield over traditional system (%)
Water management	14
Seed	40
Fertilizer	50
Seed + Fertilizer	95
Seed + Fertilizer + Water management	130

Source: Srinivasarao (2011)

Substantial variations in soil properties and nutrient and water availability exist across most fields. Thus, the ability to apply site-specific nutrient and irrigation management to match spatially and temporally variable conditions can increase application efficiencies, reduce environmental impacts, while improving yields. Precision farming technologies have now been developed to spatially vary nutrients and water prescriptions within a field based on various information sources viz. soil properties maps, terrain attributes, remote sensing, yield maps etc.

Water management in rice and wheat fields influences the extent of N losses due to the processes of nitrification, denitrification and ammonia volatilization. Upto 50% of the applied N can be lost through denitrification when alternating aerobic-anaerobic conditions prevail in the rice fields. A vast majority of soils, especially in the Indo-Gangetic Plains (IGP) are relatively coarse-textured and experience frequent alternating wetting and drying cycles. Previous research reports (Singh and Singh, 2008) indicate that the water management for rice has the implication for gaseous loss of nitrogen from the fields. Hence, water management is a key factor for N losses from the rice soils.

Irrigated rice accounts for major share of global rice production and global nitrogen consumption. The low agronomic N use efficiency (AEN, kg grain yield increase per kg N applied) of this system has become a threat to the environment (Peng *et al.*, 2006). They have found that real-time N management

(RTNM) and fixed-time adjustable-dose N management (FTNM) are the important strategies for irrigated rice-N management. A quantum leap in AEN is possible in the intensive rice-growing areas by simply reducing the current N rate and by allocating less N at the early vegetative stage.

Water management involving proper irrigation scheduling in irrigated areas and moisture conservation for rainfed agriculture is highly correlated with NUE (Prasad, 2009). In wheat, irrigation at 1.0 IW/CPE increased grain yield by 0.41 $Mg \cdot ha^{-1}$, water use efficiency by 7.5 $kg \cdot ha^{-1} \cdot cm^{-1}$ and PFPnpk by 1.7 kg grain·kg NPK^{-1} over irrigation at 0.8 IW/CPE (Verma and Singh, 2008). Similarly, construction of water harvesting structures such as tanks with open dug-wells in the upper reaches and shallow ditches in the lower reaches of a drainage line to re-harvest the seepage water from the tanks for irrigation are potential mechanisms in rainfed agriculture to increase grain yield as well as PFPnpk (Pali *et al.*, 2007). FNUE and WUE of squash increased with N fertigation compared to soil application.

Micro-irrigation and fertigation for improving NUE

Water for irrigation is becoming both scarce and expensive and necessitates to be utilized in a scientific manner. Drip irrigation has proved to be a success in terms of water saving and increased yield in a wide range of crops (Bhardwaj, 2001). With expanding area under micro irrigation in India, fertigation has great scope in fertilizer saving and increased crop yield. Studies revealed that fertilizer use efficiency up to 95% can be achieved through drip fertigation, with significant fertilizer savings of 20–60% and increase in yields of horticulture and vegetable crops by 8–41% due to fertigation (Singh *et al.*, 2010).

In Asia, yields from most crops have increased 100–400% after irrigation (FAO, 1996). Irrigation allows farmers to apply water at the most beneficial times for the crop, instead of being subject to the erratic timing of rainfall. Based on that, water resources are limited for irrigation; therefore there is a need for water-saving irrigation practices to be explored. In the process, the loss of nutrients through excess irrigation is reduced. Water saving may be achieved with drip irrigation, and even improved results seem to be possible (Shahnazari *et al.*, 2007) with partial root-zone drying, which is a new water-saving irrigation strategy. Application of urea and irrigation sequence is essential for N management (Katyal *et al.*, 1987). It is possible to increase production levels for example for a crop like potato, by well-scheduled irrigation programs throughout the growing period (Panigrahi *et al.*, 2001; Ferreira and Carr, 2002; Kashyap and Panda, 2003).

Tayel *et al.* (2006) suggested for irrigation methods as controlled surface irrigation (CSI), surface drip (SDI) and sub-surface drip (SSDI) for crops like potato, onion and peas in Cairo, Egypt for enhancing both fertilizer and water use efficiency. Tollenaar and Lee (2002) contended that yield improvement would be accompanied with an increase in both total water and fertilizer use efficiency. Waddel *et al.* (1999) found that total N uptake by potato plant was nearly 100% of that applied if the method of irrigation was sprinkler and drip. Bar-Bar-Yousef *et al.* (1989) indicated that sweet corn response to P applied via drip fertigation in calcareous soil was greater than P- application with broadcasting. Singh and Sood (1994) revealed that tuber yield of potatoes increased with increasing N-application rates under furrow, drip and sprinkler methods. On the other hand, Nimah *et al.* (2000) found no significant difference in potato yield between conventional and fertigation methods. Increasing NUE is unlikely unless a system approach is implemented that uses varieties with high harvest index, incorporated NH_4-N fertilizer, application of prescribed rate consistent with in-field variability using sensor-based systems with in production fields, low N rate applied at flowering and forage production systems (Raun and Johnson, 1999). Water limited through interval based irrigation saves water and fertilizer N at Iran (Ashouri, 2012).

Sankarnarayanan *et al.* (2011) reported that highest nutrient use efficiency (NUE) of 15 kg of seed cotton per kg of nutrients and economic nutrient use efficiency (ENUE) of 1.08 kg of seed cotton per rupee invested for nutrients were calculated with existing drip system. This was closely followed by low cost drip system of poly tubes (14.6 kg per kg of nutrients (NUE) and 1.05 kg per rupe invested for nutrients (ENUE) and microtube drip system (14.1 kg per kg of nutrients (NUE) and 1.01 kg per rupee invested for nutrients (ENUE).

In addition to proper nutrient management, other aspects of soil and crop management including the use of high yielding, nutrient-efficient cultivars, correcting soil physical and chemical problems and water management, disease and pest management (IPM), and post-harvest care and safe storage are important to achieve high fertilizer use efficiency (Prasad, 2009). Improving nutrient use efficiency (NUE) may be addressed by implementation of best management practices (BMPs) such as nutrient management planning, proper fertilizer material selection, better application timing and placement, and improved irrigation scheduling (Obreza and Sartain, 2010). Emerging technology that will aid in this effort includes increased use of enhanced efficiency fertilizers, organic soil amendments, fertigation, and foliar fertilization.

Land surface modification for managing water and reducing nutrient losses

Land levelling is the one of the most important technique to reduce water and nutrient losses. A well-leveled field is a prerequisite for good crop husbandry. When fields are not level, water may stagnate in depressions, whereas higher parts may become dry. The variability in land levelness within a field will have a major effect on crop management and crop yields. Uneven fields require more water to wet up the soil for land preparation and plant establishment. Uneven water coverage often results in uneven crop stands, weed problems, uneven ripening and uneven yields within each field. The average unevenness (i.e. the difference in height between the highest and lowest portions of the field) in Asian rice fields averages 160 mm with the range from 70 mm to 330 mm. This means that an extra 80 mm to 100 mm of water, or up to 10% of the total water requirement to grow the crop, must be stored in the field to attain complete water coverage. Crop yields are also improved by leveling the fields. Each 10 mm of variation in the surface topography reduces grain yield by approximately 280 kg ha^{-1}.

The water that continuously flows through rice fields may remove valuable (fertilizer) nutrients. This can be reduced by *field channel management.* Constructing separate channels to convey water to and from each field greatly improves the individual control of water and is the recommended practice in any type of irrigation or water management system. *Good bunds* are a prerequisite to limit seepage and underbund flows. To limit seepage losses, bunds should be well compacted and any cracks or rat holes should be plastered with mud at the beginning of the crop season. Bund heights should be high enough (at least 20 cm) to avoid overbund flow during heavy rainfall. Research reports (Bouman *et al.*, 2005) demonstrated a reduction of 450 mm in total water use in a rice field by lining the bunds with plastic. Of course, such measures are probably financially not attractive to farmers.

A number of land configurations have been tried for on-farm water management since many several decades. The Broad-bed and Furrow (BBF) land configuration has been found effective on deep Vertisols, especially under conditions of high and assured rainfall by facilitating earlier and more effective crop cover, reducing runoff and soil loss and improving yield. The system consists of a relatively flat bed or ridge and a shallow furrow. The grade under which the BBF system is normally laid out is 0.4 to 0.8%. At that grade, furrows of 100 m or less in length have performed satisfactorily with regard to discharging excess water with minimal soil erosion and the provision of adequate surface drainage even during long rainy spells. The width of one unit of the BBF is 1.5 m including furrow and the bed; of which the width of the bed is 90 cm to 1 m.

The furrow width and depth may be 25-30 cm and 12-15 cm, respectively. The BBF is generally made by bullock-drawn tropicultor or it is made by the tractor drawn BBF maker made of iron. This system provides great flexibility to fit crops either intercrops or sequence crops with widely differing row-spacing requirements. The spacing of the crop or the number of rows of crop in a bed varies with the type of crop. Precision placement of fertilizer is facilitated even when seed and fertilizer are applied separately because the wheels of the tropicultor follow the furrows. Soil compaction due to animals and equipment wheels is limited to the furrows. The results of on-station watershed at IISS, Bhopal revealed that soybean yield under BBF was greater than that under Flat on Grade (FOG) land configuration for sole soybean, soybean/maize intercropping and soybean/pigeon pea intercropping systems. On an average, BBF registered 12.7-18.0% greater yield of soybean than FOG under sole soybean (Mandal *et al.*, 2013).

Water management for rice systems and nutrient use efficiency

The system of rice intensification (SRI) is an integrated crop management technology developed by the Jesuit priest`Father Henri de Laulanie in Madagascar. SRI is characterized by the practices: transplanting 8-12 day old seedlings very carefully, transplanting single seedlings, spacing the plants widely apart in a square pattern, controlling weeds by weeding with a rotating hoe, which aerates the soil, applying compost to increase the soil organic matter, no continuous flooding during the crop growth period, applying small amounts of water regularly or alternating wet and dry (AWD) field conditions to maintain a mix of aerobic and anaerobic soil conditions. After flowering, a thin layer of water should be kept on the field, although some farmers find alternate wetting and drying of fields throughout the crop cycle to be feasible and even beneficial.

The NUE by rice crop with SRI method of rice cultivation showed that N-uptake by rice plants differed significantly according to cultivation practices and N application rates. N uptake across the four N rates was significantly higher under SRI management by 51.8 %. Agronomic N use efficiency (ANUE) was 31.3 to 44.3 kg grain kg^{-1}N under SRI, and from 23.3 to 31.6 kg grain kg^{-1}N under Transplanted Flooded Rice (TFR). ANUE with SRI was 34–40 % higher than in TFR plants. Highest ANUE was found with the application rate of 90 kgN ha^{-1} under both SRI and TFR. Partial factor productivity (PFP) from applied nitrogen was estimated to be 49% higher with SRI management when compared to TFR. Under SRI, 64 kg grain was produced with the application of 1 kg N whereas with TFR, only 43 kg grain resulted per kg N applied (Thakur *et al.*, 2013).

In saturated soil culture (SSC), the soil is kept as close to saturation as possible, thereby reducing the hydraulic head of the ponded water, which decreases the seepage and percolation. SSC in practice means that a shallow irrigation is given to obtain about 1 cm of ponded water a day or so after the disappearance of ponded water. Research results showed a water savings by 23% under SSC in transplanted and direct wet-seeded rice in puddled soil, and in direct dry-seeded rice in non-puddled soil. Mandal *et al.* (2014) evaluated rice systems irrigation and organic carbon storage potential. Raised beds can be an effective way to keep the soil around saturation. Rice plants are grown on beds and the water in the furrows is kept close to the surface of the beds. Raised beds may be 120 cm wide and separated by furrows of 30 cm width and 15 cm depth to facilitate SSC practices. Compared to flooded rice, water savings may to the tune of 20-30%.

By alternate wetting and drying irrigation water is applied to obtain flooded conditions after a certain number of days have passed after the disappearance of ponded water in this alternate wetting and drying (AWD). The number of days of nonflooded soil in AWD before irrigation is applied can vary from 1 day to more than 10 days. AWD increases water productivity with respect to total water input because the reduction in water inputs are larger than the reductions in yield. The total (irrigation and rainfall) water inputs decreases by around 15-30% without a significant impact on yield. AWD is a technology that has been widely adopted in China. It reduces seepage and percolation flows and has only a small effect on evaporation. The potential benefits of AWD are: improved root system, reduced lodging (because of a better root system), periodic soil aeration.

Aerobic method of rice culture- a fundamentally different approach to reduce water outflows from rice fields is to grow the crop like an upland crop. Unlike lowland rice, upland crops are grown in nonpuddled, nonsaturated (i.e., aerobic) soil without ponded water. When rainfall is insufficient, irrigation is applied to bring the soil water content in the root zone up to field capacity after it has reached a certain lower threshold level, such as halfway between field capacity and wilting point. The amount of irrigation water should match evaporation from the soil and transpiration by the crop. The potential water reductions at the field level when rice can be grown as an upland crop are large, especially on soils with high seepage and percolation rates. Besides a decline in seepage and percolation losses, evaporation decreases since there is no ponded water layer, and the large amount of water used for wet land preparation is eliminated altogether. Studies by Mandal *et al.* (2013b) on irrigation x N interaction revealed that a highest grain yield of 4.4 t ha^{-1} was obtained with N rate of 120 kg ha^{-1} receiving 780 mm irrigation for rice variety 'Surendra'. The next best combination viz. N rate of 80 kg ha^{-1} with 780 mm irrigation (3.84 t ha^{-1}) and N rate of 120 kg ha^{-1} and 660 mm irrigation (3.61 t ha^{-1}) were statistically similar. Hence, based on availability of irrigation water, N rate needs to be decided. The study on varietal (*viz.* 'Apo', 'Lalat' and 'Surendra') response to N rates showed that, irrespective of variety, aerobic rice with 120 N kg ha^{-1} with 780 mm irrigation gave the highest grain and straw yield of 4.24 and 6.63 t ha^{-1} with grain and straw N-uptake of 52.17 and 52.63 kg ha^{-1}, respectively .

Integrated nutrient-water management

Nutrient application to crops increases water-use efficiency by increasing the ET particularly the transpiration (T) component of ET and transpiration efficiency of the crop. Thus the T/ET ratio, which is conducive to higher WUE, is higher for the fertilized than for the unfertilised crop. The field experiments on wheat showed that, judicious application of limited irrigation water and enhancing water use efficiency (WUE) through integrated nutrient management could boost the productivity of wheat and enhancing WUE of the crop (Mandal *et al.*, 2005). Water and nutrients, and their positive interaction favourably influenced the yield, ET and WUE of wheat. In central Indian vertisols, as grain yields in three post-sowing irrigations with 100% NPK were similar to two post-sowing irrigations with 100% NPK + FYM, one pre-sowing irrigation plus two post-sowing irrigation in combination with 100% NPK and 10 t ha^{-1} farmyard manure could be a viable option in the event of limited irrigation water. For Indian mustard in central Indian vertisols, it was found that the ET-yield relationships were linear, with a lowest regression slope or marginal WUE (WUE_m) of 3.09 kg ha^{-1} mm^{-1} and elasticity of water production (E_{wp}) of 0.63 in control and considerably higher WUE_m (4.23 and 3.95 kg ha^{-1} mm^{-1}) and E_{wp} (0.71 and 0.61) in 100% NPK and 100% NPK + FYM. As the E_{wp} is positive and comparatively greater in 100% NPK, the scope of improving WUE and yield with only inorganic fertilizer is lesser, and relatively greater scope exists in the integrated management of organic manure and inorganic fertilizer, for which the E_{wp} was lesser (Mandal *et al.*, 2010). The seed yield under three irrigation with 100% NPK and two irrigation with 100% NPK+ FYM were statistically similar. This indicates that application of organic manure along with NPK fertilizers could reduce the need for one post-sowing irrigation without compromising the yield of this crop, particularly under the limited water availability situation.

Irrigation management extension for stakeholders

Over-irrigation results in leaching of NO_3 to the groundwater and reduces the efficiency of N fertilizers. Therefore, irrigation water management is essential for profitable yields and protecting water quality. Hence, scheduling irrigation guidelines may be transferred to farmers as:

- Obtain information about crop's water needs and critical growth stages, soil characteristics and irrigation system efficiency to properly schedule irrigations.
- Understand how much water the crop uses on a daily or weekly basis. This may be the ET estimated from weather data. ET rates are available.
- Determine the soil moisture content in the effective root zone and its maximum water-holding capacity by measurement or the feel method.
- The difference between the maximum water-holding capacity and the actual water content is the net amount of water to be applied.
- Determine the application efficiency of irrigation systems. Consult a qualified irrigation technician to assess irrigation system performance.
- If feasible, use irrigation systems that give higher application efficiencies.
- The gross amount of water to be applied is the net amount divided by the application efficiency of the irrigation system.
- Use measuring devices such as flumes and water meters to determine how much water is applied. When using siphon tubes or gated pipes, multiply the stream flow rate by the irrigation duration.
- Use a soil probe to monitor soil moisture. Probe the field during and after irrigation to determine depth of water penetration.
- With surface irrigation, use cutback practices to reduce deep percolation and runoff.
- Operate sprinklers at proper pressure and nozzle spacing.

Thus, it is concluded that nutrient supply should be synchronized with the soil moisture/ irrigation water availability for better root development, soil moisture extraction, water relation in plants, crop growth and nutrient uptake and enhancing efficiency of applied nutrients. A proper method of irrigation, precision irrigation techniques would enhance the saving of irrigation water as well reduce the loss of nutrients from the soil systems and in turn enhance the nutrient use efficiency.

Selected References

Alva AK, Paramasivam S, Fares A, Delgado JA, Mattos D. Jr, Sajwan K. 2008. Nitrogen and irrigation management practices to improve nitrogen uptake efficiency and minimize leaching losses. *Journal of Crop Improvement* 15 (2): 369-420.

Ashouri M. 2012. The effect of water saving irrigation and nitrogen fertilizer on rice production in paddy fields of Iran. *International Journal of Bioscience, Biochemistry and Bioinformatics* 2 (1): 56-59.

Ati AS, Iyada AD, Najim S.M. 2012. Water use efficiency of potato (*Solanum tuberosum* L.) under different irrigation methods and potassium fertilizer rates. *Annals of Agricultural Science* 57(2): 99–103.

Aulakh MS, Singh B. 1997. Nitrogen losses and fertilizer N use efficiency in irrigated porous soils. *Nutrient Cycling in Agroecosystems* 47 (3): 197-212.

Bar-Yousef B, Sagiv B, Markoitch T. 1989. Sweet corn response to surface and subsurface trickle phosphorus fertigation. *Agronomy Journal* 81: 443-447.

Bhardwaj SK. 2001. Importance of drip irrigation in Indian agriculture. *Kissan World,* 28: 32-33.

Bouman BAM, Peng S, Castaneda AR, Visperas RM. 2005. Yield and water use of irrigated tropical aerobic rice systems. *Agricultural Water Management* 74:87-105.

Cassman KG, Dobermann A, Walters DT. 2002. Agroecosystems, nitrogen use efficiency, and nitrogen management. *Ambio.* 31: 132-140.

Dobermann A, Cassman KG, Waters DT, Witt C. 2005. Balancing short- and long-term goals in nutrient management. In: Proceedings of the XV International Plant Nutrient Colloquium, Sep. 14-16, 2005. Beijing, China.

FAO 1996. Agriculture and Food Security. World Food Summit, November (1996). Food and Agricultural Organization of the United Nations, Rome, pp. 26–32.

Ferreira T, Carr M. 2002. Response of potatoes (*Solanum tuberosum* L.) to irrigation and nitrogen in a hot, dry climate: I. Water use. *Field Crops Research* 78: 51–64.

Fixen PE. 2005. Understanding and improving nutrient use efficiency as an application of information technology. In: Proceedings of the Symposium on Information Technology in Soil Fertility and Fertilizer Management, a satellite symposium at the XV International Plant Nutrient Colloquium, Sep. 14-16, 2005. Beijing, China.

Gajri PR, Prihar SS, Arora VK. 1993. Interdependence of nitrogen and irrigation effects on growth and input-use efficiencies of wheat. *Field Crops Research,* 31: 71-86.

Hati KM, Mandal KG, Misra AK, Ghosh PK, Acharya C.L. 2001. Effect of irrigation regimes and nutrient management on soil water dynamics, moisture extraction pattern, evapotranspiration and yield of wheat (*Triticum aestivum*) in Vertisol. *Indian Journal of Agricultural Science,* 71 (9): 505-509.

Kanwar JS, Rego TJ, Seetharama N. 1984. Fertilizer and water use in pearlmillet and sorghum in vertisols and alfisols of semi-arid India. *Fertiiliser News* 29 (4): 42-52.

Kashyap P, Panda R. 2003. Effect of irrigation scheduling on potato crop parameters under water stressed conditions. *Agricultural Water Management* 59: 49–66.

Katyal JC, Singh B, Vlek PLG, Buresh RJ. 1987. Efficient nitrogen use as affected by urea application and irrigation sequence. *Soil Science Society of American Journal,* 51: 366-370.

Ladha JK, Pathak H, Krupnik TJ, Six J, van Kessel C. 2005. Efficiency of fertilizer nitrogen in cereal production: Retrospects and prospects. *Advances in Agronomy,* 87: 85-156.

Mandal KG, Hati KM, Misra AK, Bandyopadhyay KK. 2010. Root biomass, crop response and water-yield relationship of mustard (*Brassica juncea* L.) grown under combinations of irrigation and nutrient application. *Irrigation Science* 28: 271-280.

Mandal KG, Hati KM, Misra AK, Bandyopadhyay KK, Mohanty M. 2005. Irrigation and nutrient effects on growth and water-yield relationship of wheat (*Triticum aestivum* L.) in central India. *Journal of Agronomy & Crop Science,*191 (6): 416-425.

Mandal KG, Kundu DK, Thakur AK, Kannan K, Brahmanand PS, Kumar A. 2013b. Aerobic rice response to irrigation regimes and fertilizer nitrogen rates. *Journal of Food, Agriculture and Environment* 11 (3&4): 1148-1153.

Mandal KG, Hati KM, Misra AK, Bandyopadhyay KK, Tripathi AK. 2013. Land surface modification and crop diversification for enhancing productivity of a Vertisol. *International Journal of Plant Production* 7 (3): 455-472

Mandal KG, Hati KM, Misra AK, Ghosh PK, Bandyopadhyay KK. 2003. Root density and water use efficiency of wheat as effected by irrigation and nutrient management. *Journal of Agricultural Physics,* 3 (1&2): 49-55.

Mandal KG, Kannan K, Thakur AK, Kundu DK, Bramhanand PS, Kumar A. 2014. Performance of rice systems, irrigation and organic carbon storage. *Cereal Research Communications* 42 (2): 346-358

Mohammad MJ. 2004. Utilization of applied fertilizer nitrogen and irrigation water by drip-fertigated squash as determined by nuclear and traditional techniques. *Nutrient Cycling in Agroecosystems* 68: 1-11.

Mosier AR, Syers JK, Freney JR. 2004. Agriculture and the Nitrogen Cycle. Assessing the Impacts of Fertilizer Use on Food Production and the Environment. Scope-65. Island Press, London.

NAAS. 2006. Low and declining crop response to fertilizers. Policy Paper No. 35, National Academy of Agricultural Sciences, New Delhi. pp. 8.

Nimah MN, Darish LI, Bashour. 2000. Potato yield response to deficit irrigation and N fertilization. *Acta Horticulturae,* 537: 823-830.

Obreza TA, Sartain JB. 2010. Improving nitrogen and phosphorus fertilizer use efficiency for florida's horticultural crops. *Horticultural Technology,* 20 (1): 23-33.

Pali G, Rajpoot RS, Sinha BL. 2007. Rainwater management strategies for drought alleviation – a success story. *Indian Farming,* 57 (8): 21–24.

Panigrahi B, Panda S, Raghuwanshi NS.2001. Potato water use and yield under furrow irrigation. *Irrigation Science,* 20: 155–163.

Peng S, Buresh RJ, Huang J, Yang J, Zou Y, Zhong X, Wang G,Zhang F. 2006. Strategies for overcoming low agronomic nitrogen use efficiency in irrigated rice systems in China. *Field Crops Research*, 96 (1): 37-47.

Prasad R. 2009. Efficient fertilizer use: The key to food security and better environment. *Journal of Tropical Agriculture* 47 (1-2) : 1-17

Raun WR, Johnson GV.1999. Improving nitrogen use efficiency for cereal production. *Agronomy Journal,* 91: 357-363.

Roberts TL. 2008. Improving nutrient use efficiency. *Turkish Journal of Agriculture and Forestry,* 32: 177-182.

Sankarnarayanan K, Nalayini P, Sabesh M, Usha Rani S, Nachane RP, Gopalkrishnan S.2011. Low cost drip- cost effective and precision irrigation tool in Bt cotton. Tech Bulletin 1/ 2011. NAIP Pub., Regional Station, CICR, Coimbatore.

Shahnazari A, Liu F Andersen MN, Jacobsen SE, Jensen CR.2007. Effects of partial root-zone drying on yield, tuber size and water use efficiency in potato under field conditions. *Field Crops Research,* 100: 117–124.

Singh N, Sood MC. 1994. Water and nitrogen needs of potato under modern irrigation methods. National Symposium held at Modipuram, pp. 142-146.

Singh RM, Singh DK., Rao KVR. 2010. Fertigation for increased crop yield and fertilizer saving. *Agricultural Engineering Today*, 34 (2): 12-16.

Singh SR. 1997. Some alternative-strategies for managing water resources to enhance agricultural production. Publication no. 4, WTCER, Bhubaneshwar.

Singh V, Singh Y. 2008. Reactive nitrogen in Indian agriculture: inputs, use efficiency and leakages. *Current Science* 94 (11): 1382-1393.

Srinivasarao Ch. 2011. Nutrient management strategies in rainfed agriculture: constraints and opportunities. *Indian Journal of Fertilizers*7(4): 12-28.

Srinivasarao Ch., Venkateswarlu B, Hedge DM, Venkateswara Rao MK, Kundu S. 2013. Use of organic fertilizers alone or in combination with inorganic ones: effects on water and nutrient use efficiency in Indian farming systems. In: Zed Rengel (Ed.) Improving Water and Nutrient- Use Efficiency in Food Production Systems, First Edition, John Wiley & Sons, Inc.

Tayel MY, El-dardiry, Ebtisam I, Abd El-Hady M. 2006. Water and fertilizer use efficiency as affected by irrigation methods. *Journal of Agriculture and Environmental Science* 1 (3): 294-301.

Thakur AK, Rath S, Mandal KG. 2013. Differential responses of system of rice intensification (SRI) and conventional flooded-rice management methods to applications of nitrogen fertilizer. *Plant and Soil* 370: 59-71

Tollenaar M, Lee EA. 2002. Yield potential, yield stability and stress tolerance in maize. *Field Crops Research* 75: 161-169.

Verma SK, Singh SB. 2008. Enhancing wheat production through appropriate agronomic management. *Indian Farming* 58 (5): 15-18.

Vijaya Kumar P, Rao SVRB, Sharma KL. 1998. A study on yield response of wheat to water and applied nitrogen using soil water balance model. *Indian Journal of Dryland Agriculture Research and Development* 13 (2): 97-102.

Waddel J, Gupta SC, Moncrief JF, Rosen CJ, Steel DD. 1999. Irrigation and nitrogen management effects on potato yield, tuber quality and nitrogen uptake. *Agronomy Journal* 91: 991-997.

14

Enhancing Nutrient Use Efficiency, pp. 179-192
Editors: K. Ramesh, A.K. Biswas, B.L. Lakaria, S. Srivastava and A.K. Patra

Enhancing Nutrient Use Efficiency through Conservation Agriculture

J. Somasundaram, K.M. Hati, R.S. Chaudhary, K. Ramesh, N.K.Sinha A.K. Biswas, A.K. Shukla and A.K. Patra

ICAR-Indian Institute of Soil Science, Nabi Bagh, Bhopal – 462038, India

Introduction

Soil resource is as important as water for agriculture as it supplies nutrients to the plants and anchorage for stand and establishment. However, people least bother about its importance despite its role in food security and ecosystem services. Good quality soil is a prerequisite for sustainable crop production. The concept of conservation agriculture (CA) has emerged in the recent past wherein conventional practices are modified to reduce farming cost in addition to sustaining soil health. Improvement in the soil health is achieved by increasing soil organic carbon (SOC), aggregation, infiltration, and also reduction in soil erosion. The key features include: (i) minimum soil disturbance by adopting no tillage (NT) and minimum traffic for agricultural operations, (ii) leave and manage the crop residues on the soil surface, (iii) adopt spatial and temporal crop sequences / crop rotations to derive maximum benefits from inputs and minimize adverse environmental impacts (Abrol and Sanger, 2006; FAO, 2008; Friedrich *et al.*, 2012; Somasundaram *et al.*, 2014a) and also sustaining agriculture in the long run.

In the conventional systems involving intensive tillage, there is a gradual decline in soil organic matter (SOM) through faster oxidation and burning of crop residues causing pollution and greenhouse gas (GHG) emission. Decline in soil fertility and biodiversity due to intensive seedbed preparation using heavy machinery further aggravates soil erosion and land degradation. When the soil surface is covered with crop residues in combination with NT, it initiates processes that lead to improved soil quality. Therefore, CA practices could lead to an overall improvement in the water use and nutrient use through their cumulative action on nutrient balance and availability. Further, an enhancement in infiltration due to residue cover and reduction in water losses due to evaporation are also achieved.

Nutrient management strategy for CA

Nutrient management strategies for CA systems would need to attend to the following four general aspects, namely (i) Strengthening the soil biological processes: to protect soil biota *vis-à-vis* SOM and soil porosity; (ii) Strengthening biomass production and biological nitrogen fixation: to keep soil energy and nutrient stocks sufficient to support higher levels of biological activity, and for covering the soil; (iii) Strengthening nutrient access to plant roots from all sources: to meet crop needs; and (iv) Maintaining soil reaction at neutral/acceptable range: to facilitate proper functioning of key soil chemical and biological processes.

Need for comprehensive nutrient management approach under CA

In the recent past, CA has emerged as a major "breakthrough" systems approach to crop and agriculture production with its change in paradigm that challenges the status quo. However, as a multi-principled concept, CA translates into knowledge-intensive practices whose exact form and adoption requires that farmers become intellectually engaged in the testing, learning and fine-tuning possible practices to meet their specific ecological and socio-economic conditions (Friedrich and Kassam, 2009). In essence, CA approach represents a highly biologically and bio-geophysically-integrated system of soil health and nutrient management for production that generates a high level of "internal" ecosystem services which reduces the levels of "external" subsidies and inputs needed. CA provides the means to work with natural ecological processes to harness greater biological productivities by combining the potentials of the endogenous biological processes with those of exogenous inputs. The evidence for the universal applicability of CA principles is now available across a range of ecologies and socio-economic situations covering large and small farm sizes including resource poor farmers (Goddard *et al.*, 2007; FAO 2008).

CA has been and is being introduced in a multitude of ecological and socio-economic situations. However each ecology imposes one or more constraints as to how fast the conversion towards CA systems can occur. In the seasonally dry tropical and sub-tropical ecologies, particularly with resource poor small farmers in drought prone zones, CA systems will take longer time to establish, and step-wise approaches to the introduction seem to show promise (Mazvimavi and Twomlow, 2006). These involve two components: the application of planting 'Zai-type' basins which concentrate limited nutrients and water resources to the plant, and the precision application of small or micro doses of nitrogen-based fertilizer. In the case of degraded land in wet or dry ecologies, special soil amendments and nutrient management practices are required to establish the initial conditions for soil health improvement and efficient nutrient management for agricultural production (Landers, 2007). What seems to be important is that whichever pathway is followed to introduce CA, there is a need for a clear understanding of how the production systems concerned should operate to sustain soil health and productivity, and how nutrient management interventions that may be proposed can contribute to the system effectiveness as a whole both in the short- and long-term.

Towards CA-based nutrient management practices

Although Integrated Soil Fertility Management (ISFM) and Integrated Natural Resources Management (INRM) approaches are in vogue in the recent years, focus is only on crop nutrient demand rather than soil health. In addition, these approaches are meant for tillage-based systems, regardless of farm size/holdings. Unless the concepts of soil health and function are explicitly incorporated in these approaches, sustainability goals and means will be weakly linked for resource poor farmers. It is often believed that CA systems have inbulit ISFM or INRM modules to optimize the use of organic plus inorganic inputs with various resources, to enhance and the system productivity in an eco-friendly responsible manner. There are empirical evidences to prove this analogy.

Focusing on soil fertility, due recognition to the tillage and cropping system, as often proposed by ISFM or INRM approaches, is an interim approach for maintaining soil health in support of sustainable production. Over the past two decades or so, evidences from the field has clearly shown that healthy agricultural soils constitute biologically active soil systems within landscapes in which both the soil resources and the landscape must operate with plants in an integrated manner to support the various desired goods and services (*e.g.*, food, feed, feedstock, biological raw material for industry, livelihood, environmental services, etc) provided by agricultural land use. Consequently, successful nutrient management strategies as a part of any ISFM or INRM approach must pay close attention to the issues of soil health management which means managing the microscopic integrity of the soil-plant system particularly as mediated by soil living biota, organic matter, physico-chemical properties, available nutrients, adapted germplasm as well as to managing the macroscopic dimensions of landscapes, socioeconomics and policy. Moreover, CA-based IFSM and INRM approaches to nutrient management and production intensification would be more effective for farmer-based innovation systems and learning processes such as those promoted through Farmer Field School networks.

CA and SOC

Long-term implementation of conservation tillage practices coupled with residue retention have been reported to increase SOM. Lower soil temperature contributes to slower rates of organic matter oxidation. A slight increase in SOM has been reported by many workers in the top 10 cm surface of soil. Once higher SOM levels are achieved, soil aggregates are stabilized resulting in good soil structure. Crop residues retained on the soil surface in CA (Fig 1), serve a number of beneficial functions, including soil surface protection from erosion, enhancing infiltration and cutting run-off rate, decreasing surface evaporation losses, moderating soil temperature and providing substrate for the activity of soil micro-organisms, and a source of SOC. This also provides organic matter necessary for soil macro-aggregate formation and fosters cellulose–decomposing fungi and thereby carbon cycling.

Fig 1. Residue retention under soybean-wheat system (left) and maize-gram system (right) (*See colour version on page 454*)

While conventional cultivation generally results in loss of SOC and nitrogen, CA has proven potential of converting many soils from sources to sinks of atmospheric C, sequestering carbon in soil as organic matter. In general, SOC sequestration during the first decade of adoption of best CA practices is 1.8 tonnes CO_2 ha^{-1} yr^{-1}. This translates to one-third of the current annual global emission of CO_2 from the burning of fossil fuels on five billion hectares of agricultural land (FAO, 2008). Lal *et al.* (1998) forecasted a widespread adoption of conservation tillage on some 400 million ha to aid in a total C sequestration of 1500 to 4900 Mg by the year 2020.

Currently, India produces a large amount of crop residues (500-550 million t yr^{-1}) that are used as animal feed, manure, thatching for rural homes and as fuel for domestic and industrial purposes. However, 30-40 % of these residues are burnt on-farm primarily to clear fields to facilitate timely planting/seeding of succeeding crops. Rice-wheat is an important cropping system followed on more than 10 million ha in the Indo-Gangetic Plains of the country. Crop residue burning under rice-wheat, soybean-wheat and maize-wheat production systems viewed as a quick, labour-saving practice to get rid of problem associated with residue retention (Prasad *et al.*, 1999), has several adverse environmental and ecological impacts in the form of CO_2 and particulate matter to the atmosphere. Further removal of soil surface cover increases the loss of mineral and organic matter–rich surface layer in run-off. If crop residues are retained, help increase in SOM levels, which facilitate better aggregation, higher infiltration rate led to higher water storage in the soil profile. Crop residues provide substrate to soil organisms which help in recycling of the plant nutrients. The annual production of crop residue is estimated to be about 3.4 billion Mg in the world. If 15% of C contained in the residue can be converted to passive SOC fraction, this may lead to C sequestration @ 0.2 x 10^{15} g yr^{-1} (Lal, 1997). Further, a reduction in tillage operation coupled with residue retention helps in maintaining the SOC (Subba Rao and Somasundaram, 2013; Somasundaram *et al.*, 2014b).

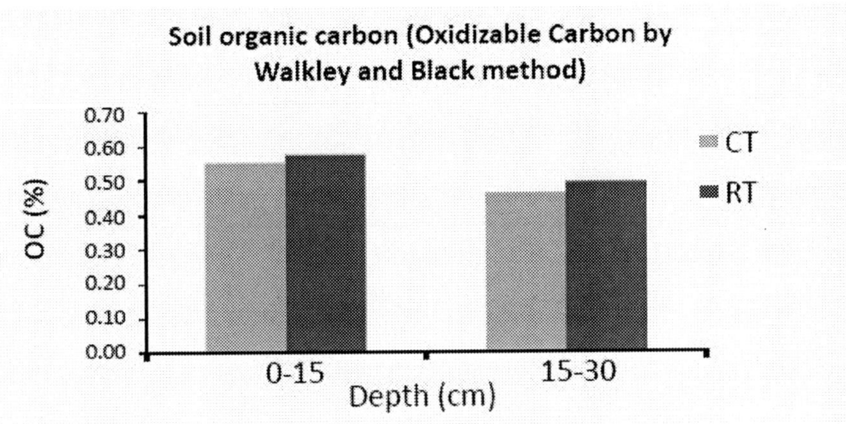

Fig 2. Effect of different tillage practices on SOC

CA on nutrient availability

Tillage, residue management and crop rotation have a significant impact on nutrient distribution and transformation in soils. Increased stratification of nutrients is generally observed, with enhanced conservation and availability (Franzluebbers and Hons, 1996). The altered nutrient availability under NT compared to conventional tillage may be due to surface placement of crop residues in comparison with incorporation of crop residues with tillage (Ismail *et al.*, 1994). Slower decomposition of surface placed residues (Kushwaha *et al.*, 2000) may prevent rapid leaching of nutrients through the soil profile, which is more likely when residues are incorporated into the soil. However, the possible development of more continuous pores between the surface and the subsurface under zero tillage may lead to more rapid passage of soluble nutrients deeper into the soil profile than when soil is tilled (Franzluebbers and Hons, 1996). Furthermore, the response of soil chemical fertility to tillage is site-specific and depends on soil type, cropping systems, climate, fertilizer application and management practices (Rahman *et al.*, 2008).

Hati *et al.* (2015), reported that SOC content at 0–15 cm depth was significantly higher in NT, reduced tillage (RT) and mouldboard tillage (MB) where wheat residues were retained after harvest than that in CT systems. The SOC, aggregate stability and saturated hydraulic conductivity were significantly higher in N150 % compared to N50 %. Similarly, Kushwa *et al.* (2016) reported from the same experiment that the highest SOC was observed in NT (8.8 g kg^{-1}) and the lowest under CT (5.9 g kg^{-1}) in 0-5 cm depth, whereas in 5-15 cm soil layer, higher SOC was observed in MB. The stratification ratio of SOC was higher in NT (2.20) followed by RT (1.93), MB (1.68) and CT (1.51). Higher available P concentration (12.8 g kg^{-1}) was recorded in NT with N50% followed by NT with N100%. They suggested that practising NT and RT systems along with residue retention and recommended rate of N would be a suitable practice for sustainable production of soybean–wheat cropping system in vertisols of central India.

In an another study, Kushwah *et al.* (2016) also reported that wheat residue either incorporated or retained on the soil surface increased the availability of P and SOC content as compared to residue burning. Residue retention or incorporation recorded higher stratification of P and soil organic carbon over burning (Fig 3). Irrespective of the nutrient treatments, greater stratification ratio of SOC and P were registered under wheat residue incorporation or retention compared to residue burning. Similarly, conservation tillage practices coupled with residue retention/incorporation have favoured the major (N, P and K) and micro-nutrients availability (Zn, Fe, Mn and Cu) in soils (Somasundaram *et al.*, 2017, unpublished data; Fig. 4).

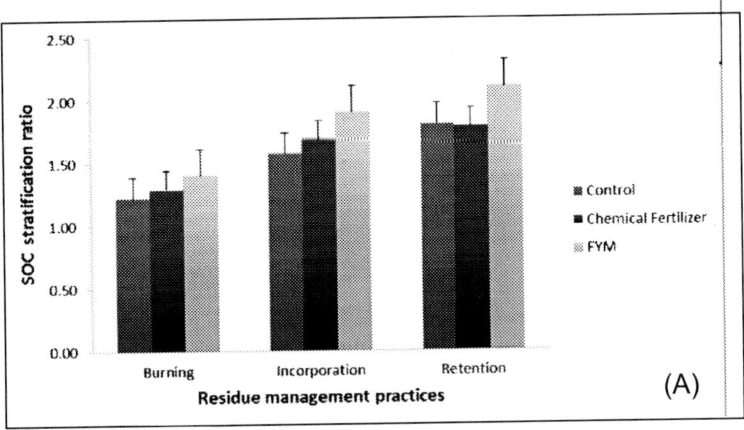

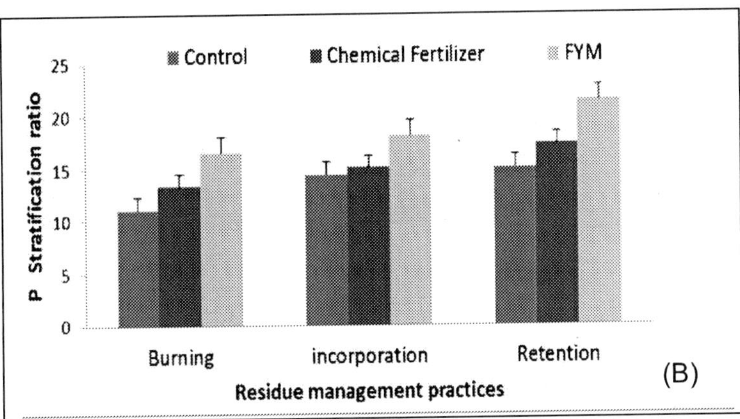

Fig 3. Effect of residue management practices (RM) and supplementary nutrient input (SNI) on SOC (A) and Phosphorus stratification (B) ratio at 0-5 cm depth (Error bar shows standard error). (*Source*: Adapted from Kushwah *et al*., 2016)

Fig 4. Crop stand under conventional tilled without residue (left) and RT with residue retained (right) under maize-gram. (*See colour version on page 454*)

Nitrogen availability

The presence of mineral soil N available for plant uptake is dependent on the rate of C mineralization. About five present higher NO-N losses with conventional tillage compared to NT has been reported by Randall and Iragavarapu (1995) as the latter, is associated with a lower N availability because of greater immobilization by the residues left on the soil surface (Bradford and Peterson, 2000), although, the impact of RT with residue retention on N mineralization is inconclusive. Greater N availability in the initial years of NT couldn't be detected in the brown soil zone in Canada (Jowkin and Schoenau, 1998), is considered as transitory by Follet and Schimel (1989), since in the long run, the higher, but temporary immobilization of N reduced the opportunity for leaching and denitrification losses of mineral N. Notwithstanding to this conclusion, Schoenau and Campbell (1996), opined that a greater immobilization of N can enhance the conservation of soil and fertilizer N in the long run, with higher initial N fertilizer requirements decreasing over time because of reduced losses by erosion and the build-up of a larger pool of readily mineralizable organic N. This was conformed by Larney *et al.* (1997) that N available for mineralization was greater in NT under continuous spring wheat after eight years since the N mineralization generally increased in the top 0-5 cm soil layer (Wienhold and Halvorson, 1999). Kushwaha *et al.* (2000) reported that incorporated crop residues decompose 1.5 times faster than surface placed residues. However, also the type of residues and the interactions with N management practices also determine C and N mineralization. Evidently, organic matter becomes more accessible to soil microorganisms due to aggregate disruption by tillage and increasing mineral N release from active and physically protected N pools (Six *et al.*, 2002). Govaerts *et al.* (2006) found after 26 cropping seasons in a high-yielding, high input irrigated production system that the N mineralization rate was higher in permanent raised beds with residue retention than in conventionally tilled raised beds with all residues incorporated, and also that N mineralization rate increased with increasing rate of inorganic N fertilizer application. Permanent raised beds, a more comfortable method under Vertisols with residue retention resulted in more stable macro aggregates and increased protection of C and N in the micro aggregates within the macro aggregates compared to conventionally tilled raised beds (Lichter *et al.*, 2008).

Phosphorus availability

Accumulation of P at the surface of continuous NT (Matowo *et al.*, 1999) is commonly observed, largely due to reduced mixing of the fertilizer P with the soil, leading to lower P-fixation has been reported. After 20 years of NT, extractable P was 42% greater at 0-5 cm, but 8-18% lower at 5-30 cm depth compared with conventional tillage in a silty loamy soil (Ismail *et al.*, 1994). When fertilizer P is applied on the soil surface, a part of P will be directly fixed by soil particles. When P is banded as a starter application below the soil surface, authors ascribed P stratification partly to recycled P by plants. This may either be a benefit when P is a limiting nutrient, or a threat when P is an environmental problem because of the possibility of soluble P losses in runoff water (Duiker and Beegle, 2006). Deeper placement of P under NT may be profitable if the surface soil dries out frequently during the growing season as suggested by Mackay *et al.* (1987). In that case, injected P may be more available to the crop. However, if mulch is present on the soil surface in NT, the surface soil is likely to be moist and deep placement may not be necessary, especially in humid areas. Kushwah *et al.* (2016) also reported that wheat residue either incorporated or retained on the soil surface increased the availability of P and SOC than residue burning. Residue retention or incorporation increased stratification of P and soil organic carbon over the residue burning.

Potassium availability

NT conserves and increases availability of nutrients, such as K, near the soil surface where crop roots proliferate (Franzluebbers and Hons, 1996). According to Govaerts *et al.* (2007b), permanent raised beds had a concentration of K 1.65 times and 1.43 times higher in the 0-5 cm and 5-20 cm layer, respectively, than conventionally tilled raised beds, both with crop residue retention. In both tillage systems, K accumulated in the 0-5 cm layer, but this was more accentuated in permanent than in conventionally tilled raised beds. Other studies have found higher extractable K levels at the soil surface as tillage intensity decreases (Lal *et al.*, 1990). Du Preez *et al.* (2001) observed increased levels of K in NT as compared to conventional tillage, but this effect declined with depth. Some authors have observed surface accumulation of available K irrespective of tillage practice (Duiker and Beegle, 2006). Follett and Peterson (1988) reported either higher or similar extractable K levels under NT as compared to mould board tillage, while Roldan *et al.* (2007) couldn't detect any effect on available K.

Micronutrients

Micronutrient cations (Zn, Fe, Cu and Mn) tend to be present in higher levels under NT with surface residue retention, especially extractable Zn and Mn near the soil surface (Franzluebbers and Hons, 1996). In contrast, Govaerts *et al.* (2007b) could not observe the effect of tillage practice on the concentration of extractable Fe and Cu, except Zn in the 0-5 cm layer of permanent raised beds with residue retention with a reduction in Mn. Similar results were reported by Du Preez *et al.* (2001) and Franzluebbers and Hons (1996). Mn concentrations were reported to increase with higher SOM contents Peng *et al.* (2008).

Conclusions

Conservation agricultural practices not only improve soil aggregation but also influence on nutrient availability over conventional tillage systems. The combination of RT with residue retention increases the SOC in the topsoil. Moreover, C-cycle is greatly influenced by CA, and N cycle too. Numerous studies have reported that CA increases availability of nutrients near the soil surface where crop roots proliferate. Adoption of CA systems with crop residue retention may result in initial N immobilization. However, rather than reducing N availability, CA may stimulate a gradual release of N in the long run and can reduce the susceptibility to leaching or denitrification, when no growing crop is able to take advantage of the nutrients at the time of their release. Also crop diversification, an important component of CA, has to be seen as an important strategy to govern N availability through rational sequences of crops with different C/N ratios. Tillage, residue management and crop rotation have a significant impact on micro- and macronutrient distribution and transformation in soils. The altered nutrient availability may be due to surface placement of crop residues in comparison with incorporation of crop residues with tillage. Slower decomposition of surface placed residues prevents rapid leaching of nutrients through the soil profile. However, the response of soil chemical fertility to tillage is site-specific and depends on soil type, SOC content, cropping systems, climate, fertilizer application and management practices..

Selected References

Abrol IP, Sanger S. 2006. Sustaining Indian Agriculture-Conservation agriculture the way forward. *Current Science* 91 (8): 1020-1024.

Balota EL, Colozzi A, Andrade DS, Dick RP. 2004. Long-term tillage and crop rotation effects on microbial biomass and C and N mineralization in a Brazilian Oxisol. *Soil and Tillage Research* 77:137-145.

Blanco-Canqui H, Lal R. 2007. Impacts of long-term wheat straw management on soil hydraulic properties under no-tillage. *Soil Science Society of American Journal* 71:1166-1173.

Bot A, Benites J 2001. Conservation agriculture: Case studies in Latin America and Africa. Rome, Italy: FAO. CTIC Conservation Technology Information Centre. (1996a). Facilitating Conservation Farming Practices and Enhancing Environmental Sustainability with Agricultural Biot e chnology 1 - 2 ; (ht tp://www. c t i c .purdue . edu/medi a/pdf/ Biotech Executive_Summary.pdf).

Bot A, Benites J 2001. Conservation agriculture: Case studies in Latin America and Africa. Rome, Italy: FAO.

Bradford JM, Peterson GA. 2000. Conservation tillage. In *Handbook of soil science*, ed. M. E. Sumner, G247-G269. Boca Raton, FL, USA: CRC Press.

CTIC. 1996b. 17th Annual Crop Residue Management Survey Report. West Lafayette, In: Conservation Technology Information Center.

Baker JM, Ochsner TE, Venterea RT, Griffis TJ. 2007. Tillage and soil carbon sequestration - What do we really know? *Agriculture, Ecosystems and Environment* 118:1-5.

Dolan MS, Clapp CE, Allmaras RR, Baker JM, Molina JAE. 2006. Soil organic carbon and nitrogen in a Minnesota soil as related to tillage, residue and nitrogen management. *Soil and Tillage Research* 89:221-231.

Du Preez CC, Steyn JT, Kotze E. 2001. Long-term effects of wheat residue management on some fertility indicators of a semi-arid Plinthosol. *Soil Till. Res.* 63:25-33.

Duiker SW, Beegle DB. 2006. Soil fertility distributions in long-term no-till, chisel/disk and moldboard plow/disk systems. *Soil and Tillage Research* 88:30-41.

Ellert BH, Bettany JR. 1995. Calculation of organic matter and nutrients stored in soils under contrasting management regimes. *Canadian Journal of Plant Science* 75:529-538.

Etana A, Hakansson I, Zaga E, Bucas S. 1999. Effects of tillage depth on organic carbon content and physical properties in five Swedish soils. *Soil and Tillage Research* 52: 129-139.

Eve MD, Sperow M, Howerton K, Paustian K, Follett RF 2002. Predicted impact of management on soil carbon storage for each cropland region of the conterminous US, *Journal of Soil and Water Conservation* 57: 196–204.

FAO. 2008. Investing in Sustainable Agricultural Intensification: The Role of Conservation Agriculture – A Framework for Action' FAO Rome, August 2008 (available at www.fao.org/ag/ca/).

Follett RF, Peterson GA. 1988. Surface Soil Nutrient Distribution As Affected by Wheat-Fallow Tillage Systems. *Soil Science Society of American Journal* 52:141-147.

Follett RF, Schimel DS. 1989. Effect of tillage on microbial biomass dynamics. *Soil Science Society of American Journal* 53:1091-1096.

Franzluebbers AJ, Hons FM. 1996. Soil-profile distribution of primary and secondary plant available nutrients under conventional and no tillage. *Soil and Tillage Research* 39:229-239.

Friedrich T, Derpsch R, Kassam A. 2012. Overview of the Global Spread of Conservation Agriculture, http://factsreports.revues.org/1941 Published 12 september 2012.

Friedrich T, Kassam AH. 2009. Adoption of Conservation Agriculture Technologies: Constraints and Opportunities. Invited paper, IV World Congress on Conservation Agriculture, 4-7 February 2009, New Delhi, India.

Galantini JA, Landriscini MR, Iglesias JO, Miglierina AM, Rosell RA. 2000. The effects of crop rotation and fertilization on wheat productivity in the Pampean semiarid region of Argentina 2. Nutrient balance, yield and grain quality. *Soil and Tillage Research* 53:137-144.

Goddard T, Zoebisch M, Gan Y, Ellis W, Watson A, Sombatpanit S. 2007 (Eds.). No-Till Farming Systems. WASWC Special Publication No. 3, Bangkok, 544 pp.

Govaerts B, Sayre KD, Ceballos-Ramirez JM, Luna-Guido ML, Limon-Ortega A, Deckers J, Dendooven L. 2006. Conventionally tilled and permanent raised beds with different crop residue management: Effects on soil C and N dynamics. *Plant Soil* 280:143–155.

Govaerts B, Sayre KD, Lichter K, Dendooven L, Deckers J. 2007b. Influence of permanent raised bed planting and residue management on physical and chemical soil quality in rain fed maize/wheat systems. *Plant Soil* 291:39-54.

Govaerts B, Verhulst N, Sayre KD, Dixon J, Dendooven L. 2009b. Conservation Agriculture and Soil Carbon Sequestration; Between Myth and Farmer Reality. *Critical Reviews in Plant Science* 28(3):97-122.

Hati KM, Chaudhary RS, Mandal KG, Bandyopadhyay KK, Singh RK, Sinha NK, Mohanty M, Somasundaram J, Saha R. 2015. Effects of tillage, residue and fertilizer nitrogen on crop yields, and soil physical properties under soybean–wheat potation in Vertisols of central India. *Agricultural Research* 4 (1): 48-56.

Hulugalle NR, Entwistle P. 1997. Soil properties, nutrient uptake and crop growth in an irrigated Vertisol after nine years of minimum tillage. *Soil and Tillage Research* 42:15-32.

Ismail I, Blevins RL, Frye WW. 1994. Long-Term No-Tillage Effects on Soil Properties and Continuous Corn Yields. *Soil Science Society of American Journal* 58:193-198.

Jantalia CP, Resck DVS, Alves BJR, Zotarelli L, Urquiaga S, Boddey RM. 2007. Tillage effect on C stocks of a clayey Oxisol under a soybean-based crop rotation in the Brazilian Cerrado region. *Soil and Tillage Research* 95:97-109.

Jowkin V, Schoenau JJ. 1998. Impact of tillage and landscape position on nitrogen availability and yield of spring wheat in the Brown soil zone in southwestern Saskatchewan. *Canadian Journal of Plant Science* 78:563-572.

Kushwa V, Hati KM, Sinha NK, Singh RK, Mohanty M, Somasundaram J, Jain RC, Chaudhary RS, Biswas AK, Patra AK. 2016. Long-term conservation tillage effect on soil organic carbon and available phosphorous content in Vertisols of central India. *Agricultural Research* 5: 353-361.

Kushwah SS, Damodar Reddy D, Somasundaram J, Srivastava S, Khamparia SA. 2016. Crop Residue Retention and Nutrient Management Practices on Stratification of Phosphorus and Soil Organic Carbon under Soybean-Wheat System in Vertisols of Central India. *Communications in Soil Science and Plant Analysis* 47:2387-2395.

Kushwaha CP, Tripathi SK, Singh KP. 2000. Variations in soil microbial biomass and N availability due to residue and tillage management in a dryland rice agroecosystem. *Soil and Tillage Research* 56:153-166.

Lal R.1998. Soil quality and sustainability. In: Methods for Assessment of Soil Degradation. (Eds: Lal, R.; Blum, W. H.; Valentine, C.; Stewart, B. A.) CRC Press, New York, 17-30.

Lal R. 1997. Residue management, conservation tillage and soil restoration for mitigating greenhouse effect by CO -enrichment. *Soil and Tillage Research* 43: 81-107.

Lal R, Logan TJ, Fausey NR. 1990. Long-Term Tillage Effects on a Mollic Ochraqualf in North-West Ohio.3. Soil Nutrient Profile. *Soil and Tillage Research* 15:371-382.

Landers J. 2007. Tropical Crop-Livestock Systems in Conservation Agriculture: The Brazilian Experience. *Integrated Crop Management* Vol. 5. FAO, Rome.

Larney FJ, Bremer E, Janzen HH, Johnston AM, Lindwal CW. 1997. Changes in total, mineralizable and light fraction soil organic matter with cropping and tillage intensities in semiarid southern Alberta, Canada. *Soil and Tillage Research* 42:229-240.

Lichter K, Govaerts B, Six J, Sayre KD, Deckers J, Dendooven L. 2008. Aggregation and C and N contents of soil organic matter fractions in a permanent raised-bed planting system in the Highlands of Central Mexico. *Plant and Soil* 305:237-252.

Limon-Ortega A, Sayre KD, Drijber RA, Francis CA. 2002. Soil attributes in a furrow-irrigated bed planting system in northwest Mexico. *Soil and Tillage Research* 63:123-132.

Mackay AD, Kladivko EJ, Barber SA, Griffith DR. 1987. Phosphorus and Potassium Uptake by Corn in Conservation Tillage Systems. *Soil Science Society of American Journal* 51:970-974.

Matowo PR, Pierzynski GM, Whitney D, Lamond RE. 1999. Soil chemical properties as influenced by tillage and nitrogen source, placement, and rates after 10 years of continuous sorghum. *Soil and Tillage Research* 50:11-19.

Mazvimavi K, Twomlow S. 2006. Conservation Farming for Agricultural Relief and Development in Zimbabwe. In: No-Till Farming Systems (Goddard, T. *et al.*, Eds.), pp. 169-175. WASWC Special Publication No. 3, Bangkok.

Mohamed A, Hardtle W, Jirjahn B., Niemeyer T, von Oheimb G. 2007. Effects of prescribed burning on plant available nutrients in dry heathland ecosystems. *Ecology* 189:279-289.

Peng KJ, Luo CL, You WX, Lian CL, Li XD, Shen ZG. 2008. Manganese uptake and interactions with cadmium in the hyper-accumulator - *Phytolacca americana L. Journal of Hazardous materials* 154:674-681.

Prasad R, Gangaiah B, Aipe KC. 1999. Effect of crop residue management in rice-wheat cropping system on growth and yield of crops and on soil fertility. *Experimental Agriculture*, 35:427-435.

Qin RJ, Stamp P, Richner W. 2004. Impact of tillage on root systems of winter wheat. *Agr. J.* 96:1523-1530.

Rahman MH, Okubo A, Sugiyama S, Mayland HF. 2008. Physical, chemical and icrobiological properties of an Andisol as related to land use and tillage practice. *Soil and Tillage Research* 101:10-19.

Randall GW, Iragavarapu TK. 1995. Impact of Long-Term Tillage Systems for Continuous Corn on Nitrate Leaching to Tile Drainage. *Journal of Environmental Quality* 24: 360-366.

Roldan A, Salinas-Garcia JR, Alguacil MM, Caravaca F. 2007. Soil sustainability indicators following conservation tillage practices under subtropical maize and bean crops. *Soil and Tillage Research* 93:273-282.

Sakala WD, Cadisch G, Giller KE. 2000. Interactions between residues of maize and pigeonpea and mineral N fertilizers during decomposition and N mineralization. *Soil Biology and Biochemistry.* 32:679-688.

Sanger LJ, Whelan MJ, Cox P, and Anderson JM. 1996. Measurement and modelling of soil organic matter decomposition using biochemical indicators. In *Progress in Nitrogen Cycling Studies*, ed. O. Van Cleemput, G. Hofman, and A. Vermoesen, 445-450. Netherlands: Kluwer Academic Publ.

Schoenau JJ, Campbell CA. 1996. Impact of crop residues on nutrient availability in conservation tillage systems. *Canadian Journal of Plant Science* 76:621-626.

Sisti CPJ, dos Santos HP, Kohhann R, Alves BJR, Urquiaga S, Boddey RM. 2004. Change in carbon and nitrogen stocks in soil under 13 years of conventional or zero tillage in southern Brazil. *Soil and Tillage Research* 76:39-58.

Six J, Conant RT, Paul EA, Paustian K. 2002. Stabilization mechanisms of soil organic matter: Implications for C-saturation of soils. *Plant and Soil* 241:155-176.

Somasundaram J, Chaudhary RS, Hat, KM, Coumar VM, Sinha NK, Jha P,Ramesh K, Neenu S, Blaise D, Saha R, Ajay, Biswas AK, Maheswari M, Rao DLN, Subba Rao A, Venkateswarlu B. (2014b). Conservation Agriculture for Enhancing Soil Health and Crop Productivity. Published by Indian Institute of Soil Science, Bhopal in collaboration with National Initiative on Climate Resilient Agriculture (NICRA), CRIDA, Hyderabad, 64 p.

Somasundaram J, Chaudhary RS, Subba Rao A, Sinha NK, Coumar V. 2014a. Conservation Agriculture for carbon sequestration and sustaining soil health. New India Publishing Agency, New Delhi. P. 1-528.

Subba Rao A, Somasundaram J. 2013. Conservation Agriculture for carbon sequestration and sustaining soil health, In: Agriculture Year Book 2013, Published by Agriculture Today (The National Agricultural Magazine) pp:184-188.

Vanden BAJ, Angers DA. 2006. Towards accurate measurements of soil organic carbon stock change in agroecosystems. *Canadian Journal of Plant Science* 86:465-471.

Vanlauwe B, Dendooven L, Merckx R. 1994. Residue Fractionation and Decomposition – the Significance of the Active Fraction. *Plant and Soil* 158:263-274.

Welch RM.2002. The impact of mineral nutrients in food crops on global human health. *Plant and Soil* 247:83-90.

West TO, Post WM. 2002. Soil organic carbon sequestration rates by tillage and crop rotation: global data analysis. *Soil Science Society of American Journal* 66:1930-1946.

Wienhold BJ, Halvorson AD. 1999. Nitrogen mineralization responses to cropping, tillage, and nitrogen rate in the Northern Great Plains. *Soil Science Society of American Journal* 63:192-196.

15

Enhancing Nutrient Use Efficiency, pp. 193-203
Editors: K. Ramesh, A.K. Biswas, B.L. Lakaria, S. Srivastava and A.K. Patra

Soil Microbial Diversity in Different Sustainable Agricultural Management Systems

A.K. Patra, S.R. Mohanty and K. Bharati

ICAR-Indian Institute of Soil Science, Nabi Bagh, Bhopal – 462 038, India

Introduction

Soil is a complex system consisting of physical, chemical and biological components. These components dynamically interact making it functional to sustain life. All life forms rely on soil microbial processes for their survival. Microbial diversity in the soil is greater than the diversity of any other group of organisms existing in the earth. Microbes as inhabitant of soil are responsible for diverse metabolic functions that affect soil and plant health (Table 1). Because of their activity as single or community as whole is responsible for performing many critical ecosystem functions and can be important biotic indicators of soil and ecological health (Doran *et al.*, 1996; Barrios, 2007). For centuries, biologists have studied patterns of plant and animal diversity at different spatial scales. However microbial diversity, to a large extent, is un-explored. Because of the limitation of techniques that the spatial variation of soil microbial diversity has long been considered as 'noise' in microbiological studies (Ettema and Wardle, 2002), although the composition and diversity of microbial communities are thought to have a direct influence on a wide range of ecosystem processes (Madsen, 1996; Balvanera *et al.*, 2006).

Table 1: Major characteristics of soil with microorganisms as inhabitant

Sl No	Soil characteristics with microorganisms as inhabitant
1.	Only 5% of the soil space is occupied by organisms.
2.	Large microbial biomass and huge microbial diversity for eco-specific response
3.	Complex trophic interactions for nutrient channeling
4.	Soil colloids adsorbing important biological molecules (DNA, proteins, etc.)
5.	Domination of the solid phase. Vary in micro and macroaggregates
6.	Abiotic or enzyme-like reactions for different ecosystem services

Microbial community is under constant threat and is pronounced due to intensive human activity. Any detrimental effects of agricultural management systems on soil microbial communities would damage the functions that they perform and hence impact the ecological services provided by soils, such as nutrient cycling and crop protection. Thus, it is important to gain a better mechanistic understanding of the functional diversity, composition, and dynamics of soil microbial communities under various management systems. This would help to identify the ecological consequences of various agricultural practices and the development of beneficial management strategies.

This paper summarized (1) the recent progress in techniques for exploring soil microbial communities that provides insight into the relationship between structure and function. (2) The effects of agricultural practices (cropping, cultivation and nutrient management) on soil microbial community.

Microbial diversity and ecosystem functioning

The relationship between microbial species composition and ecosystem functioning is difficult to quantify. When species disappear, others can become more dominant and take over a link in the process. It is possible that a process will continue while species composition has changed or degraded. So the preservation of biodiversity cannot be guaranteed solely by measuring process values. Many processes are too general or insensitive as an early warning indicator. The most obvious conflict is between works on natural and synthesized ecosystems. Important works showed that ecosystem processes were determined much more by the functional characteristics of component organisms than by species number (McGrady-Steed *et al.,* 1997). Other studies showed that the greatest deterioration in ecosystem processes occurs as diversity declines from moderate to very low value (Tillman *et al.*, 1996). A large body of ecological literature was developed thinking about vegetal and animal communities and only tentatively adapted to the microbial ones. In fact, limitations due to the cultivation methods and to the extremely high diversity of soil microbial communities, makes it difficult to use approaches based on determining the distribution of different types of organisms. Although utilization of molecular techniques are increasing the capability to analyze microbial communities, we still have conceptual troubles in categorizing the constituents of a communities or in understanding the functional abilities of each individual organism type (Franklin and Mills, 2006). Relationship between microbial community structure and function presents another complication in the physiological versatility of many microorganisms. Functional redundancy of microorganisms was found being quite high in many experimental systems after artificial reduction in species diversity (Franklin and Mills, 2006; Wertz *et al.*, 2006). However, different ecosystem functions have to be considered in evaluating impact by microbial diversity, since many experimental study, for instance, showed an high positive correlation between diversity and invaders survival (Matos *et al.*, 2005; Kennedy *et al.*, 2002). Certainly, biodiversity may represent a form of biological insurance against the loss of selected species and in turn to the maintaining of specific functions. If this is the case, for instance, heavy pollution or disturbances select for a few resistant species. In such situations the ecological basis for processes may become very narrow. When the resistant species also disappear, or are inhibited, as a result of future and yet unknown human activities a process stops and the life support function is permanently affected. Due to redundancy of species, a process is assumed to continue to exist with fewer species, when species disappear, in which case the risk of instability and uncontrolled fluctuations will increase. Research is still necessary to understand the threshold afterwards further species losses could impair ecosystem functioning and then soil quality.

Microbial diversity estimation

Soil microbial communities are among the most difficult to study and fully characterize, because of their immense phenotypic and genotypic diversity. The bacterial population in soil top layers can rise up to more than 10^9 cells per gram soil and has been postulated to represents the world's greatest reservoir of biological diversity. Hence, it is a challenge to access this largely unexplored microbial "wealth". Traditional approaches to study microbial communities in soils are based on culturable protocols. These approaches are useful for isolation purposes, but are very limited in their scope to understand microbial communities and diversity as less than 1% of the bacteria present in soil may be readily isolatable (Torsvik and Ovreas, 2002).

A range of novel methods based on rRNA and rDNA analysis have revealed parts of soil microbial diversity. The key problem posed by the link between microbial diversity and soil function is to understand the relations between genetic diversity or community structure and community function. A better understanding of the relations requires not only the application of more accurate assays for taxonomical and functional characterization of DNA and RNA extracted from soil, but also high resolution techniques with which to detect inactive and active microbial cells in the soil matrix. Nucleic acid-based methods offer the possibility to assess the total microbial diversity by bypassing the limitation of cultivation-based studies. "The PCR-DGGE, TRFLP, highthroughput sequencing, phylochip, functional arrays etc are" is among the most widely used methods for studying the microbial communities in environmental samples. The analysis of data from PCR-DGGE banding patterns is a powerful and relatively easy way to compare microbial communities and diversities from different samples (Muyzer *et al.*, 1993). Terminal restriction fragment length polymorphism (TRFLP) also has been widely employed to study microbial diversity in terrestrial samples (Avis *et al.*, 2006). This technique is characterized as high resolution molecular technique to study microbial diversity at the species level. Recently, the application of DNA microarray (Mohanty, 2012) and stable isotope probe (Mohanty *et al.*, 2006) has been provided huge amounts of new data. This approach has potentially improved our understanding of the structure and function of soil microbial community, and the interactions that occur within. We have discussed the two molecular techniques which has better perspective understanding microbial diversity in the soil.

Microbial diversity and agricultural practices

One of the basic functions of soil microorganisms is the decomposition and transformation of organic materials, which are mostly derived from above and below-ground plant residues. Thus, soil microbial communities play a critical role in ecosystem processes, such as carbon cycling, nutrient turnover, or the production of trace gases. Soil microbial activities, populations and communities are governed by environmental variables and agricultural system, as conventional and organic system (Parton *et al.*, 1988; Araújo *et al.*, 2009). Soil quality is the capacity of soil to maintain some key ecological functions, such as decomposition and formation of soil organic matter (Doran *et al.*, 1996). Microbial processes are important for the management of farming system and improvement of soil quality. Microbial respiration of soil has received considerable attention because it can be used as a soil quality indicator (Haynes, 2000) and it is one important variable to quantify soil microbial activity (Jandl *et al.*, 2007). Soil microbial biomass carbon is considered the most dynamic and labile component of soil organic carbon. Thus, the pool of soil microbial carbon, its activity and composition are key parameters in soil processes (Davidson and Schimel, 1995; Kemmitt *et al.*, 2008).

Information on the comparison of microbial population and activity (processes) in organic and conventional systems is limited. In Brazil, a long-term experiment was carried out to evaluate such microbial response in conventional and organic farming. This was to test a hypothesis that there would be changes in soil microbial activity by practices in organic farming due high input of organic C. Result revealed that different soil management approaches resulted in rapid changes in certain soil properties like soil respiration and bulk density. Organic farming practices resulted in lower soil bulk densities and higher soil microbial activity (measured by soil respiration, and organic carbon). This effect was due to the higher inputs of organic matter for the soil microbial communities. These activated microbial biomass effected turnover of applied nutrients (Fig 1).

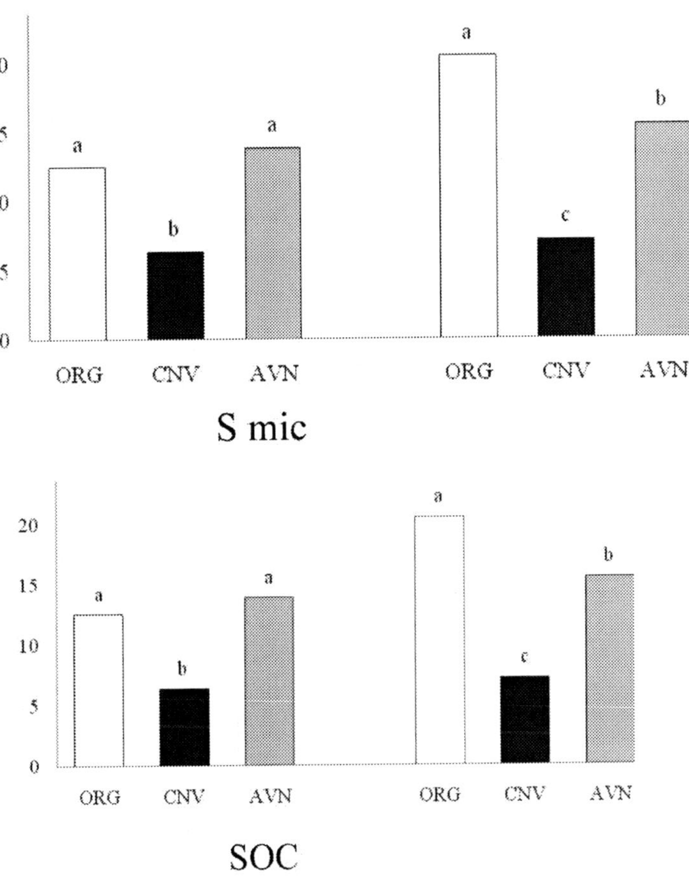

Fig. 1. Soil microbial biomass in the left panel and organic carbon in the right panel in plots under conventional (CNV) and organic farming (ORG) and native vegetation (AVN). Y axis of left panel represents microbial biomass in ug per g soil. Y axis of right panel represents soil organic carbon in mg per g soil. X axis represents different agricultural practices (Araújo *et al.*, 2009)

Crop diversity enhances belowground microbial diversity

The densely populated, intensively cropped subtropical highlands of the world have agricultural sustainability problems from soil erosion and fertility decline. Increasing dependence on a small number of agricultural crops like soybean, wheat or corn may lead to reductions in agricultural biodiversity. Reductions in the number of crops in rotation or the replacement of rotations by monocultures are responsible for this loss of biodiversity. The belowground implications of simplifying agricultural plant communities remain unresolved; however, agroecosystem sustainability will be severely compromised if reductions in biodiversity reduce soil C and N concentrations, alter microbial communities, and degrade soil ecosystem functions as reported in natural communities. In an experiment crop rotation was evaluated on its influence on total soil C and N concentrations, microbial biomass C and N pools. Meta analysis revealed that adding one or more crops in rotation to a monoculture increased total soil C by 3.6% and total N by 5.3%, but when rotations included a cover crop (i.e., crops that are not harvested but produced to enrich the soil and capture inorganic N), total C increased by 8.5% and total N 12.8%. Rotations substantially increased the soil microbial biomass C (20.7%) and N (26.1%) pools, and these overwhelming effects on microbial biomass were not moderated by crop type or management practices. Crop rotations, especially those that include cover crops, sustain soil quality and productivity by enhancing soil C, N, and microbial biomass, making them a cornerstone for sustainable agroecosystems (McDaniel *et al.*, 2014).

Crop rotation and tillage practice also influence soil microbial diversity. In an experiment microbial diversity was assessed using BIOLOG™ system. This system is used for detection of specific patterns of substrate utilization by microbes like bacteria. Soil was sampled (0–7.5 cm) in the wheat phase of different cropping rotations which had been established under zero tillage or conventional tillage on a Gray Luvisol in northern Alberta. Data indicated that tillage significantly ($P<0.05$) reduced the diversity of bacteria by reducing both substrate richness and evenness. The influence of tillage on microbial diversity was more prominent at the flag-leaf stage than at planting time and more prominent in bulk soil than in the rhizosphere at the flag-leaf stage. Microbial diversity was significantly higher under wheat preceded by red clover green manure or field peas than under wheat following wheat (continuous wheat) or summer fallow. The substrate utilization patterns of the bacterial communities also revealed that the bacterial community assemblages under conventional tillage had more similar structures than those under zero tillage. Study highlighted that conservation tillage and legume-based crop rotations support diversity of soil microbial communities and may affect the sustainability of agricultural ecosystems (Lupwayi *et al.*, 1998).

In 1991, the International Maize and Wheat Improvement Center (CIMMYT) initiated a long-term field experiment at its semi-arid highland experiment station in Mexico to investigate the long-term effects of tillage/seeding practices, crop rotations, and crop residue management on soil microbial biomass (SMB) (substrate-induced respiration (SIR) and chloroform fumigation incubation (CFI)) and micro-flora physiological and catabolic diversity (BIOLOG™). SMB-C (CFI, SIR) was significantly and respectively 1.2 and 1.3 times higher for residue retention (average 387 mg C kg^{-1} dry soil and 515 mg C kg^{-1} dry soil, respectively) compared to residue removal. SMB-C (CFI) was significantly higher for wheat (369 mg C kg^{-1}dry soil) compared to maize (319 mg C kg^{-1} dry soil). SMB-N (CFI) was significantly 1.3 times higher for residue retention (average 28 mg N kg^{-1} dry soil) compared to residue removal. BIOLOG ecoplate essay indicated there were large differences in the catabolic capability of soil microbial communities after 15 years of contrasting management practices. While maize and wheat rotation under conventional tillage with residue retention showed a significantly higher overall microbial activity compared to the other treatments, Microbial activity was significantly higher for residue retention compared to residue removal and for wheat as compared to maize. For maize, the management practices were divided into two groups; zero tillage with residue removal was separate from all other treatments. For wheat, conventional tillage was separate from all zero tillage treatments. This study suggests that in the target area, a cropping system that includes zero tillage, crop rotation, and crop residue retention can increase overall biomass and micro-flora activity and diversity compared with common farming practices. In the long term, zero tillage combined with residue retention creates conditions favourable for the development of antagonists and predators, and fosters a new ecological stability. Zero tillage without residue retention is an unsustainable practice that leads to poor soil health in the long run (Govaerts *et al.*, 2007).

Both no-till and crop rotations have been widely adopted in many agricultural settings, and although it is generally accepted that these practices have the potential to increase microbial biomass and activity (Franzluebbers *et al.*, 1995), the specific impacts that these practices have on microbial community composition are largely unknown. It is reported that no consistent effects on bacterial abundance or biomass in a 30-year tillage plot (Frey *et al.*, 2008). Other studies have indicated that tilled soil may or may not contain greater bacterial diversity than non-tilled soil (Ferreira *et al.*, 2000). Tillage typically has exhibited more consistent, predictable effects on soil fungal populations, often resulting in the inhibition of active soil hyphae and arbuscular mycorrhizal fungi (Mozafar *et al.*, 2000; Galvez *et al.*, 2001; Kabir, 2005). Long-termstudies (25–30 years) have found that fungal biomass and hyphal lengths increase with no-tillage compared to conventional tillage (Frey *et al.*, 2008; Helgason *et al.*, 2009). However, reported results have been mixed as to whether no-tillage practices have a greater impact on bacterial or fungal populations (Feng *et al.*, 2003; Helgason *et al.*, 2009). Similarly, varying outcomes have been reported with respect to the effects of monoculture production versus crop-rotation on soil microbial communities. Some studies suggest that monocultures select for less diverse microbial communities, with research indicating that monocultures of wheat (*Triticum aestivum* L.), maize (*Zea mays* L.), or soybean (*Glycine max* L.) may lead to reduced levels of metabolic diversity (Lupwayi *et al.*, 1998), the dominance of certain genotypes of Rhizobium (Depret *et al.*, 2004), or the decline of fungal species (Meriles *et al.*, 2009), respectively. In addition, it has been reported that potato (*Solanum tuberosum* L.) monocultures result in an increased abundance of plant pathogens, leading to a decline in both crop yield and quality (Hide and Read 1991; Honeycutt *et al.*, 1996). In contrast, other studies have reported that long-term rotations of wheat and maize do not appear to significantly impact total bacterial populations but may alter diversity by specifically decreasing the abundance of fluorescent *Pseudomonas* spp. (Govaerts *et al.*, 2007; Hungria *et al.*, 2009).

In an experiment bacterial and fungal relative abundance estimated by qPCR targeting 16S rRNA gene was used to examine the effects of long-term crop-management practices (no-till vs. conventional tillage, and continuous wheat (*Triticum aestivum* L.) vs. sorghum-wheat-soybean rotation (*Sorghum bicolor* L. *Moench-Triticum aestivum* L.-*Glycine max* L. Merr) in Texas, USA (Ng *et al.*, 2012). Bacterial copy numbers were significantly higher in no till plots with continuous wheat than with rotation wheat (sorghum-wheat-soybean) and showed a decreasing, but not significant, trend in plots with conventional tillage (Fig 2).

Shannon's diversity index values were slightly higher in CW than RW samples, and no consistent differences were observed with respect to tillage treatment. In contrast, richness estimates were higher in NT than CT samples, but there were not consistent differences in comparison with the rotation. The CT-RW treatment combination resulted in both the lowest diversity and richness estimates. Results indicate that long-term tillage practices and cropping rotations can impact soil microbial populations, diversity, richness, and community composition. Interestingly, cropping sequence had a larger impact on bacterial community composition and relative levels of bacteria and fungi than did the tillage treatment.

The economic value of ecosystem services carried out microorganisms is not exactly evaluated but it can be at least many tens of billions of dollars per annum. Microbial communities are potentially affected by natural and anthropogenic activities such as agriculture, pesticides, and environmental pollution. It is not yet known how changes in microbial diversity can influence ecosystems. Knowledge of the composition of microbial communities and how these communities are affected by land use measures will find wider application in land use sustainability programs and contribute to future global policies.

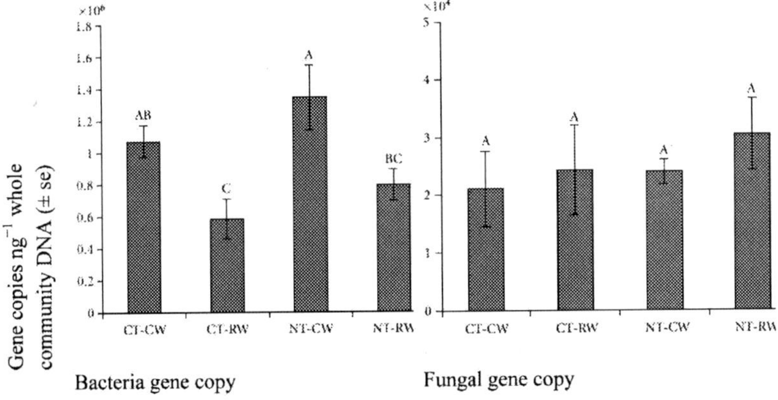

Fig 2. Abundance of bacteria (a) and fungi (b) as determined by qPCR, in soil under different longtermtillage and rotation treatments. Error bars represent ± standard error of three replicate samples. NT: no tillage; CT: conventional tillage; CW: continuous wheat; RW: sorghum-wheat-soybean rotation. Different uppercase letters above bars within a figure indicate significant differences at P = 0.05. (Ng *et al.*, 2012)

Selected References

Araújo AS, Leite LF, Santos VB, Carneiro RF. 2009. Soil microbial activity in conventional and organic agricultural systems. *Sustainability* 1:268–276.

Avis PG, Dickie IA, Mueller GM. 2006. A "dirty" business: testing the limitations of terminal restriction fragment length polymorphism (TRFLP) analysis of soil fungi. *Molecular Ecology,*15:873–882.

Balvanera P, Pfisterer AB, Buchmann N, *et al.* 2006. Quantifying the evidence for biodiversity effects on ecosystem functioning and services. *Ecology Letters* 9:1146–1156.

Barrios E.2007. Soil biota, ecosystem services and land productivity. *Ecological economics* 64:269–285.

Davidson EA, Schimel JP .1995. Microbial processes of production and consumption of nitric oxide, nitrous oxide and methane. In: Matson, PA and Harris RL (Eds) Biogenic trace gases: measuring emissions from soil and water 327–357.

Depret G, Houot S, Allard MR, *et al.* 2004. Long term effects of crop management on Rhizobium leguminosarum biovar viciae populations. *FEMS Microbiology Ecology* 51:87–97.

Doran JW, Sarrantonio M, Liebig M .1996. Soil health and sustainability. In: Sparks, DL (Ed), Advances in Agronomy, *Academic Press, San Diego* 56:1 54.

Ettema CH, Wardle DA .2002. Spatial soil ecology. *Trends in Ecology & Evolution*17:177–183.

Feng Y, Motta AC, Reeves DW, *et al.* 2003. Soil microbial communities under conventional-till and no-till continuous cotton systems. *Soil Biology and Biochemistry* 35:1693–1703.

Ferreira MC, de S Andrade D, de O Chueire LM, *et al.* 2000. Tillage method and crop rotation effects on the population sizes and diversity of bradyrhizobia nodulating soybean. *Soil Biology and Biochemistry* 32:627–637.

Franzluebbers AJ, Hons FM, Saladino VA.1995. Sorghum, wheat and soybean production as affected by long-term tillage, crop sequence and N fertilization. *Plant and Soil* 173:55–65.

Frey B, Pesaro M, Rudt A, Widmer F .2008.Resilience of the rhizosphere Pseudomonas and ammonia-oxidizing bacterial populations during phytoextraction of heavy metal polluted soil with poplar. *Environmental Microbiology* 10:1433–1449.

Galvez L, Douds Jr DD, Drinkwater LE, Wagoner P.2001. Effect of tillage and farming system upon VAM fungus populations and mycorrhizas and nutrient uptake of maize. *Plant and Soil* 228:299–308.

Govaerts B, Mezzalama M, Unno Y, *et al.* 2007. Influence of tillage, residue management, and crop rotation on soil microbial biomass and catabolic diversity. *Applied Soil Ecology* 37:18–30.

Haynes RJ .2000.Labile organic matter as an indicator of organic matter quality in arable and pastoral soils in New Zealand. *Soil Biology and Biochemistry* 32:211–219.

Helgason BL, Walley FL, Germida JJ 2009. Fungal and bacterial abundance in long-term no-till and intensive-till soils of the Northern Great Plains. *Soil Science Society of America Journal* 73:120–127.

Hide GA, Read PJ 1991. Effects of rotation length, fungicide treatment of seed tubers and nematicide on diseases and the quality of potato tubers. *Annals of Applied Biology* 119:77–87.

Honeycutt CW, Clapham WM, Leach SS 1996. Crop rotation and N fertilization effects on growth, yield, and disease incidence in potato. *American Potato Journal* 73:45–61.

Hungria M, Franchini JC, Brandão-Junior O, *et al.* 2009. Soil microbial activity and crop sustainability in a long-term experiment with three soil-tillage and two crop-rotation systems. *Applied Soil Ecology* 42:288–296.

Jandl R, Lindner M, Vesterdal L, *et al.* 2007. How strongly can forest management influence soil carbon sequestration? *Geoderma* 137:253–268.

Kabir Z (2005) Tillage or no-tillage: impact on mycorrhizae. *Canadian Journal of Plant Science* 85:23–29.

Kemmitt SJ, Lanyon CV, Waite IS, *et al.* 2008.Mineralization of native soil organic matter is not regulated by the size, activity or composition of the soil microbial biomass—a new perspective. *Soil Biology and Biochemistry* 40:61–73.

Lupwayi NZ, Rice WA, Clayton GW. 1998. Soil microbial diversity and community structure under wheat as influenced by tillage and crop rotation. *Soil Biology and Biochemistry* 30:1733–1741.

Madsen EL 1996. A critical analysis of methods for determining the composition and biogeochemical activities of soil microbial communities *in situ*. In: Stotzky G, Bollag JM (eds) Soil Biochemistry, Vol 9. Marcel Dekker, 270 Madison Ave, New York, NY 10016, pp 287–370

McDaniel MD, Tiemann LK, Grandy AS 2014. Does agricultural crop diversity enhance soil microbial biomass and organic matter dynamics? A meta-analysis. *Ecological Applications* 24:560–570.

Meriles JM, Vargas Gil S, Conforto C, *et al.* 2009. Soil microbial communities under different soybean cropping systems: Characterization of microbial population dynamics, soil microbial activity, microbial biomass, and fatty acid profiles. *Soil and Tillage Research* 103:271–281.

Mohanty SR .2012. 16S rRNA gene microarray analysis of microbial communities in ethanol-stimulated subsurface sediment. *Microbes Environment* 26(3):261-5.

Mohanty SR, Bodelier PL., Floris V, Conrad R 2006. Differential effects of nitrogenous fertilizers on methane-consuming microbes in rice field and forest soils. *Applied and Environmental Microbiology* 72:1346.

Mozafar A, Anken T, Ruh R, Frossard, E. 2000.Tillage intensity, mycorrhizal and nonmycorrhizal fungi, and nutrient concentrations in maize, wheat, and canola. *Agronomy Journal* 92: 1117–1124.

Muyzer G, Dewaal EC, Uitterlinden AG. 1993. Profiling of complex microbial populations by denaturing gradient gel electrophoresis analysis of polymerase chain reaction-amplified genes coding for 16S rRNA. *AEM* 59:695–700.

Ng JP, Hollister EB, González-Chávez M del CA, *et al.* 2012. Impacts of cropping systems and long-term tillage on soil microbial population levels and community composition in Dryland agricultural setting. *ISRN Ecology* 2012:

Parton WJ, Stewart JWB, Cole CV.1988. Dynamics of C, N, P, and S in grassland soils[mdash]a model. *Biogeochemistry* 5:109–131.

Torsvik V, Ovreas L. 2002. Microbial diversity and function in soil: from genes to ecosystems. *Current Opinion on Microbiology* 5:240–245.

16

Enhancing Nutrient Use Efficiency, pp. 205-215
Editors: K. Ramesh, A.K. Biswas, B.L. Lakaria, S. Srivastava and A.K. Patra

Nano Science and Nanotechnology for Nutrient Use Efficiency

Tapan Adhikari

ICAR-Indian Institute of Soil Science, Nabi Bagh, Bhopal – 462 038, India

Introduction

Nano science is the study of the properties of structures of the size smaller than several hundreds of nano meter (nm). Nanotechnology consists in techniques for designing and manufacturing these structures as well as applications arising from these. The development of nanoscience and nanotechnology is in line with the trend towards miniaturisation. As nanoscience and nanotechnology developed, they make the barriers between traditional scientific and technological disciplines more permeable. Today, there is increasing interaction between electronics, chemistry, physics, biology, medical sciences, information and communication sciences. Scientists describe this as a process of "convergence" of disciplines. Smaller objects or devices are usually easier to transport, require less energy and produce less waste. When things are miniaturised to nanometric sizes, it enables scientists to understand and exploit certain phenomena that are difficult to explain, in varied fields such as chemistry, physics and biology. Nature abounds with many effects linked to nanometric size. The surface of certain leaves and petals are covered with micropillars of water-repellent, hydrophobic wax. The droplets slide off carrying dust with it and leading to a self-cleaning effect is known as the Lotus effect. Similarly the Gecko effect depicts the toe pads of tokay geckos have microscopic hairs, lined with thousands of spatulae of with a width of approximately 200 nm. These spatulae can make close contact with surfaces at nanometric distances. At such distances, there are forces, or bonds between the molecules that are very weak but in very large quantities. They add up and enable the gecko to move on smooth vertical surfaces and even on ceilings. Although quantum physics is already the basis of such areas as electronics (transistors, computers *etc.*) optics (lasers, *etc.*), quantum effects are sometimes "hidden" at the macroscopic scale. With the nanometric scale, they can be observed and exploited in their "pure" form. Usually, the colour of an object remains the same whether it is big or small.

This is not the case on a nanoscale. Some nanoparticles, when lighted appropriately, for example, under ultraviolet light, display different colours depending on their size. It is a quantum effect that is observed only for nanometric sizes. Several nanometres of nanocrystals of semiconductors such as cadmium selenide (CdSe) are dissolved.

When they are exposed to ultraviolet light, they emit a fluorescent light with colours depending on the size of the crystals: blue at 2.3 nm, yellow at 4.5 nm and red at 5.5 nm.

This makes it possible to manufacture probes that can monitor chemical reactions or biological processes.

Nutrient Use Efficiency

Nutrient efficiency can be classified into three groups *viz.*, agronomic efficiency, physiological efficiency and apparent recovery efficiency. The agronomic efficiency is defined as is economic production obtained per unit of nutrient applied. It can be calculated with the help of following equation:

$$\text{Agronomic efficiency } (\text{kg kg}^{-1}) = \frac{(\text{Grain yield of fertilized crop in kg-grain yield in unfertilised crop in kg})}{\text{Quantity of fertilizer applied in kg}}$$

Sometime, agronomic efficiency is also called economic efficiency. If the efficiency is determined under greenhouse conditions, the agronomic efficiency may be expressed in g g^{-1} or mg mg^{-1}.

Physiological efficiency is defined as the biological production obtained per unit of nutrient observed. Sometimes, it is also known as biological efficiency or efficiency ratio.

It can be calculated with the help of the following equation:

$$\text{Physiological efficiency } (\text{kg kg}^{-1}) = \frac{\text{Total dry matter yield of fertilized crop in kg} - \text{Total dry matter yield of unfertilized crop in kg}}{(\text{Nutrient uptake by fertilized crop in kg} - \text{Nutrient uptake by unfertilized crop in kg})}$$

The apparent recovery efficiency is defined as the quantity of nutrient absorbed per unit of nutrient applied. It can be calculated with the help of the following expression:

$$\text{Apparent recovery efficiency} = \frac{\text{Nutrient uptake by fertilized crop} - \text{Nutrient uptake by unfertilized crop}}{\text{Quantity of fertilizer applied}} \times 100\ \%$$

Physiological and recovery efficiency can be combined to obtain the nutrient use efficiency: Nutrient use efficiency = Physiological efficiency × Apparent Recovery efficiency.

To reduce dependence on import of fertilizers, the indigenously available nutrient sources like low grade rock phosphate, waste mica and phosphogypsum need to be promoted as a source of phosphorus, potassium and sulphur, respectively. In addition to agricultural intensification on the best arable land, management practices need to be identified for rational utilization of marginal lands for agriculture for enhancing sustainable crop production in developing countries (Lal, 2000). In this context, possibilities of harnessing newly emerging concepts, mechanisms and techniques in cellular and molecular biology need to be explored for better understanding of tolerance mechanisms to stress so that appropriate strategies could be developed for identification of crop genotypes with superior resource use efficiency. Integration of such crops into crop rotations in conjunction with improved management practices will play an increasingly important role in enhancing crop production especially under conditions of low availability of nutrients and other stresses. Significant progress has been made in understanding the physiological traits of several crop species responsible for tolerance to a wide range of nutrient deficiencies in soils. Preliminary studies have shown that inclusion of such promising crop species into crop rotations or to intercropping would be a useful strategy to improve nutrient use efficiency, crop nutrition and yields (Hocking, 2001). Fertilizers have played a stellar role in improving crop productivity and production and would continue to do so in future as well. However, presently there is a growing concern about the low use efficiency of nutrients which range from 2 to 50%. Such a low efficiency increases the cost of production and leads to severe environmental consequences. It is estimated that just by raising the nutrient-use efficiency by 10%, the country can save almost 20 million ha of land at the current level of productivity.

The impaired soil health and the declined productive potential is primarily due to imbalanced fertilizer use coupled with low use of organic manures. The soils are not being adequately replenished even with the macro-nutrients, but also secondary and micro-nutrients. The improper nutrient management has, therefore, led to multi-nutrient deficiencies in Indian soils. The deficiencies range from 3% of copper to 89% of nitrogen with other elements falling in the range. The deficiencies are becoming more critical for sulphur, zinc and boron. About 47 million ha in major cropping systems are deficient in sulphur. The zinc deficiency is rampant in alluvial soils of Indo-Gangetic plain, black soils of Deccan Plateau and red and other associated soils. The boron deficiencies are showing up in red, lateritic and calcareous soils of Bihar, Orissa and West Bengal. The limiting nutrients by not allowing the full expression of other nutrients, lower fertilizer responses and crop productivity. The site-specific integrated nutrient management encompassing conjunctive use of inorganic and organic fertilizers is the most ideal system for maintaining soil health and enhancing nutrient-use efficiency. The country will require about 45 MT of nutrients to produce 300 MT of foodgrains by 2025. Therefore, the fertilizer industry has to augment fertilizer production substantially from the present level of about 22 million tonnes of nutrients to keep pace with the growing food demands of the country. At the same time, the Government should have adequate provisions for setting up of units of compost and biofertilizers in rural and urban areas of the country. There is near stagnation in capacity and investment in fertilizer sector since 2000, which has adversely affected the production of fertilizers in the country.

Nitrogen, which is a key nutrient source for food, biomass, and fibre production in agriculture, is by far the most important element in fertilizers when judged in terms of the energy required for its synthesis, tonnage used and monetary value. However, compared with amounts of nitrogen applied to soil, the nitrogen use efficiency (NUE) by crops is very low. Between 50 and 70% of the nitrogen applied using conventional fertilizers plant nutrient formulations with dimensions greater than 100 nm is lost owing to leaching in the form of water soluble nitrates, emission of gaseous ammonia and nitrogen oxides, and long-term incorporation of mineral nitrogen into soil organic matter by soil microorganisms. Numerous attempts to increase the NUE have so far met with little success, and the time may have come to apply nanotechnology to solve some of these problems. Carbon nanotubes were recently shown to penetrate tomato seeds and zinc oxide nanoparticles were shown to enter the root tissue of ryegrass. This suggests that new nutrient delivery systems that exploit the nanoscale porous domains on plant surfaces can be developed. The potential use of nanotechnology to improve fertilizer formulations, however, may have been hindered by reduced research funding and the lack of clear regulations and innovation policies. Current patent literature shows that the use of nanotechnology in fertilizer development remains relatively low (about 100 patents and patent applications between 1998 and 2008) compared with pharmaceuticals (more than 6,000 patents and patent applications over the same period). A nanofertilizer refers to a product that delivers nutrients to crops in one of three ways. The nutrient can be encapsulated inside nanomaterials such as nanotubes or nanoporous materials, coated with a thin protective polymer film, or delivered as particles or emulsions of nanoscale dimensions. Owing to a high surface area to volume ratio, the effectiveness of nanofertilizers may surpass the most innovative polymer-coated conventional fertilizers, which have seen little improvement in the past ten years. Ideally, nanotechnology could provide devices and mechanisms to synchronize the release of nitrogen (from fertilizers)with its uptake by crops; the nanofertilizers should release the nutrients on-demand while preventing them from prematurely converting into chemical/gaseous forms that cannot be absorbed by plants. This can be achieved by preventing nutrients from interacting with soil, water and microorganisms, and releasing nutrients only when they can be directly internalized by the plant. Examples of these nanostrategies are beginning to emerge. Zinc–aluminium-layered double-hydroxide nanocomposites have been used for the controlled release of chemical compounds that regulate plant growth. Improved yields have been claimed for fertilizers that are incorporated into cochleate nanotubes (rolled-up lipid bilayer sheets). The release of nitrogen by urea hydrolysis has been controlled through the insertion of urease enzymes into nanoporous silica. Although these approaches are promising, they lack mechanisms that can recognize and respond to the needs of the plant and changes in nitrogen levels in the soil. The development of functional nanoscale films and devices has the potential to produce significant gains in the NUE and crop production. In addition to increasing the NUE, nanotechnology might be able to improve the performance of fertilizers in other ways. For example, owing to its photocatalytic property, nanosize titanium dioxide has been incorporated into fertilizers as a bactericidal additive. Moreover, titanium dioxide may also lead to improved crop yield through the photoreduction of nitrogen gas. Furthermore, nanosilica particles absorbed by roots have been shown to form films at the cell walls, which can enhance the plant's resistance to stress and lead to improved yields. Clearly, there is an opportunity for nanotechnology to have a profound impact on energy, the economy and the environment, by improving fertilizer products. New prospects for integrating nanotechnologies into fertilizers should be explored, cognizant of any potential risk to the environment or to human health. With targeted efforts by governments and academics in developing such enabled agri-products, we believe that nanotechnology will be transformative in this field.

Nano-fertilizer

Nano-fertilizer may be defined as the nano particles, which can supply essential nutrients precisely for maximum plant growth, have higher use efficiency and can be delivered in a timely manner to a rhizospheric target or by foliar spray (Adhikari *et al.*, 2010). Nano-fertilizers are molecular modified or synthesized materials, with the help of nano technology used to improve the fertility of soil for a better yield and increased crop quality. The nano-fertilizer technology is an innovative strategy. Nano fertilizers play an important role in increasing the quantity of agricultural products and in removing the soil and environmental hazards. The purity of elements is very high in such fertilizers. One of the advantages of such fertilizers is that they can be used in tiny amounts, to the extent that an adult plant requires only 40-50 kg of fertilizer while an amount of 150 kg would be required for ordinary fertilizers, means it saves cost by less expenditure on fertilizer which is economical for farmers point of view. In general advantages of application of nano fertilizers are (i) increased yield of crop by an average of 20% (ii) enhanced nutritional value of the produce and (iii) overall health of the crop is enhanced, making it more resistant to severe weather and extreme atmospheric conditions. Immunological response is heightened allowing the plant to fight disease and prevent infections.

In contrast to conventional P fertilizers, nano rock phosphate fertilizer gave a better effect if band placed for localized effect near the root zone. As nano fertilizers required in less quantity, the nano fertilizers can be mixed well with finely ground soil or farm yard manures before localized placement surrounding the root zone of the plants. The reason in support of this practice is that two phenomena will occur after placement of nano fertilizers viz. (i) more number of rock phosphate nano particles could come in intimate contact with soil particles and thereby will be subjected to greater action of soil solvents and (ii) nano particles directly interact with the plant root surface and penetrate through the cell wall and enter inside the cell. Other methods such as foliar application with effective surfactants, seed coating *etc.* of nano fertilizers can be adopted to enhance the use efficiency of the fertilizers. Developed new products such as coatings with nano material for controlled release, incorporation of other essential nutrients, liquid and suspension forms etc. have the potential to markedly improve nutrient use efficiency.

Nano phosphatic fertilizer

Adhikari *et al.* (2009) conducted a series of laboratory experiments to know the effect of rock phosphate nano particle on germination and growth of seeds of soybean and mustard and found that, germination was not affected up to 200 ppm P Udaipur rock phosphate nano particles in both the crops, and an increasing trend was observed in root growth of the crops. Adhikari *et al.* (2011) showed the application of nano rock phosphate particles enhanced the growth of maize plant in a solution culture study. The experimental results revealed that the maize plant showed response to nano rock phosphate application. The dry weight of shoot and root was increased due to application of phosphorus through KH_2PO_4, and nano rock phosphate particles. But the highest dry matter weight of maize shoot (3.10g) and root (0.28 g) was recorded at 31 ppm P through nano rock phosphate particles. It was also observed that application of P through nano rock phosphate particle increased the plant height from 45 cm to 80 cm. Experimental results indicated that plant root might have got the unique mechanism of assimilating nano rock phosphate particle and using for its growth and development. Adhikari *et al.* (2011) observed that the biomass yield of maize plant harvested after 60 days of sowing on different soils in a pot culture study responded distinctly different to P applied in the form of SSP, nano rock-phosphate and micron sized phosphate. This exploratory investigation clearly indicated relatively higher yield response to nano-rock phosphate in all the soils as compared to micron sized phosphates. Further the results suggested that crop utilization of P from nano rock-phosphate was at par with that of P from SSP in Vertisol and Inceptisol, while the yield response to P from nano-rock phosphate was marginally lower than to P from SSP in Alfisol and Aridisol. Adhikari *et al.* (2013) evaluated the efficacy of different nano rock phosphates, in a field experiment taking maize as a test crop. The results showed that the shelling percentage, 1000-grain weight, grain and stover yield etc. were significantly higher in the maize plant treated with Udaipur nano rock phosphate (34 % P_2O_5) as compared to the control, which was being at par with single super phosphate treated plants. These results showed that crop utilization of P from nano rock phosphate was at par with that of P from SSP while yield response to P from nano rock phosphate was marginally lower than to P from SSP but much more economical. This might be due to the fact that nano form of rock phosphate particles increased dissolution rate. The finer size of nano rock phosphate enhanced the degree of contact between rock phosphate and soil, which caused greater rate of dissolution. Hence more solubility of P, released P slowly and provide a constant supply of P to the plant for entire growth period and thereby more utilization of the P by the plants led to better plant growth, which may in turn to higher biomass yield resulted into more P uptake. Nano-particles can absorb on to the clay lattice thereby preventing fixation while releasing into the soil solution than can be utilized by the plants. This process improved the soil health and nutrient use efficiency by the crops. However, the use of Udaipur nano rock phosphate (34%) gave less but almost similar effect as using of single super phosphate. The residual effect of one time application of the P sources was also able to sustain successive cropping of other crops.

As economic point of view, the return profit against invested in fertilizer, was higher for nano rock phosphates. Because the cost of super phosphate was almost three to four times more than that of rock phosphates, assuming that the fertilizer prices are farm gate prices and that all other costs are equal for all phosphate sources. Based on the economy point of view, this practice of utilizing nano rock phosphate as a source of phosphate fertilizer will open a new avenue to the farmers. The experimental results have an agronomic importance for cereal cultivation and crop yield. Because cereals require a greater amount of phosphorus fertilizers, especially when grown in alkaline soil, as some proportion of these fertilizers become unavailable for plant. Finally, it can be inferred that rock phosphates when applied in the form of nano particles, synthesized from a cheap source of phosphorus, could be a potential and economic alternative source to the costly super phosphate. Utilization of nano rock phosphate will help in reducing the quantity of fertilizer application and saving the input cost of farmers. In the near future, this practice may assist in solving the problems encountered with food shortage and crop production economy.

Nano zinc fertilizer

In India, zinc (Zn) has been recognized as the fourth most important yield-limiting nutrient after N, P and K. Zinc deficiency affecting crop yields severely. It is difficult to eliminate because (1) there is a large temporal and spatial variation in availability of soil Zn, (2) a relatively large portion of soil-applied Zn may be fixed into plant-unavailable-forms, and (3) soil-applied Zn is not readily transported down the soil profile. Therefore, the use efficiency of Zn fertilizer to the soil is very low 2 to 5%. Supplementing the Zn requirement of agricultural crops through water soluble $ZnSO_4$ fertilizer is a costly management option whereas, utilization of ZnO (water insoluble and a cheaper material) as a source of Zn could be an alternative cost effective option to encourage farmers for wider adoption. Recently, nano particles have received considerable attention due to their increased uptake by plants as they are small in size and have high rate of penetration through plant cell membrane. Adhikari *et al.* (2011) reported that better germination percentage (93-100%) due to nano ZnO (<50nm) coating with Pine Oleo Resin (POR) as compared to uncoated seeds (80%) and also observed in a pot culture experiment conducted with coated seeds that the crop growth with ZnO coated seeds were similar to that observed with soluble Zn treatment applied as $ZnSO_4.7H_2O$ (@ 2.5 ppm Zn). Seed coating is an on-seed delivery mechanism that provides all the advantages and more of a starter fertilizer. The ingredients wrapped around the seed include plant nutrients. The POR containing nano ZnO particles increased seed germination and improved seedling growth throughout the growing season. POR forms a compact protective film on the seed surface, which can delay the release of fertilizer elements, reduce nutrient losses and significantly improve fertilizer efficiency. The Zn fertilizer can increase the enzymatic activity of the seeds, which is helpful to the transformation of protein into amino acids and starch into simple sugars or fat to provide abundant nutrient sources for germs to grow. It will produce obvious economic and environmental benefits when this novel seed coating technique, using POR is used in agriculture. The most important advantage of seed coating with ZnO (both micron/nano scale) is that it did not exert any osmotic potential at the time of germination of the seed, thus, the total requirement of Zn of the crop can be loaded with the seed effectively through nano scale ZnO particle. Seed coating with nano ZnO (<50nm) and POR can be used by the seed producing agencies to produce customized seed for Zn deficient areas of the country.

Nano copper fertilizer

Copper is an essential micronutrient for plants, which acts either as the metal component of enzymes or as a functional structural or a regulatory co-factor of a large number of enzymes. Studies showed that the effects of nano-particles on plants can be beneficial (seedling growth and development) (Zhu *et al.*, 2008). The propensity of the NPs to cross barriers and their interaction intracellular structures owing to their small size and high surface reactivity govern the growth of plants. In industry nano-CuO has been extensively used for several decades. But its application in agricultural field is still not emphasized and practiced so as to increase the nutrient use efficiency. Results showed that Cu nano particles can enter into the plant cell through roots and leaves. Bioaccumulation increased with increasing concentration of Cu nano particles (NPs), and agglomeration of particles was observed in the cells using transmission –electron microscopy. Application of Cu nano particles through solution culture as well as spray enhanced the growth (51%) of maize plant in comparison to control. Foliar nutrient absorption is not a biological process that requires time. It is very fast process because of it is governed by physical and chemical process. Chlorophyll production and photosynthesis rate were enhanced due to application of foliar fertilizer, which in turn increases the uptake of soil applied fertilizers in response to increased need for water by the leaf bringing more nutrients to the plant via the vascular system. However, accumulation of nano particles on photosynthetic surface cause foliar heating which results in alterations to gas exchange due to stomatal obstruction that produce changes in various physiological and cellular function of plants. Studies on the mechanism of nano particle uptake by plants still need more clarification.

Oleoresin coated urea fortified with nano-particles

Adhikari *et al.* (2011) developed a protocol to fortify the Urea granules with a consortium of nano-particles of Zn, as ZnO (<100 nm), Fe as Fe_3O_4 (50 nm) Cu as CuO (50 nm) and Si as SiO_2 (<20 nm) using oleoresin. The protocol can successfully be used to deliver nano-particles of micronutrients along with urea. The nano-particles coated urea contained 43.84% N, 2.20 mg Zn/g Urea, 1.10 mg Fe /g Urea, 0.66 mg Cu / g Urea and 1.06 mg Si / g Urea. Application of such urea @ 200 kg/ha will supply , 440 g Zn, 220 g Fe, 132 g Cu and 212 g Si along with 87.68 kg N/ha to the crops.

Conclusion

The decreasing availability and increasing cost of fertilizer as well as to preserve the soil and underground water resources, the strategies for improving the nutrient use efficiencies must be devised. The use of slow-released fertilizer has become a new trend to save fertilizer consumption and to minimize environmental pollution. Due to its high surface energy and chemical activity, the application domain of nano materials has significantly expanded. It is the beginning of a new advanced era and there is a great need of agricultural technique modification to fulfill the requirement of this and upcoming generation. Therefore more research and industrial setup should be emphasized in the field of nano-fertilizers. What policy or technology alternatives are there to nanotechnology for the sustainable use of fertilizers and micronutrient supplements in soil? There is an urgent public policy need to acquire the data necessary to determine the environmental fate of agri-nanotechnology applications.

Selected References

Abdzad Gohari, A, Noorhosseini Niyaki. S.A. 2010. Effects of Iron and Nitrogen Fertilizers on Yield and Yield Components of Peanut (*Arachis hypogaea* L.) in Astaneh Ashrafiyeh, *Iran Journal of Agriculture and Environmental Sciences* 9: 256-262.

Adhikari T, Kundu S, Biswas, A.K. 2009. Annual Report (2009-2010). Indian Institute of Soil Science (ICAR), Nabi Bagh, Bhopal. p. 26-27.

Adhikakri T, Kundu S, Biswas, A.K. 2011. Annual Report (2011-2012). Indian Institute of Soil Science (ICAR), Nabi Bagh, Bhopal. p.30.

Adhiakri T, Kundu S, Biswas A.K. 2013. In: Nanotechnology in soil science and plant nutrition: (Tapan Adhikari, S. Kundu and A.Subba Rao, Eds.) pp. 9-21 (2013).

Bozorgi HR.2012. *ARPN* Effects of foliar spraying with marine plant Ascophyllum nodosum extract and nano iron chelate fertilizer on fruit yield and seaweed attributes of eggplant (*Solanum melongena* L.). *Journal of Agriculture and Biological Sciences* 7: 233 -237.

Brittenham GM. 1994. New advances in iron metabolism, iron deficiency and iron overload. *Current Opinion in Hematology* 1: 549-556.

Khanna PK Gaikwad S, Adhyapak PV, Singh N, Marimuthu R. 2007. Synthesis and characterization of copper nanoparticles. *Material Letters* 61: 4711- 4714.

Miller GW, Huang IJ, Welkie GW, Pushmik J.C.1995. In: Abadia, J., (Ed.), Iron nutrition in soils and Plants. Kluwer Academic Publishers, Dordrecht. pp. 19-28.

Mohapatra M, Anand S. 2010. Synthesis and applications of nano-structured iron oxides/ hydroxides – a review. *International Journal of Engineering Science and Technology* 2: 127-146.

Nazari A. 2012. Morphological traits of sweet basil (*Ocimum basilum* L.) as influenced by foliar application of methanol and nano-iron chelate fertilizers. *Annals of Biological Research,* 3: 5511-5514.

Pal S, Saimbhi MS, Bal SS. 2002. Effect of nitrogen and phosphorus levels on growth and yield of brinjal hybrid (*Solanum melongena* L.). *Journal of Vegetable Science* 29: 90-91

Sat P, Saimbhi MS. 2003. Effect of varying levels of nitrogen and phosphorus on earliness and yield of brinjal hybrids. *Journal of Research on Crops* 4: 217-222

Shah MA. 2010. Preparation of MgO nanoparticles with water. Preparation of MgO Nanoparticles With Water. *African Physics Review* 4 (0004): 21-23.

Singh AL, Dayal BD. 1992. Foliar application of iron for recovering groundnut plants from lime induced iron deficiency chlorosis and accompanying losses in yield. *Journal of Plant Nutrition* 15: 1421-1433. .

Yadav BC, Srivastava R, Kumar A. 2007. Characterization of ZnO nanomaterial synthesized by different methods. *International Journal of Nanotechnology and Applications* 1: 1-11.

Zareie S, Golkar P, Mohammadi Nejad GH. 2011. Effect of nitrogen and iron fertilizers on seed yield and yield components of safflower genotypes. *African Journal of Agricultural Research* 6: 3924-3929.

17

Enhancing Nutrient Use Efficiency, pp. 217-225
Editors: K. Ramesh, A.K. Biswas, B.L. Lakaria, S. Srivastava and A.K. Patra

Machinery for Precise Nutrient Application as a Means of Increasing Nutrient Use Efficiency with Special Reference to Fertigation

K.V. Ramana Rao

ICAR- Central Institute of Agricultural Engineering, Bhopal – 462038, India

Introduction

Fertilizers in India, are mostly applied either through broadcasting, foliar application or using seed cum fertilizer drills. With the introduction of micro irrigation systems, the application of fertilizers along with the irrigation water became promising and is gaining momentum. The goal of fertilizer application through irrigation is to maximize the uptake rate of fertilizers in order to minimize the losses.

What is fertigation

Fertigation is the method of application of soluble fertilizer with irrigation water. Fertigation uses either granular or liquid fertilizers, which are dissolved in water, and injected into the irrigation system. Nutrients can be applied through drip or spray systems, and can vary in concentration and composition. Fertigation provides uniform and relative ease of distribution of nutrients and can be fine-tuned to the nutritional requirements of a particular crop. In general, application of fertilizer with irrigation water gives a better crop response than either broadcast or foliar application. Fertigation is a prerequisite for drip irrigation. Since the wetted soil volume is limited, the root system is confined and concentrated. The nutrients from the root zone are depleted quickly and a continuous application of nutrients along with the irrigation water is necessary for adequate plant growth. Fertigation offers precise control on fertilizer application and can be adjusted to the rate of plant nutrient uptake.

Advantages of fertigation

The advantages of fertigation over conventional application methods could be:

- The supply of nutrients can be more carefully regulated and monitored.
- The nutrients can be distributed more evenly throughout the entire root zone or soil profile.
- The nutrients can be supplied incrementally throughout the season to meet the actual nutritional requirements of the crop.
- Nutrients can be applied to the soil when crop or soil conditions would otherwise prohibit entry into the field with conventional equipment.
- Soil compaction is avoided, as heavy equipment never enters the field.
- Less equipment, labour and energy is required for application.
- Minimizes the water and land pollution.
- Safe application without damage to foliage.

All the chemical fertilizers applied through irrigation systems must meet the following criteria.

- Avoid corrosion, softening of plastic pipe and tubing, or clogging of any component of the system
- Safe for field use
- Soluble in water
- Should not react adversely to salts or other chemicals in the irrigation water.

Proper management and efficient application of these nutrients offer significant saving of these costly inorganic fertilizers, better scheduling as per crop need, improvement in yield and nutrient use efficiency and less impact on environment which is an important component of sustainable soil management. The main drawbacks associated with fertigation are the initial set-up costs and the need to monitor the operation carefully to ensure that irrigation and injection systems are working correctly. Water quality can limit the use of fertigation; irrigation waters that are high in salts are not suitable for fertigation. Generally the concentration of salts in the fertigation solution should not exceed 3000 μs/cm. The most significant risk when utilizing fertigation is for water source contamination due to back siphoning. Fertigation may favour NO_3^- -N leaching, which requires a careful calculation of the fertilizer dose to minimize the risk of groundwater contamination.

Liquid fertilizers are best suited for fertigation. In India, inadequate availability and high cost of liquid fertilizers restrict their uses. Fertigation using granular fertilizers poses several problems namely, their different levels of solubility in water, compatibility among different fertilizers and filtration of undisclosed fertilizers and impurities. Different granular fertilizers have different solubility in water. When the solutions of two or more fertilizers are mixed together, one or more of them may tend to precipitate if the fertilizers are not compatible with each other. Therefore, such fertilizers may be unsuitable for simultaneous application through fertigation and would have to be used separately. This chapter reports on the various issues of fertigation i.e. advantages and limitations, selection of water soluble fertilizers (granular and liquid), fertigation scheduling in various crops and fertigation system for efficient fertigation programme and response of plants to fertigation and its economic use.

Selection of fertilizers

Effective fertigation requires an understanding of plant growth behavior including nutrient requirements and rooting patterns, soil chemistry such as solubility and mobility of the nutrients, fertilizer chemistry (mixing compatibility, precipitation, clogging and corrosion) and water quality factors including pH, salt and sodium hazards, and toxic ions. The granular and liquid fertilizers used in fertigation are available in various chemical formulations, solubility and different types of coatings; therefore selection of fertilizer is an important issue in fertigation. The selection of granular fertilizer will depend on the nutrient that is applied, fertilizer solubility and ease of handling. An essential pre-requisite for the solid fertilizer use in fertigation is its complete dissolution in the irrigation water. Examples of highly soluble fertilizers appropriate for their use in fertigation are: ammonium nitrate, potassium chloride, potassium nitrate, urea, ammonium monophosphate and potassium monophosphate. The fertilizer solutions stored during the summer form precipitates when the temperature decrease in the autumn, due to the diminution of the solubility with low temperatures. Therefore, it is recommended to dilute the solutions stored at the end of the summer. Fertilizer solutions of smaller degree specially formulated by the manufacturers are used during the winter.

Table 1: Fertilizers solubility and temperatures (g/100 g water) (Wolf *et al.*, 1985).

Temperature	KC_1	K_2SO_4	KNO_3	NH_4NO_3	Urea
10°C	31	9	21	158	84
20°C	34	11	31	195	105
30°C	37	13	46	242	133

To ensure the fertilizers selected will not precipitate in irrigation pipe, mix the fertilizer solution with sample of irrigation water in the same proportions as and when they are mixed in the irrigation system. If the chemical stays in the solution, then the product is safe to use under fertigation. Liquid fertilizers offer many intrinsic advantages over granular fertilizers for fertigation. It is considered as most suitable for fertigation under micro and sprinkler irrigation. These are available as fertilizer solutions and suspension, both of which may contain single or multi nutrient materials. To avoid damage to the plant roots due to high fertilizer concentrations, the fertilizer concentration in irrigation water should not exceed 5%. Although, the susceptibility to root burning from concentrated fertilizers varies with crops, fertilizers and accompanying irrigation practices, therefore it is safer to keep the fertilizer concentration of 1-2% in the irrigation water during fertigation.

Fertigation uses either granular or liquid fertilizers, which are dissolved in water, and injected into the irrigation system The quantity of fertilizer that can be dissolved in unit quantity of water is called the solubility. The solubility is greatly affected by the temperature variations. Normally these values are taken for 20°C.

Nitrogenous fertilizers used in fertigation

Most of the commercially available forms of nitrogenous fertilizers are readily soluble in water and hence have got good scope for fertigation.

Fertilizer	N content (%)	Solubility (g/l) at 20°C
Ammonium Sulphate	21	750
Urea	46	1100
Ammonium nitrate	34	1920
Calcium nitrate	15.5	1290

Source: Rajput and Patel, 2002.

Phosphatic fertilizers used in fertigation

Most of the phosphate fertilizers marked for field applications can't be suitable for fertigation, as they are too low in their solubility both at low and high pH. The maximum solubility of phosphorus is occurs in the range of pH from 6.5 to 7.0 (Beegle, 1990). Mono Ammonium Phosphate (MAP) contains 12% of Nitrogen and 61% of phosphorous and Mono Potassium Phosphate (MKP) contains 52% of phosphorus and 34% of potassium. Phosphate fertilizers like MAP, MKP, though recommended under fertigation, still they have precipitation problems if the water is hard or contains high calcium contents or with the presence of bicarbonates. Use of Phosphoric acid (52%) with the drip irrigation system not only helps in providing phosphatic fertilizers to the plant but also minimizes the problems of drip emitters clogging.

Potassic fertilizers used in fertigation

Potassic fertilizers are most suitable fertilizers for fertigation with drip system.

Fertilizer	K content (%)	Solubility (g/l) at 20°C
Potassium sulphate	50	110
Potassium chloride	60	340
Potassium nitrate	44	133

Source: Rajput and Patel, 2002.

Micronutrients that can be used in fertigation

Apart from major nutrients, water soluble micro nutrients are also readily available in the market for application through drip irrigation system. Some of them are:

Fertilizer	Content (%)	Solubility (g/l) at 20°C
Solubor	20 B	220
Copper sulphate	25 Cu	320
Iron sulphate	20 Fe	160
Magnesium sulphate	10 Mg	710
Ammonium molybdate	54 Mo	430
Zinc sulphate	36 Zn	965
Manganese sulphate	27 Mn	1050

Source: Rajput and Patel, 2002.

Points to be considered during fertigation

- Avoid high irrigation water having high pH values
- In waters having high calcium content and bicarbonates, avoid use of sulphate fertilizers
- Presence of high concentrations of calcium and magnesium and high pH values lead to the precipitation of calcium and magnesium phosphates.
- When irrigation water has an EC > 2 dS/m (with high salinization hazard), and crop is sensitive to salinity, decrease the amount of accompanying ions added with the N or K.
- Avoid mixing of $(NH_4)_2SO_4$ and KCl in the fertilizer tank as the solubility of the mixture reduces due to the K_2SO_4 formation.

- Other forbidden mixtures are: Calcium nitrate with any phosphates or sulphates; Magnesium sulphate with di or mono-ammonium phosphate; Phosphoric acid with iron, zinc, copper and manganese sulphates.

Fertilizer applicators

A number of techniques are used to introduce the fertilizers into the irrigation system. Generally, fertilizers are injected into irrigation systems by three principal methods namely, (1) fertilizer tank (the by-pass system), (2) the venturi pump and (3) the injection pump (piston or dozotron pump). Non-corrosive material should be used for the fertilizer containers and for the injection equipment.

Equipment and methods for fertilizer injection

Injection of fertilizer and other agrochemicals such as herbicides and pesticides into the drip irrigation system is done by i) Fertilizer tank method (By-pass system) ii) Venturi system and iii) Direct injection system.

Fertilizer tank method (By-pass system)

This method employs a tank into which the fertilizer solution is filled. The tank is connected to the main irrigation line by means of a by-pass so that a part of the irrigation water flows through the fertilizer tank and dilutes the fertilizer solution. Water flows because of pressure gradient between the entrance and exit of the tank, created by a control valve. The concentration of fertilizer in the tank decreases gradually until it reaches the level of the irrigation water, as governed by following equation:

$$C_t = C_0 (Q / Q_i) e^{(Q_i / V_t) t_f}$$

where,

C_t = Change in the concentration of nutrients in fertilizer solution after t hour of irrigation duration, mg/L

C_0 = Initial concentration of nutrients in the fertilizer tank, mg/L

Q = Flow rate of irrigation water, Lph

Q_i = Injection rate of fertilizer solution, Lph

V_t = Capacity of fertilizer tank, L

t_f = Fertigation duration, h

The rate of flow through the by-pass is determined by the pressure head difference between the entrance and exit, which is usually of the order of 1-5 m of head of water. The choice of tank size (available sizes, 50-1000 litres) is related to the area to be fertilized. The pressure difference needed in order to empty the tank during one irrigation needs to be determined empirically.

In case of liquid or already dissolved fertilizers, the amount of water that equals 4 times the fertilizer tank volume must be passed through the tank to transfer out 98% of the fertilizer. But if granular fertilizers are directly put in the fertilizer tank, higher amounts of water should be passed through the tank. A special valve regulates the water flow velocity through the tank. Important advantages and disadvantages of a by-pass tank system are listed below:

Advantages

- Simple in construction and operation.
- No need for external power supply.
- Relative insensitivity to changes in pressure or flow rate.
- It enables a high fertilizer injection application rate.
- Hydraulic head loss is low.

Disadvantages

- The tank must be refilled with fertilizer solution for each irrigation cycle, thus the system is not suitable for automatic or serial fertigation.
- Fertilization is not proportional, therefore, control over the fertilizer concentration in the water is limited.

Venturi injector

Venturi consists of a converging section, a throat and a diverging section, like in parshall flumes. When the flowing water passes through the constricted throat section, its velocity increases and the pressure reduces, creating a suction effect. Due to suction, liquid fertilizer enters into the drip system through a tube from the fertilizer storage tank connected to the throat. To start the venturi system the desired pressure difference across the venturi is created by using pressure regulating valves to enable the flow of fertilizer into the drip system. The rate of flow is regulated by means of the valves. The venturi system works because of differential pressure in the system (usually 20 %) from one side of the device to the other. Injection rate depends on the pressure difference and pressure fluctuations in the system change the injection rate. The advantages and disadvantages of using a venturi system for fertigation are given below:

Advantages

- Ease in operation
- Relatively inexpensive.
- Control over the fertilizer concentration with the irrigation water

Disadvantages

- High pressure loss (i.e. about one third of the total operating pressure).
- Relatively low discharge rate system.
- Precise regulation of flow is difficult because the rate of injection is very sensitive to the pressure and rate of flow of water in the main system.
- The irrigation system needs to be operated at full capacity prior to injecting the fertilizer solution.

Direct injection system

With this method, a pump is used to inject fertilizer solution into the irrigation line. The type of pump used is dependent on the power source. The pump may be driven by an internal combustion engine, an electric motor or hydraulic pressure. The electric pump can be automatically controlled and is thus the most convenient to use. However its use is limited by the availability of electrical power. The use of a hydraulic pump, driven by the water pressure of the irrigation system, avoids this limitation. The injection rate of fertilizer solution is proportional to the flow of water in the system (Tiwari, 2007). A high degree of control over the injection rate is possible, no serious head loss occurs and operating cost is low. Another advantage of using hydraulic pump for fertigation is that if the flow of water stops in the irrigation system, fertilizer injection also automatically stops. This is the most perfect equipment for accurate fertigation. Two injection points should be provided, one before and one after the filter for fertigation. This arrangement helps in by-passing the filter, if filtering is not required and thus avoids corrosion damage to the valves, filters and filter-screens or to the sand media of sand filters.

The capacity of the injection system depends on the concentration, rate and frequency of application of fertilizer solution.

Advantages

- Flexible and high discharge rate.
- Do not add to head loss in drip irrigation system.

- Maintains constant concentration of nutrients throughout the fertigation period.

Disadvantages

- Expensive.
- Need skilled personnel to operate and maintain the device.
- Changes in water flow rate, power failure or mechanical failure may cause serious deviations from planned concentrations.
- Need external power source to operate the device.

Selected References

Rajput TBS, Patel N. 2002. Fertigation Theory and Practices. Technical Bulletin.. Water Technology Centre, IARI, New Delhi:

Singh PK. 2012. Chemigation under protected cultivation. In training programme on "Diseases and management of crops under protected cultivation.

Tiwari KN. 2007. Pressurized Irrigation. PFDC/IITKGP/1/2007.

18

Enhancing Nutrient Use Efficiency, pp. 227-234
Editors: K. Ramesh, A.K. Biswas, B.L. Lakaria, S. Srivastava and A.K. Patra

Socio-Economic and Technological Issues Associated with the Field Level Adoption of Nutrient Management Technologies

Shinogi, K.C. and Sanjay Srivastava

ICAR-Indian Institute of Soil Science, Nabi Bagh, Bhopal – 462 038, India

Introduction

Land serves as storage for water and nutrients for plants and other living organisms, but the land resources are limited. India has about 18% of world's population and 15% of livestock population to be supported from only 2% geographical area and 1.5% of forest and pasture lands (Rao *et al.*, 2015). The per capita availability of agricultural land has declined from 0.48 hectare in 1951 to 0.16 hectare in 1991 and is likely to decline further to 0.08 hectare in 2035. This decline in per capita land availability in the country is mostly on account of rising population (GOI, 2010). Agriculture is subjected to high risk because of the volatile nature of the factors of production. Since its development, agriculture has expanded vastly in geographical coverage and new technologies are integrated throughout this expansion. Development of agricultural technologies has steadily increased agricultural productivity and the widespread diffusion of these technologies revolutionized county's agricultural scenario. However, for food, energy and other human requirements, preservation and improvement of the productivity of land is inevitable.

In the recent past, land based livelihoods of small and marginal farmers are becoming unsustainable, because of the inequitable access to productive resources, such as land, water, improved inputs and technologies, microfinance etc. This is keeping most of the farmers in the poverty trap. Hence, to develop a better agricultural growth that ensure an improved livelihood to the farming community, at some point of time it is necessary to think about how to manage the existing systems for a better agricultural growth than spending on new initiatives (Naik, 2010).

Agricultural technology

The use of the term "technology" has changed significantly over the last 200 years. Before the 20th century, the term technology was uncommon in English. The *Merriam-Webster Dictionary* offers a definition of the term: "the practical application of knowledge especially in a particular area" and "a capability given by the practical application of knowledge". Technology can be tools and machines that may be used to solve real-world problems, and also a collection of techniques. In this context, it is the current state of humanity's knowledge of how to combine resources to produce desired products, to solve problems, fulfill needs, or satisfies wants through technical methods, skills, processes, techniques, tools and raw materials.

Still, there are several confusions regarding what is meant by a technology in agriculture. Any agriculture related technology is a factor that changes the production function though existence of some uncertainty is there. The uncertainty diminishes over the time through the acquisition of experience and information, and the production function itself may change as the adopters become more efficient in the application of the technology (Feder and Umali, 1993). Andrei Parwan explains that a technology is assumed to be a new scientifically derived and complex input supplied to farmers by organization with deep technical expertise. However, a project or a technique can be still considered as a technology even if the science in many steps removed from the eventual implementer. For instance, a project where a field level extension agent motivate a farmer to adopt crop rotation or any other agronomic interventions in the farm can be treated as a "low-tech", but explaining the chemistry behind the nitrogen fixation while encouraging the adoption of biofertilizers is "extensive-tech" (Parwan, 2011).

Measurement of technology adoption

'Centre for International and Area Studies' report that the major challenge for agricultural researchers in developing countries is to understand how and when new technologies are used by farmers. For this task, agricultural scientists have turned to social scientists, asking for improved understanding of the mechanisms underlying technology adoption (Caswell *et al.,* 2001). Over the years, researchers have worked to answer changing questions about agricultural technology adoption. Initially, policy makers and researchers sought simple descriptive statistics about the use and diffusion of new seed varieties and associated technologies such as fertilizer and irrigation. Concerns arose later about the impact of technology adoption - on commodity production, on poverty and malnutrition, on farm size and input use in agriculture, on genetic diversity, and on a variety of social issues (Doss, 2006). Numerous researchers (many associated with the CGIAR centers) have developed innovative methodologies for addressing such concerns, carried out surveys and collected enormous amounts of data to describe and document the adoption of new agricultural technologies. Yet many questions remain. At the simplest level, we still have considerable gaps in our knowledge of which technologies are being used, where, and by whom.

The available literature on technology adoption is spread over three categories; 1) innovative econometric and modeling methodologies to understand adoption decisions, 2) examinations of the process of learning and social networks in adoption decisions and 3) micro-level studies based on local data collection intended to shed light on adoption decisions in particular contexts for policy purposes (Doss, 2006). In fact, understanding the adoption pattern of different technologies helps to know to what extent they are socially benefited and thus contributing to the economic development of the nation. Many of the socially valuable and potential technologies are slow in the diffusion process because of the information constraints and other externalities. There are so many approaches to analyze the technology adoption pattern and the three basic empirical methods out of those are *Time-Series Studies, Cross-Sectional Studies, and Panel Data Studies* (Besley and Case, 1993).

Time-series studies are only an aggregate measure of adoption, for example percentage of farmers adopted the new technology over a period of time is studied and gives a rough idea about the major driving force behind adoption. Time-series data explains the rate of change in adoption of a particular technology over a time period but does not attend to the fundamental reasons for adoption of technology.

Cross-sectional studies at the micro level answer the questions regarding the decision making process of the farmer in the choice of technologies and other farm management practices. The method analyzes the adoption pattern in two ways, the first one is a probit analysis model where the profit of a farmer by using a new technology is measured at some date (snapshot analysis) and the second method based on the recall of the time when the farmers adopted the technology (recall analysis). Cross-section analysis at the micro-level can answer important questions about technology use and also provide some information on patterns of adoption and disadoption (Doss, 2006). However, the researchers often make unrealistic assumptions to convert the cross sectional data obtained at a single point of time to a larger time frame to explain dynamics of technology adoption. This is the main shortfall of cross sectional studies compared to the time-series studies. Similarly, cross-section data cannot tell us much about the impact of a new technology on the well-being of farmers or farm communities.

Panel studies bring together time-series and cross-sectional data to understand the dynamics of adoption decisions, rather than just developing static descriptions of particular areas. Technology adoption decisions are inherently dynamic. Farmers do not simply decide whether or not to permanently adopt an improved variety, but instead they make a series of decisions: whether or not to try planting an improved variety, how much land to allocate to the improved variety, whether or not to continue to grow it, whether to try a different improved variety. Decisions about other input use and management techniques are similarly complex. Decisions in one period depend critically on decisions made in previous periods. To understand these decisions, farmers' decisions need to be followed over a period of time. This is best done with panel data sets of farmers. Panel data would also help us to understand the distributional impacts of new technology. Since many things change within rural communities, both in response to new agricultural technologies and in response to changes in outside forces, panel data are needed to sort out the effects.

Nutrient management technologies

World bank report in the late 90s showed that the nutrient capacity of soil was boosted by about 2000 kg of nitrogen (N), 700 kg of phosphorus (P) and 1000 kg of potassium per hectare of arable land in Europe and North America over a period of 30 years due to the over application of organic and inorganic fertilizers (World Bank, 1996). But, leaching and run-off of surplus N and P into rivers and lakes also caused eutrophication. There are a number of technologies generated for the improvement of soil fertility and maintaining soil health with optimum use of external inputs like chemical fertilizers. However, the adoption of any one of these technologies depends on a number of factors like the socio-economic profile of the farmer, risk involved with the technology and relative advantage and cost involved with the adoption of a particular technology compared to the rest of the available technologies. Though many studies assessed the benefits of adopting input use efficiency to maintain sustainable soil systems, no study is available regarding the assessment of factors that affect the adoption decision of farmers and extent of adoption of nutrient management technologies (Grazhdani, 2013).

Existing international research literature on the adoption of technology mostly deals with those of green revolution i.e. high yielding variety (HYV) seeds, the irrigation schedule and fertilizer use associated with these crop varieties (Niell and Lee, 2001) and these literatures are ultimately concerned with understanding the choices of rural communities regarding the farming methods and technologies. Moreover, mostly fertilizer use is bundled up as one of the essential input of HYVs and presented before the farmers as a single choice. So, there is no evidence to understand the decision making pattern regarding the adoption of input use pattern (Parwan, 2011).

Factors influencing the adoption of innovative agricultural/nutrient management technologies

There are not much recent studies available dealing with the assessment of factors affecting the nutrient management technologies. The decision to adopt a new technology is similar to an investment decision and the farmer need to think about the monetary cost as well as the nonmonetary cost involved while adopting. The five attributes of innovation identified by Rogers viz., relative advantage, compatibility, complexity, trialability, and observability are still valid as the major factors affecting the adoption of a technology (Rogers, 1983). Profitability of the technology is only one factor to motivate for adoption as effectiveness of any innovative technology depends on where and when it is used (Stoneman and David, 1986). In the adoption of organic fertilizer technologies like biofertilizers, animal manures, green manures and crop residues the major constraints are requirement of considerable labour input, substantial risk involved when farmer need to produce food crops on all arable land and so these technologies often may not meet the criteria of farmers (Snapp *et al.*, 1998). In general, the different factors can be categorized as.

I. Techno-Economic Factors

a. **Uncertainties associated with the technology:** When a technology is first introduced to a farmer or a group of farmers, uncertainty about the performance of the technology in the field condition will be high even after the scientist or extension agent convinces them by showing the superior result in the research station. However, these kinds of perceptions are subjective and mostly change over time once they learn more about the benefits associated from those who already adopted the technology in the nearby area. Once all the uncertainties are over, some farmers may become ready to adopt it in a portion of the farmland.

b. **Limited access to resources:** Size of land holding, income of the farm family, irrigation water availability of the farm, availability of institutional credit and even education level of the farmer have a direct positive correlation with farmers' decision making process while adopting any new agricultural technology (Ahmed, 2014). Larger farms are more likely to adopt the technology that comes in bundles or as a complete package sooner than small farms. However, the effect of farm size may not have the same effect for all the technologies. Still, the size of the farm may act as a socio-economic indicator for the increased access to credit facilities.

c. **Preventive innovations:** Innovations that are hard to demonstrate the advantages of adoption since the benefits will occur only at some future, unknown time are known as preventive innovations (Rogers, 1983). Most of the resource conservation and nutrient management technologies belong to this category and hence, they have always a low rate of adoption.

d. **Inability to adopt:** Sometimes farmers may not be able to adopt the technology even if they are interested to do because of lack of sufficient information or poor accessibility to the information sources, limited availability of supporting resources, inadequate managerial skills, or no control over the adoption decision (Nowak, 1992). Farmers are always looking for different options to combine inputs in order to maximize the output with minimum cost (Thomas and Sumberg, 1995). But, for the use of both organic and inorganic sources of soil fertility, information is largely incomplete. Whether it is the optimum fertilizer use or the best combination of the organic and inorganic input combinations, the information sources that are easily accessible to the rural farmers fail to provide the relevant information. This prevents them to adopt those good nutrient management technologies available in the country.

II. Socio-psychological factors

Unwillingness to adopt: Several farmers show unwillingness to adopt beneficial technologies because of the belief in the traditional practices they are following over a period of time in the farmland, conflicts or inconsistency in the information regarding the technology, or increased risk of negative outcomes. The U.S. Department of Agriculture Area Studies Project report says that more experienced farmers are less likely to adopt modern nutrient management practices like N-testing, and micro nutrients application since they may feel they already have sufficient knowledge to determine appropriate fertilizer needs or perhaps they are more reluctant to switch from technologies they have used for years (Caswell *et al.,* 2001). Adherence to traditionalism and ethnicity of some rural farming community make them less responsive to the adoption of new agricultural technologies especially that of nutrient management practices like soil testing and method of fertilizer or pesticide use.

Strategies for improving the technology adoption process

The existing gaps between the research findings and reality in farmers' fields, only partly explains the low adoption rates of nutrient management technologies that enhance the soil productivity under field condition (Schlecht *et al.*, 2007). Still, there are a number of opportunities before the researchers, extension agents and policy makers to disseminate the innovative technologies in an effective way. This will improve the adoption rate of agricultural/nutrient management technologies by the farmers.

a. **Farmers' participation in technology dissemination:** Disseminating the information through key farmers or master farmers is the emerging concept in the technology diffusion process and many agencies like Honey Bee Network supported by Society for Research and Initiatives for Sustainable Technologies and Institutions (SRISTI) and Vegetable and Fruit Promotion Council Keralam (VFPCK) are already adopted this method. The master farmers will be local progressive farmers who have the complete knowledge about the village and the rural community. They have an influential power over the farmers because of their position in the society knowledge in farming. Once these local leaders are convinced with the benefits of new technology, the further diffusion of the technology in the rural society will be easy.

b. **Ensure Financial Support:** To promote any innovative agricultural technology among the small or marginal farmers it is essential to support them financially through specific schemes by providing inputs at least for a minimum period. Because, small and marginal farmers who practice agriculture to support their livelihood may not be ready to sacrifice their assured profit they are gaining through their own farming methods. The financial support may motivate these neo-converts to adopt the new technology since there is a compensation for their losses.

Conclusion

Spreading of nutrient management technologies among the traditional rural farmers is always a tedious job as they are more conservative than the educated progressive farmers. The time lag associated with most of the nutrient management technologies using on-farm organic resources to show a visible impact may be the major challenge here. Because the extension principle mostly works to deal with the rural farming community is "*seeing and believing*". A visible effect of technology like that of HYV crops can motivate them to adopt it but, mostly they miss the high nutrient requirement of those crops to give the potential yield. However, the bottom-up approach of technology dissemination may be more successful to deal with these traditional rural farmers. Still, farmers' trust on the information source or the researcher or extension agent who introduce the technology before them and financial support from the promoting agency in the initial years of adoption play the major role in convincing the farmers to adopt the technology in a proper manner.

Selected References

Ahmed S. 2014. Factors and constraints for adopting new agricultural technology in Assam with special reference to Nalbari district: An empirical study. *Journal of Contemporary Indian Polity and* Economy, Available online:http://jocipe.com/jocipe1.html, Accessed on 3 August 2014.

Beseley T, Case, A. 1993. Modeling technology adoption in developing countries, *The American Economic Review* 83(2): 396-402

Caswell M, Fuglie K, Ingram C, Jans S, Kascak C. 2001. Adoption of Agricultural Production Practices: Lessons Learned from the U.S. Department of Agriculture Area Studies Project. Agricultural Economic Report No. (AER-792) 116 pp. United States Department of Agriculture. Available:http://www.ers.usda.gov/publications/aer-agricultural-economic-report.

Doss C R. 2006. Analyzing technology adoption using microstudies: Limitations, challenges, and opportunities for improvement. *Agricultural Economics* 34(3):207-219.

GOI. 2010. *India 2010*. Publication division, Ministry of information and broadcasting, Government of India.1286p.

Grazhdani D. 2013. An analysis of factors affetcting the adoption of resource conserving agricultural technologies in Al-Prespa Park. *Natura Montenegrina*, 12(2):431-443.

Naik G. 2010. Indian agriculture - Issues and reforms.*tejas@iimb: an IIMB Management Review Intitiative*, Indian Institute of Management, Bangalore. Available: http://www.tejas.iimb.ac.in/interviews.

Neill S, Lee D R. 2001. Explaining the adoption and disadoption of sustainable agriculture: The case of cover crops in northern Honduras, *Economic Development and Cultural Change* 49: 793-820.

Novak PJ. 1992. Why farmers adopt production technology, *Journal of Soil and Water Conservation*, 47(1):14-16.

Parwan A. 2011. Agricultural technology adoption: Issues for consideration when scaling up. *The Cornell Policy Review* 1(1).

Rogers E M. 1983. *Diffusion of Innovations*,Third Edition, The Free Press, New York.

Rao MV, Babu SV, Chandra KS, Chary GA. 2015. Integrated land use planning for sustainable agriculture and rural development. Apple Academic Press Inc. Canada. 329 pp.

Schlecht E, Buerkert A, Tielkes E, Bationo A. 2007. A critical analysis of challenges and opportunities for soil fertility restoration in Sudano-Sahelian West Africa. In: Bationo, A., Waswa, B., Kihara, J., Kimetu, J. (eds.). *Advances in Integrated Soil Ferttility Management in Sub-Saharan Africa:Challenges and Opportunities*. 1077p.

Stoneman PL, David PA. 1986. Adoption Subsidies vs. Information Provision as Instruments of Technology Policy, *The Economic Journal*, 96:142-150.

World Bank. 1996. Natural resource degradation in Sus-saharan Africa: Restoration of Soil fertility. A Concept Paper and Action Plan, Washington DC.

Enhancing Nutrient Use Efficiency in Crops and Cropping Systems

19

Enhancing Nutrient Use Efficiency, pp. 235-262
Editors: K. Ramesh, A.K. Biswas, B.L. Lakaria, S. Srivastava and A.K. Patra

Dryland Crops and Cropping Systems

K. Sammi Reddy, Sumanta Kundu, K.L. Sharma and Ch. Srinivasarao

ICAR-Central Research Institute for Dryland Agriculture, Santoshnagar Saidabad (P.O.), Hyderabad, 500 059, Andhra Pradesh, India

Introduction

India has predominantly agrarian economy with just 2.5% of the world's geographical area, but has about 17% of its population. India's population has increased from 361 million in 1951 to 1270 million at present: a three and half-fold increase in a span of 63 years. By 2020, India will need about 294 million tonnes food grains (Kumar, 1998). However, only 259.29 m t was produced in 2011–12 (Directorate of Economics and Statistics, Ministry of Agriculture, 2012), implying that about 34 m t additional food grains have to be produced from the same or even lesser land area.

Out of the estimated 141 m ha of net cultivated land in India, 85 m ha is rainfed which produces 40 percent of the food grains in the country. Low and erratic rainfall, high temperature, degraded soils with low available water content and multinutrient deficiencies are important factors contributing to low crop yields in these regions. In India, the predominant rainfed production systems are upland rice, sorghum, maize, pearlmillet, fingermillet, cotton, groundnut and soybean besides several pulses and oilseeds, which are characterized by diverse climate of arid, semi-arid and subhumid, with mean annual rainfall ranging from 412 mm to 1400 mm and spatial variability of soils consisting of Entisols, Inceptisols, Alfisols, Vertisols, Aridisols and Oxisols. Drylands are not only thirsty but also hungry. Major reasons for low fertility of dryland soils are low in organic matter, low biomass generation, low left over crop residue and its rapid oxidation due to high temperatures (Srinivasarao *et al.*, 2013). Nitrogen is universally deficient in these soils, P is deficient in black soils while some of red soil regions observed build up of available P. Again K application was never considered seriously especially in rainfed areas. The deficiency of secondary and micro nutrients also cropped up in extensive rainfed regions as supplementation of nutrients was seldom practiced. Potassium along with secondary nutrients, like sulphur, calcium and magnesium are emerging as constraints for rainfed production. Among micro nutrients, zinc and boron deficiencies are wide spread (Rego *et al.*, 2007).

Fertilizer consumption increased 326 times in India during the 1950–51 to 2011–12 period. The yield curve in most of the crops in irrigated areas has touched the plateau because of stagnated response to the added inputs. But there is a huge yield gap in rainfed agriculture. Considering the gap between possible potential yield of rainfed crops and their actual level realized yields, there is tremendous scope to enhance the yield levels in rainfed agriculture. Data from on farm trials conducted under the Simple Fertilizer Trials Scheme of ICAR show that response ratio was the highest for N (11.6 to 16.7 kg grain kg N^{-1}), followed by P (5.5 to 12.5 kg grain kg $P_2O_5^{-1}$), and the least for K (3.6-6.2 kg grain kg K_2O^{-1}) (Prasad, 2008). Furthermore, response to NP, NK, or NPK was not additive of their individual responses, which made the farmers to apply mostly N alone.

Constraints of dryland farming

The major constraints in improving the productivity and returns from dryland farming in India are: (i) erratic and uncertain rainfall, leading to moisture scarcity, droughts and failure of crops, especially annual crops, (ii) soil degradation and poor soil quality (iii) fragmented and small holding size, leading to constraints in mechanization, (iv) poverty among growers and constraints in availability and purchase of essential inputs, such as seeds and fertilizers, bullock-drawn small seed-cum-fertilizer drills, etc., (v) lack of assured credit and financial support and marketing, (vi) inadequate infrastructure for post-harvest value-addition and storage of produce, (vi) low procurement prices of agricultural commodities, in general, and (vii) inadequate earnings for livelihood from the farming profession because of low volume of business due to small holding size, low productivity and low produce prices, etc. The consequences of these constraints are likely to lead the marginal and small-farming communities towards distraction from agriculture, migration to cities to look for alternate assured wages, suicides, etc. To mitigate these constraints and transform the rainfed farming to an attractive option, there is a strong need for strategic planning and policy changes in a phased manner.

Dryland production systems

Historically dryland farmers practice high diversity in cropping systems with livestock integration which is an inbuilt risk management strategy. The cropping patterns have evolved based on the rainfall, length of the growing season and soil types. However, due to changed consumer preferences and market demand, farmers are now rapidly shifting to crops and cropping patterns which are more remunerative. Data in Fig.1 shows that how sharp increase in area under maize and cotton took place in few years at the cost of coarse cereals like sorghum, pearl millet mainly due to higher returns. Such changes will have implications on fodder availability to livestock. This trend is likely to further increase by 2030.

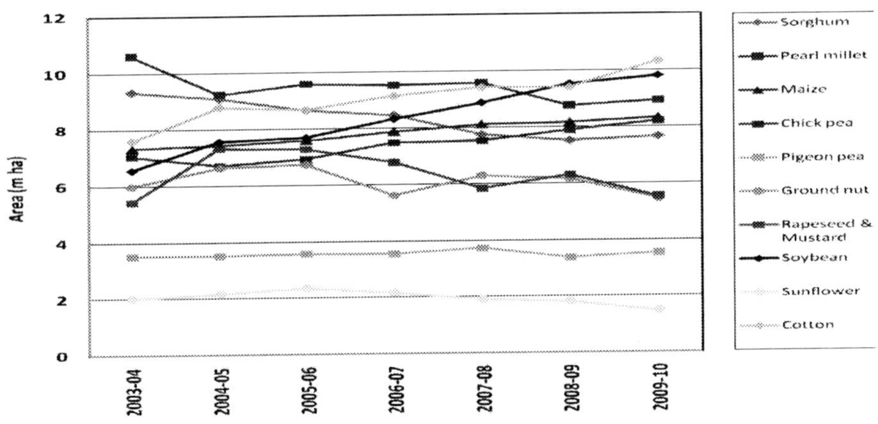

Fig.1. Recent trends in shift in cropped area under major rainfed crops (*Source*: CRIDA Vision, 2030)

The change in cropping pattern will have implications on the resource use. Continuous mono cropping increases vulnerability of farmers to weather risks, depletes soil fertility, ground water and leads to build up of pests and diseases. This issue has to be dealt both through technology and policy. For example, we have to evolve management practices for farmers choice of remunerative crops without degradation of the natural resource base and also define agro-ecological zones where such cropping systems can be adopted sustainably. Simultaneously, need based policy incentives are required to encourage farmers adopt agro-ecology-compatible cropping systems so that the farmers' income is maintained and the natural base of the country is not degraded.

Yield gap in dryland crops

It is in the rainfed belt where most of the coarse cereals (91%), pulses (91%), oilseeds (80%) and cotton (65%) are produced. It emphasizes the importance of dryland agriculture in the country's economy and food security. The yield curve in most of the crops in irrigated areas has touched the plateau because of stagnated response to the added inputs. Therefore, it is anticipated that if at all another green revolution is expected in Indian agriculture, it would come from these gray areas only. Considering the gap between possible potential yield of rainfed crops and their actual realized yields, there is tremendous scope to enhance the yield levels in rained agriculture.

An analysis of the critical yield gaps in important dryland crops indicate the large difference in yield levels between national demonstrations and national average (Fig. 2). Similarly in different rainfed pulses, there is a wide gap among experimental yields, front line demonstration yields and national average yields (Table 1). This suggests that there is enormous scope for enhancing the production and productivity of pulse crops in India with improved management.

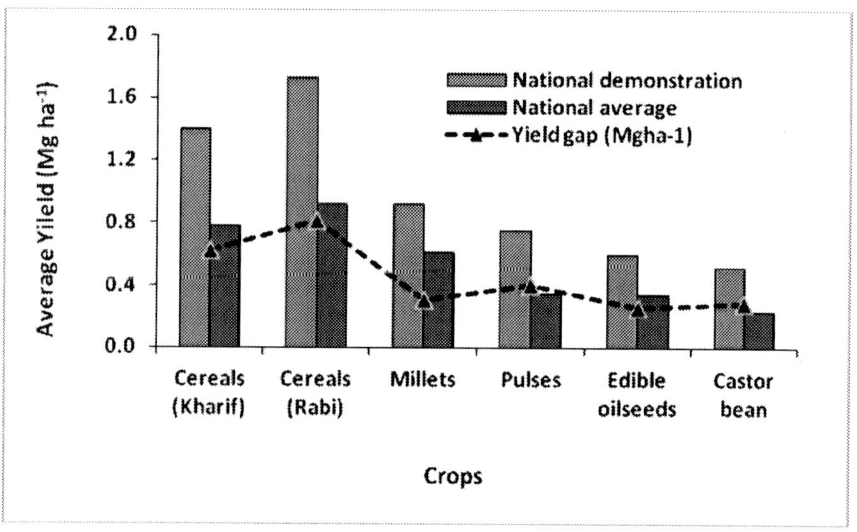

Fig. 2. Yield gaps in dryland agriculture in India (*Source*: Srinivasarao *et al.*, 2013)

Table 1: Critical yield gaps (q ha^{-1}) in major pulse crops under rainfed conditions

Crop	Yield potential on research plots	Yield in FLDs at farmers fields	National average
Chickpea	20-22	15-18	8.06
Pigeon pea (Early)	15-17	12-15	7.97
Pigeon pea (Late)	20-25	20-22	-
Mung bean	11-12	9-10	3.81
Urd bean	10-12	8-9	4.40
Field pea	20-22	15-18	10.34
Lentil	15-18	12-14	7.32

Source: Chaturvedi and Ali (2002)

The projected area, production and yield of cereals under various production systems of India for 2030 and 2050 are shown in Table 2. These projections assume little impact of technology in rainfed production systems. While irrigated systems can contribute an additional yield of 15%, the rainfed systems could remain the same. Thus, there would be a need for a strategic mix of better technology adoption, institutional innovations and incentive system to enhance productivity of rainfed cereal systems.

Table 2: The projected area, yield and production of cereals under different production systems of India

Cereals area	2030	2050
Rainfed area (million hectare)	40	36
Rainfed yield (tonnes/hectare)	1.8	2.0
Rainfed production (million metric tonnes)	73	72
Irrigated area (million hectare)	57	62
Irrigated yield (tonnes/hectare)	4.3	4.6
Irrigated production (million metric tonnes)	248	285
Total area (million hectare)	97	98
Total yield (tonnes/hectare)	3.3	3.7
Total production (million metric tonnes)	321	357

Source: CRIDA Vision, 2050

Eroding demand for some of the rainfed commodities is another challenge while other crops are enjoying favourable market demand. Therefore, technologies are to be developed for those crops which have expanding markets but are difficult to be grown in rainfed conditions. Wherever appropriate, steps must be taken to contribute to formulation of policies for promoting crops that are not so resource-demanding but at the same time have better food, nutrition and fodder value.

Nutrient deficiencies in rainfed agriculture

Cropping systems of present day are heavy feeders and are bound to extract heavily the nutrients from the soil. Hence nutrient deficiencies are inevitable unless steps are taken to restore fertility levels. Deficiency of essential elements in Indian soils and crops started emerging since 1950s' after the governmental efforts of independent India to give a fillip to food production through intensification. Extent of deficiency in major and micro nutrients in predominant rainfed regions of India are presented in Tables 3 and 4. Major reason for low fertility of dryland soils is low organic matter. Low biomass generation, lower left over crop residue and its rapid oxidation with high temperatures are causes for low organic matter in dryland soils (Srinivasarao *et al.*, 2009). Nitrogen is deficient in most of these soils. P is deficient in black soils while some of red soil regions like Bangalore and Anantapur showing build up of available P. However, K deficiency is one of the emerging constraint which limits crop productivity. The deficiency of secondary and micro nutrients also creeped in extensive rainfed regions as supplementation of nutrients was seldom practiced. Summary of emerging nutrient deficiencies at 21 locations of rainfed regions of India is given in Table 5 indicating that most of the rainfed soils have multi-nutrient deficiencies. Potassium along with secondary nutrients, like sulphur, calcium and magnesium are emerging as constraints for rainfed production. Among micronutrients, zinc and boron are severely deficient in soils of rainfed regions. (Srinivasarao and Vittal, 2007).

Table 3: Fertility status of N. P and K in predominant rainfed states of India

State	No of samples analysed	Nitrogen			Phosphorus			Potassium		
		L	M	H	L	M	H	L	M	H
Madhya Pradesh	1,38,553	40	41	19	39	38	23	10	32	58
Uttar Pradesh	8,07,424	80	15	5	7186	26	3	12	55	33
Maharashtra	93,142	67	26	7	57	12	2	8	18	74
Andhra Pradesh	3,12,521	62	21	17	31	29	14	9	30	61
Karnataka	3,17,213	29	37	34	59	48	21	7	32	61
Orissa	2,51,196	60	23	17	24	28	13	33	41	26
Tamil Nadu	4,91,657	75	16	9	42	41	35	12	36	52
India	36,50,004	63	26	11		38	20	13	37	50

(*Source*: Motsara, 2002) L: Low; M: Medium, H:High

Table 4: Micronutrient deficiencies in predominant rainfed states of India

State	No. of samples (range of samples analyzed for different states)	Percentage deficiency of available micronutrients			
		Zn	Fe	Cu	Mn
Andhra Pradesh	5219-6563	51	2	1	2
Bihar	17802-19078	54	6	3	2
Gujarat	29532	24	8	4	4
Madhya Pradesh	11204-12000	63	3	1	3
Tamil Nadu	19559-20580	53	15	3	8
Uttar Pradesh	24425-25122	45	6	1	3
Karnataka	24411-25542	78	39	5	19

Source: Takkar (1996)

Table 5: Emerging nutrient deficiencies in dryland soils (0-15 cm) in India

Location	Limiting Nutrient (Low/Deficient)
Varanasi	N, Zn, B
Faizabad	N
Phulbani	N, Ca, Mg, Zn, B
Ranchi	Mg, B
Rajkot	N, P, S, Zn, Fe, B
Anantapur	N, K, Mg, Zn, B
Indore	-
Rewa	N, Zn
Akola	N, P, S, Zn, B
Kovilpatti	N, P
Bellari	N, P, Zn, Fe
Bijapur	N, Zn, Fe
Jhansi	N
Solapur	N, P, Zn
Agra	N, K, Mg, Zn, B
Hisar	N, Mg, B
SK. Nagar	N, K, S, Ca, Mg, Zn, B
Bangalore	N, K, Ca, Mg, Zn, B
Arjia	N, Mg, Zn, B
Ballowal-Saunkri	N, K, S, Mg, Zn
Rakh-Dhiansar	N, K, Ca, Mg, Zn, B

Source: Srinivasarao and Vittal (2007)

Fertilizer consumption

With the increased fertilizer consumption, there was an increase in food grain production. There was a five-fold increase in cereal production, i.e., from 42.4 m t in 1950–51 to 215.6 m t in 2011–12. With a consumption of 14.4 m t N, 5.5 m t P_2O_5, and 2.6 m t of K_2O in 2007–08, India now occupies the second position after China in N and P consumption. In K consumption, India occupies the fourth position.

The role of fertilizers in augmenting food production in India was realized only with the introducing of high yielding dwarf varieties of wheat and other cereals. However, there are wide disparity in fertilizer consumption between irrigated and rainfed agriculture, primarily due to uncertainty associated with rainfed crops. In India, about 80% fertilizer is concentrated in the irrigated areas, and remaining 20% is used 70% of the cropped area. Average NPK fertilizer consumption increased to about 120 kg ha^{-1} in irrigated conditions the same is about 55 kg ha^{-1} under rainfed agriculture (Fig. 4). Even among rainfed conditions major share of nutrients goes to cotton, sunflower and other commercial crops (Fig. 5). However, fertilizer use efficiency in rainfed crops in rather low.

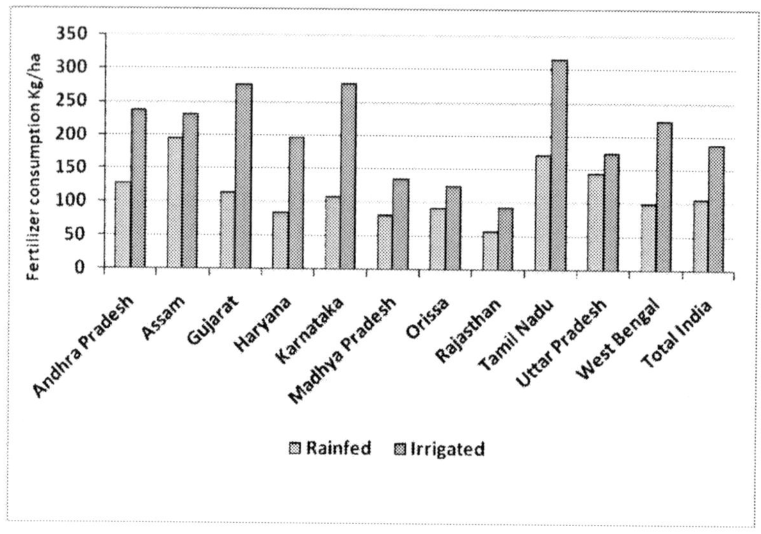

Fig. 4. Fertilizer consumption in irrigated and rainfed areas in major states in India (*Source*: FAI, 2012)

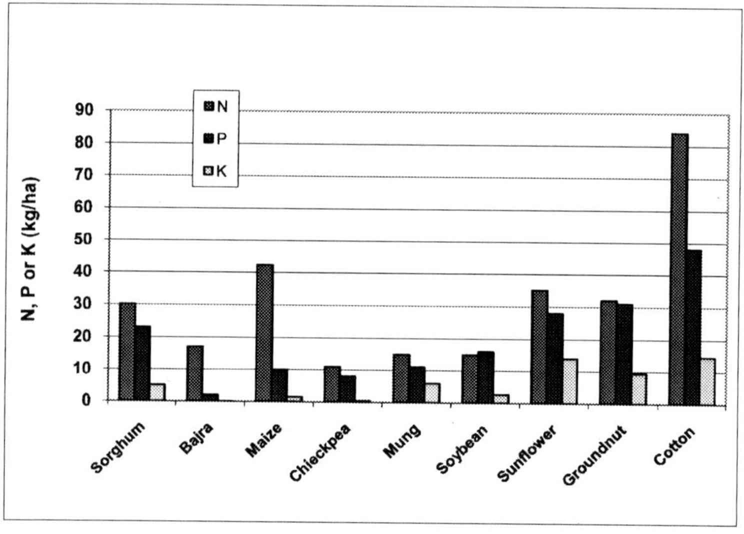

Fig. 5. Consumption of NPK in major rainfed crops of India (*Source*: Srinivasarao *et al.*, 2011)

Nutrient removals

Intensive cropping invariably results in heavy withdrawal of nutrients from soils and its sustenance largely depends upon the judicious application of inputs commensurate with nutrient uptake. Nutrient uptake values generally provide a reliable estimate of nutrient requirements under varying agro-ecological regions which would form the basis for the development of a sound fertilizer recommendation strategy for realizing higher productivity and maintaining soil fertility. N, P and K uptake per tonne of grain yields have indicated that NPK requirements are as high as 67, 28 and 105 kg ha^{-1} in some important rainfed crops (Table 6). Similarly, uptake of secondary and micronutrients in several rainfed crops (Table 7) shows that sulphur removal goes up to 7 kg ha^{-1} while the uptake of B and Zn were as high as 52 g and 192 g ha^{-1}.

Table 6: Nutrient uptake of some important rainfed crops in India

Crop	Produce	kg nutrient/tonne produce		
		N	P_2O_5	K_2O
Sorghum	Grain	22.4	13.3	34.0
Pearl millet	Grain	42.3	22.6	90.8
Rice	Grain	20.1	11.2	30.0
Chickpea	Grain	46.3	8.4	49.6
Groundnut	Grain	58.1	19.6	30.1
Soybean	Grain	66.8	17.7	44.4
Sunflower	Grain	56.8	25.9	105.0
Cotton	Seed	44.5	28.3	74.7

Source: Srinivasarao and Venkateswarlu, 2009

Table 7: Mean yield and uptake of nutrients by crops grown in water several sheds in Andhra Pradesh under rainfed conditions

Crop	Stover (t ha^{-1})		Grain (t ha^{-1})		Total Nutrients Removed (g ha^{-1})					
					Control			Treated		
	Control	Treated	Control	Treated	S	B	Zn	S	B	Zn
Mungbean	0.73	1.00	0.77	1.11	2325	20	46	4009	30	68
Maize	3.46	4.29	2.73	4.56	4536	16	112	7014	19	192
Groundnut	1.99	2.49	0.70	0.94	4355	40	50	6418	52	81
Pigeonpea	1.31	2.10	0.54	0.87	1619	22	27	2649	36	45
Castor	0.82	1.19	0.59	0.89	2216	18	40	3650	26	62

Source: Rego *et al.* (2007)

Nutrient use efficiency: Present scenario

Nutrient use for agricultural production is essential to increase food production but it is a major cause of pollution. With the discovery that plant growth requires nutrient elements in addition to the basic factors of soil, water and air, agricultural production has increased by leaps and bounds. The urgency for higher agricultural production and the greed for higher profits have made nutrient applications in agriculture unscientific with more wastage leading to pollution of soil, air and water. Although fertilizer consumption is increasing quantitatively, the corresponding yield increase per unit of nutrient is substantially diminishing over the years. NUE depends on several agronomic factors including tillage, time of sowing, appropriate crop variety, proper planting or seeding, sufficient irrigation, weed control, pest/disease management, and balanced and proper nutrient use. These factors largely influence NUE, either individually or collectively. For example, selection of proper planting material, population density, and balanced fertilization could collectively improve NUE by 25 to 50% (Rao, 2007). Soil degradation is occurring due to inadequate and imbalanced fertilization leading to nutrient mining and development of second generation problems in nutrient management. The supply-demand gap is of concern that the partial factor productivity of fertilizers has been continuously declining. The efficiency of fertilizer N is only 30-40% in rice and 50-60% in other cereals, while the efficiency of fertilizer P is 15-20% in most crops. The efficiency of K is 60-80%, while that for S is 8-12%. As regards the micronutrients, the efficiency of most of them is <5% (NAAS, 2006). The developed economies of the world (France, Japan, USA, UK) operate at potential yields of crops, with highest rate of fertilizers per unit area (300 to 600 NPK kg/ha). In USA, the fertilizer-N efficiency in maize has increased by >30% over the last 20 years (Fixen and West, 2002). In Japan, fertilizer N use has decreased by about 30% from 1960 to 2000, while keeping a little increase in cereal yields. In our country operating at the lowest productivity of most crops, the fertilizer consumption has reached 107 kg ha^{-1}. Yet, our advantage to achieve higher use efficiency or recovery even at this low rate of application is not realized due to the limiting agro-ecological conditions of crop production. Large areas of cropping under rainfed condition, imbalanced fertilization, tropical climate, cultivation of traditional crops and varieties/genotypes, above all, general low investment capacity of farmers result in low NUE and profitability.

Strategies for enhancing NUE

Fertilizer-use efficiency can be increased through two approaches: (i) Product strategy, wherein *per-se* fertilizer material- carrying nutrients to plants are improved that can resist losses or fixation and match the crop uptake (coated fertilizers, slow-release fertilizers, nitrification inhibitors, urease inhibitors, use of nanotech materials etc.). Use of these new products will be expensive besides their limited availability; and (ii) Management strategy, wherein for the existing fertilizer or material, higher use efficiency is achieved through improved application methods, *viz.* split application, N rate based on plant analysis or variable rate (time and space, precision farming). Bottlenecks and strategies for increasing nutrient-use efficiency (NUE) can be well understood by analyzing the pathways of nutrients reaching the plant roots and its uptake and incorporation in plant tissues.

Time and method of nutrient application

Time of application

Under irrigated conditions split application of N is a well accepted method of increasing NUE, and plenty of literature is available on the subject (Prasad, 2007). Split application of N is highly desirable because crop plants take up small amounts of N per ha per day. Excess N not used by crops is subject to various losses (Prasad *et al.*, 1999b; Adhya *et al.*, 2007; Pathak *et al.*, 2008). But in rainfed condition, where it is difficult to predict the rainfall event and most of the times it does not match with plant's physiological stages, application of fertilizer in splits matching with crop's requirement is difficult. Recent research has shown that for determining the proper time of post transplant/sowing application of N, use of new tools such as chlorophyll meters and leaf color charts holds promise.

Most P and K are applied at sowing. The simple agronomic practice of split-application of nutrients does not increase production costs by much, but it significantly improves NUE. Experiments on horticultural crops such as banana and mulberry have shown that split application of nutrients, especially N and K, improved NUE (Fig. 6) (Rao, 2007).

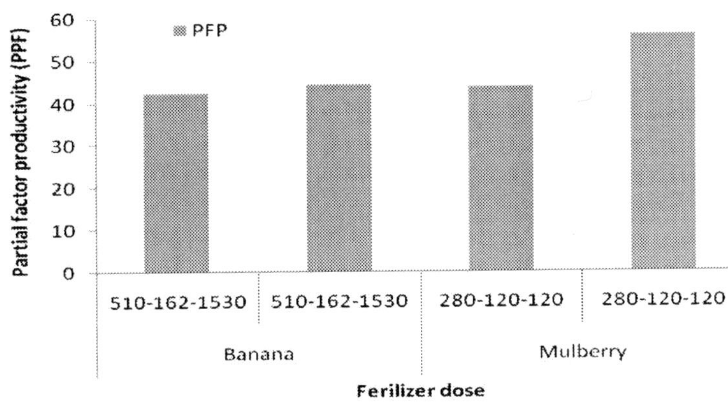

Fig. 6. Effect of split application of nutrients on partial factor productivity (*Source*: Rao, 2007)

Method of application

Considerable literature exists in India on the advantage of deep placement of P for increasing its efficiency for crops other than rice (*e.g.* Srinivasarao *et al.*, 2003). However, only in areas where agriculture is mechanized, deep placement of P is practiced; elsewhere, it is still broadcast, depriving the farmers of the full benefits of P fertilization.

Regarding N, deep placement increases use efficiency considerably. Panda *et al.* (2007) reported that band furrow placement of N doubled N, use efficiency compared to its broadcast application in rainfed lowland rice. Lately, the International Fertilizer Development Centre, USA, has developed a machine for deep placement of urea, which is performing well in Bangladesh (Sharma, 2008). In addition, deep placement of N has definite advantages in dryland crops such as oilseeds (Hegde, 1995).

Foliar application of N is desirable in dryland agriculture because the farmers in these areas apply fertilizers only after rains, and these are often delayed. Under such conditions foliar application of N is the best choice. Kumar and Kumar (2007) recommended foliar application of K in banana. Foliar application of KNO_3 in addition to soil K application was recommended for potato in Ludhiana (Brar and Kumar, 2006).

Use of improved materials

There are various products available which supply plant nutrients, offer a variety of nutrient contents, physical forms, and other properties to meet individual needs. Several researches are going on to find new sources of fertilizers and modifiers to minimize losses, thereby achieve higher use efficiency. Much of the work has been done on N with major source as urea containing high N (46%) and its proneness to losses. The approaches have been to manipulate the granule size variations, coatings with neem or coal tar, modifiers or additives to control the nutrient release rate. Application of controlled release fertilizers (CRFs) is a new approach for minimizing non-point contamination in agriculture. The CRFs have higher N-use efficiency, thus reducing N loss through leaching and volatilization while keeping higher yields (Mao *et al.*, 2005). The CRF applied at 1/3 rd of the rate of common urea fertilizer (CUF) also increased the yield by 15.1% and raised NUE up to 51.2%. The CRF showed advantages in controlling release of N, increasing grain yield and reducing N application. The N uptake efficiency of sesame, upon application of slow-release fertilizers such as cyclodiurea and latex- coated urea, increased by 10-25% in polyethylene film mulching and 3-10% in non-mulching compared with that of urea in silty clay-loam soils of Korea (Lee *et al.*, 2006). The SSNF could be especially useful in drought-prone areas where the availability of water is insufficient. The gelbased CRFs showed significant positive influence on agronomical, physiological and biochemical characteristics of maize plants (Hegde *et al.*, 2007). These increased dry biological yield of maize by 26.8-42.3%, and improved N-use efficiency by 17.0 -31.7%, P-use efficiency by 8.0-16.0%, and K-use efficiency by 4.6-18.3%. Besides, the nutrients (N, P and K) in gel-based fertilizers were leached more slowly into the soil than common fertilizers.

The IARI's urea coating technology employing neem oil emulsion needing 0.5-1.0 kg neem oil per tonne of urea was found superior to prilled urea (Prasad *et al.*, 2001). The neem coated urea of National Fertilizers Limited recorded better shelf-life, slow dissolution as well as nitrification inhibition property. With marginal additional cost for coating, the product showed increased NUE in Punjab, Uttar Pradesh and Himachal Pradesh (Mangat, 2004). Thus, NUE could be increased through suitable modification of urea by compacting it with acid and nonacid producing fertilizers such as NH_4Cl, KCl, $ZnSO_4$ and DAP. The development of thin polymer coatings has improved the opportunity to coat fertilizer granules and increased the predictability of time of availability of nutrients from the controlled release product.

Use of organic materials

Organic materials have a major role to play in maintaining buffering capacity of soil and are important for maintaining soil physical and biological properties. Various factors such as soil temperature, moisture and chemical composition of the organic material influence N release from organic sources in soil. The build-up of soil organic matter is required to increase the potential for N mineralization. The challenge in optimizing crop N uptake in organic and cover crop-based systems does not entirely rely on developing organic matter pools but is more important to influence the rate and timing of N mineralization. It is also well-known that N from many organic fertilizers often shows little effect on crop growth in the year of application because of the slow release characteristics of organically bound N. Nitrogen immobilization after application can occur, leading to enrichment of the soil N pool. This process increases the long-term efficiency of organic fertilizers. In wheat–soybean cropping system conducted for seven years during 1995-2002 under the fine-textured Vertisols, the highest productivity was obtained when the organic sources were applied along with 100% NPK (Behera and Pandey, 2006). Soil fertility parameters showed conspicuous improvement over the initial status under FYM and poultry manure. The combined application of crop residue with 60 kg N ha^{-1} and 13.1 kg P ha^{-1} in sunflower – groundnut cropping system further improved the yield potential of sunflower by 10-11% and of succeeding groundnut by 10-24% over that of fertilizer N and P (Aulakh, 2007). In groundnut - maize cropping system, incorporation of groundnut stover improved the subsequent maize yield than surface application. A higher N recycling efficiency was observed when residues were incorporated. The seed yield of sunflower in recommended fertilizer treatment was 1.49 t/ha which increased to a maximum of 1.87 t/ha due to incorporation of pigeonpea stalks treated with *Aspergillus awamori* along with recommended fertilizer (Qureshi, 1991). Srinivas *et al.* (2005) showed that agronomic efficiency and apparent recovery of applied N were higher with

Glyricidia. It was proved that *Glyricidia* could be used as source of N to crops with partial inorganic N fertilizer. Laxminarayana and Patiram (2005) showed that conjunctive use of organic manure (15 t FYM or 5 t pig manure or 5 t poultry manure/ha) along with inorganic chemical fertilizers resulted in enhanced fertilizer-use efficiency in groundnut. The higher NUE achieved through mulching was due to weed suppression and increased soil moisture conservation. In Anantapur, application of groundnut shell @4t ha^{-1} along with 50% RDF showed higher NUE.

Biological nitrification inhibition

Nitrification results in substantial losses to agricultural systems. As much as 50-70% of the fertilizer N can be lost because of nitrification-associated processes, contributing significantly to global warming, destruction of the ozone layer in the stratosphere through nitrous oxide emissions, and serious nitrate pollution of surface and groundwater bodies. Nitrification inhibition by the ethylene diaminebased chelating agents such as ethylene diamine tetraacetic acid (EDTA), diethylene triamine pentaacetic acid (DTPA), and ethylene diamine (EDA) inhibited ammonium oxidation but did not inhibit nitrite oxidation (Hu *et al.*, 2003). The EDA appears to inhibit nitrifying activity by a different operative mechanism than EDTA and DTPA. Operationally, the only way to regulate the rate of nitrification in agricultural systems is through application of synthetic nitrification inhibitors or slow-release N fertilizers. Subbarao (2006) found that some forage grasses e.g. *Brachiaria* have the ability to regulate nitrification in soils by releasing inhibitors through root exudates. This biological mechanism has evolved to minimize N losses associated with nitrification, thus improving NUE. The inhibitory activity produced/released from plants is termed as biological nitrification inhibition (BNI). Wide variation in BNI capacity was recorded among the 18 species tested for it. The BNI capacity could either be managed and/or introduced into pasture/crops with an expression of this phenomenon via genetic improvement approaches that combine high productivity along with some capacity to regulate soil nitrification process (Subbarao *et al.*, 2007). Sustained release of BNI compounds occurred only in the presence of NH_4^+ in the root environment. This is of practical utility that there is no need to change the present sources of N application wherein the BNI activity will be rather triggered.

Use of biofertilizers

Microbes harbouring rhizosphere of crops provide benefits to crops through better nutrient availability by way of atmospheric N_2 fixation or solubilising fixed mineral forms of nutrients. The symbiotic and non-symbiotic N_2 fixation is well-known phenomenon involving legumes and nonlegumes. Rice-based cropping system offers a lot of scope for the effective utilization of a wide range of biofertilizers such as *Azolla, Cyanobacteria, Azospirillum, Rhizobium, Azorhizobium, Acetobacter* and other heterotrophic N_2 fixing bacteria as well as leguminous green manures and phosphate solubilizing bacteria, which help in increasing NUE and yield by reducing the cost of cultivation. Phosphorus use efficiency of frenchbean on vertisol was improved when rock-phosphate was applied with FYM or vermicompost at 1 : 2 ratio along with phosphate solubilizing bacteria (Manjunath *et al.*, 2006). Utilization of soybean residues and *Azospirillum brasilense* inoculation with or without inorganic N fertilizer on poor fertility sandy soils showed significant increase of N uptake by either shoot or grains of wheat plants (Galal and Thabet, 2003). Mycorrhyza is a soil fungus and effective arbuscular mycorrhizal (AM) associations may help to improve early season P nutrition in crops. The external hyphae of AM extend from the root surface to the soil beyond the P depletion zone and so access a greater volume of undepleted soil than the root alone. Therefore, a root system that has formed a mycorrhizal network will have a greater effective surface area to absorb nutrients and explore a greater volume of soil than non-mycorrhizal roots. The VAM endophytes are not host-specific in general, although preferential associations are observed with some host plants (Bagyaraj *et al.*, 2002). Mycorrhizal inoculation significantly increased plant growth, P and N contents, acid and alkaline phosphatases and total soluble protein of soybean (Fattah, 2002). In cowpea, significant genotypic variations in N_2 fixation were observed. A better root infection by arbuscular mycorrhizal fungi in genotype 'IT90K-59' and root morphological and physiological characteristics in 'IT89KD-391' were the most important factors for increasing P uptake (Martin *et al.*, 2006). Mycorrhizal plants can absorb more P at lower concentration in the soil solution than non-mycorrhizal plants. Mycorrhizal inoculation stimulated the growth of soybean. The external P requirements were 0.11 ppm for mycorrhizal and 0.15 ppm for non-mycorrhizal soybeans to obtain 80% of the maximum yield, which corresponded to a saving of 97 kg P ha^{-1} (Plenchette and Morel, 1996). Incorporation of VAM along with starter dose of P in P-deficient acid soils improved the total uptake of P, K, Ca, Mg and Zn by 2.5-6.0 fold in sorghum and cowpea when compared to only P application (Bagayoko *et al.,* 2000). The VAM fungal hyphae also play an important role in soil aggregation by creating skeletal structure of macro-aggregates through physical entanglement of soil particles and organic materials (Tisdall, 1994). CRIDA tested Zinc solubilizing bacteria in some dryland crops like maize in Zn deficient soil and found positive results.

Nutrient-use efficiency in cropping systems

Individual crop nutrient management relies on the crop requirement matching the soil fertility. Cropping systems with differential requirement and contribution to modifying the rhizosphere by different crops provide newer challenge as well as opportunity for management to achieve higher NUE. Pigeonpea substitution for rice during monsoon season in rice – wheat cropping system improved wheat yield by 16%, and total N and P uptake by 31 and 10%, respectively. The better N and P-use efficiency in pigeonpea–wheat system was due to improvement in soil physical properties (Singh *et al.*, 2006b). Garcia *et al.* (2006) summarizing the nutrient-use efficiencies in major cropping systems of Argentina reported increased agronomic nutrient-use efficiencies and yield differences over time as a consequence of residual effects on soil fertility.

Inclusion of legumes in the cropping systems

The beneficial effect of important legumes on increasing productivity and NUE in various systems was recently reviewed by Ghosh *et al.* (2007). In alley cropping system, *Leucaena leucocephala* (subabul) prunings provide N to the extent of 75 kg, which benefit castor and sorghum. Legumes with indeterminate growth are more efficient in N_2 fixation than determinate types. Fodder legumes in general are more potent in increasing the productivity of succeeding cereals. Symbiotic N_2 fixation with compatible rhizobial strains resulted in carryover of N for succeeding crops to the tune of 60-120 kg in berseem, 75 kg in Indian clover, 75 kg in clusterbean, 35-60 kg in fodder cowpea, 68 kg in chickpea, 55 kg in urdbean, 54-58 kg in groundnut, 50-51 kg in soybean, 50 kg in *Lathyrus*, and 36-42 kg ha^{-1} in pigeonpea. Direct and residual effect of partially-acidulated material and mixture of rock-phosphate + single superphosphate were observed to be better when these were applied to mungbean in winter season than to rice in rainy season simply because of legume effect.

Tillage and weed management

Reducing tillage and optimizing N fertilization are important strategies for soil and water conservation and N use efficiency. In dry years with scarce rainfall, N should not be applied in any tillage system (Angas *et al.*, 2006). Though tillage systems did not significantly affect the plant tissue contents of wheat (at tillering) and cotton (at flowering), tillage and fertilizer treatments had a positive effect on nutrient uptake by wheat (Ishaq *et al.*, 2001). Conservation tillage and deep tillage increased N, P and K uptake compared to minimum tillage. Soil, water and nutrients play an important role in increasing yields of crops in black soils of semi-arid tropics during post-rainy season. Tillage practices along with organic material further affected the moisture and nutrient availability to crops. Availability of nutrients and water at different growth stages of sorghum were increased by deep tillage (Patil and Sheelavantar, 2006). Nitrogen-use efficiency and N uptake efficiency were greater with conventional tillage than with no-tillage (López-Bellido and López-Bellido, 2001). In general, crop yields under no-till practices are more stable than under tilled systems with greater efficiency in the use of nutrients (Martin, 2006). Improper land leveling is the serious cause for losses in water, nutrients and improper population resulting in low yields and profits. A reduction in the yield of seed cotton up to 20.1% was observed on traditionally-leveled fields compared to precision leveling.

Efficient genotypes for higher nutrient-use

Development of varieties with better nutrient-use efficiency is becoming a necessity both to allow the maintenance of a sufficient profit for the farmer and preserve groundwater and water bodies from pollution. The efficiency of a genotype to utilize nutrients arises from a balance of several processes associated with acquisition and usage. Genetic variation in nutrient efficiency may be attributed to two multi-factorial components: (i) genotypes may differ in the efficiency with which the nutrients in the plant are utilized to produce yield (utilization efficiency), and/or (ii) they may differ in their effectiveness in absorbing nutrients from the soil (uptake efficiency). Experiments conducted at Tamil Nadu Agricultural University comparing hybrid and non-hybrid plant ability to use P and K nutrients indicate that although non-hybrids and hybrids of rice/cotton have large responses to P and K application, the degree of agronomic response is greater in hybrid crops. For instance, the cotton hybrid TCHB 213 produces more yield per unit of K fertilisation (Agronomic Efficiency of K or AEK) than the non-hybrid MCU 5 (Fig. 7) (Rao, 2007).

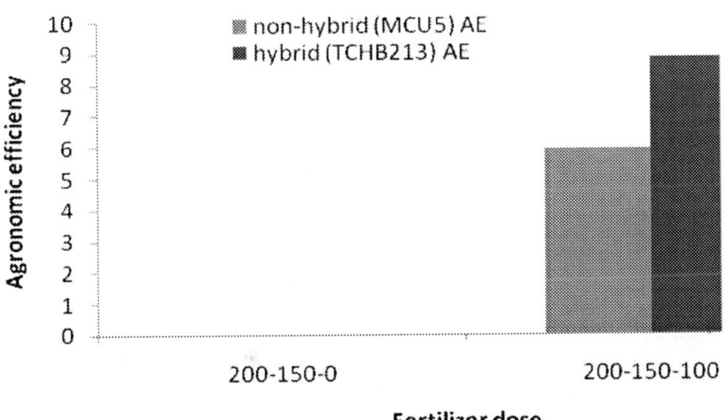

Fig. 7. Agronomic efficiency of K for hybrid and non-hybrid cotton (*Source*: Rao, 2007)

Site specific nutrient management (SSNM)/Balanced fertilization

Balanced fertilizer use is not only the first requirement; it is rather a pre-requisite since no amount of agronomic manipulation can produce high efficiency out of an imbalanced fertilizer dose. In fact, efficient nutrient use is essentially an offspring of balanced fertilizer use and sound management practices and decisions. With the adoption of exhaustive cropping systems, widespread deficiencies of S and Zn, sporadic deficiencies of Fe, Mn, and B have been noticed in intensively cultivated areas. The best strategies for sustainable crop production and environmental safety with respect to P are to produce optimum crop yields by: (a) alternative P application only to winter crops, (b) periodic reduction in P rates according to accumulated P, and (c) occasionally without fertilizer P to enable the utilization of residual fertilizer P that had accumulated in the soil over the years (Aulakh, 2006, 2007; Roberts, 2006). Application of rock-phosphate along with green leaf manure and P solubilizing fungi gave significantly higher grain yield compared to application of rock-phosphate alone (Srinivasamurthy *et al.,* 2006). The total dry matter, P and Ca uptake in soybean was significantly improved with the use of low-grade rock-phosphate in conjunction with phosphate solubilizers and FYM. *Aspergillus awamorii* was found to be efficient P-solubilizer, followed by *Bacillus striata* in solubilizing North Carolina and Mussoorie rock-phosphates (Qureshi and Narayanasamay, 1999). About 64-71% equivalent of water-soluble P requirement of crops could be met through rock-phosphates, seed inoculation of P solubilizing microbes and FYM (Qureshi *et al*., 2005). Potassium, an essential nutrient required by plants in large quantities is present in soils in four pools, of varying availability to plants, in dynamic equilibrium. Potassium deficiencies have been observed in crops, such as cotton, maize and sorghum in vertisols (Kathryn *et al.,* 2006). Application of K fertilizers in many of the soils exhibits no yield response to applied K despite an increase in K uptake. Riley *et al.* (2002) reported that sulphate was highly mobile and prone to leaching under the experimental conditions, whereas the slow release characteristics of elemental S led to smaller leaching losses and larger residual effects. Inclusion of limiting nutrients in fertilization programme has shown excellent results. Balanced and adequacy of all limiting nutrients for different cropping systems in alluvial and lateritic soils of West Bengal ensured high relative agronomic efficiency (RAE). Significant reduction in RAE compared to soil test and yield target-based treatment was observed under inadequate nutrient application. High levels of P application may induce Zn deficiency in plants grown on Zn-deficient soils. On P- and Zn-deficient typic *Ustochrept* soils, it was found that increasing P up to 13.5 mg kg^{-1} soil increased Zn concentration, while further increase led to decreased Zn concentration. Iron concentration decreased with increasing P levels. Up to 13.5 mg kg^{-1} P application, Cu concentration increased, and thereafter decreased. Manganese concentration gradually increased with increasing P levels (Srinivasarao *et al.*, 2006). Potassium improved N recovery efficiency of cotton (Fig. 8) (Rao, 2007).

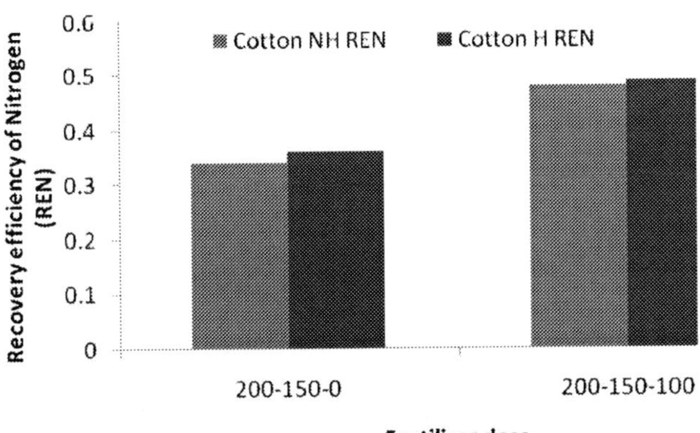

Fig. 8. Influence of K on N recovery efficiency (REN) (*Source*: Rao, 2007)

Integrated nutrient management - towards improving nutrient use efficiency

Organic matter affects crop growth and yield either directly by supplying nutrients or indirectly by modifying soil physical properties such as stability of aggregates and porosity that can improve the root environment and stimulate plant growth (Darwish *et al.*, 1995). Incorporation of organic matter either in the form of crop residues or farmyard manures has been shown to improve soil structure and water retention capacity (Bhagat and Verma, 1991), increase infiltration rates and decrease bulk density (Khaleel *et al.*, 1981). Therefore, any nutrient management practice that can improve organic matter status of soil is important. However, neither inorganic nor organic amendments alone can maintain organic matter status of soil and sustain the productivity in the semi-arid tropics (Prasad, 1996). A judicious and combined use of organic and inorganic sources of nutrients is essential to maintain soil health and to augment the efficiency of nutrients (Lian, 1994). Additionally, such integration plays an important role in better penetration and establishment of plant roots, which helps the plant to utilize water from deeper layers and to maintain high relative plant water content under a soil moisture stress condition, which is quite common in rain-fed farming.

In a study conducted at Bhopal on a Vertisol, Hati *et al.* (2006) reported that manured and fertilized plots (NPK + FYM) produced significantly higher seed yield than did the plots receiving either inorganic fertilizer alone or no fertilizer in all the three years of study. The combined use of inorganic fertilizers and organic manures enhanced the inherent nutrient supplying capacity of the soil and also improved the physical properties of the soil, which promoted better rooting, higher nutrient and water uptake by the crops, and increased the seed yield as well as water-use efficiency of soybean. In a 20 years old experiment on a Aridisol, with average annual rainfall of 550 mm consisting of pearl millet-castor-cluster bean rotation indicated that the highest agronomic efficiency varied from 3.0 (50 kg N-Urea) to 5.1 (25 kg N-Urea + 25 kg N-FYM) (Fig. 9). At similar N levels, N use efficiency (AE_N) is higher with integrated use of N these organic and inorganic sources. Similarly highest AE_N (9.7) was obtained in INM treatment (25 kg N-Urea + 25 kg N-FYM) in *rabi* sorghum at Solapur (Fig. 10) (Srinivasarao *et al.* 2011).

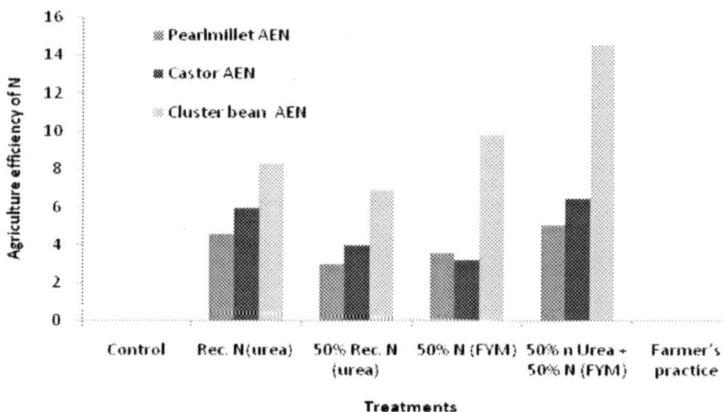

Fig. 9. Agronomic efficiency of N in pearlmillet, castor and cluster bean on Aridisol at SK Nagar (Gujarat) during 1988-2006 (*Source*: Srinivasarao *et al.*, 2011)

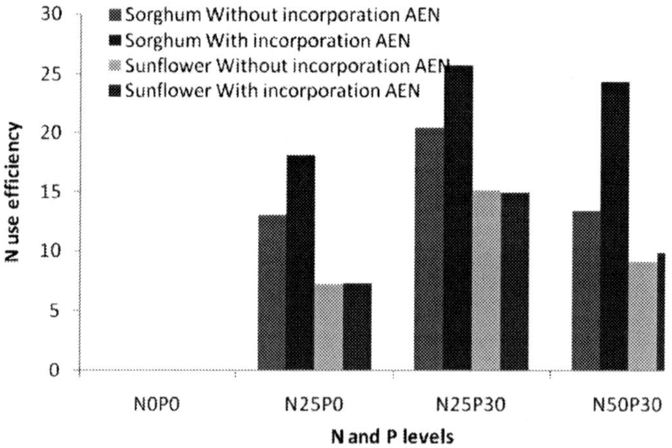

Fig. 10. Nitrogen use efficiency during 22 years sorghum cropping on Vertisols of semi-arid, Solapur (*Source*: Srinivasarao *et al.*, 2011)

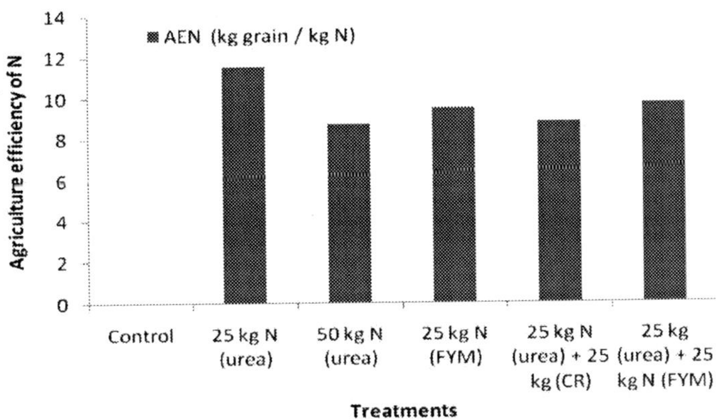

Fig. 11. N use efficiency in sorghum and sunflower at different NP levels as influenced by horsegram biomass incorporation during 10 years experiment on Alfisols, Hyderabad (*Source*: Srinivasarao *et al.*, 2011)

Better water-nutrient interaction for improving NUE

Water management involving in situ or ex situ moisture conservation in rain-fed agriculture is highly correlated with NUE. Construction of water-harvesting structures such as tanks with open dug wells in the upper reaches and shallow ditches in the lower reaches of a drainage line to re-harvest the seepage water from the tanks for irrigation are potential mechanisms in rainfed agriculture to increase grain yield as well as NUE (Pali *et al.*, 2007).

Crop residues and manures are often used in temperate regions to control soil erosion, maintain soil organic matter and improve moisture-holding capacity. In 1991 and 1992, a field experiment was conducted to study the effect of crop residue incorporation in a loamy sand (Typic Camborthid) in Rajasthan, India. Residues of clusterbean [*Cyamopsis tetragonoloba* (L) Taub], mung bean [*Vigna radiata* (L) R. Wilczek] and pearl millet [*Pennisetum glaucum* (L) R. Br.], and farmyard manure (FYM) were incorporated by disking. Crop residues increased soil water content at seeding in the order pearl millet > mungbean = farmyard manure > clusterbean. Addition of crop residues and FYM generally enhanced soil fertility status (N and P availability, organic matter, enzyme activity) 10 to 20%. Hence, incorporation of crop residues and FYM in arid tropical soils benefits soil water storage, soil nutrient availability, and crop yields (Aggarwal *et al.*, 1997). Soil amendment through tank silt and biochar for higher water retention in light textured soil can help in improving NUE.

Conservation agriculture

Adoption of conservation agriculture (CA) in rainfed farming is an emerging dimension for recycling biomass into soil system and improving soil organic carbon as well as improving NUE through better water nutrient synergy. Kundu *et al.* (2013) reported an increase in use efficiency of N, P, K, S and Zn to the extent of 11, 16, 14, 13 and 21% through adoption of CA practices in maize-horsegram sequence at Alfisols of Hyderabad. In arid (<500 mm rainfall) region, low tillage is comparable to conventional tillage and weed problem is manageable. In semi arid region (500–1000 mm), conventional tillage can be superior to reduced tillage, and successful crop production depends on water infiltration and conserving soil moisture in the profile. A long-term study comprising of tillage conventional (CT) and reduced (RT) and conjunctive nutrient use (fertilizers and low cost farm-based organics) treatments conducted in Alfisol (Typic Haplustalf) at Hyderabad, India, involving sorghum and mungbean cropping system, indicated significant improvements in grain yields and SYI of several crops (Sharma *et al.,* 2004). The results of this and other experiments have raised the expectations of improving rainfed farming on Alfisols by using some kind of CA system. But the potential for CA exist in rainfed crops like maize, pigeonpea, castor, cotton, sunflower etc. where CRs are not used as feed and for other competing purposes. Being the only source of water, interactions between rainwater conservation and CA must be studied in an integrated manner. With canopy cover of maize grown during the rainy season, there is a possibility of growing a post-rainy legume (*i.e.*, horsegram or *Macrotyloma uniflorrus*) crop in degraded Alfisols of Southern India, which is otherwise a mono cropped area.

Future line of work

1. In India, concept of nutrient use efficiency remains largely confined to the scientific community. Because nutrient use in India is largely subsidy driven and not science-based, the practical benefits of high use efficiency are distorted since N, the cheapest or most subsidized nutrient, is used most while other nutrients are presently ignored. Therefore, there is a need to create awareness among the farming community about the benefit of higher NUE.

2. Developing recommended dose for balanced fertilization on cropping system basis and exploiting positive interactions among inputs for sustainable high NUE and environmental protection.

3. Soil fertility research should also focus climate change influence on soil that has a marked influence on nutrient dynamics in soil and plant uptake.

4. Nitrogen-nutrient augmentation through improvements and exploitation of alternative sources such as BNF and potential of BNI to control nitrification in cropping systems should be intensified.
5. Identifying and developing superior plant types and genotypes for high nutrient-use efficiency.
6. Developing precision farming methodology for higher NUE.
7. Results of long-term fertility experiments across rainfed agro-ecosystems have to be critically examined for their trends of responses for guiding the management strategies.
8. Development of sustainable farming systems with higher nutrient-use efficiency and specific guidelines for good agricultural practices.

Conclusion

Declining use efficiency of applied nutrients is a major cause of concern for increasing crop production, cost of cultivation and environmental pollution. Several techniques of increasing nutrient use efficiency like time and method of nutrient application, use of nanoparticles/materials in slow-release fertilizers, organic materials, biological nitrification inhibition, use of soil amendments and microbial sources, proper tillage and weed management, and precision agriculture are used globally. But these technologies should be tested in Dryland conditions for their efficacies in a particular situation. Our on-farm experience in India suggests that balanced fertilization and site-specific nutrient management or integrated nutrient management can be effective in increasing nutrient-use efficiencies in poorly fertile soils in arid and semi-arid climates, thus improving productivity and profitability of the farms. Locally-available organic manure can be used successfully for improvement of nutrient use efficiencies.

Selected References

Acharya CL, Sharma AR. 2008. Integrated input management for improving nitrogen-use efficiency and crop productivity. *Indian Journal of Fertilizers* 4(2): 33–40, 43–50.

Al-ali MA, Morris T. 2006. In-season nitrogen fertilization to improve nutrient-use efficiency in maize. 18th World Congress of Soil Science, held during 9–15 July, Philadelphia, Pennsylvania, USA.

Alam MM, Ladha JK, Foyjunnessa RZ, Khan SR, Harun-ur-Rashid KAH, Buresh RJ. 2006. Nutrient management for increased productivity of rice–wheat cropping system in Bangladesh. *Field Crops Research* 96(2- 3): 374–386.

Angas P, Lampurlanes J, Cantero MC. 2006. Tillage and N fertilization: effects on N dynamics and barley yield under semiarid Mediterranean conditions. *Soil and Tillage Research* 87(1): 59–71.

Arnall DB, Tubana BS, Girma K, Raun WR, Freeman KW, Teal RK. 2006. Relationship between nitrogen-use efficiency and response to fertilizer in winter wheat. 18th World Congress of Soil Science, held during 9-15 July, Philadelphia, Pennsylvania, USA.

Aulakh MS. 2006. Assessment and use of accumulated fertilizer phosphorus in a 30-year study in subtropical region for sustainable crop production and environmental safety. 18th World Congress of Soil Science, held during 9-15 July, Philadelphia, Pennsylvania, USA.

Aulakh MS. 2007. Yields and nutrient use-efficiency in groundnut sunflower cropping system in Punjab, India. *Journal of Sustainable Agriculture* 31(1):1.

Bagayoko M, George E, Romheld V, Buerkert A. 2000. Effects of mycorrhizae and phosphorus on growth and nutrient uptake of millet, cowpea and sorghum on west African soil. *Journal of Agricultural Science* 135(4): 399–407.

Bagyaraj DJ, Mehrotra VS, Suresh CK. 2002. Vesicular arbuscular mycorrhyza biofertilizers for tropical forest plants, pp. 299-311. In: *Biotechnology of Biofertilizers*. Kannayaiyan, C. (Ed.) Narosa Publishing House, New Delhi.

Baligar VC, Bunce JA, Bailey BA, Machado RC, Pomella AW. 2004. Carbon dioxide flux and photosynthetic photon flux density effects on growth and mineral uptake parameters of cacao. *Journal of Food, Agriculture and the Environment* 3(2): 142–147.

Behera UK, Pandey HN. 2006. Sustaining productivity of wheat-soybean cropping system through integrated nutrient management practices in the vertisols of central India. 18th World Congress of Soil Science, held during 9-15 July, Philadelphia, Pennsylvania, USA.

Bijay-Singh, Tiwari MK, Abrol YP. 2008. Reactive nitrogen in Agriculture, Industry and Environment in India. IGBP-WCRP-SCOPE-Report Series, 42, Indian Ntional Science Academy, New Delhi.

Cassman KG, Dobermann A, Walters DT. 2002. Agroecosystems, nitrogen-use efficiency, and nitrogen management *Ambio* 31(2): 132–140.

Chang Ki Woon, Pil Joo Kim, Chang Hoon Lee, and Yong Bok Lee. 2006. Enhancing nitrogen efficiency of rice (*Oryza sativa* L.) by silicate application in Korean paddy soil. 18th World Congress of Soil Science, held during 9-15 July, Philadelphia, Pennsylvania, USA.

CRIDA, Vision, 2030. Central research Institite for Dryland Agriculture, Hyderabad.

CRIDA, Vision, 2050. Central research Institite for Dryland Agriculture, Hyderabad.

De Datta SK, Buresh RJ, Mamaril CP. 2005. Increasing nutrient-use efficiency in rice with changing needs. *Nutrient Cycling in Agroecosystems* 26(1-3): 157–167.

Fageria NK. 2006. Nutrient efficient plants in improving crop yields in the twenty first century. 18th World Congress of Soil Science, held during 9-15 July, Philadelphia, Pennsylvania, USA.

Fageria NK, Baligar VC. 2005. Enhancing nitrogen-use efficiency in crop plants. *Advances in Agronomy* 88: 97–185.

FAI. 2012.Fertilizer Statistics 2001-2012. The Fertilzier Association of India, New Delhi, p 240

Fattah Abdel GM. 2002. Measurement of viability of VAM fungi colonized of roots using three different stains and its relation to growth and metabolic activities of soybean plants. *Agrochimica* 46(5): 177–189.

Fixen FE, West FB. 2002. Nitrogen fertilizers: meeting contemporary challenges. *Ambio.* 31(2):169–176.

Fudao Z. Rutang, W. Qiang, X. Yajun, W, Jianfeng, Z. 2006. Effects of slow / controlled release fertilizer cemented by nano materials on biology. II. Effects of slow /controlled release fertilizer cemented and coated by nano materials on plants. *Nanoscience* 11(1): 18–26.

Gallais A, Coque M. 2005. Genetic variation and selection for nitrogen-use efficiency in maize: a synthesis. *Maydica* 50(3- 4): 531–547.

Garcia F, Miguel B, Jorge M, Hugo B, Raul H, German DM, Angel B. 2006. Direct and Residual Effects of Balanced Fertilization in Field Crops of the Pampas of Argentina. 18th World Congress of Soil Science, held during 9-15 July, Philadelphia, Pennsylvania, USA.

Ghosh PK, Bandyopadhyay KK, Wanjari RH, Manna MC, Misra AK, Mohanty M, Rao AS. 2007. Legume effect for enhancing productivity and nutrient use-efficiency in major cropping systems - an Indian perspective: a review. *Journal of Sustainable Agriculture* 30(1): 61–86.

Hegde DM, Murthy IYLN. 2005. Management of secondary nutrients: achievements and challenges. *Indian Journal of Fertilisers* 1(9): 93–100, 105 110.

Hegde DM, Sudhakara Babu SN. 2004. Role of balanced fertilization in improving crop yield and quality. *Fertilizer News* 49(12): 103–110, 113–114, 131.

Hirel B, Le Gouis J Ney, B, Gallais A. 2007. The challenge of improving nitrogen-use efficiency in crop plants: towards a more central role for genetic variability and quantitative genetics within integrated approaches. *Journal of Experimental Botany* 58(9): 2369-2387.

Hodge A. 2006. Plastic plants and patchy soils. *Journal of Experimental Botany* 57(2): 401-411.

Hong D, Zhang, Yu-shu. 2006. Effect of Gel-based controlled release fertilizers on crop yield and nutrient-use efficiency. 18th World Congress of Soil Science, held during 9- 15 July, Philadelphia, Pennsylvania, USA.

Hu Z, Kartik C, Domenico, G , Smets BF. 2003 Nitrification Inhibition by Ethylenediamine-Based Chelating Agents. *Environmental Engineering Science* 20(3): 219-228.

Ishaq M, Ibrahim, M and Lal, R. 2001. Tillage effect on nutrient uptake by wheat and cotton as influenced by fertilizer rate. *Soil and Tillage Research* 62(1-2): 41–53.

John KS. 2006. Efficiency of soil and fertilizer phosphorus: reconciling changing concepts of soil phosphorus chemistry with agronomic information. 18th World Congress of Soil Science, held during 9-15 July, Philadelphia, Pennsylvania, USA.

Ju Jing, Yamamoto Y, Wang Y, Shan Y, Dong D, Yoshida T, Miyazaki A. 2006. Genotypic differences in grain yield, and nitrogen absorption and utilization in recombinant inbred lines of rice under hydroponic culture. *Soil Science and Plant Nutrition* 52(3): 321–330.

Kathryn T, Singh B, Schwenke G. 2006. Potassium dynamics in vertisols following potassium fertilization and plant uptake. 18th World Congress of Soil Science, held during 9-15 July, Philadelphia, Pennsylvania, USA.

Katyal JC, Gadalla AM. 1990. Fate of urea-N in floodwater. I. Relation with total N loss. *Plant and Soil* 121(1): 21–30.

Kundu S, Srinivasarao Ch, Mallick RB, Satyanarayana T, Prakash NR, Johnston A, Venkateswarlu B. 2013. Conservation agriculture in maize (*Zea mays* L.)-horsegram (*Macrotyloma uniflorum L.*) system in rainfed Alfisols for carbon sequestration and climate change mitigation. Journal of Agrometeorology, 15 (Special issue I): 144-149.

Laxminarayana K, Patiram. 2005. Influence of inorganic, biological and organic manures on yield and nutrient uptake of groundnut (*Arachis hypogaea*) and soil properties. *Indian Journal of Agricultural Sciences* 75(4): 218–221.

Lee CH, Park CY, Park KD, Jeon WT, Kim PJ. 2004. Long-term effects of fertilization on the forms and availability of soil phosphorus in rice paddy. *Chemosphere* 56(3): 299–304.

Liu JL, Zhang F, Liao WH. 2005. Transformation of phosphorus in rhizosphere of different wheat varieties as influenced by VA mycorrhiza (VAM). *Plant Nutrition and Fertilizer Science* 7(1): 23–30.

Liu-Ming Q, Yu-Zhen R, Liu-Yun H. 2005. Modeling the relationship between summer maize NPK uptake and yield on the basis of soil fertility indices. *Agricultural Sciences in China* 4(10): 774–780.

López-Bellido RJ, López-Bellido L. 2001. Efficiency of nitrogen in wheat under Mediterranean conditions: effect of tillage, crop rotation and N fertilization. *Field Crops Research* 71(1,5): 31–46.

Mangat GS. 2004. Relative efficiency of NFL [National Fertilizers Limited] - Neem coated urea and urea for rice. *Fertilizer News* 49(2): 63–64.

Manjunath MN, Patil PL, Gali SK. 2006. Effect of organic amended rock phosphate and P solubilizer on P-use efficiency of French bean in a Vertisol of Malaprabha right bank command of Karnataka. *Karnataka Journal of Agricultural Sciences* 19(1): 36–39.

Mao XiaoYun, Sun K, Wang D, Liao Z. 2005. Controlled-release fertilizer (CRF): a green fertilizer for controlling non-point contamination in agriculture. *Journal of Environmental Sciences* 17(2): 181–184.

Martin, D. 2006. Innovations for Improving productivity and nutrient-use efficiency: no-till grain cropping systems of South America. 18th World Congress of Soil Science, held during 9-15 July, Philadelphia, Pennsylvania, USA.

Martin J, Robert CA, Christian N, Walter J 2006. Genotypic variation for phosphorus uptake dinitrogen fixation in cowpea on low-phosphorus soils of southern Cameroon. *Journal of the Science of Food and Agriculture* 58(1): 1–8.

NAAS. 2006. Low and declining crop response to fertilizers. *Policy Paper No. 35, National Academy of Agricultural Sciences, New Delhi.*

Nair K, Prabhakaran P. 2006. Sustaining Crop Production in the Developing World through "The Nutrient Buffer Power Concept". 18th World Congress of Soil Science, July 9-15th, Philadelphia, Pennsylvania, USA.

Pal SS, Jat ML, Subba Rao AVM. 2003. Laser land leveling for improving water productivity in rice-wheat system. PDCSR Newsletter, Modipuram, Meerut.

Pandu KH, Angadi VV, Chittapur BM. 2005. NUE and N uptake of upland rice as influenced by integrated nitrogen management through leaf colour chart. *Karnataka Journal of Agricultural Sciences* 18(1): 124–126.

Panneerselvam P, Bheemaiah G. 2005. Effect of integrated nutrient management practices on productivity, nutrient uptake and economics of sunflower intercropped with *Azadirachta indica* and *Melia azadirach* trees. *Journal of Oilseeds Research* 22(1): 231–234.

Pathak H, Prasad R. 2008. Fate of nitrogen in Indian Agriculture and its environment impact. NAAS News 8(2): 1-4.

Patil SL, Sheelavantar MN. 2006. Soil water conservation and yield of winter sorghum as influenced by tillage, organic materials and nitrogen fertilizer in semi arid tropical India. *Soil and Tillage Research* 89(2): 246–257.

Plenchette C, Morel C. 1996. External phosphorus requirements of mycorrhizal and non-mycorrhizal barley and soybean plants. *Biology and Fertility of Soils* 21(4): 303–308.

Prasad R. 2009. Enhancing nutrient use efficiency- Environmentaal benign strategies. Souvenir, 67-74, The Indian Society of Soil Science, New Delhi.

Prasad R. 2008. Fertilizer use, food security, health and the environment. *Current Science* 75: 677-684.

Prasad R. 2009. Efficient fertilizer use: The key to food security and better environment. *Journal of Tropical Agriculture* 47: 1-17.

Prasad R, Sharma SN, Singh S, Saxena VS, Shivay YS. 2001. Pusa neem emulsion as an ecofriendly coating agent for urea quality and efficiency. *Fertiliser News* 46 (7):73– 74.

Premlatha PR, Angadi VV. 2005. Crop need based N management through LCC in direct dry seeded rainfed lowland rice. *Mysore Journal of Agricultural Sciences* 39(1): 140–143.

Purakayastha TJ, Katyal JC. 1998a. Evaluation of compacted urea fertilizers prepared with acid and non-acid producing chemical additives in three soils varying in pH and cation exchange capacity I. NH3 volatilization. *Nutrient Cycling in Agroecosystems* 51(2): 107–115.

Qureshi AA. 1991. Effect of incorporation of crop residues and fertilizer levels on soil properties and crop yields of sunflower and chickpea. M.Sc. thesis, University of Agricultural Sciences, Dharwad.

Qureshi AA, Narayanasamy G. 1999. Direct effect of rock phosphates and phosphate solubilizers on soybean growth in a Typic Ustochrept. *Journal of Indian Society of Soil Science* 47(3): 475–478.

Qureshi AA, Narayanasamy G, Chhonkar PK, Balasundaram V.R. 2005. Direct and residual effect of phosphate rocks in presence of phosphate solubilizers and FYM on the available P, organic carbon and viable counts of phosphate solubilizers in soil after soybean, mustard and wheat. *Journal of Indian Society of Soil Science* 53 (1): 97–100.

Rao KV, Kavitha B, Rama Prasad AS, Surekha K, Gandhi G. 2006. Genotypic variability in nitrogen utilization efficiency in wetland rice. 18th World Congress of Soil Science, held during 9-15 July, Philadelphia, Pennsylvania, USA.

Rao TN. 2007. Improving nutrient-use efficiency: the role of beneficial management practices. *Better Crops: India* 1(1): 6–7.

Raun WR. 2006. What new equipment is available to improve N use efficiency? 18th World Congress of Soil Science, held during 9-15 July, Philadelphia, Pennsylvania, the USA.

Rego TJ, Sahrawat KL, Wani SP, Pardhasaradhi G. 2007. Widespread deficiencies of sulphur, boron and zinc in Indian semi arid tropical soils: on farm crop responses, *Journal of Plant Nutrient* 30.

Riley NG, Zhao FJ, McGrath S.P. 2002. Leaching losses of sulphur from different forms of sulphur fertilizers: a field lysimeter study. *Soil Use and Management*. 18(2): 120–126.

Sharma KL, Kusuma GJ, Mandal UK, Gajbhiye PN, Srinivas K, Korwar GR, Himabindu V, Ramesh V, Ramachandran K, Yadav SK. 2008. Evaluation of long-term soil management practices through key indicators and soil quality indices using principal component analysis and linear scoring technique in rainfed Alfisols. *Australian Journal of Soil Research* 46: 368-377.

Sharma KL, Srinivas K, Mandal UK, Vittal KPR, Kusuma GJ, Maruthi Sankar G. 2004. Integrated Nutrient Management Strategies for Sorghum and Green gram in Semi arid Tropical Alfisols. *Indian Journal of Dryland Agricultural Research and Development* 19 (1): 13-23.

Singh B, Singh Y, Singh, Mharban SGPS, Ladha JK, Vethaiya B. 2006a. Plant need based real time nitrogen management in rice grown by small farmers in Asia. 18th World Congress of Soil Science, held during 9-15 July, Philadelphia, Pennsylvania, USA.

Singh U, Giller KE, Palm CA, Ladha JK, Breman H. 2001. Synchronizing N release from organic residues: opportunities for integrated management of N. 2nd International Nitrogen Conference, held during 14-18 October 2001 at Potomac, Maryland, the USA,

Singh VK, Dwivedi BS, Shukla AK. 2006b. Yields and nitrogen and phosphorus use efficiency as influence by fertilizer NP additions in wheat under rice – wheat and pigeon pea–wheat system on a Typic Ustochrept soil. *Indian Journal of Agricultural Sciences* 76(2): 92–97.

Singh Y, Singh B, Ladha JK, Gupta R, Pannu R. 2006c. Tillage and residue management effects on yield and nitrogen-use efficiency in wheat following rice in the Indo-Gangetic plains of India. 18th World Congress of Soil Science, held during 9-15 July, Philadelphia, Pennsylvania, USA.

Sinha SP, Prasad SM, Singh SJ. 2005. Nutrient utilization by winter maize and weeds as influenced by integrated weed management. *Indian Journal of Agronomy* 50(4): 303-304.

Srinivas K, Sridevi S, Sharma KL. 2005. Effect of sole and conjunctive applications of plant residues and inorganic nitrogen on growth, yield and N uptake of sorghum. *Indian Journal of Dryland Agricultural Research and Development* 20(1): 62–69.

Srinivasamurthy CA, Sunil Kumar MV, Ravi S, Bhaskar, Siddaramappa R. 2006. Use of indigenous Indian rock phosphates as cheap source of P to increase rice production. 18th World Congress of Soil Science, held during 9-15 July, Philadelphia, Pennsylvania, USA.

Srinivasarao Ch, Vittal KPR (2007) Emerging nutrient deficiencies in different soil types under rainfed production systems of India. *Indian Journal of Fertilisers* 3: 37-46.

Srinivasarao C, Ganeshamurthy AN, Ali M, Venkateswarlu B. 2006. Phosphorus and micronutrient nutrition of chickpea genotypes in a multi-nutrient deficient typic Ustochrept. *Journal of Plant Nutrition* 29(4): 747–763.

Srinivasarao Ch., Chary GR, Venkateswarlu B, Vittal KPR, Prasad JVNS, Kundu S., Singh SR, Gajanan GN, Sharma RA, Deshpande AN, Patel JJ, Balaguravaiah G. 2009. Carbon sequestration strategies in rainfed production systems of India. Central Research Institute for Dryland Agriculture, Hyderabad. P 102.

Srinivasarao Ch, Venkateswarlu B, Lal R, Singh AK, Kundu S. 2013. Sustainable Management of Soils of Dryland Ecosystems of India for Enhancing Agronomic Productivity and Sequestering Carbon. In Donald L. Sparks, editor: *Advances in Agronomy*, 121, Burlington: Academic Press, 2013, pp. 253-329. ISBN: 978-0-12-407685-3.

Srinivasarao Ch, Venkateswarlu B, Sharma KL, Ramachandrappa BK, Patel JJ, Deshpande AN. 2011. Analyzing Nitrogen Use Efficiency in Long Term Experiments in Rainfed Conditions. *Indian Journal of Fertilizers*, 7: 36–47

Srinivasarao Ch, Wani SP, Sahrawat KL, Rego TJ, Pardhasaradhi G. 2008. Zinc, boron and sulphur deficiencies are holding back the potential of rainfed crops in semi-arid India: Experiences from participatory watershed management. *International Journal of Plant Production (Iran)*, 2: 89-99.

Subbarao GV. 2006. Biological nitrification inhibition (BNI): is it a widespread phenomenon? *Plant and Soil* 294(1-2): 5–18.

Subbarao GV, Tomohiro B, Masahiro K, Osamu I, Samejima H, Wang HY, Pearse SJ, Gopalakrishnan S, Nakahara K, Zakir AKM, Sujimoto HHT, Svecnjak Z, Rengel Z. 2006. Canola cultivars differ in nitrogen utilization efficiency at vegetative stage. *Field Crops Research* 97(2-3): 221–226.

Takkar PN. 1996. Micronutrient research and sustainable agricultural productivity. *Journal of the Indian Society of Soil Science* 44: 563-581.

Tisdall JM. 1994. Possible role of soil microorganisms in aggregation in soils. *Plant and Soil* 159(1): 115–121.

Venkateswarlu B, Srinivasarao Ch, Ramesh G, Venkateswarlu S, Katyal JC.2007. Effects of long term legume cover crop incorporation on soil organic carbon, microbial biomass, nutrient build up and grain yields of sorghum/sunflower under rainfed conditions. *Soil Use and Management*, 23:100-107.

20

Enhancing Nutrient Use Efficiency, pp. 263-278
Editors: K. Ramesh, A.K. Biswas, B.L. Lakaria, S. Srivastava and A K. Patra

Cereal Based Cropping Systems

R.H. Wanjari

ICAR-Indian Institute of Soil Science, Nabi Bagh, Bhopal – 462 038, India

Introduction

Out of the total cultivated area of 142 million hectares (M ha) in India during 2011–12, 100.2 M ha was under cereals (the word cereal derives from Ceres, the name of the Roman goddess of harvest and agriculture) and 24.8 M ha under pulses (splitting seed), the staple food grains in the country (Economic Survey 2013). There is very little scope of bringing additional area under food grains; on the contrary this may decline in future due to the land needed for civil amenities and industrial purposes. Further, a lot of good agricultural land around villages and towns is being lost to urban development. At the 1950–51 level of productivity, when very little fertilizer was used, 100.2 M ha under cereals would have produced only 54.2 million tonnes (Mt) of cereals, and 24.8 M ha under pulses would have produced 10.9 Mt of pulses. As against this during 2011–12, 240 Mt of cereals and 17.2 Mt of pulses were produced. This was possible due to improved agricultural technology that included high-yielding crop varieties, use of fertilizer, an increase in irrigated area and introducing plant protection. As regards fertilizers, 27.57 M t of fertilizer ($N + P_2O_5 + K_2O$) was used in 2011–12 as compared to a mere 0.069 M tonnes in 1950–51. This brings out the fact that the soil can supply only limited amount of productivity level of 1950–51, and for each additional tonne of food grain produced, adequate amounts of additional plant nutrients must be externally applied as fertilizer. Thus, the fertilizer has been the key input for realizing yield potential of responsive, high-yielding crop varieties. Role of fertilizer in food grain production became more evident after the green revolution and NPK consumption increased from 1.1 Mt in 1966–67 to 28 Mt in 2011–12, and the food grain production increased from 74 Mt in 1966–67 to 257 Mt in 2011–12; about 70% of the total NPK was consumed in food grains production.

Pulses have major role to maintain soil fertility and nutrient use efficiency. The beneficial effect of pulse crops in improving soil health and sustaining productivity has long been realized. On account of biological nitrogen fixation, addition of considerable amount of organic matter through root biomass and leaf fall, deep root systems, mobilization of nutrients, protection of soil against erosion and improving microbial biomass, they keep soil productive and alive by bringing qualitative changes in physical, chemical and biological properties. As a result of this, the productivity of cereals following a preceding grain legume often increases and corresponds to 40-60 kg N equivalent. Moreover, India has key place in global pulses production and contributes about 25% to the total pulse basket (Singh *et al.*, 2009).

Fertilizer is one of the key inputs in crop production and India has made rapid growth in fertilizer consumption, particularly after the introduction of high yielding varieties (HYVs) in mid sixties (Chander, 2013). Soil fertility depletion due to imbalanced and inefficient use of nutrients has become serious constraint in improving yields. Continuous nutrient mining and non-recycling of crop residues have aggravated the problem of multi-nutrient deficiencies in Indian soils. The nutrient use efficiency is low and declining. The low nutrient use efficiency not only affects crop yields and farmer's profit but also poses a great threat to the environment. Thus, nutrient application plays key role in maintaining productivity and sustainability of the crops/ cropping system. Amongst them is the fertilizer. According to Majumdar *et al.* (2013) for best fertilizer use one should adopt and follow the principle of "4 R" – as the right fertilizer source, at the right rate, right time and right place to achieve economics, social and environmental goals. This holds true not only for fertilizer but also for nutrient in totality. Thus, nutrient use efficiency represents a key indicator to assess progress towards better nutrient management.

Nutrient scenario

To meet food grains requirement of 300 Mt by 2025, 45 Mt of $N + P_2O_5 + K_2O$ is estimated to be required per annum (Prasad, 2012). Out of this, 35 Mt is proposed to be met from the chemical fertilizers and the rest from organic manures. A similar estimate of one-and-half times to that in year 2007– 08 consumption of 23 Mt of NPK was made by Tiwari (2007). The present installed capacity of fertilizers is only 12.3 Mt of N and 5.7 Mt of P_2O_5. All potash is imported. During 2007–08, 6.9 Mt of urea, 3.0 Mt of DAP/MAP and 4.4 Mt of muriate of potash (1KCl) were imported. Obviously, imports will increase in the coming years to sustain increased food production unless adequate incentives are provided to the fertilizer industry to increase the installed fertilizer production capacity.

Table 1: Available nutrients from organic manure

Component	Potential availability (Mt)	Actual availability (Mt)	Nutrient value (Mt)
Crop residue	603	201	5
Animal dung	791	287	4
Green manure	4.5*	NA	0.2
Rural compost	184	184	2.6
City compost	12.2	12.2	0.4
Biofertilizer	0.01	Negligible	0.4
Others	96.6	NA	0.9
Total			**12.8**

(*Source*: Bhattacharya, 2007)

Some estimates of N, P and K removed per tonne of food grain produced (Table 2) imply that both cereals and pulses are heavy feeder of nutrients especially N, P and K. However, majority of pulses derive nitrogen from atmosphere in the form of biological nitrogen fixation and requires only few N through fertilizer. Plant nutrient (N + P_2O_5 + K_2O) availability from organic sources such as farmyard manure, compost, vermicompost and green manure is estimated at 13 Mt (9 Mt net) (Table 1). These estimates exclude secondary and micronutrients added, which are sizeable. Organic manures and crop residues can play a major role in recycling K. Concomitant use of organic manure and fertilizer-N can reduce leaching losses of N. Thus Integrated Nutrient Management is vital for increasing food production.

Table 2: Removal of NPK (kg/t grain) of major food grain crops

Crop	N	P (P_2O_5)	K (K_2O)
Rice	20.4	3.6 (8.2)	20.4 (24.5)
Wheat	22.4	3.8 (8.7)	28.2 (33.8)
Maize	24.3	6.4 (14.6)	18.3 (22.0)
Sorghum	26.1	4.5 (10.3)	21.5 (25.8)
Pearl millet	27.1	8.2 (18.8)	39.7 (47.6)
Chickpea	50.6	8.6 (19.7)	29.7 (35.8)
Pigeonpea	92.1	8.2 (18.8)	30.7 (36.8)

At present, India is the second largest producer of fertilizer N and the third largest producer of phosphate fertilizers in the world (Table 3). Potash is totally imported. As regards consumption, India is second only to China in nitrogen and phosphorus. However, the fertilizer consumption in India is quite skewed. The average fertilizer consumption of 120 kg/ha (in 2007–08) masks more than it reveals. During 2007–08, fertilizer NPK consumption (kg/ha) was maximum in Andhra Pradesh (205) followed by Punjab (196), Tamil Nadu (184), Haryana (182) and Uttar Pradesh (154), but was less than 2 kg/ha in Arunachal Pradesh and Nagaland (Prasad, 2012).

Table 3: Production and consumption of nitrogen and phosphate fertilizers (2007–08)

Country	Production (Mt)		Consumption (Mt)	
	N	P_2O_5	N	P_2O_5
China	35.3	12.6	31.3	11.5
India	10.9	3.7	14.4	5.5
USA	8.5	8.9	11.6	4.1

Although it is reasonable to assume that, on a global scale, at least 50% of the fertilizer N applied is lost from agricultural systems and most of these losses occur during the year of fertilizer application. However, these losses could be minimized by adoption of good agronomic practice and best fertilizer management options. In case of P fertilizer applications typically result in cereal yield increases by 20 to more than 50 kg grain/kg P applied. Under favorable growth conditions, most agricultural crops recover 20-30% depending upon the growth stage of P applications. A large portion of the unused P accumulates in the soil and is eventually recovered by subsequent crops over, a much smaller fraction of P losses as runoff or through leaching that can be secondary off-site impacts (Dobermann, 2007). On the contrary, it has been reported that in developing countries, K input output budgets in agriculture are highly negative (Sheldrick *et al.*, 2002). In countries like India, annual K losses hover around 20-40 kg ha^{-1} and those have been increased steadily during the past 40 years (Majumdar *et al.,* 2013). Therefore, it is expected that there will be high nutrient use efficiency for the K in such areas. Today economic and environmental challenges are demanding more attention to nutrient use efficiency. Moreover, higher prices for both crops and fertilizers have highlighted interest in efficiency improving technologies and practices that also improve productivity. In addition, nutrient losses that contaminate air and water bodies. This contamination can be reduced by improving use efficiencies of nutrients particularity for nitrogen (N) and phosphorus (P).

Agricultural cropping systems contain complex combinations of components like soils, plants, water, soil microbes, crop rotations etc. Improvements in the efficiency of one component may or may not be effective in improving the efficiency of the cropping system. Nutrient inputs may include fertilizer alone or other sources including manures, nutrients in the soil, biological activities (*e.g.* N-Fixation). Outputs may consider the specific nutrient in question. Because a cropping system includes multiple inputs and outputs, its overall efficiency depends on the impact of economics. To maximize profit is to obtain the maximum values of outputs per unit values of all inputs. At the rate where the net return to the use of one input peaks, the input is making its maximum contribution to increase the efficiency of all other inputs involved. Rates of nutrient application optimal for economic yields often minimize nutrient losses (Hong *et al.*, 2007)

Concept and principles for boosting the NUE

All plants require at least 17 essential elements to complete their life cycle. Nutrient availability in many native soils is too low in at least one or more of the essential nutrients to allow crops to express their genetic potential for growth. Each plant nutrient has specific function within the plant. Some are relatively simple while others take part in extremely complex biochemical reactions. To enhance NUE one should remember and follow the concept that fertilizer must be applied as right source at the right rate, time and place.

Applying the right source of plant nutrients at the right rate at the right time, and in the right place is the case concept of '4 R' nutrient stewardship. In this regard following principles are need to be addressed in nutrient management approaches:

Principles supporting practices

Specific scientific principles guide the development of practices determining right source, rate, time and place. The principles are the same globally, but how they are put into practice varies locally depending on specific soil, crop, climate, weather, economics and social condition. Farmers and crop advisers make sure the practices they select and apply locally are in accord with these principles.

Right source

The idea of selecting the most appropriate nutrient source seems simple in concept, but many factors need to be considered when adopting this choice. Factors influencing selecting the right source are namely, fertilizer delivery, issues, environmental concerns, product price and economic constraints. Important considerations for high NUE are:

- Consider rate, time and place of application
- Supply nutrients in plant available forms
- Suit soil physical and chemical properties
- Recognize synergisms among nutrient elements and sources
- Recognize blend compatibility

- Recognize benefits and sensitivities to associated elements
- Control effects of non-nutritive elements.

Right rate

The core scientific principles that define right rate for a specific set of conditions are the following:

- Consider source, time and place of application
- Assess plant nutrient demand
- Use adequate methods to assess soil nutrient supply
- Assess all available nutrient sources
- Predict fertilizer use efficiency
- Consider soil resource impacts
- Consider rate-specific economics.

Right time

The core scientific principles that define right time for a specific set of conditions are the following:

- Consider source, rate and place of application
- Assess timing of plant uptake
- Assess dynamics of soil nutrient supply
- Recognize dynamics of soil nutrient loss
- Evaluate logistics of field operations.

Many examples (Majumdar *et al.*, 2013) of timing fertilizers applications based on stage of crop growth can be given but few are referred here: (i) *N application to small grains such as wheat* : Most wheat recommendations calls for some N applied at planting, with the majority top-dress applied by (before) jointing. By the time wheat begins heading later in the season the majority of N has been taken up, and if good N a management practices were not previously used, then yield will suffer. (ii) *Ca for groundnut*: Groundnuts are especially sensitive to Ca deficiency. High levels of available Ca are needed in the soil zone where groundnuts pods are developing and thus pre-bloom applications of soluble Ca materials (*i.e.* calcium sulfate or calcium nitrate) are sometimes made to groundnuts. (iii) *Mn for soybean*: Early season foliar applications of Mn are often made to soybean in areas when deficiency symptoms appear on the plant tissue.

Right place

The core scientific principles that define right time for a specific set of conditions are the following:

- Consider source, rate and place of application
- Consider where plants roots are growing
- Consider soil reactions
- Suit the goal of the tillage system
- Manage spatial variability.

Measures of nutrient use efficiency

In general, four terms are used in relation to NUE. These are: Agronomic Efficiency (AE), Recovery Efficiency (RE), Physiological Efficiency (PE), and Partial Factor Productivity of Fertilizers (PFP_f). The following expressions are used for determining these NUE measures (Prasad, 2009):

AE (kg grain /kg nutrient applied) = $(Y_f - Y_c)/ N_a$

RE (% of nutrient taken up by a crop) = $(NU_f - NU_c)/Na \times 100$

PE (kg grain/kg nutrient taken up by a crop) = $(Y_f - Y_c)/(NU_f - NU_c)$

PFP_f (kg grain/kg nutrient applied) = Y_f/ N_a

where, Y_f, Y_c = Yields (kg ha^{-1}) in fertilized (f) and control (c) plots

NU_f, NU_c = Nutrient uptake by a crop (kg ha^{-1}) in fertilized (f) and control (c) plots

N_a = nutrient applied

AE is the same as "crop response ratio" or productivity index used by FAO (1989) and can be determined for a single nutrient (N, P, or K) or for a combination of nutrients (NP, NK, PK, or NPK), or for a fertilizer material *per se*. PFP_f can also be determined for a single or a combination of nutrients or for a fertilizer *per se*. PFP_f which was recently introduced does not ask for a 'no-fertilizer control' plot yield. This term permits comparison of fertilizer use efficiency in different countries or in different regions of a country. The term is useful in comparing the advantages of fertilizer use in experiments on tillage, irrigation, weed control etc., where a 'no fertilizer control' is typically not provided. Further, RE may be apparent recovery efficiency (RE) or true recovery efficiency (RE_t). RE_t is determined with the help of ^{15}N for N and ^{32}P for P. RE is used by soil and environment scientists in finding out the part of nutrient taken up by crop and the part causing environmental pollution. PE is used by plant physiologists and plant breeders in studying the efficiency of different crops or cultivars of a crop in utilizing the absorbed nutrients. PE is actually AE/ RE.

Impact of nutrient management on NUE

AE or crop response ratio

Cereals

In the pre-Green Revolution era tall rice and wheat varieties showed highest response ratio for N (11.6 to 16.7 kg grain/kg N), followed by P (5.5 to 12.5 kg grain/kg P_2O_5), and the least for K (3.6-6.2 kg grain/kg K_2O) (Prasad, 2009). Furthermore, response to NP, NK, or NPK was not additive of their individual responses, which made the farmers to apply mostly N alone. Similarly, after Green Revolution era the increase in yield due to fertilizer was much higher (1.1 to 2.6 Mg ha^{-1} compared to 0.47 to 1.25 Mg ha^{-1} for tall wheat), the response ratio to NPK application was not much due to higher NPK doses and ranged from 4.7 to 10.9 kg grain/kg nutrient.

Pulses

The AE for soybean found around 10 kg grain/kg N, 5 for P, 16 for K and 9 for combined use of NPK in long term fertilizer experiments at Jabalpur (Vertisols) (Singh and Wanjari, 2009). However, in Alfisols of Ranchi the AE_n was negative due to adverse effect of N alone treatment than control. But application of P and K has enhanced the crop response ratio in Alfisols to 29, 34 and 21 kg grain /kg P, K and NPK, respectively.

Recovery efficiency

Cereal

It is documented that rice-wheat cropping system is the backbone of food security in India (Prasad, 2005) and values of true recovery efficiency from some experiments in India showed that in rice the values ranged from 26 to 35.8%, while in wheat these ranged from 25.6 to 44%. The global average values for rice, wheat, and maize clearly show that the values of all the terms associated with nitrogen use efficiency (N_iUE) declined as the rate of N applied increased. At similar N levels, values of all N_iUE terms in rice were lower in India compared to the global values. On the other hand, values of all terms of N_iUE in wheat were higher in India than the global values, showing that in India

N is more efficiently utilized for wheat than rice. Thus, in rice there is considerable scope to increase N_iUE. RE of P varied from 20–37%, while that of K varied from 40–56% in the rice-wheat cropping system.

Pulses

It has been estimated that 668,000 tonnes of nitrogen can be incorporated in the soil through the inclusion of legumes in cropping systems. The intrinsic nitrogen fixing capacity of pulse crops enables them to meet large proportion of their nitrogen requirement and also helps in economizing nitrogen in succeeding non-legume crops due to the residual effect. Different legumes have different capacity to leave behind varying amounts of N for use by the succeeding crops. In sequential cropping involving pulses, the preceding pulse may contribute 18-70 kg N/ha to the soil and thereby considerable amount of nitrogen to succeeding crop (Ali and Mishra, 2000). The beneficial effect of pulses was more pronounced in maize as compared to sorghum after chickpea and pigeonpea whereas after lentil and peas the higher N equivalent benefit was observed after pearlmillet. Growing of short duration legumes such as green gram and cowpea in widely spaced crops and ploughing back the same in the soil after picking the grains resulted in an advantage of 30 kg N/ha on fertilizer basis in Alfisol of Hyderabad. Rekhi and Meelu (1983) found that incorporation of crop residue of mungbean in rice-wheat system not only added 100 kg N/ha to the soil but also maintained high availability of N during various growth stages of rice.

Approaches to improve NUE

Crop yield directly or indirectly is the numerator in all the terms of FUE/NUE and the crop, soil and agronomic factors that increase crop yield may therefore increase FUE/NUE (Prasad, 2009 and 2012). Data on potential yield, on-station, and on-farm yields shows a gap of 37–52% between potential and on-station yields and a 35–70% gap between potential and on-farm yields. The gap between on-station and on-farm yields varied from 6–44%. In general the gaps are wider in rice than in wheat. The available farm technology can at least reduce on-station–on-farm gap and this can increase rice and wheat production by 15–20%. However, in rice the on-station–on-farm gap is zero in Ludhiana region of Punjab and in wheat it is even slightly negative in Pantnagar, Uttaranchal (Prasad, 2009). This shows that the farmers have already applied the available technology in these regions. Thus, with good extension efforts it can be replicated in other parts of the country. Information on benefits of improved technology as compared to farmers' practices in increasing crop yield is also available for pulses (Ali *et al.,* 2002).

Soil management

Both chemical amendments such as lime and gypsum and physical management involving tillage are important for increasing crop yields and in doing so, they improve NUE.

Liming acid soils: Nearly 51 million ha of soils in India have pH 5.5 or less (Mahapatra and Pattanayak, 2008). Long-term fertilizer experiments from Ranchi (Nambiar, 1994) further show that over a period of 13 years, maize yield was higher under NPK + lime compared to no-lime treatments. Continuous application of FYM also maintained reasonably good yields (71.7% of that obtained with NPK + lime).

Gypsum application in sodic soils: Crop yields are low on sodic soils and can be largely increased by gypsum application. Data from Karnal (Singh and Abrol, 1988) illustrated that PFP_{npk} in wheat almost doubled following gypsum application.

Tillage: Puddling rice paddies reduces percolation of water and leaching of fertilizers, especially N, besides helping in weed control. The net result is increased rice yield and PFPn (Dwivedi *et al.*, 2003). Several new tillage implements such as laser aided land leveller, mechanical rice transplanter, and drum seeders have recently become available (Tomar *et al.*, 2006) and their use in rice cultivation will increase NUE. It is now possible to sow wheat soon after rice harvest without primary cultivation with zero-till machines, which permits timely sowing, besides ensuring increased grain yields and PFP_{npk} (Yadav *et al.*, 2005; Singh *et al.*, 2008). Advantage of zero tillage has also been reported for maize after rice in Telangana region of Andhra Pradesh (Reddy and Veeranna 2008). In arid regions, off-season tillage can help in storing soil moisture, which increases crop yield and PFP_{npk} (Samra, 2003).

Crop management

Crop varieties and cropping systems: The Green Revolution in India was initiated with the introduction of high yielding fertilizer responsive dwarf varieties of wheat (Swaminathan, 2006), which not only gave higher yields but also higher NUE. The introduction of rice hybrids (Chang *et al.,* 1988; Siddiq, 2006) promises one tonne additional yield over that obtained with current high yielding varieties and an increase in AE_n and RE_n (Kumar and Prasad, 2004; Kumar *et al.,* 2007).

Timely sowing/transplanting: Delayed transplanting of rice reduces grain yield and PFP_{npk} as is obvious from the data from Bhubaneswar (Nayak *et al.,*

2003). Late sown wheat in rice-wheat cropping system results in16–24% less grain yields and 21.9 to 16–19.4% lower PFP_{npk} (Tripathi *et al.,* 2002).

Plant population: Sub-optimal plant population is one factor that reduces crop yields in India more than any other factor. Lower seed rates associated with wider spacing and fewer seedlings per hill in the case of rice and seedling mortality due to diseases and pests are the major factors that affect plant population in other crops. In rice transplanting two seedlings is advantageous from the viewpoint of grain yield as well as PFP_{npk} (Nayak *et al.,* 2003). Increasing seed rate from 100 to 125 kg ha^{-1} increased rice yield by 240 kg ha^{-1} and PFP_{npk} by 2% (Tripathy and Mohapatra, 2007). Further increase in seed rate, however, reduced grain yield and PFP_{npk}. A large volume of data exists on seed rate, spacing, thinning etc. on most crops in India but it has not been linked with NUE.

Weed management: Weeds compete with crop plants for nutrients, water, and sunlight and as a consequence reduce crop yield and NUE. Examples are the menace of Phalaris minor in wheat in India (Sharma, 2007) and weedy rice (a natural hybrid of *Oryza sativa* and wild rices *O. rufipogon* and *O. nivara*) in South and Southeast Asia (Anonymous, 2007). Considerable information exists in India on effective weed control through mechanical methods and herbicides (Gupta, 1984), albeit it is not linked with NUE. Saha *et al.* (2007), however, showed that effective weed control in rainfed rice in Cuttack raised grain yield by 1.34 Mg ha^{-1} and PFP_{npk} by 11.2%.

Water management: Water management involving proper irrigation scheduling (irrigated areas) and moisture conservation (rainfed agriculture) is highly correlated with NUE. For example, in wheat, irrigation at 1.0 IW/CPE increased grain yield by 0.41 Mg ha^{-1}, water use efficiency by 7.5 kg $ha^{-1}cm^{-1}$ and PFP_{npk} by 1.7 kg grain kg NPK^{-1} over irrigation at 0.8 IW/CPE (Verma and Singh 2008). Similarly, construction of water harvesting structures for irrigation is potential mechanisms in rainfed agriculture to increase grain yield as well as PFP_{npk} (Pali *et al.,* 2007).

Fertilizer management

Materials: These are of two kinds slow release nitrogen fertilizers: the coated conventional fertilizers such as sulphur coated urea, polymer coated urea, neem coated urea, and the inherently less soluble materials, which are mostly urea-aldehyde products, such as ureaform (urea-formaldehyde), isobutylidene diurea (IBDU), and crotonaldehyde diurea (CDU). However, the cost of N in these materials is twice or thrice or even more than the conventional fertilizers, making them beyond the reach of common farmers. Another approach has been to use nitrification inhibitors to retard nitrification of applied NH_3 or urea-N and to reduce leaching and de-nitrification losses (Prasad, 2005). The most widely tested and used nitrification inhibitors are Nitrapyrin or N-Serve, AM (2-amino-4-chloro, 6-methyl pyridine), and dicyandiamide. On-farm trials in Delhi, Punjab, Haryana, and Uttar Pradesh have shown that neem cake coated urea (NCCU) results in 6 to 11% increase in rice yield. PFPn for NCU ranged from 41 to 43% compared to 36 to 41% for prilled urea (Prasad, 2007). Yield benefits with urea super granules (USG) over prilled urea varied from 0.2 to 1.2 Mg ha^{-1} at the same level of N (Kumar *et al.,* 1989). Production of USG in India, however, did not take off. Nonetheless, it was more popular and well received by the rice farmers of Bangladesh (Balasubramanian *et al.,* 2004). There has been not much research on slow-release P and K fertilizers. As regards to phosphate fertilizers, the phosphate rock is insoluble but on acid soils it can be directly used. However, good response have tested in lab to nano phosphorus which now being taken to field for their recommendation as it is low volume and highly efficient material from NUE.

Time of application: Split application of N is highly desirable since crop plants take up very small amounts of N $ha^{-1}day^{-1}$. For example, Prasad (2006) reported that rice removed just 1–1.2 kg N ha day^{-1}. Excess N not used by crops is subject to various mechanisms of losses (Adhya *et al.,* 2007; Pathak *et al.,* 2008). Recent research has shown that for determining the proper time of post transplant/sowing application of N, use of new tools such as chlorophyll meters and leaf colour charts holds promise (Pathak and Ladha, 2007; Bijai Singh, 2008). Most P and K are applied at sowing/transplanting, although it is reported that in wheat P may be applied after the first irrigation, in case it is not available or applied at sowing (Singh, 1985). Likewise, some reports on the advantage of split application of K in rice are available (Meena *et al.,* 2002).

Method of application: It is known that there is advantage of deep placement of P for increasing its efficiency for crops other than rice (Rao *et al.,* 2003). However, only in areas where agriculture is mechanized, deep placement of P is practised; elsewhere, it is still broadcast depriving the farmers the full benefits of P fertilization. Regarding N also, deep placement increases nitrogen use efficiency (NiUE). Panda *et al.* (2007) reported that band furrow placement of N doubled NiUE compared to its broadcast application in rainfed lowland rice. Foliar application of N is desirable in dryland agriculture, because the farmers in these areas apply fertilizers only when rains come, and these are often delayed. Thanunathan *et al.* (2004) recommended foliar application of N and K in flooded rice.

Balanced NPK Fertilization and Site Specific Nutrient Management: Balanced NPK fertilization has received considerable attention in India (Ghosh *et al.,* 2004). Farmers, specially the marginal and dryland farmers, generally, tend to apply only N. However, the AE_n of applied N can be largely increased by adequate P and K fertilization. Adequate application of S and Zn in the soils deficient in these nutrients automatically increases the AE_{npk} (John *et al.*, 2006). Widespread deficiencies of S, Zn, and B have led to the evolution of site specific nutrient management (Singh *et al.,* 2008). The SSNM increases the AE of all nutrients applied as it involves analyzing the soils for all essential plant nutrients and developing fertilizer recommendations based on soil analysis.

Integrated Plant Nutrient Supply System (IPNS): The IPNS demands a holistic approach to nutrient management for crop production and it involves judicious combined use of fertilizers, biofertilizers, organic manures (FYM, compost, vermicompost, biogas slurry, green manures, crop residues *etc.*), and growing of legumes in the cropping systems (Prasad, 2008). IPNS also encompasses balanced fertilization and SSNM. Considerable research on IPNS has been done in India (Rao *et al.*, 2002). Moreover, long-term fertilizer experiments have shown that addition of organic manures in addition to NPK (add-on series) results in high yields over a long period of time as compared to a decline in yield over time when only inorganic fertilizers were applied (Swarup, 2002). Sarkar and Singh (2002) reported that for soybean-wheat cropping system in the acidic soils of Ranchi (pH < 5.4), soybean yield (averaged over 28 years) was 0.33 Mg ha^{-1} and wheat yield, 0.43 Mg ha^{-1} for plots receiving N alone as compared to 1.59 Mg ha^{-1} in soybean and 2.65 Mg ha^{-1} in wheat when NPK was applied. Application of FYM with NPK increased the soybean yield to 1.86 Mg ha^{-1} and that of wheat to 3.19 Mg ha^{-1}. Further, the effects of NPK + FYM were at par with NPK + lime, implying that in acid soils continuous application of FYM can also partially offset soil acidity.

Green manure crops have the intrinsic potential to recycle considerable quantities of organic materials and nutrients. Productivity of rice-wheat cropping could be raised by 1.2 Mg ha^{-1} with 80 kg N ha^{-1} (with Sesbania and cowpea GM) applied to rice over 120 kg N ha^{-1} applied in control plots (Misra and Prasad, 2000). Green manures contribute 60–120 kg N ha^{-1} to the succeeding crop (Sharma *et al.,* 1996, Palaniappan *et al.,* 1997). Legumes fix 50–500 kg N ha^{-1} depending upon the crop and its growth period, and leave a residual N varying from 30–70 kg N ha^{-1} to the succeeding crop (Venkatesh and Ali, 2007). Organic manures supply small amounts of micronutrients (Mishra *et al.,* 2006) and also improves the soil physical, chemical and biological properties (Misra and Saha, 2008, Vineela *et al.* (2008). Biofertilizers [*Rhizobium*, *Azotobacter*, *Azosprillum*, blue green algae (BGA), *azolla*, phosphate solubilizing organisms

(PSO, PSB, PSF), vescicular arbuscular mycorrhiza (VAM)] can become an important component of IPNS (Swarnalakshmi *et al.,* 2006, Tewatia *et al.,* 2007) specially under low-land rice cultivation and dryland agriculture, where only low levels of fertilizers are applied. Organisms accelerating the decomposition of crop residues also have a role (Sharma and Prasad, 2002).

Conclusion

Crop yield directly or indirectly is the numerator in all the terms of nutrient use efficiency. These yields are influenced by the crop, soil and agronomic factors. Now suitable agricultural technologies are available for enhancing the productivity to meet the yield gap between potential yield, on station and on farm yield. In this context, an effective nutrient management has to play key role in the development of site specific nutrient recommendations including balanced NPK doses, timely application of fertilizers using appropriate source, methods at right time and place. In addition to this, development and production of slow-release N fertilizers and indigenous nitrification inhibitors, and developing and practicing an integrated plant nutrient supply system (IPNS) using chemical fertilizers, organic manures, crop residues, and biofertilizers. Along with above strategies on nutrient management, other aspects of soil and crop management including the use of high yielding, nutrient-efficient cultivars, correcting soil physical and chemical problems and water management, disease and pest management (IPM), and post-harvest care and safe storage are important to achieve high nutrient use efficiency.

Selected References

Adhya TK, Pathak H , Chhabra A. 2007. N-fertilizers and gaseous- N emission from rice-based cropping systems. *In:* Agricultural Nitrogen Use and Its Environmental Implications. Abrol, Y.P., Raghuram N., and Sachdev, M.S. (eds). I.K. Pub. House Pvt. Ltd., New Delhi, pp.459–476.

Ali M, Mishra JP. 2000. Nutrient management in pulses and pulse based cropping systems. *Fertilizer News* 45(4): 57-69.

Anonymous. 2007. Weedy rice attacks Asia's direct-seeded rice. *Ripple* 2(2): 5.

Balasubramanian V, Alves B, Aulakh MS, Bekunda M, Cai ZC, Drinkwater L, Mugendi D, Van Kessel C, Oenema O.2004. *Crop environmental factors affecting N use efficiency.* Agriculture and the Nitrogen Cycle: Assessing the Impacts of Fertilizer Use in Food Production and the Environment. Mosier AR, Syers JK and Freney JR (eds). Island Press. Scientific Committee on Problems of the Environment (SCOPE) Series Vol. 65, Paris, France, pp.19–33.

Bhattacharya P. 2007. Prospects of organic nutrient resource utilization in India. *Indian Journal of Fertilizers* 3: 93–107.

Bijai Singh. 2008. Crop demand driven site specific nitrogen applications in rice and wheat: some recent advances. *Indian Journal of Agronomy* 53: 1–30.

Chang Q, Hu N, Jing N and Huang F. 1988. Physiolo-gical and biochemical characters of hybrid rice in Dongting Lake region, China. In: Hybrid Rice, International Rice Research Institute, Los Banos, Manila, pp. 279–286.

Dobermann A. 2007. IFA International Workshop on Fertilizer Best Management Practices. 7-9 March, Brussels, Belgium

Dwivedi BS, Shanker AK and Singh VK. 2003. Annual Report 2002-2003. Project Directorate for Cropping Systems Research (PDCSR), Modipuram.

Economic Survey.2013. Economic Survey 2012-2013, Government of India.

Gupta OP. 1984. Scientific Weed Management. Toady and Tomorrow's Printers & Publishers, New Delhi.

Hong N, Scharf PC, Davis JG, Kitchen NR, Sudduth KA. 2007. *Journal of Environmental Quality* 36: 354- 362.

Kumar N, Prasad R .2004. Nitrogen uptake and apparent N recovery by a high yielding variety and a hybrid of rice as influenced by levels and sources of nitrogen. *Fertiliser News* 49(5): 65–67.

Kumar N, Prasad R ,Zaman FU. 2007. Relative response of high yielding variety and a hybrid of rice to levels and sources of nitrogen. *Proceedings of Indian National Science Academy* 73: 1–6.

Mahapatra IC, Pattanayak SK .2008. Acid soils and their management – A case study with paper mill sludge. *NAAS News* 8(4): 6–7.

Majumdar Kaushik, Johnston AM, Dutt Sudarshan, Satyanarayana T, Roberts TL .2013. Fertilizer best management practices concept, global perspectives and application. *Indian Journal of Fertilizer* 9 (4) : 14-31.

Meena SL, Singh S, Shivay YS. 2002. Response of hybrid rice (Oryza sativa) to nitrogen and potassium application. *Indian Journal of Agronomy* 47: 207–211.

Mehla RS, Verma JK, Gupta RK , Hobbs PR. 2000. Stagnation in the productivity of wheat in the Indo-Gangetic plains: zero-till-seed-cum-fertilizer drill as an integrated solution. Rice-wheat Consortium Paper Series 8, Rice-Wheat Consortium for the Indo-Gangetic Plains, New Delhi, 12p.

Mishra BN, Prasad R, Gangaiah B, Shivakumar BG. 2006 Organic manures for increased productivity and sustained supply of micro nutrients Zn and Cu in a rice-wheat cropping system. *Journal of Sustainable Agriculture* 28: 55–66.

Misra BN , Prasad R (2000) Integrated nutrient management for sustained production in a rice-wheat cropping system. *Acta Agronomica Hungarica* 48:257–262.

Nambiar KKM (1994) Soil Fertility and Crop Productivity under Long Term Fertilizer Use, Indian Council of Agricultural Research, New Delhi, 144p.

Nayak BC, Dalei BB, Choudhury BK 2003. Response of hybrid rice (*Oryza sativa*) to date of planting, spacing and seedling rate during wet season. *Indian Journal of Agronomy* 48: 172–174.

Pathak H, Ladha JK (2007) Improving nitrogen Use efficiency: strategies, tools,management and policy options. *In:* Agricultural Nitrogen Use and Its Environmental Implications. Abrol, Y.P., Raghuraman, N., and Sachdev, M.S. (eds). I.K. Int. Pub. House, New Delhi, pp. 279–302.

Pathak H, Bhatia A, Jain N, Mohanty S, Chandrasekharan S (2008) Nitrogen, phosphorus and potassium budgets in Indian Agriculture. *In* Indian Science Congress, Andhra University, Visakhapatnam, 3-7 January, 2008, 45p.

Prasad R.2005. Rice-wheat cropping systems. *Advances in Agronomy* 86: 255–339.

Prasad R .2006.) Nitrogen uptake pattern as a guide for nitrogen application practices in rice. *Indian J Fertiliser* 2(6): 39–41.

Prasad R .2012. Fertilizers and manures. Special Section. *Current Science* 102(6):894-898

Prasad R .2009.Efficient fertilizer use: The key to food security and better environment. *Journal of Tropical Agriculture* 47 (1-2): 1-17.

Rao AS, Chand S, Srivastava S. 2002. Opportunities for integrated plant nutrient supply for crops/cropping systems in different agro-ecosystems. *Fertiliser News* 47(12): 75–90.

Reddy RR, Veeranna G 2008. Rice-zero tillage in maize proved economical in Andhra Pradesh. *Indian Farming* 58(5): 8–9.

Rekhi RS, Meelu OP. 1983. Effect of complementary use of mung straw and inorganic fertilizer-N on N availability and yield of rice. *Oryza* 20: 125-129.

Sharma R. 2007. Integrated weed management in wheat and rice crop. *Indian Farming* 57(8): 29–34.

Sharma SN, Prasad R, Singh P. 2000. On-farm trials of the effect of introducing a summer green-manure of mungbean on the productivity of rice wheat cropping system. *Journal of Agricultural* 134: 169–172.

Sheldrick WF, Syers JK, Lingard J .2002. *Nutrient cycling in Agroecosystem* 62: 61-72.

Siddiq EA. 2006. Genetic improvement of rice in perspective. *Indian Farming* 56(7): 10–16.

Singh G, Singh OP, Kumar V , Kumar T. 2008. Effect of methods of establishment and tillage practices on productivity of rice – wheat cropping system in low-lands. *Journal of Agricultural* 78: 163–166.

Singh KK, Ali M, Venkatesh MS. 2009. Pulses in Cropping Systems. *Technical Bulletin*, IIPR, Kanpur.

Singh RP. 1985. Report of the All India Wheat Improvement Project, Wheat Agronomy 1984-85. Project Directorate for Wheat, New Delhi.

Singh SB, Abrol IP. 1988. Long-term effects of gypsum application and rice-wheat cropping on changes in soil properties and crop yield. *Journal of Indian Society of Soil Science* 36: 316–319.

Swaminathan MS. 2006. Sustainable Agriculture – Towards an Evergreen Revolution. Konark Pub. Pvt. Ltd., New Delhi, 217p.

Swarnalakshmi K, Dhar DW, Singh PK. 2006. Blue-green algae: A potential biofertilizer for sustainable rice cultivation. *Proceedings of Indian National Science Academy* 72: 167–178.

Swarup A. 2002. Lessons from Long term fertilizer experiments in improving fertilizer use efficiency and crop yields. *Fertilizer News* 47(12): 59-73.

Tewatia RK, Kalve SP, Chaudhary RS. 2007. Role of biofertilzers in Indian agriculture. *Indian Journal of Fertilizers* 3(1): 111–118.

Thanunathan K, Kandasamy S, Vaiyapuri V, Imayavaramban V, Singarvel R .2004. Conjunctive use of N and K through foliage for augmenting the yield of flood affected rice. *J Potassium Research* 20: 116–117.

Tiwari KN. 2007. Reassessing the role of fertilizers in maintaining food, nutrition and environmental security. *Indian Journal of Fertilizers* 3: 33–50.

Tomar RK, Sahoo RN, Garg RN, Gupta VK. 2006. Resource conservation technologies – potential tools for attaining food, nutritional and livelihood security. *Indian Farming* 56(9): 24–30.

Tripathi SC, Chauhan DS, Sharma RK, Kharub AS, Chhonkar RS, Singh S. 2002. Effect of sowing time and temperature on wheat productivity in NWPZ and NEPZ of India. *Indian Wheat Newsletter* 8(2): 6–7.

Tripathy SK, Mohapatra S. 2007. Maximizing direct-sown rice production. *Indian Farming* 57(2):3–6.

Vineela C, Wani SP, Srinivasarao Ch, Padmaja B, Vittal KPR .2008. Microbial properties of soils as affected by cropping and nutrient management practices in several long-term manual experiments in the semi-acid tropics of India. *Ecology* 40: 165-173.

Yadav DS, Shukla RP, Sushant, Kumar B .2005. Effect of zero tillage and nitrogen level on wheat (*Triticum aestivum*) after rice (*Oryza sativa*). *Indian Journal of Agronomy* 50: 52–53.

21

Enhancing Nutrient Use Efficiency, pp. 279-289
Editors: K. Ramesh, A.K. Biswas, B.L. Lakaria, S. Srivastava and A.K. Patra

Rainfed Pulses Based Cropping Systems

P.M. Shanmugam and S.P. Sangeetha

Tamil Nadu Agricultural University, Coimbatore, Tamil Nadu-641 003, India

Introduction

India is the largest producer, importer and consumer of pulses in the world, accounting for 25 percent of the global production, 15 per cent trade and 27 per cent consumption, being an inseparable ingredient in the diet of the vast majority of population. Since time immemorial, pulses have been cultivated on marginal and sub-marginal lands, which are characterized by poor soil fertility and moisture stress, and consequently their yield potentials have not been realized. Further, more than 90 per cent areas under pulses are rainfed. Drought and heat stress may reduce seed yields by 50%, especially in arid and semi-arid regions. Pulses are predominantly grown under resource poor and harsh environments frequently prone to drought and other biotic and abiotic stresses. As a result, the productivity of the pulses in India is quite low even less than 1 tonne per hectare compared to wheat and rice (Johonson *et al.*, 2000). To meet the demand of pulses, India is presently importing about 3 million tons of pulses. To increase the pulse production to the tune of about 18 million tons from existing 15 million tonnes, rainfed rice fallow lands offer a huge potential niche for pulses production. Nearly 82% of the rainfed rice fallow lands are located in the states of Assam, Bihar, Chhattisgarh, Jharkhand, Madhya Pradesh, Odisha, and West Bengal. These rainfed ricc fallow lands offer a huge potential niche for pulses production. Farmers are often pressed to use inadequate quantity of manures and fertilizers to rainfed pulses for economic reasons. Therefore, there is a great scope of increasing the production in rainfed areas through efficient nutrient management.

Major nutrient management issues in rainfed pulses

- Inadequate and unbalanced use of fertilizers
- Emerging multi-nutrient deficiencies particularly of secondary and micronutrients
- Low fertilizer use efficiency
- Declined crop response ratio.

Crop responsiveness to fertilizer is maximized and environmental impact of fertilizers reduced, and often eliminated, when crops are managed for improved nutrient use efficiency through best management practices which balance production inputs at the appropriate levels.

Nutrient requirements of pulses

Of the 16 essential elements required for the nutrition of plants, pulses specially need adequate amount of P, Ca, Mg, S and Mo. Phosphorus is required for proper root growth and growth of rhizobia. Calcium and magnesium are required to stimulate growth and to increase the size of the nodules, pod formation and grain setting. Sulphur is required for nodulation and protein synthesis, molybdenum for nitrogen fixation and assimilation and boron for reproduction.

The nutrient removal by different pulse crops for an average yield level are furnished in Table 1 and 2.

Table 1: Macro and secondary nutrient removal (kg /ha) by different pulse crops

Crops	N	P	K	Ca	Mg	S
Blackgram	48.5	7.1	42.5	24.5	5.8	6.5
Greengram	41.0	8.6	40.5	21.5	6.5	5.3
Chickpea	46.3	8.4	49.6	NA	NA	NA
Pigeonpea	63.0	11.5	62.5	NA	NA	NA
Soybean	55.5	8.8	76.5	57.5	12.4	6.3

Table 2: Micronutrient removal (g /ha) by different pulse crops

Crops	Fe	Mn	Zn	Cu	B
Blackgram	350	150	210	17	30
Greengram	150	78	130	10	21
Soybean	190	75	55	21	18

Inadequate use of fertilizers

Pulses remove from soil a good amount of nitrogen and phosphorus along with other nutrients. When N supply from soil becomes limited, they meet their N requirement by symbiotic fixation. One tonne of legume biomass removes nearly 30-50 kg N, 2.7 kg P, 12-30 kg K, 3-20 kg Ca, 1-5 kg Mg, 1-3 kg S, 200-500 g Mn, 5 g B, 1 g Cu and 0.5 g Mo. Pulses require a high amount of P and Ca. Besides, they have a greater demand for micro nutrients such as Mo, B, Co, Cu and Zn. Co and Mo are particularly beneficial for nodulation.

Pulses have the unique ability to trap atmospheric nitrogen in root nodules in association with *rhizobium* bacteria and consequently only a small amount of nitrogen (20 kg/ha) as starter dose is recommended. The process of nitrogen fixation peaks at 2-3 weeks before the onset of flowering and then starts degenerating. The scarcity of soil moisture further restricts biological nitrogen fixation. Consequently pulses in rainfed areas often experience nitrogen deficiency at the peak reproductive stages viz., at flowering and pod formation. Phosphorus deficiency in soils is wide spread and most of the pulse crops have shown good response to 20–60 kg P_2O_5/ha depending upon nutrient status of soil, cropping system and moisture availability. Method of supplying this nutrient need to be further explored for increasing the efficiency of phosphorus application. Super phosphate is recommended and is costly and hence to reduce the cost, alternate sources like rock phosphate can be used. The efficiency of rock phosphate can be increased by bio-amending with Farm Yard Manure (FYM) and *phosphobacteria.* Response to potassium application is location specific.

In the recent years, use of sulphur (20–30 kg/ha) and some of the micronutrients such as Zn, B, Mo and Fe have improved productivity of pulse crops considerably in many pockets. B and placement of phosphatic fertilizers and use of bio-fertilizers enhance the efficiency of applied as well as native P. Foliar nutrition of some micronutrients proved quite effective. The amount and mode of application is determined by indigenous nutrient supply, moisture availability and genotypes. Balanced nutrition is indispensable for achieving higher productivity. At the same time, in view of increasing nutrient demand, there is immense need to exploit the alternate source of nutrients viz., organic materials and bio-fertilizers to sustain the productivity with more environment friendly nutrient management systems. The environmental issues and other hazards emerging out of the imbalanced use of nutrients should also be addressed properly.

Lack of organic manure application

Intensive cultivation of pulses without addition of organic matter resulted in multiple nutrient deficiency especially for nutrients like zinc and sulphur. Declining trend in the addition of organic matter is a serious concern in general and in particular for pulse crop production. With the reduced generation of FYM in Indian farms, other methods of addition like residue management and incorporation of green manures needs attention in maintaining soil health and improving the productivity. There is a need to deliver in-situ pulse crop residue management for improving soil organic matter and soil fertility. The nutrient use efficiency of pulses is given in Table 3.

Table 3. Current status of nutrient use efficiency

Nutrient	Efficiency percentage
N	30-50
P	10-20
K	<80
S	8-12
Zn	2-5
Fe	1-2
Cu	1-2
Mn	1-2

(Chowdhury and Gupta, 2012)

Nutrient management technologies for pulses involve

i) Integrated nutrient management with major and micro nutrients

ii) Foliar nutrition of pulses

iii) Balanced nutrition.

Input supply (micronutrients and fertilizer application)

Legumes fix atmospheric nitrogen. However, availability of quality *Rhizobium* inoculum is limiting. Phosphorous is becoming a limiting macro-nutrient which will affect the pulses production. A common difficulty in recovering P from the soil is that it is not readily available to plants because P reacts with aluminium, iron and calcium in the soil to form complexes. These nutrients are essentially insoluble resulting in very little movement of P in the soil solution, and none of the complexes can be taken up directly by roots (Sinclair and Vadez 2002). The use of phosphate solubilizing bacteria (strains from the genera of *Pseudomonas*, *Bacillus* and *Rhizobium* are among the most powerful P solubilizers) as inoculants simultaneously increases P uptake by the plant and thus crop yields (Khan *et al*, 2009).

A recent study by ICRISAT indicated that soils in many states in India are deficient in micro-nutrients such as boron, sulfur, zinc and magnesium (Wani *et al*., 2012). Application of small quantities (0.5 to 2 kg ha^{-1}) of micronutrients has resulted in 40-120% increase in grain yield. Hence, making these micro-nutrient fertilizers easily available to smallholder farmers in remote areas will go a long way in enhancing productivity and production of pulses. Under a mission to boost productivity of rainfed agriculture through science-led interventions in Karnataka, the improved management practices (including application of micronutrients) have increased the yield by 31-57% in green gram, 26-38% in pigeonpea and 27- 39% in chickpea during 2010-11. Similarly in 2011-12, black gram and green gram grain yields increased by 33-42% in response to improved management when compared to farmers management (Wani *et al.*, 2012).

Foliar spray of nutrients

There is no possibility of basal application of fertilizers for rice fallow pulses, since the pulses are sown prior to harvest of rice crop. Therefore, fertilizer incorporation becomes impossible. Hence, foliar fertilization with the spraying of 2% DAP and 1% KCl is recommended at 30 and 45 DAS of the pulse crop both in rainfed and irrigated conditions. To mitigate the drought effect on the pulse crop, cycocel spray at 100 ppm (100 mg/lit) is recommended. Cycocel spray will enhance the root development, which facilitates the crop to get moisture from deeper layer. To check the flower dropping during drought situation, Alpha naphthaleneacetic acid NAA (4 ml in 4.5 lit) spray is advocated, first spray at floral initiation and another spray at 15 days thereafter.

Techniques to improve nutrient use efficiency in rainfed pulses

Blackgram

1. Seed treatment

- Treat the seeds with 3 packets (600 g/ha) of Rhizobial culture CRU-7 + 3 packets (600 g/ha) of PGPR and 3 packets (600 g/ha) of Phosphobacteria using rice kanji as binder. If the seed treatment is not carried out apply 10 packets of *Rhizobium* (2000 g/ha) + 10 packets of PGPR (2000 g/ha) and 10 packets (2000 g) of Phosphobacteria with 25 kg of FYM and 25 kg of soil before sowing.
- Seed coating with biofertilizers and micronutrients *viz*., Zn, Mo and Co @ 4, 1, 0.5 g/kg of seed is recommended.

2. Fertilizer application and foliar spray

- Apply fertilizers 12.5 kg N + 25 kg P_2O_5 + 12.5 kg K_2O +10 kg S ha basally before sowing under rainfed condition. (S is applied in the form of gypsum if single super phospate is not applied as a source of phosphorus)

- Soil application of micronutrient mixture @ 5 kg/ha as Enriched FYM (Prepare enriched FYM at 1:10 ratio of MN mixture and FYM; mix at friable moisture and incubate for one month in shade)
- For yield improvement through increasing the physiological, biochemical attributes, foliar spray of urea 1% on 30 and 45 days after sowing is recommended. For rice fallow pulses in cauvery delta area, the present recommendation of foliar spray of 2% DAP may be continued.
- To mitigate stress foliar spraying of 2% KCl + 100 ppm Boron during dry spell as mid season management practice in blackgram during *Rabi* season is recommended to increase the yield over KCl spray alone .
- 50 per cent nitrogen can be substituted through organic source (850 kg of vermicompost per hectare). Lime application is recommended for pulses with soil pH less than 6.0.

In a field experiment conducted at Coimbatore, it was found that fertilization with 25 kg N, 50 kg, P_2O_5, 10 kg Mg per ha, 6.5 ppm of Fe and 0.1 ppm Mo along with rhizobium increased the yield of blackgram and resulted in high N fixation. Among the forms of N applied, the total dry matter production, grain yield and N balance were maximum with NH_4 form and NO_3 form performed better for the number, weight and N content of nodule and acetylene reduction per plant.

The usefulness of enriched biocompost made out of distillary spent wash and solid wastes *viz*., press mud, yeast sludge and bagasse ash, were tested in CO_5 blackgram to investigate the possibility of converting them into biocompost by adding rock phosphate and micronutrients such as $ZnSO_4$, $FeSO_4$ and Borax. The composts matured between 60th and 90th day and had higher nutrient contents. The results showed that application of enriched biocompost along with 100% NPK increased the yield of blackgram significantly and superior to 100% NPK alone or control.

The crop response studies conducted at the AICRP Micronutrient research unit of TNAU, Coimbatore centre on soil application of micronutrients have shown that (i) Fe application has enhanced the yield of pulses by 3.0 to 5.8 q/ ha, (ii) application of 0.50 kg sodium molybdate /ha for green gram and black gram, enhanced the yield by 0.5 q/ha in blackgram and 2.1q/ha in greengram, the yield increase being 6.3 and 27 per cent respectively. (iii) seed coating with Zn, Mo and Co at 4.0, 1.0 and 0.5 g/kg seed increased the grain yield of black gram and green gram by 261 kg (30.9 per cent) and 264 kg /ha (27.7 per cent) respectively and (iv) soil application of 5 kg Zn +1.5 kg B+ 0.5 Mo+ 40 kg S ha^{-1} increased the yield of black gram by 179 kg/ha (23.6 per cent) and green gram by 156 kg/ha (20.4 per cent) over check. Rhizobium application increased nodulation and grain yield of blackgram (Table 4).

Table 4: Influence of organic amendments on nodulation and grain yield of blackgram

Treatments	Nodule No./plant	Grain yield (kg/ha)	% Increase over control
Control	4.0	835	-
Rhizobium	25.0	995	22.6
Biodigested slurry	17.0	970	16.1
BDS + Rhizobium	22.0	1060	26.9
Sheep manure	15.0	965	15.2
SM + Rhizobium	22.0	1015	21.6
FYM	1.0	940	12.5
FYM + Rhizobium	23.0	1005	20.1

(Prabakaran and Ravi, 1998)

The micronutrient fertilizers singly or in combination significantly increased the yield of black gram over control to the tune of 10-17% depending on nutrient combination. Soil application of combinations of 5 kg Zn, 40 kg S, 1.5 kg B, 0.5 kg Mo ha^{-1} recorded the highest grain yield of 735 kg ha^{-1} which was 17.2% higher than NPK control. The availability of these nutrients had nearly doubled as a consequence of addition in comparison to control.

Greengram

1. Seed treatment

- Treat the seeds with 3 packets (600 g/ha) of Rhizobial culture CRU-7 + 3 packets (600 g/ha) of PGPR and 3 packets (600 g/ha) of Phosphobacteria using rice kanji as binder. If the seed treatment is not carried out apply 10 packets of Rhizobium (2000 g/ha) + 10 packets of PGPR (2000 g/ha) and 10 packets (2000 g) of Phosphobacteria with 25 kg of FYM and 25 kg of soil before sowing.
- Seed coating with biofertilizers and micronutrients viz., Zn, Mo & Co @ 4, 1, 0.5 g/kg of seed is recommended.

2. Fertilizer application and foliar spray

- Apply fertilizers 12.5 kg N + 25 kg P_2O_5 + 12.5 kg K_2O +10 kg S*/ha basally before sowing under rainfed condition.
- Soil application of micronutrient mixture @ 5 kg/ha as Enriched FYM.
- Foliar spray of urea 1% on 30 and 45 days after sowing is recommended.
- Under rice fallow condition foliar spray of NAA 40 mg/litre and salicylic acid 100 mg/litre once at pre-flowering and another at 15 days thereafter

- Foliar spray of pulse wonder @ 5 kg/ha once at flowering or DAP 20 g/lt once at flowering and another at 15 days thereafter.

Field experiments conducted by the AICRP-Micronutrients scheme of TNAU in the farmer's holdings have shown that the soil application of 5 kg Zn + 40 kg S + 1.5 kg B + 2.5 kg Cu + 0.5 kg Mo ha^{-1} had registered the highest grain and haulm yields of greengram, the increase being 10-15 per cent over NPK check. The uptake of added micronutrients were also got increased resulting in yield advantage in greengram. Soil application of $ZnSO_4$ @ 25 kg ha^{-1} for pulses (greengram and blackgram) was found to enhance the yield to the tune of 10 to 15%.

A field experiment was conducted in TNAU wetlands to assess direct, residual and cumulative effect of Mussoorie rock phosphate and DAP in rice-greengram rotation involving IR 60 rice in Rabi followed by CO_3 greengram during summer as test crop in Noyyal series soil low in available N and P and high in K. Application of MRP at 75 kg P_2O_5 / ha with GLM at 10 t / ha to first crop of rice benefited the succeeding greengram. Cumulative effect was better than residual effect indicating that the indigenously available MRP in combination with GLM can be effectively and economically utilized as an alternate source of P in rice-pulse cropping system.

Pigeonpea

1. Seed treatment

- Using rice kanji as binder, Rhizobium should be given as seed treatment only. For PSB and PGPR, if the seed treatment is not carried out, apply 10 packets (2 kg) of Phosphobacteria (*Bacillus megaterium*) and 10 packets (2 kg) of PGPR (*Pseudomonas* sp.) with 25 kg of FYM and 25 kg of soil before sowing.

2. Fertilizer application and foliar spray

- Apply fertilizers 12.5 kg N + 25 kg P_2O_5 + 12.5 kg K_2O +10 kg S*/ha basally before sowing under rainfed condition.

- Foliar spray of NAA 40 mg/l once at pre-flowering and another at 15 days thereafter.

- Foliar spray of DAP 20 g/l or urea 20 g/l once at flowering and another at 15 days thereafter.

- Foliar spray of salicylic @100 mg/litre once at pre flowering and another at 15 days thereafter.

Studies on effect of plant growth regulators on redgram and soybean in Agricultural Research Station (ARS) Bhavanisagar revealed that foliar spray of growth regulator NPGR 3 at 500 ml / ha in red gram enhanced the leaf area index, total chlorophyll content, pod numbers, seed yield (34% over control) and the protein content. A considerable increase in yield (400 kg ha^{-1}) in pigeon pea was observed due to foliar spray of 0.5 per cent zinc sulphate at flower initiation stage.

Cowpea

1. Seed treatment

- The improved rhizobial strain COC 10 is more effective in increasing the yield. Treat the seeds with 3 packets (600 g/ha) of Rhizobial culture COC 10 and 3 packets (600 g/ha) of Phosphobacteria using rice gruel as binder. If the seed treatment is not carried out apply 10 packets of Rhizobium (2000 g/ha) and 10 packets (2000 g) of Phosphobacteria with 25 kg of FYM and 25 kg of soil before sowing.

2. Fertilizer application and foliar spray

- Apply fertilizers 12.5 kg N + 25 kg P_2O_5 + 12.5 kg K_2O +10 kg S*/ha basally before sowing under rainfed condition.

Zinc fertilisation had synergistic effect on P availability. Significant residual effect was observed in the treatment on cowpea crop and zinc availability in soil. In AICRP long term fertilizer experiments on Maize-Cowpea cropping sequence in Coimbatore in Vertisols, integrated use of NPK and FYM was found to positively influence the grain yield of cowpea.

Effect of micronutrient on the productivity of cowpea was studied under the All India Co-ordinated Research Project on arid legumes showed that 0.5percent $FeSO_4$ + 0.5 percent $ZnSO_4$ spray both at 25 and 45 DAS proved most effective and increased the seed yield by 27.7 per cent when compared with control ICAR, 2002.

Soybean

1. Seed treatment

- Treat the seeds atleast 24 hours before sowing.
- Treat the seeds with 3 packets (600 g/ha) of Rhizobial culture (COS-1) and 3 packets (600 g/ha) of Phosphobacteria using rice kanji as binder. If the seed treatment is not carried out apply 10 packets of Rhizobium (2000 g/ha) and 10 packets (2000 g) of Phosphobacteria with 25 kg of FYM and 25 kg of soil before sowing.

2. Fertilizer application and foliar spray

- Apply 20 kg N and 40 kg P_2O_5 and 20 kg K_2O and 20 kg of S as gypsum /ha as basal dressing under rainfed.
- Foliar spray of NAA 40 mg/l and salicylic acid 100 mg/l once at pre-flowering and another at 15 days thereafter.
- Foliar spray of DAP 20 g/l or urea 20 g/l once at flowering and another at 15 days thereafter.

Studies on the interaction of P and K in irrigated soybean in 4 major soil groups of Tamil Nadu in CO1 variety was carried out. The results showed that the application of 100 kg P_2O_5 and 50 kg K_2O was optimum for higher yields of soybean. Among the soils, laterite soil recorded higher dry matter yield than red, black and alluvial soils at initial stage of crop but at later stages, the difference between soils was not marked. The black soil recorded highest seed yield and the lowest was in the red soil.

Application of 20 : 80 : 20 kg N, P_2O_5, K_2O ha^{-1} with FYM 12.5 t ha^{-1} and Rhizobium @ 2 kg ha^{-1} in Typic Ustivertepts Periyanaickenpalyam series in CO 1 soybean recorded 43% increase in grain yield over control, with highest value cost ratio of 2.6. Phosphorus incorporated as SSP with organics and biofertilizer improved the oil content and DAP enhanced the grain and protein yield. In the same soil, the crop responded very well to the application of micronutrients and cytozyme. The combined nutrient spray and cytozyme spray super imposed over NPK significantly increased the crop yield.

In black calcareous soil deficient in micronutrients, soil application of $MnSO_4$ was found to be the best in increasing seed yield of soybean followed by foliar spray at one per cent $MnSO_4$ thrice during the crop growth at 20, 30 and 40 DAS. The other treatments such as application of sodium molybdate and $ZnSO_4$ as foliar spray also produced desirable effects such as increased crude protein content of seeds and nodules for the former treatment.

The critical soil S was found to be 9.5 ppm and 11.5 ppm for soybean and blackgram crops respectively with an increased seed yield of 112.6 and 69.0 per cent respectively over control for the combined application of FYM + *Thiobacillus* + 20 kg S ha^1.

The impact of foliar application of KCl 0.5 per cent and NAA 40 ppm was compared and found that crop growth and yield of soybean can be increased by KCl sprays under water stress conditions by maintaining tissue water potential and preventing water loss. It also increased the oil content significantly. Foliage applied macro and micronutrients at critical stages of the crop were effectively absorbed and translocated to the developing pods, producing more number of pods and better filling in soybean.

Garden lablab and field lablab

- Fungicide treated seeds should be again treated with bacterial culture. There should be an interval of atleast 24 hours between fungicidal and bacterial culture treatments. Three packets of bacterial culture are sufficient for treating seeds required for one hectare. The bacterial culture may be prepared with rice kanji. Dry the inoculated seeds in shade for 15 minutes, before sowing.
- Apply fertilizers 12.5 kg N + 25 kg P_2O_5 + 12.5 kg K_2O +10 kg S*/ha basally before sowing under rainfed condition.

Selected References

Chowdhury T, Gupta SB. 2012. Importance of soil health for sustainable pulse production. In: Model Training Course on Technologies of *rabi* pulses. Directorate of Extension Services, Indira Gandhi Krishi Vishwavidyalaya, Raipur 09-16 January, 2012.

Johansen C, Ali M, Gowda A, Ramakrishna S, Nigam N, Chauhan YS. 2000. Regional opportunities for warm season grain legumes in the Indo-Gangetic Plain. In Legumes in rice and wheat cropping systems of the Indo-Gangetic Plain-Constraints and opportunities. Patancheru 502 324, Andhra Pradesh, India. pp. 185-199.

Khan AA, Jilani G, Akhtar MS, Naqvi SMS, Rasheed M. 2009. Phosphorous solubilizing bacteria: Occurance, Mechanisms and their role in crop production. *Journal of Agriculture and Biological Sciences*. 1: 48-58.

Prabakaran J, Ravi KB. 1996. Response of soybean to *Rhizobium* and organic amendments in acid soil. *Madras Agricultural Journal*, 83: 132-133.

Sinclair TR, Vadez V. 2002. Physiological traits for crop yield improvement in low N and P environments. *Plant and Soil*. 245: 1–15.

Wani SP, Sarvesh KV, Krishnappa K, Dharmarajan BK, Deepaja SM. 2012. Bhoochetana: Mission to boost productivity of rainfed agriculture through science-led interventions in Karnataka. International Crops Research Institute for the Semi-Arid Tropics, Patancheru 502 324, Andhra Pradesh, India. ISBN 978-92-9066-548-9. pp.84.

22

Enhancing Nutrient Use Efficiency, pp. 291-341
Editors: K. Ramesh, A.K. Biswas, B.L. Lakaria, S. Srivastava and A.K. Patra

Soybean Based Cropping Systems

S. D. Billore

ICAR- Indian Institute of Soybean Research, Indore – 452 001, India

Introduction

Soybean (*Glycine max* L. Merril) is the world's most important oil yielding legume which contributes 25% of the global edible oil, about two thirds of the world's protein concentrate feeds for livestock, poultry and fish. India ranks fifth in the area and production in the world after USA, Brazil, Argentina and China. The contribution of India in the world soybean area is 10% but to total world soybean production is only 4% indicating the poor levels of productivity of the crop in India (1.1 t/ha) as compared to other courtiers (World average 2.2 t/ha) which is a major cause of concern.

The commercial cultivation of soybean crop in India began in late sixties; thereafter it made phenomenal growth, which is unparallel in the history of the country. Starting from just 30, 000 ha in 1970, area under soybean has increased to 10.69 million ha in 2012. The production and productivity levels of 14000 tones and 0.43 t/ha in 1970 have increased to 12.67 million tones and 1.19 t/ha in 2012, respectively. Soybean is predominantly grown as rainfed crop in Vertisols and associated soils with an average crop season rainfall of 900 mm which varies greatly across locations and years. Introduction of soybean in these areas has led to a shift in cropping system from rainy season fallow followed by post rainy season wheat or chickpea system fallow-wheat/chickpea) system to soybean followed by wheat or chickpea (soybean-wheat/chickpea) system. This has resulted in an enhancement in the cropping intensity and resultant increase in the profitability per unit land area. In fact, soybean is one of the most resilient crops for the rainfed *kharif* season as despite aberrant weather conditions in recent past, the crop has maintained its performance.

The area under soybean is spread mainly in the states of Madhya Pradesh, Maharashtra, Rajasthan, Chhattisgarh, Andhra Pradesh and Karnataka. These states together contribute to about 98% of the total soybean production in the country.

Soybean and soybean based cropping systems

Soybean offers good potentials to get involved in the cropping sequences or intercropping systems. It is a short duration (85 to 130 days depending on the latitude) leguminous energy rich crop. It is relatively tolerant to drought and excessive moisture and extends benefits of 45 to 60 kg residual nitrogen per hectare to the succeeding crop in addition to improvements in physico-chemical environment in the soil for crop growth. It is also instrumental in sustaining soil organic matter status through substantial recycling of foliage/ rhizosphere root mass. Experimental evidences have established that soybean could fit aptly in any of the traditional cropping systems in all the five agro climatic zones of India specified for soybean (Table 1). Hence, the inclusion of soybean in any of the existing cropping systems generates much more dividends than any other cropping sequence. The productive and remunerative soybean based cropping systems has been presented in Table 2 (Billore and Joshi, 2004) and well reviewed by Vyas *et al.* (2008)

Table 1: Soybean agro-ecological zones of India

Zone	Area Covered
Northern Hill	H.P. and Northern Hills of Uttar Pradesh.
Northern Plain	Punjab, Haryana, Delhi and North-eastern Plains of Uttar Pradesh and Western Bihar
Central	M.P. Bundelkhand Region of Uttar Pradesh, Rajasthan, Gujarat, North-Western parts of Maharashtra and Orissa
Southern	Karnataka, Tamil Nadu, Andhra Pradesh., Kerala and Southern parts of Maharashtra
North-Eastern	Assam, West Bengal, Bihar and Meghalya

Table 2: Soybean based cropping sequences and intercropping

Sl. No.	Zone	Cropping sequence	Inter-/mixed-/companion cropping
	Central	***Major*** :Soybean-chickpea (rainfed) Soybean-wheat (irrigated) ***Others*** : Soybean-potato (I) Early soybean-garlic (I) Soybean-pigeon pea (R) Early soybean-safflower (R) Early soybean-rapeseed/mustard	***Major***: Soybean + pigeon pea Soybean+corn ***Others***: Soybean + sorghum Soybean + cotton Soybean + groundnut Soybean + pearl millet Soybean in fruit (mango/guava) orchards
	Southern	***Major***: Soybean-wheat-groundnut (I) Soybean-finger millet-beans (I) Wheat-soybean-finger millet-peas (I) ***Others***: Oat-cow pea-barley-soybean (I)	***Major***: Soybean-pigeon pea Soybean-finger millet Soybean-sugarcane Soybean-sorghum ***Others***: Soybean+wheat(Rabi) Soybean in fruit (mango/guava) orchards
	Northern Plain	***Major***: Soybean-wheat (I) Soybean-potato (I) Soybean-chickpea (R)	***Major***: Soybean + pigeon pea Soybean-corn Soybean-sorghum ***Others***: Soybean in fruit (mango/guava) orchards
	Northern Hill	***Major***: Soybean-wheat (I) Soybean-peas (I) ***Others***: Soybean-lentil (R) Soybean-toria (R)	***Major***: Soybean+corn Soybean+pigeon pea

I = Irrigated; R= Rainfed

Nutrient use efficiency

Efficiencies are generally calculated as ratios of outputs to inputs in a system. Agricultural cropping systems contain complex combinations of components, including soils, soil microbes, roots, plants, and crop rotations. Improvements in the efficiency of one component may or may not be effective in improving the efficiency of the cropping system. Efficiency gains in the short term may sometimes be at the expense of those in the long-term. Short-term reductions in application rates increase nutrient use efficiencies, even when yields decline. However, in the long-term, lower yields reduce production of crop residues, leading to increased erosion risks, decreased soil organic matter, and diminished soil productivity. Sustainable system efficiency demands attention to the long-term impacts.

Efficiency is defined as the amount of product produced per unit of resource used. This means nutritional efficiency is the amount of dry matter produced per unit of nutrient applied or absorbed. However, the nutrient efficiency has been defined in so many ways in literature available. Cooke (1987) defined NUE as the increase in yield of the harvested fraction of the crop per unit of nutrient supplied by fertilizer. Israel and Rufty (1988) reported that, traditionally efficiency of nutrient utilization has been defined as the ratio of biomass of the total amount of nutrient in the biomass. NUE is obtained by dividing the whole plant biomass by the nutrient concentration in the plant. The nutrient utilization efficiency can be separated into two components- the utilization quotient and biomass production (Israel and Rufty, 1988).

Estimation of NUE in Plants

The evaluation of NUE is useful to differentiate plant species or genotypes for their ability to absorb and utilize nutrients for maximum yields. The NUE depends on (a) uptake efficiency (b) incorporation efficiency (transports to shoot and leaves are based on shoot parameters) and (c) utilization efficiency (remobilization, whole plant i.e. root and shoot parameters). Craswell and Godwin (1984) classified nutrient efficiency in to *agronomic efficiency*, *physiological efficiency* and *apparent recovery efficiency* (Table 3). The AE and RE terms are most informative about single sources. The PFP and PNB terms can over estimate efficiency when applied only to a single source, if other sources are also significant. In the short term, all four component efficiencies increase as rates of fertilizer application are decreased below economic optimum. This might cause one to falsely conclude that the lowest fertilizer rate would result in the most efficient cropping system. This is untrue, as production depends on many non-nutrient inputs, including seed, fuel, crop protection products, machinery, labor, land, and capital. Also, in the long term, limiting nutrient rates in a deficient situation reduces production of crop residues, leading to increased erosion risks, decreased soil organic matter, and diminished soil productivity.

Table 3: Definitions of the different types of nutrient efficiencies

NUE terms	Unit	Formula
Partial factor productivity	PFP (kg/kg)	= Yield/ amount of nutrient applied (single nutrient or total nutrients)
Agronomic efficiency / economic efficiency	AE (kg/kg)	= yield in fertilized crop – yield in unfertilized / Amount of nutrient applied
Agro-physiological efficiency/ Physiological efficiency	APE/ PE(kg/kg)	= (Yield in fertilized crop – Yield in unfertilized crop) / (Nutrient uptake by fertilized crop – uptake in unfertilized crop)
Apparent recovery efficiency	ARE (%)	= (Nutrient uptake in fertilized crop – nutrient uptake in unfertilized crop) / Amount of nutrient applied x 100

(Contd.)

Nutrient efficiency ratio	NER (kg/kg)	= (Units of Yields, kg) / (Unit of elements in tissue, kg)
Partial nutrientbalance (removalto use ratio)	PNB (kg/kg)	= nutrient content of harvested portion of crop/ amount of nutrient applied
Nutrient use efficiency		=Physiological efficiency x Recovery efficiency
Nutrient(s) use efficiency		= total DM / total amount of nutrient(s) in plants
Nutrient(s) uptake efficiency	NUE (kg/kg)	= total amount of nutrient(s) in plants / root DM
Nutrient(s) translocation efficiency	NTE (%)	= (amount of nutrient(s) in shoot / total amount of nutrient(s) in plants) x 100
Nutrient use efficiency		Uptake efficiency x Utilization efficiency
Utilization efficiency		= Harvest index x Nutrient biomass production efficiency or Harvest index x Inverse of total nutrient
Grain yield response index	GI (kg/kg)	=(Yield in non- stress nutrient soil) -(Yield in nutrient stress soil)/ (Differences in applied nutrient levels between non-stress and stress)

Nutrient uptake = concentration x dry matter / yield.

Efficient does not necessarily mean effective

Improving NUE is an appropriate goal for all agricultural producers. However, effectiveness cannot be sacrificed for the sake of efficiency. Much higher nutrient efficiencies could be achieved simply by sacrificing yield, but that would not be economically effective or viable for the farmer, or the environment.

Relationship between efficiency and effectiveness was further explained when Fixen (2006) suggested that the value of improving NUE is dependent on the effectiveness in meeting the objectives of nutrient use, objectives such as providing economical optimum nourishment to the crop, minimizing nutrient losses from the field, and contributions to system sustainability through soil fertility or other soil quality components. He concluded that the highest recovery efficiency of P was associated with lowest P rate and at the lowest soil test value however, this practice would be ineffective because yield was sacrificed.

Agronomic measures for enhancing nutrient use efficiency

Effectiveness is maximized when the most appropriate nutrient sources are applied at the right rate, time, and place in combination with conservation practices within intensively managed cropping systems that achieve both increasing yields and diminishing nutrient losses. This approach ensures that improvements to the NUE of the components contribute toward improving the efficiency of the entire system. Because a cropping system includes multiple inputs and outputs, its overall efficiency depends on the science of economics.

To maximize profit is to obtain the maximum value of outputs per unit value of all inputs. At the rate where the net return to the use of one input peaks, the input is making its maximum contribution to increasing the efficiency of all other inputs involved. Rates of nutrient application optimal for economic yields often minimize nutrient losses.

There exists enough evidence that fertilizers are not as effective as they should have been for many reasons, such as the quality of fertilizers, use of wrong fertilizers, sub-optimal use levels, absence of complementary inputs such as improved seed varieties and adequate water, bad cultural practices, and similar factors. The losses of nutrients caused not only an environmental hazard but also a substantial economic loss. Reducing nutrient losses is a critical step toward improving soil fertility and agricultural productivity for the poor farmers. It makes sense from every perspective agronomic, economic and environmental. Thus improving the FUE is of paramount importance and needs particular attention.

Improved management practices, product and crop attribute all lead to increased NUE. The integrated use of mineral fertilizers and recycled waste products not only reduces the amount of needed fertilizer but also improves nutrient and water use efficiency. The key is to synchronize the nutrient delivery from soils, biological nitrogen fixation, organic materials and mineral fertilizers with crop requirements (Singh, 2002). In order to address the key concerns such as food security, profitability in agriculture, and environmental quality, improving the efficiency, effectiveness and sustainability of fertilizer use is the fundamental challenge for the region. In the region, the low nutrient absorption rate of 30 - 40% by crops for N, P and K fertilizers are inefficient and must be improved.

Crop yield directly or indirectly is the numerator in all the terms of *FUE/NUE* and the crop, soil and agronomic factors that increase crop yield may therefore increase *FUE/NUE*.

External factors such as soil management, climatic factors, allelopathy, diseases, and weeds profoundly affect the plants ability to absorb and utilize nutrients more effectively (Baligar and Bennett, 1986a and b, Fageria *et al.*, 1990 and 1997a, Fageria, 1992, Baligar and Fageria, 1997). Soil temperature and moisture greatly influence nutrient transformation (release) from organic forms, their uptake by roots and their subsequent translocation and utilization by plants. Plant health is influenced by diseases, insects and weeds that compete for nutrients and water resources and lower NUE (Barber, 1995 and Fageria *et al.*, 1997a).

Soil management

About half of the world's soils are deficient in macro- and micro-nutrients. If new cultivars that have higher yields are developed, the dynamics of nutrients could change due to larger removal of these elements from the cropping systems in the harvested portions of the crops. In such a case, nutrients will have to also be monitored for these soils to ensure that higher yields and NUE are maintained.

Adverse soil physical properties affect the longitudinal and radial root growth, root distribution, morphology by stunting, thickening, reduction of second and third order lateral roots and root anatomical changes (Bennie, 1996). High mechanical impedance leads to loss of root caps and reduction in radial thickening primarily due to shorter and wider cells with the same volume in the cortex (Camp and Lund, 1964) and a thicker cortex. This may also cause changes in cell structure of the endodermis and pericycle (Bennie, 1996). Such changes in the size and internal and external morphology of roots due to the adverse soil physical conditions will influence the root's ability to explore larger soil volume and reduce nutrient and water availability and uptake, leading to low NUE and lower yields.

Crop management

Soybean yield potential has been defined as the maximum yield of a crop cultivar grown in an environment to which it is adapted, with nutrients and water non-limiting, and pests and diseases effectively controlled. Soybean yield have steadily increased in the past 30 years due to a combination of genetic and management improvement. The annual rate of yield increase was on an averages 31 kg/ha in the US (Specht *et al.*, 1999) and 28 kg/ha globally (Wilcox, 2004). Hence, yield increases will become the major source for sustaining further increases in soybean production. Across a 60 year period, cultivar development efforts by soybean breeders have resulted in a 21-31 kg/ha/yr increase in soybean yield (Wilcox, 2001). It can be approximated that recent yield gains are about 50% due to cultivar genetic improvement and 50% to improved cultural practices. Potential gains from improved cultural practices for any given locale are usually determined by comparing farmer yields with those done using recommended practices (Foulkes *et al.*, 2009). Yield potential studies of the world show yield ranging from 60 to 80% of the optimal level (Foulkes *et al.*, 2009). This yield gap is attributed to a suboptimal physical environment coupled with inadequate application of fertilizer and pest control. Thus, improvement of cultural practices can be expected to increase yield anywhere from 25 to 66%.

Soybean varieties

Plant genetic variability can be defined as the heritable characters of a particular crop species or cultivars that show differences in growth or production in comparison with other species or cultivars of the same species under favorable or unfavorable growth conditions. It has been shown that large variations exist among species or cultivars of the same species, in absorption and utilization of mineral nutrients. The differential requirement of nutrients by different crop species or cultivars of the same species may be related to nutrient absorption, translocation and metabolism. According to Graham (1984) better nutrient use efficiency in crop plants may be related to following mechanisms: (i) Better root geometry (ii) Faster rate of absorption at low concentration of nutrients (iii) Chemical modification of the root- soil interface to solubilize more of the limiting nutrients (iv) Improved internal redistribution (v) Superior utilization or lower functional requirement for the nutrient.

Genetic variability has been reported to explain the differences in NUE and the parameters of nutrient uptake. Such differences in growth and NUE in plants have been related to differences in absorption, translocation, shoot demand, dry matter production per unit of nutrient absorbed, and environmental interactions (Baligar and Duncan, 1990 and Clark and Duncan, 1991). Overall NUE in plants is governed by the flux of ions from the soil to the root surface and by the influx of ions into roots followed by their transport to the shoots and remobilization to plant organs. Various soil and plant mechanisms and processes that contribute to such differences were reviewed by Epstein (1972), Barber (1995) and Baligar and Fageria (1997).

Those cultivars that have the ability to absorb large amounts of nutrients and convert them in to useful dry matter of highly enriched soils, in which a less efficient cultivar reaches a yield plateau have been described as the nutrient efficient species or cultivar (Arnon, 1974). More emphasis is now being given to that plant cultivar that should produce more on soils having a low fertility (Clarkson and Hanson, 1980). Variety that produced higher yields under low nutrient supply has evolved one or more of the following characteristics: an efficient internal economy, which may result from efficient redistribution within the plant, or lower requirements at functional sites (Clarkson and Hanson, 1980). It is the growing need today that nutrient uptake and utilization of crop plants should be as efficient as possible, to reduce the cost of production and achieve a higher profit for the farmers.

A total of more than 105 soybean varieties have been developed and recommended to the farmers as per their suitability to different agroclimatic zones of India. Yielding ability of soybean genotypes differs significantly within zone and across the zones. A significant variation has been observed in fertilizer PFPf (6.02 to 17.28kg/kg) in soybean genotypes (Table 4). The highest PFPf was associated with SL 444 in north plain, JS 95 60 in central, RAUS 5 in north eastern and MACS 450 in southern zones and the corresponding lowest values was recorded with genotype DS 98 14, MAUS 71, RKS 18 and Lsb1, respectively. It is also noted that the genotypes behaved differently in different zones. Genotype RKS 18 and JS 93 05 showed higher PFPf in southern and central zones as compared to north eastern zone. These variations may be due to the variations in climatic and soil factors existed in the different zones. The differential response among the various genotypes is might be due to variations in their genetic makeup (Epstein, 1972; Duncan and Carrow, 1999).

Table 4: Genotypical variation in fertilizer partial factor productivity (2009-2011)

Zone and Variety	Yield (kg/ha)	PFPf (kg/kg)	Zone and Variety	Yield (kg/ha)	PFPf (kg/kg)
North plain (NPZ)	1566	12.04	North eastern (NEZ)	1835	10.19
PS 1347	1375	10.58	RKS18	1401	7.78
SL 525	1561	12.00	JS 97 52	1737	9.65
DS 9814	1303	10.02	JS 335	1341	7.45
SL 744	2050	15.76	RAUS 5	2365	13.14
Bragg	1542	11.86	BSS 2	2150	11.94
Central (CZ)	1799	12.85	JS 93 05	2014	11.19
JS 95 60	1935	13.82	Southern (SZ)	1748	11.65
JS 97 52	1887	13.48	RKS18	2447	16.31
MAUS 71	1604	11.45	MAUS 61	1612	10.74
JS 93 05	1769	12.63	MACS 450	2592	17.28
			Lsb 1	903	6.02
			JS 335	1908	12.72
			Co 3	1024	6.82

Source: Modified from Billore and Srivastava (2013)

Soybean genotype PS 1347 in north plan, JS 9752 in north eastern and RKS 18 in southern zone were found to be more nutrient use efficient (PFPf and AE) than other varieties. Genotype JS 95 60 showed higher PFP and lower AE than JS 97 52 in central zone. Overall, north plain zone genotypes were posses highest PFP and lowest AE, while the central zone genotypes were more efficient with regard to AE (Table 5 and 6). Billore *et al.* (2009) concluded that genotype JS 335 showed higher K use efficiencies except AE and internal K use efficiency which were higher in JS 93 05 (Table 7). Two years results revealed that the long duration genotype NRC 37 was more efficient with regards to sulphur use efficiencies as compared to variety JS 95 60 (Table 8).

Table 5: Impact of integrated nutrient management schedule on yield and nutrient use efficiencies of soybean varieties in north plain and north eastern zones (2009-11)

Treatment	North plain zone						North eastern zone								
	PS 1347			SL 525			JS 97 52			RKS 18			JS 93 05		
	Yield	PFPf	AE	Yield	PFPf	AE	Yield	PFPf	AE	Yield	PFPf	AE	Yield	PFPf	AE
75 % RDF+ FYM @ 5 t/ha	2135	21.89	2.49	2130	21.85	1.63	1567	11.60	3.60	1222	9.04	2.30	1440	10.67	1.90
75 % RDF	2127	21.82	2.41	2072	21.25	1.04	1707	12.64	4.64	1387	10.27	3.25	1620	12.00	3.24
100 % RDF	2124	16.34	1.78	2196	16.89	1.73	1849	10.27	4.27	1509	8.38	3.12	1725	9.58	3.01
100 % RDF + FYM @ 5 t/ha	2148	16.52	1.97	2169	16.68	1.52	1961	10.89	4.89	1641	9.12	3.85	1773	9.85	3.28
125 % RDF	2266	13.94	2.30	2192	13.48	1.36	1965	8.73	3.92	1514	6.73	2.52	1938	8.61	3.36
125 % RDF + FYM @ 5 t/ha	2306	14.19	2.55	2158	13.28	1.15	2052	9.12	4.32	1617	7.19	2.97	2077	9.23	3.97
Control	1892			1971			1081			948			1183		
Mean		17.45	2.25		17.24	1.41		10.54	4.27		8.46	3.00		9.99	3.13

Table 6: Impact of integrated nutrient management schedule on yield and nutrient use efficiencies of soybean varieties in central and southern zone (2009-11)

Treatment	Central zone						Southern zone					
	JS 95 60			JS 97 52			RKS 18			MAUS 61		
	Yield	PFPf	AE	Yield	PFPf	AE	Yield	PFPf	AE	Yield	PFPf	AE
75% RDF 105	1603	15.27	3.53	1457	13.88	3.38	1857	16.50	4.30	1457	12.95	2.94
75% RDF+FYM @ 5 t/ha	1659	15.80	4.07	1660	15.80	5.31	2000	17.78	5.57	1558	13.85	3.84
100% RDF 140	1811	12.93	4.13	1784	12.74	4.87	1923	12.82	3.67	1489	9.93	2.42
100% RDF+ FYM @ 5 t/ha	1922	13.72	4.93	1818	12.98	5.11	2035	13.57	4.41	1594	10.63	3.12
125% RDF 175	1852	10.58	3.54	1807	10.32	4.03	1909	10.18	2.86	1508	8.04	2.03
125% RDF + FYM @ 5t/ha	1927	11.01	3.97	1894	10.82	4.53	2114	11.27	3.95	1637	8.73	2.73
Control	1232			1102			1373			1126		
Mean		13.22	4.03		12.76	4.54		13.69	4.13		10.69	2.85

Table 7: Genotypic variations for potassium use efficiency

Genotype	Soybean yield (kg/ha)	Wheat yield (kg/ha)	SEY (kg/ha)	System total K uptake (kg/ha)	K-HI (%)	PFP (kg/kg)	AE (kg/kg)	RE (%)	PE (kg seed/ kg K uptake)	Internal K use efficiency (kg seed/kg K)
JS 335	2349	5034	7383	87.91	32.35	98.92	14.57	22.78	89.91	83.98
JS 93 05	2350	4860	7210	82.37	36.03	96.96	14.76	20.51	75.94	87.53
NRC 7	2305	4913	7218	87.13	35.71	96.76	13.76	20.25	69.56	82.84
CD	NS	114.50	NS	2.29	0.98	-	-	-	-	1.28

The existence of considerable genotypic variations, techniques and selection criterion could enhance the feasibility of breeding crop cultivars for improved mineral nutrient use efficiency (Fageria and Baligar, 1994 and Graham, 1984). Identification of cultivars with greater tolerance to sub-optimal soil nutrient levels offer considerable promise for increasing the crop production potential of marginal low fertility lands throughout the world (Clark and Duncan, 1991, Baligar and Fageria, 1997 and Duncan and Carrow, 1999). When nutrient supply from soil is sub-optimal, the efficiency with which mineral nutrients are used by plants is important in overall nutrient efficiency. The nutrient efficient varieties must be possesses plant characteristics such as the ability to produce near maximum yields at low nutrient levels, and extensive root systems efficient in exploring large soil volumes to produce cultivars with high NUE that can contribute to sustainability and environmental protection (Sattelmacher *et al.*, 1994). Most efficient and most inefficient nutrient efficiency ratios (NER) in different genotypes within species have been reported (Fageria *et al.*, 1988a and b).

Table 8: Genotypic variation for sulphur use efficiency

Treatment	Seed yield (kg/ha)	Total S uptake (kg/ha)	PFP (kg/kg)	AE (kg/kg)	PE (kg/kg)	RE (%)
			Variety			
JS 95 60	2041	11.62	63.73	12.03	108.58	0.11
NRC 37	2130	12.88	66.52	13.53	130.48	0.10
SEm	17.36	0.72	7.86	0.61	7.43	0.004
CD (P=0.05)	50.33	2.08	22.72	1.77	21.51	0.010

In plant uptake and utilization, efficiency of nutrients is governed by different physiological mechanisms and their response to deficiency, tolerance and toxicity of element(s) and climatic variables (Baligar and Duncan, 1990, Baligar and Fageria, 1997 and Duncan and Carrow, 1999). Genetic improvement in tolerance to toxicities of Al, Mn, H, Na, trace elements, salts and to nutrients deficiencies, drought, temperature extremes, aeration and high soil bulk density, will enhance the plants' ability to absorb and utilize nutrients more effectively. The numerous nutritional differences among cultivars and strains of plants indicate genetic control of inorganic plant nutrition. Genetic variation for NUE has been widely reported within and among crop species. Gene factors and inheritance of traits related to NUE have been well documented.

Root density (length, mass) or the amount of roots per unit volume of soil, plays a central role in any consideration of the absorption of water and nutrients from soil. Rate and pattern of root growth vary with plant genetic potential. A more extensive root system potentially has access to larger reserves of soil water and nutrients. Plant species adapted to low P soils usually have larger, more fibrous root systems, which allow more efficient mining of the soil volume. Root parameters such as length, thickness, surface areas, density, root hairs and root growth rate expressed as dry mass and/or root: shoot ratios are affected by deficiencies of essential minerals and/or excess of minerals (Baligar and Duncan, 1990, Kafkafi and Bernstein,1996, Fageria *et al.*, 1997a, and b and Baligar *et al.*, 1998). Clark (1970) reported that in solution culture studies, reducing the supply of essential nutrients from full strength to none increased root: shoot ratio in P, Ca, S and Zn treatments; however, root: shoot ratios decreased in NO_3^-N, Mg, Mn, and Cu treatments.

Table 9: Influence of tillage systems on yield and partial factor productivity of soybean

Tillage	Soybean yield (kg/ha)	PFPf (kg/kg)
No till	1594	13.28
Minimum till	1552	12.93
Conventional till	1550	12.92

Source: Modified from Billore *et al.* (2013)

Results from a long term experiment (1995-2001) showed that the soybean productivity did not influence by the different tillage systems under Vertisols of Madhya Pradesh (Table 9). The no till system slightly improved the PFPf as compared to minimum and conventional tillage systems. Minimum tillage improved nutrient PFPf in soybean–wheat and soybean chickpea cropping systems than conventional and no till systems (Table 10).

Minimum tillage, no tillage, conservation tillage and traditional tillage can bring profound changes in soil quality, soil organic matter and nutrients throughout different soil horizons (Mahboubi *et al.*,1993). Rooting pattern, water holding capacity, water penetration, aeration, soil compaction, and soil temperature are also influenced by type of tillage practices. Changes in the soil nutrient reserve and alteration in root systems under different tillage systems might have direct bearing on the nutrient availability and uptake by crops. Tillage practices are known to bring changes in soil organic matter, nutrient concentrations, bulk density, water holding capacity and soil temperature among others. Higher contents of available P, Ca, K and organic C and N have been reported for no tillage than for conventional tillage (Mahboubi *et al.*, 1993 and Ismail *et al.*, 1994). Minimum tillage increases root growth in the top 12 cm of soil and increase root weight, length, and density, increasing the nutrient and water use efficiencies. Baligar *et al.* (1998) reported that shoot dry matter yields and root length in no-till were significantly higher than in conventional tillage. Such improved root parameters contributed to higher yields and uptake efficiencies of N, P, Ca, S, Cu, Fe and Zn. Improved tillage equipment and practices need to continue being developed to increase NUE across different agro ecosystems.

Table 10: Influence of tillage systems on yield of soybean-wheat and soybean –chickpea cropping system (pooled data 1995-2001)

Tillage	Yield (kg/ha)		PFPf (kg/kg)	Yield (kg/ha)		PFPf (kg/kg)
	Soybean	Wheat		Soybean	Chickpea	
No till	1584	2425	12.52	1562	1018	10.75
Minimum till	1594	2739	13.54	1508	1228	11.40
Conventional till	1550	2709	13.30	1550	1165	11.31

Time of sowing

Sowing date affects significantly on different growth stages as one of the important its cultivation issues and it is one of the important factors in determining harvest of maximum cultivar yield in one or another regions. Appropriate planting date causes optimal utilization of the climatic resources such as temperature, humidity, day length and also anthesis time adaptation with proper temperature (Hashemi, 2001). Early planting of soybean was found to be more productive than later ones (Shafigh *et al.*, 2006). With delay in planting due to high sensitivity of soybean to light period duration and temperature affects yield negatively by reducing the duration of vegetative and reproductive growth and yield components drop (Kazemi *et al.*, 2005).

The regression analysis showed that the soybean yields declined by 7 kg/ha/day in North Plain, 9 kg/ha/day in Central, 20.5 kg/ha/day in North East, 25.47 kg/ha/day in South zones (Billore and Srivastava, 2013). The results of multi location (8) and year (3) experiments revealed that the planting of soybean in between the 20th June to 5th July in north plain and central zone and 15th June to 30th June in north eastern and southern zones exhibited maximum fertilizer PFP as compared to either early or late planting of soybean which drastically reduced the fertilizer PFPf (Table 11). Late planting of soybean showed lower potassium use efficiencies (K-HI, PFP, AE, RE, PE and Internal K use efficiency) in Vertisols (Table 12). The increasing levels of K showed declining trend in K use efficiencies in soybean –wheat cropping system (Billore *et al.*, 2009).

Table 11: Soybean yield and fertilizer partial factor productivity as influenced by sowing date (2009-2011)

Sowing date	North Plain zone		Central zone		Sowing date	North eastern zone		Southern zone	
	Yield (kg/ha)	PFP_f (kg/kg)	Yield (kg/ha)	PFP_f (kg/kg)		Yield (kg/ha)	PFP_f (kg/kg)	Yield (kg/ha)	PFP_f (kg/kg)
05th Jun	1646	12.66	1755	12.53	15th June	2166	12.03	2214	14.76
20th Jun	1616	12.43	2118	15.13	30th June	2206	12.25	2086	13.90
05th Jul	1630	12.54	2046	14.61	15th July	1761	9.78	1584	10.56
20th Jul	1305	10.04	1347	9.62	30th July	1289	7.16	1107	7.38

Source: Modified from Billore and Srivastava (2013)

Table 12: Effect of planting time on productivity and potassium use efficiency in soybean

Planting time	Yield (kg/ha)			Total K uptake (kg/ha)	K-HI (%)	PFP (kg/kg)	AE (kg/kg)	RE (%)	PE (kg/kg K uptake)	IUE (kg/ kg K)
	Soybean	Wheat	SEY							
Normal	2544	4908	7452	91.70	35.25	100.05	14.03	22.28	60.11	81.26
Late	2115	4938	7030	80.38	33.95	94.76	15.52	20.28	76.15	87.45
CD	152.57	NS	241.60	3.96	1.09	-	-	-	-	2.20

Source: Billore *et al.* (2009)

Crop geometry and plant population

It is very much clear that the plant population which was determined by the seed rate and row spacing are the most important factor among the cultural practices of soybean and both the factors are responsible for the light interception and micro climate alteration which ultimately affect the yield of soybean. Data accrued from multi location (10) and year (3) presented in Table 13 revealed that the a seed rate of 65 to 75 kg/ha planted at 45 and 30 cm row spacing resulted in maximum fertilizer PFPf in north plain, north eastern, central and southern zones, respectively. Plant population studies conducted under short-season conditions also have outlined a yield-control mechanism very similar to those described for narrow vs. wide row spacing and shade (Purcell, 2002). Yield losses started occurring when average light interception across this period falls 14% below that for full coverage canopies. This yield response to reduced light interception corresponds very closely to that shown by Board *et al.* (1996) for wide vs. narrow row spacing. Row spacing and plant population also affects soybean yield.

Table 13: Effect of seed rate and row spacing on productivity of soybean in different agro-climatic regions of India (2009-11)

Treatment	North plain zone		North eastern zone		Central zone		Southern zone	
	Yield (kg/ha)	PFPf (kg/kg)	Yield (kg/ha)	PFPf (kg/kg)	Yield (kg/ha)	PFPf (kg/kg)	Yield (kg/ha)	PFPf (kg/kg)
Seed rate (kg/ha)								
55	2161	16.62	1716	9.53	1652	11.80	2385	15.90
65	2249	17.30	1819	10.10	1683	12.02	2576	17.17
75	2219	17.06	1861	10.33	1684	12.03	2605	17.37
CD	461.38		537.07		144.34		126.63	
Row spacing (cm)								
30	2178	16.75	1815	10.08	1691	12.08	2578	17.19
45	2281	17.54	1944	10.80	1741	12.44	2469	16.46
60	2169	16.68	1637	9.09	1619	11.56	-	
CD	93.69		83.87		51.80		87.93	

Source: Modified from Billore and Srivastava (2014).

Fertilizer management: There is generally an over application of N compared with P and K that contradicts with the tenets of balanced fertilization and thereby, influences soil fertility, crop yields and quality and the sustainability of agriculture in the long run. In India, the main consumer of all three macronutrients, the share of application of N fertilizers in total is around 65% where as the share of P is around 25% and that of K is 10% as against the recommended ratio of 4:2:1 for N, P and K, respectively (58% of N, 28% of P and 14% of K). These imbalances in fertilizer use are attributed to several factors including high prices of P and K compared with N, low in-country production base of P and K fertilizers, excessive promotion of urea by local producers, and policies such as subsidy, decontrol of fertilizers favoring increased use of fertilizers particularly N.

Many soils are severely depleted of some nutrients and organic matters (OM) because of intensive farming and the non-return of organic residues; crop residues are widely used as fuel and fodder and usually not returned to the soil. The widespread deficiency of P, S and OM severely affects crop production. Such deficiency can lead to low crop quality and imperfect plant morphological structure along with lower efficiency of major macronutrients, such as N, P and K and thereby reduced crop yield. Therefore, in order to gain optimal productivity it is important to know both soil type and plant requirement and then to apply fertilizers. At present soil-testing facilities are limited and farmers are still not aware of the plant requirement and are facing difficulty in getting the appropriate nutrients in the right amount and time. Often the delay in fertilizer supply leads to untimely application of fertilizer. Inadequate warehousing capacity, delay in subsidy payment to fertilizer, poor infrastructure are some of the major bottlenecks in the timely availability of fertilizers during the peak seasons.

Rate, source, time and application methods of fertilizer

In case of mobile nutrients, such as N, application of higher doses at one time may result in nutrient loss and N efficiency is greatly reduced. In case of low P conditions, banded P is usually more effective than broadcast application. Seed or row placement of P has been the most effective on low P soils. The NUE *viz.*, AE, ARE and PFP are more significantly enhanced by the application of tablet forms of NPK sources than other slow release forms as well as standard fertilizer materials (Jagadeeswaran *et al.*, 2005). Slow release N fertilizers are now becoming popular among the farmers (Prasad and Power, 1995).There is the need for additional research with these slow and controlled release fertilizers and their interaction with different management situations, soil types and cropping systems.

Nitrogen use efficiency

In order to achieve high yield potential, soybean must sustain high photosynthesis rates and accumulate large amounts of N in seeds. Their nitrogen demand by Biological N_2 fixation (BNF) and mineral soil or fertilizer N is the main sources of meeting the N requirement of high yielding soybean. However, antagonism between nitrate concentration in the soil solution and the BNF process in the nodules is the main constraint the crop faces in terms of increasing N uptake when no other abiotic stress that reduce BNF activity occurs, e.g. soil moisture (Purcell *et al.*, 2004), soil pH or soil temperature. Maximum BNF occurs between the R3 and R5 stages of soybean and any gaps between crop N demand and N supply by BNF must be met by N uptake from other sources. If the overall N supply does not meet soybean requirements, the crop will remobilize N accumulated in leaves to the seed, which diminishes the photosynthetic capacity of the canopy and thus limits yield potential. Kessel and Hartley (2000) suggested that BNF will increase in high yielding environments since the nitrogenase, located in the nodules, will adjust its activity to the demand of the legume. However, the generally observed reduction in BNF activity between the R5 and R7stages could lead to as shortage of N during seed-filling in high yielding environments.

Applying fertilizer–N has been proposed as an aid for increasing available N in the soil. Studies of nodulated soybean showed significant yield response to frequent N additions when the BNF apparatus could not meet N demand. However, soybean yield response to fertilizer N has been inconsistent at economically acceptable levels. Those studies reporting no increase in grain yield assumed that the crop simply substitutes the N it ordinarily would have derived from BNF with N from fertilizer, or that more N translocation from vegetative reserves occurs when applied N lowers the rate of BNF (Herridge *et al.*, 1984). Although it is generally believed that late N applications at reproductive stages (*i.e.*, R3 to R5) should theoretically increase yields in high yielding environments, the empirically measured responses in seed yield to fertilizer-N applied at late R-stage is not universal. Whereas early application of even small amounts of N often results in temporary suppression of nodule establishment and subsequent activity (Hungaria *et al.*, 2005), an early season N deficiency may delay early crop growth and thus the development to an efficient nodulation system. Overall, the contradictory results obtained in N fertilization studies do not provide clear evidence as to whether N fertilization is required to complement the N supply from BNF to achieve soybean yield that approach yield potential levels. A number of reviews have been published on BNF in soybean in particular (Hungaria *et al.*, 2005 and 2006,).

Approximately 55 to 60% of the total N demand is fulfilled by nitrogen fixation in soybean and remaining through soil N and additional applied N. A small starter dose is recommended for fulfilling the initial nitrogen demand up to nodule formation. A four years trial was conducted at Directorate of soybean Research, Indore to study the effect of various levels of nitrogen and their application time on productivity and nitrogen efficiency in soybean. A mean linear increase in soybean seed yield was of 0.013 Mg/kg increase in N accumulation in above-ground biomass. The gap between crop N uptake and N supplied by BNF tended to increase at higher seed yield for which the associated crop N demand is higher. A negative exponential relationship was observed between N fertilizer rate and N fixation when N was applied on the surface or incorporated in the top most soil layers. Deep placement of slow-release fertilizer below the nodulation zone, or late N applications during reproductive stages, may be promising alternatives for achieving a yield response to N fertilization in high yielding environments. The results from many N fertilization studies are often confounded by insufficiently optimized BNF or other management factors that may have precluded achieving BNF-mediated yields near the yield potential ceiling (Salvagiotti *et al.*, 2008).

Agronomic efficiency of applied N varied widely due to large variation in indigenous soil N supply, N rates, application methods, and other factors affecting yield responses to N. In N-responsive experiments, soil application of N fertilizer before the R3 stage mostly resulted in AE of 2–10 kg grain/kg N. The largest AE was observed at N rates less than 50 kg/ha. It is noteworthy that AE averaged 8.7 kg grain/kg N when less than 50 kg N/ha was applied before R3, but averaged a substantive three times greater (24 kg grain/kg N) when N was applied after R3 stage. The latter value is comparable to the AE of applied N in well-managed cereal crops (Ladha *et al.*, 2005). Generally speaking, little quantitative in formation exists about the relative contributions of RE and PE to the AE of applied N in soybean. RE values were ranging from 0.12 to 0.96 kg plant N/ kg N. There was no clear differentiation in RE between early or late applications of N, although in one study RE increased from 0.18 to 0.74 kg/kg when N was applied late as compared to early.

At comparable N rates, AE of foliar-applied N was generally lower than that achieved with soil-applied fertilizer delivered after the R3 stage, and it ranged from 2.9 to 10.6 kg grain/kg N. A major limitation of foliar fertilization is the difficulty of applying larger quantities of N because of risk of leaf injury. Considering this risk of injury and the low AE, it is unlikely that foliar application of N can be reliably contribute to significant yield increases in soybean unless these limitations can be alleviated with improved foliar N formulations.

An experiment was conducted at Directorate of soybean Research, Indore to study the effect of basal and split application of nitrogen in soybean and results revealed that the highest PFP and AE was recorded with recommended dose of nitrogen i.e. 20 kg/ha applied as basal. The additional increase in N rates (10 kg/ha) which was top dressed at R1 and further splitted 5 kg each at R1 and R5 stage reduced the nitrogen use efficiency drastically. The split application has very little effect on the soybean yield as compared to recommended dose (Table 14). The results across the country showed that the increase in nitrogen levels (40 kg N/ha) from recommended dose (20 kg N/ha), the partial factor productivity was just 50% of the recommended one in all the zones. However, AE was approximately three times lesser in north plain zone and tow times lower in north east, central and southern zones (Table 15).

Table 14: Nitrogen nutrition in soybean (1995-98)

Treatment	Seed yield (kg/ha)	PFP(kg/kg)	AE(kg/kg)
Control	1266	-	-
20 Kg N basal	1460	73.00	9.70
20 kg N basal+10 kg N at R1(TD)	1443	48.10	5.90
20 kg N basal+10 kg N at R5 (TD)	1418	47.27	5.07
20 Kg N basal +5+5kg N at R 1 &R5 (TD)	1461	48.70	6.50
20 kg N basal+10 kg N at R1(F)	1445	48.17	5.97
20 kg N basal+10 kg N at R5 (F)	1400	46.67	4.47
20 Kg N basal +5+5kg N at R 1 &R5 (F)	1430	47.67	5.47

Table 15: Effect of nitrogen on productivity of soybean under different agroclimatic zones of India

N level (kg/ha)	North plain zone			North eastern zone			Central zone			Southern zone		
	Yield (kg/ha)	PFP (kg/kg)	AE (kg/kg)	Yield (kg/ha)	PFP (kg/kg)	AE (kg/kg)	Yield (kg/ha)	PFP (kg/kg)	AE (kg/kg)	Yield (kg/ha)	PFP (kg/kg)	AE (kg/kg)
20	2897	144.85	28.95	1946	97.30	34.65	1597	79.85	20.55	1848	92.40	19.20
40	2717	67.93	9.98	1934	48.35	17.03	1701	42.53	12.88	1868	46.70	10.10
CD	83.29			NS			74.46			51.28		
Control	2318			1253			1186			1464		

Phosphorus use efficiency

The recovery of P by crops that are planted immediate after the soluble fertilizer application amounts to only 10 to 30% of the quantity applied to the soil (Baligar and Bennet, 1986). The remaining 70 to 90% has been accounted for by absorption by soil microbes, precipitation by cations in soil solution, or by adsorption on the clay matrix. The P use efficiency is depends on soil reaction, soil P status, specific needs of crop and certain pedological differences. The most important factors controlling the solubility and the availability of P of soils are pH of the soil solution, content of cations and clay and organic matter (Baligar and Bennet, 1986).

The application of phosphorus significantly enhanced the soybean yield in all the zones, while just reverse trend was observed with regards to partial factor productivity (Table 16). Application of 40 and 60 kg P_2O_5/ha showed more or less equal/ similar agronomic efficiency and further increase in P levels decreased the AE substantially in north plain and Central zone. The increased levels of phosphorus failed to alter the agronomic efficiency in north eastern zone, however, the higher levels showed little bit higher AE as compared to lowest level in north eastern zone. In southern zone, a decreasing trend of AE was observed as the levels of phosphorus increases.

Table 16: Effect of phosphorus on productivity of soybean under different agro-climatic zones of India

P level (kg/ha)	North plain zone			North eastern zone			Central zone			Southern zone		
	Yield (kg/ha)	PFP (kg/kg)	AE (kg/kg)	Yield (kg/ha)	PFP (kg/kg)	AE (kg/kg)	Yield (kg/ha)	PFP (kg/kg)	AE (kg/kg)	Yield (kg/ha)	PFP (kg/kg)	AE (kg/kg)
40	2762	69.05	11.10	1695	42.38	11.05	1525	38.13	8.48	1818	45.45	8.85
60	2776	46.27	11.45	1948	32.47	11.58	1693	28.22	8.45	1863	31.05	6.65
80	2884	36.05	7.08	2178	27.23	11.56	1730	21.63	6.80	1894	23.68	5.38
CD	102			156			91			63		
Control	2318			1253			1186			1464		

Potassium use efficiency

In tropical soils, the efficiency of applied K is anywhere from 20 to 40% (Baligar and Bernett, 1986). Most of the adsorbed K in crop plants remains in the tops and a small amount is translocated to the grains.

The three years results of AICRP on soybean revealed that the application of potassium @ 20 kg/ha showed maximum PFP and AE and further increase in potassium levels substantial reduction was observed in both the efficiencies (Table 17). In another experiment in which the different levels of potassium were tried in soybean-wheat cropping system which indicated that the productivity of both the crops and K-HI enhanced with the increased levels of K (Table 18). The highest PFP was recorded with lowest dose of K *i.e.* 49.8 kg

Table 17: Effect of potassium on productivity of soybean under different agroclimatic zones of India

K level (kg/ha)	North plain zone			North eastern zone			Central zone			Southern zone		
	Yield (kg/ha)	PFP (kg/kg)	AE (kg/kg)	Yield (kg/ha)	PFP (kg/kg)	AE (kg/kg)	Yield (kg/ha)	PFP (kg/kg)	AE (kg/kg)	Yield (kg/ha)	PFP (kg/kg)	AE (kg/kg)
20	2815	140.75	24.85	1900	95.00	32.35	1629	81.45	22.15	1859	92.95	19.75
40	2800	70.00	12.05	1979	49.48	18.15	1668	41.70	12.05	1857	46.43	9.83
CD	83.29			127.06			74.46			51.28		
Control	2318			1253			1186			1464		

Source: Modified from Billore and Srivastava (2013)

Table 18: Effect of potassium levels, soybean genotypes and planting time on productivity, K uptake and K use efficiencies under soybean and wheat cropping system (2006-08)

K levels – *Kharif* + *Rabi* (kg/ha)	Soybean yield (kg/ha)	Wheat yield (kg/ha)	SEY (kg/ha)	Total K uptake (kg/ha)	K-HI (%)	PFP (kg/kg)	AE (kg/kg)	RE (%)	PE (kg/kg K uptake)	IUE (kg /kg K)
0 + 0=0	2006	4366	6372	72.85	33.27	-	-	-	-	86.90
16.6 + 33.2=49.8	2230	4778	7008	82.89	34.29	116.80	10.59	16.73	67.43	84.55
33.2 + 33.2=66.4	2461	5139	7600	90.35	35.40	95.00	15.11	21.88	72.51	84.11
49.8 + 33.2=83.0	2643	5403	8046	98.26	35.43	80.46	14.78	25.41	69.97	81.88
CD	73.42	141.38	173.06	2.84	0.79	-	-	-	-	1.58

Source: Billore *et al.*, 2009.

K/ha, while the AE and PE was highest with 66.4 kg K/ha. The highest IUE was associated with control (Billore *et al*., 2009). Potassium AE was found to be higher in soybean than wheat in a LTFE (Table 19) conducted at Jabalpur and Ranchi while the just reverse was true with per kg response of K at Jabalpur only (Samra and Swarup, 2005). K use efficiency increased with K application and 8.25 kg grain for each kg K_2O applied when 40 kg K_2O/ha which was 1.5 times higher than 20 K_2O/ha (Dixit *et al.* 2011).

Table 19: Average grain yield of crops (t/ha) and K response (kg/ha) over the years in LTFE

Centre	Crops	NP (100%)	NPK (100%)	K response (kg/ha)	AE (kg/kg)
Jabalpur(1972-99) Vertisol	Soybean	1.9	2.1	200	11.7
	Wheat	3.9	4.2	300	9.0
Ranchi(1973-99) Alfisols	Soybean	0.9	1.6	700	21.2
	Wheat	2.3	2.6	300	9.0

Source: Samra and Swarup (2005)

Table 20: Effect of sulphur and boron levels on yield, partial factor productivity and agronomic efficiency of soybean in different agro-climatic zones of India (2007-08)

Treatment	North plain zone		North eastern zone		Central zone		Southern zone	
	Yield (kg/ha)	AE (kg/kg)	Yield (kg/ha)	AE (kg/kg)	Yield (kg/ha)	AE (kg/kg)	Yield (kg/ha)	AE (kg/kg)
S level (kg/ha)								
0	1995	-	1594	-	1892	-	1171	-
10	1942	-	1792	19.80	2045	15.30	1290	11.90
20	2040	2.25	2031	21.85	2165	13.65	1404	11.65
30	2106	3.70	2281	22.90	2172	9.33	1490	15.95
40	2107	2.80	2361	19.18	2218	8.15	1415	6.10
B level (kg/ha)								
0.0	2026	-	1701	-	1939	-	1188	-
0.5	1976	-	1974	546.00	2088	298.00	1531	686.00
1.0	2043	17.00	2076	375.00	2171	232.00	1488	300.00
1.5	2035	6.00	2138	291.33	-	-	1383	130.00
2.0	1878	-	2072	185.50	-	-	1181	-

Soybean yield was found to increase as levels of S and B increased up to 30 kg/ha in north plain, central and southern zones while yield increased up to 40 kg S/ha and 1.5 kg B/ha in north eastern zone. The highest sulphur –AE was recorded under north eastern zone conditions followed by central, southern and north plain zones (Table 20). The maximum S-AE was associated with 30 kg S/ha in north plain, north east and southern zone, while it was maximum with 10 kg S/

ha in central zone. The highest response of boron was noted in southern zone followed by north eastern, central and lowest with north plain zone. The maximum B-AE was associated with 0.5 kg B/ha in north eastern, central and southern zone whiles it was maximum with 1 kg B/ha in north plain zone. The application of 40 kg S/ha showed higher values of AE, PE and RE, however, PFP was highest with 20 kg S/ha in Vertisols of Madhya Pradesh (Table 21). Three years results indicated that the maximum sulphur- PFP and AE was noted with lowest sulphur level in all the zones except north plain zone, where highest S-PFP and AE was associated with 5 and 20 kg S/ha, respectively (Table 22).The maximum soybean yield was recorded with 20 kg S/ha in north plain zone, with 30 kg S/ha in central and southern zone and with 35 kg S/ha in north eastern zone. S use efficiencies (PFP and AE) decreased as the levels of S increases (Table 22).

A substantial soybean yield improvement was observed due to application of Zn, B and Mo (Table 23). The further yield enhancement was recorded when these nutrients applied in conjunction with FYM in central and southern zone only. The PFP and AE trend was almost similar as was observed in yield. The values of PFP and AE seem to be very much larger because the application rates were very low which indicated that these two efficiencies overestimate the effect of micronutrients. A similar trend was also recorded in another experiment (Table 25).

Table 21: Effect of sulphur levels on yield, uptake and their use efficiencies of soybean (2011-12)

S level (kg/ha)	Seed yield (kg/ha)	Total S uptake (kg/ha)	PFP (kg/kg)	AE (kg/kg)	PE (kg/kg)	RE (%)
0	1713	9.18	-	-	-	-
20	1971	11.66	98.55	12.90	104.03	0.120
40	2308	13.38	57.70	14.88	141.66	0.105
60	2348	15.32	39.13	10.58	103.42	0.102
CD	119	2.95	32.14	2.50	30.41	0.014

Table 22: Response of sulphur to soybean in north plain and north eastern zones of India

S level (kg/ha)	North plain zone			North eastern zone			Central zone			Southern zone		
	Yield (kg/ha)	PFP (kg/kg)	AE (kg/kg)	Yield (kg/ha)	PFP (kg/kg)	AE (kg/kg)	Yield (kg/ha)	PFP (kg/kg)	AE (kg/kg)	Yield (kg/ha)	PFP (kg/kg)	AE (kg/kg)
0	1422	-	-	1505		-	1639		-	2230		-
5	1449	289.80	5.40	1601	320.20	19.20	1756	351.20	23.40	2372	474.40	28.40
10	1468	146.80	4.60	1670	167.00	16.50	1805	180.50	16.60	2340	234.00	11.00
15	1359	90.60	-	1669	111.27	10.93	1886	125.73	16.47	2387	159.13	10.47
20	1644	82.20	11.10	1729	86.45	11.20	1920	96.00	14.05	2398	119.90	8.40
25	1358	54.32	-	1802	72.08	11.88	2023	80.92	15.36	2448	97.92	8.72
30	1498	49.93	2.53	1814	60.46	10.30	2064	68.80	14.17	2500	83.33	9.00
35	1554	44.40	3.77	1873	53.51	10.51	2007	57.34	10.51	2495	71.28	7.57
40	1430	35.75	0.20	1856	46.40	8.78	2004	50.10	9.13	2430	60.75	5.00
CD	118			138			88			155		

Table 23: Effect of integrated micronutrient management in soybean (1992-1994)

Treatment	North plain zone			Central zone			Southern zone		
	Yield (kg/ha)	PFP (kg/kg)	AE (kg/kg)	Yield (kg/ha)	PFP (kg/kg)	AE (kg/kg)	Yield (kg/ha)	PFP (kg/kg)	AE (kg/kg)
Control	2139	-	-	1768	-		2231	-	
Zn @ 5 kg/ha	2547	509.40	81.60	1927	385.40	31.80	2369	473.80	27.60
Zn @ 10 kg/ha	2567	256.70	42.80	1925	192.50	15.70	2283	22.83	5.20
$NaMoO_4$ @ 4 g/kg seed	2339	7796.67	666.67	1935	6450.00	566.67	2287	7623.33	186.67
B @ 0.5 kg/ha	2329	4558.00	380.00	1841	3682.00	148.00	2303	4606	144.00
B @ 1.0 kg/ha	2277	2277.00	138.00	1907	1907.00	139.00	2409	2409	178.00
Zn@ 5 kg + FYM	2448	489.60	61.80	1997	399.40	45.80	2471	494.20	48.00
Zn @ 10 kg + FYM	2469	246.90	33.00	2037	203.70	26.90	2358	23.58	12.70
$NaMoO_4$ + FYM	2437	8123.33	993.33	1904	6346.67	453.33	2430	8100.00	663.33
B @ 0.5 kg + FYM	2297	4594.00	316.00	2006	4012.00	476.00	2420	4840.00	378.00
B @ 1.0 kg+ FYM	2328	2328	189.00	1968	1968.00	200.00	2389	2389.00	158.00

Table 24: Effect of integrated micronutrient management in soybean (1995-1999)

Treatment	North plain zone			Central zone			Southern zone		
	Yield (kg/ha)	PFP (kg/kg)	AE (kg/kg)	Yield (kg/ha)	PFP (kg/kg)	AE (kg/kg)	Yield (kg/ha)	PFP (kg/kg)	AE (kg/kg)
Control	1658	-	-	1628	-	-	1967	-	-
Zn @ 5 kg/ha	1907	381.40	49.80	1795	359.00	33.40	2119	423.80	30.40
$NaMoO_4$ @ 4 g/kg seed	1928	6426.67	900.00	1760	5866.67	400.00	2117	21.17	500.00
B @ 0.5 kg/ha	1877	3754.00	438.00	1754	3508.00	252.00	2064	4128.00	194.00
Zn + FYM	1904	380.80	492.00	1979	395.80	70.20	2311	462.20	68.80
B + FYM	1842	3684.00	368.00	1902	3804.00	548.00	2234	4468.00	534.00
$NaMoO_4$ + FYM	1985	6616.67	1090.00	1862	6206.67	780.00	2189	7296.66	740.00
$ZnSO_4$ @5 kg + lime @ 2.5 kg (Foliar)	1814	362.80	31.20	1751	350.20	24.60	2106	421.20	27.80
Zn + B soil application	1849	336.18	34.72	1817	330.36	34.36	2147	390.36	32.72

Time of application: Soybean being an oil yielding legume, in general all the recommended nutrients are applied as basal because of soybean requires only starter dose of nitrogen while phosphorus and potassium are not easily lost from the soil. The application of phosphorus and potassium either below the seed or band placement were found better than broadcasting which helps greater uptake and utilization. However, there are some research evidences are available which indicated that the split application of nitrogen and potassium enhanced their use efficiency in soybean (Vyas *et al.*, 2008).

Placement: Placement of nutrients and fertilizer rates are important factors to be considered to produce maximum yield of crops. Particularly deep placement of nutrients might be beneficial to crop growth. Increased early growth has been observed with deeper P placement as well as by deep band placement of K when compared to broadcast application. The method of N, P and K placement has typically been found effective over broadcasting on the top of the soil, and it is also influenced by the amount of water used for irrigation. Normally fertilizers are broadcasted (Radhika *et al.*, 2013).

Integrated Plant Nutrient Supply System (IPNS)

Small farmers are facing problems of declining soil fertility due to intensive cultivation, nutrient removal via crop harvest and soil erosion. Use of commercial fertilizers to address the declining soil fertility remains minimal due to farmer's low income which limits their ability to purchase fertilizers. High costs of fertilizer, lack of credit, delays in delivery of fertilizer due to poor transport and marketing infrastructure, and lack of know-how about their usage have individually or jointly constrained fertilizer optimal use (Makokha *et al.*, 2001). One current proposition towards solving agro-environmental problems is INM, which does not aim to remove fertilizer totally in the short run but to reduce the negative impacts of overuse of fertilizers containing N, P, and other elements. The INM system promotes low chemical input but improved NUE by combining natural and manmade sources of plant nutrients in an efficient and environmentally prudent manner. This will not sacrifice high crop productivity in the short term nor endanger sustainability in the long term (Adesemoye *et al.*, 2008). It was demonstrated that PGPR-elicited plant growth promotion resulted in enhanced N uptake by plant roots which might have resulted from increased fertilizer N utilization efficiency in an INM system.

Organic amendments help in improving the soil physical and biological environment besides supplying nutrients for crop growth. Results from different studies revealed that continuous application of FYM and green manure improved the soil organic carbon under different soils and cropping systems (Biswas *et al.,* 1971 and Swarup, 1998). Neither inorganics nor organics alone can maintain organic matter status of soil and sustain productivity (Prasad, 1996). So judicious uses of organic and inorganic fertilizers is essential to safeguard soil health and augment the nutrient use efficiency. Integrated use of organic and inorganic sources of nutrients can improve the root growth of soybean facilitating efficient utilization of stored soil moisture and nutrients from different layers in the profile. The positive effect of integrated use of FYM and inorganic fertilizers on productivity of soybean has been reported by many workers (Singh *et al.*, 1999, Hati *et al.*, 2000 *and* Mandal *et al.*, 2000).

Besides growing legumes and short-duration pulse crops in crop rotations, the effects of integrated use of organics (farmyard, piggery, poultry and green manures and crop residues) with chemical fertilizers, and impacts of long-term use of INM on enhancing crop productivity were studied. The results clearly demonstrated that INM enhances the yield potential of crops over and above achievable yield with recommended fertilizers, and results in better synchrony of crop N needs due to (a) slower mineralization of organics, (b) reduced N losses via denitrification and nitrate leaching, (c) enhanced NUE and recovery by crops, and (d) improvements in soil health and productivity, and hence could sustain high crop yields in various cropping systems ensuring long-term sustainability of the system.

Application of FYM @ 4 t/ha with NPK significantly improved the seed yield of soybean by14 and 51.8% over NPK and control, respectively. The application of NPK and NPK + FYM @ 4 t/ha enhanced the PFP to the tune of 7.26 and 8.04 kg/kg and AE 1.66 and 2.40 kg/kg, respectively. The root mass of soybean was mostly confined to 15 cm soil depth. Combined application of NPK and FYM recorded 5.5 and 18.8% higher root length density of soybean in the 0-15 cm soil layer at flowering stage over NPK and control, respectively. The root mass density and root volume density followed the trend similar to root length density.

Three year experimentation results indicated that the integration of organic and inorganic sources of nutrient improved the soybean yield as well as nutrient use efficiencies than their lone application and management (Table 25). Sufficient evidence indicates that the combined application of NPK fertilizers and organic fertilizers can generally achieve higher crop yields than any of them applied alone (Zhao *et al.*, 2010).

A long-term field experiment from 1973 to 2004 indicated significant downward trend in yield of soybean for both the control and NPK treatment compared with a significant upward trend with time for the NPK + FYM (Bhattacharyya *et al.*, 2008). Similar, but not statistically significant, trends were observed for wheat yields. They concluded that the different combinations of mineral

Table 25: Impact of integrated nutrient management schedule on yield and nutrient use efficiencies of soybean in different agro-climatic zones of India (2009-11)

Treatment	North plain zone			North eastern zone			Central zone			Southern zone		
	Yield (kg/ha)	PFP (kg/kg)	AE (kg/kg)	Yield (kg/ha)	PFP (kg/ha)	AE (kg/ha)	Yield (kg/ha)	PFP (kg/ha)	AE (kg/ha)	Yield (kg/ha)	PFP (kg/kg)	AE (kg/kg)
75 % RDF	2133	21.88	2.06	1410	10.44	2.51	1530	14.57	3.46	1657	14.73	3.62
75 % RDF+ FYM @ 5 t/ha	2100	21.54	1.72	1571	11.63	3.70	1660	15.81	4.69	1779	15.84	4.70
100 % RDF	2160	16.62	1.75	1694	9.41	3.46	1798	12.84	4.51	1706	11.37	3.04
100 % RDF+ FYM @ 5 t/ha	2159	16.61	1.75	1792	9.96	4.01	1870	13.36	5.02	1815	12.10	3.77
125 % RDF	2229	13.72	1.83	1806	8.02	3.27	1830	10.45	3.79	1709	9.11	2.45
125 % RDF + FYM @ 5 t/ha	2232	13.73	1.85	1915	8.51	3.75	1911	10.92	3.79	1876	10.00	2.81
Control	1932			1071			1167			1250		

N, P and K fertilizers could not achieve yield sustainability under soybean-wheat system. The higher soybean and wheat yields obtained with FYM + NPK were possibly caused by other benefits of organic matter exceeding N,P and K supply, such as improvements in microbial activities, better supply of macro- and micro-nutrients such as S, Zn, C and B, which are not supplied by inorganic fertilizers, and lower losses of nutrients from the soil. They also stated that improved soil physical properties on the FYM-treated plots were observed and may have contributed to the improvement in crop yields. In another 28-year long-term experiment, Hati *et al.*, (2007) found that the combined application of NPK and FYM not only produced significantly higher soybean and wheat yield than NPK treatment, but also increased SOC by 56.3% compared with the initial SOC value. The treatment increased soil electrical conductivity, SOC, aggregation, water retention, micro porosity and available water capacity, but decreased soil bulk density over all other treatments.

Biofertilizers

Microbial inoculants are promising components for integrated solutions to agro-environmental problems because inoculants possess the capacity to promote plant growth, enhance nutrient availability and uptake, and support the health of plants (Adesemoye *et al.*, 2008). Microbial inoculants include three major groups: (1) arbuscular mycorrhiza fungi (AMF), (2) PGPR, and (3) the nitrogen-fixing rhizobia, which are usually not considered as PGPR.

Enhanced beneficial microbes such as rhizobia, diazotrophic bacteria, and mycorrhizae in the rhizosphere have improved root growth by BNF, suppressing pathogens, producing phytohormones, enhancing root surface area to facilitate uptake of less mobile nutrients such as P and micronutrients, and mobilization and solubilization of unavailable organic/inorganic nutrients. Ladha *et al.*, (1996) reported that free-living and/or associated phototrophs and heterotrophs can fix from 50 to 100 kg N/ha, contributing to the increased supply and efficiency of N.

Rhizobia and Biological N Fixation (BNF)

Biological nitrogen fixation by legume species plays an important role in increasing N supply to growing crops in rotation or in legume grass mixtures. The overall contributions of legumes are most effective in humid and sub-humid regions and N_2 fixed ranges from 24 to 267 kg/ha/yr (Fageria, 1992). The amount of N_2 fixed varies, depending on crop species/cultivars, soil acidity, temperature, drainage, and the timing of harvest (Fageria, 1992). Legumes supply substantial N to the soil, but the amount of N fixed is highly variable. Soybean is able to fix atmospheric N in between 170 to 217 kg/ha (Thurlow and Hiltabold, 1985).

Nitrogen is a critical limiting element for plant growth and production. Soybean is categorized as good nitrogen fixers and will fix all of their nitrogen needs other than that absorbed from the soil. It may fix up to 250 kg N/ha and are not usually fertilized. In fact, they usually don't respond to nitrogen fertilizer as long as they are capable of fixing nitrogen. Nitrogen fertilizer is applied at planting to soybean to supply nitrogen i.e. 20 kg N/ha to the plant before nitrogen fixation starts. When large amounts of nitrogen are applied, the plant literally slows or shuts down the nitrogen fixation process. It is easier and less energy consuming for the plant to absorb nitrogen from the soil than to fix it from the air.

Inoculation of soybean is a significant agency for the manipulation of rhizobia for improving crop productivity and soil fertility. It can lead to the establishment of a large rhizobial population in the plant rhizosphere and to improved nodulation and BNF, even under adverse soil N conditions (Peoples *et al.*, 1995). Sometimes, yield is not increased by inoculation, but N concentration in seed or plant parts may be increased over that of non-inoculated plants (Wani *et al.*, 1995).

Mycorrhizae: The term mycorrhiza describes structures formed by the association of plant roots with fungi. According to Moser and Wandter (1983) that more than 80% of the higher plants, the root system and hence, at least the mode of nutrient uptake, is modified by symbiosis with fungi. Two major groups of mycorrhizae are ecto and endo mycorrhizae. Two sub groups occur within the endomycorrhizae: those produced by septate fungi and those produced by non-septate fungi. The later are known as Vesicular-arbuscular mycorrhizae (VAM). The plant roots infected with VAM fungal mycelium exploit a greater soil volume and this helps in more absorption of nutrients, especially P. Moreover, an increase in uptake of P, VAM infected roots, increase in S, K, Ca and Zn are also reported (Moser and Wandter, 1983).

Arbuscular mycorrhizal (AM) fungi forms a beneficial symbiosis with roots, thereby increasing root surface area which assists roots in exploring larger soil volumes thereby bring more ions closer to roots and contributing to higher nutrient inflow (Smith *et al.*, 1993). Primary benefits of AM are enhanced acquisition of mineral nutrients, plant tolerance to soil chemical constraints such as acidity, salinity, alkalinity, and increases the ability of the host plants to withstand or have reduced acquisition of elements toxic to plant growth. Other benefits for the host-plants are the improvement of water uptake and the withstanding of drought. The AM infection may act as general modifiers of NUE regardless of the extent to which the plant roots are infected and extent of infection appears to be under genetic control and shows considerable variability between crop species/cultivar and AM isolates. (Smith *et al.*, 1993). Such an interaction could be used to select crop cultivars and effective AM isolates to enhance nutrient uptake from nutrient deficit or low input systems. The P uptake in plant is most affected by AM interaction, but AM can also directly increase the uptake of Zn, Cu, N, and the cation/ anion ratio. The capacity of AMF to influence plant growth, water and nutrient content has been widely reported over the years (Giovannetti *et al.*, 2006). The AMF have a high-affinity P-uptake mechanism that enhances P nutrition in plants.

Other microbes: The continued use of chemical fertilizers and manures for enhanced soil fertility and crop productivity often results in unexpected harmful effects. INM systems are needed to maintain agricultural productivity and protect the environment. Microbial inoculants are promising components of INM. The using microbial inoculants are plant growth-promoting rhizobacteria (PGPR) and AMF for increasing the use efficiency of fertilizers. Studies with microbial inoculants and nutrients have demonstrated that some inoculants can improve plant uptake of nutrients and thereby increase the use efficiency of applied chemical fertilizers and manures.

Plant Growth Promoting Bacteria (PGPR): Benefits to plants from plant-PGPR interactions have been shown to include increases in seed germination rate, root growth, yield, leaf area, chlorophyll content, nutrient uptake, protein content, hydraulic activity, tolerance to abiotic stress, shoot and root weight, bio-control and delayed senescence (Bakker *et al.*, 2007). Other beneficial effects of PGPR strains include enhancing P availability, BNF (Bashan *et al.*, 2004), sequestering Fe for plants by production of siderophores (Bakker *et al.*, 2007), producing plant hormones, auxins and synthesizing the enzymes, which lowers plant levels of ethylene, thereby reducing environmental stress on plants.

Adesemoye *et al.* (2008) confirmed that inoculations with mixed strains were more consistent than single strain inoculations. Inoculants, PGPR and AMF, are playing significant roles in the solubilization of inorganic phosphate and mineralization of organic phosphates (Mahmood *et al.*, 2001 and Tawaraya *et al.*, 2006). Inoculants have been shown to influence the uptake of many other elements in addition to N and P (Adesemoye *et al.*, 2008). Han and Lee (2005) reported an increased uptake of P and K when soil was fertilized with rock P and K and coinoculated with P solubilizing bacteria *B. megaterium* and K solubilizing bacteria *B. mucilaginosus*. Liu *et al.* (2000) reported an increase in acquisition of Fe, Zn, Cu, and Mn by mycorrhiza. Sulfur and Fe uptake have been achieved through sulfur-oxidizing bacterial inoculant and siderophore-producing bacteria, respectively (Banerjee *et al.*, 2006 and Bakker *et al.*, 2007).

Table 26: Effect of phosphate solubilizing micro-organisms inoculation on soybean yield and nutrient use efficiency

Treatment	Seed yield (kg/ha)	PFP(kg/kg)	AE (kg/kg)
Un-inoculated	2220	-	-
Aspergillus awamori	2260	-	-
Psudomonas strata	2400	-	-
SSP @ 30 kg P_2O_5/ha	2460	82.00	8.00
RP @ 30 kg P_2O_5/ha	2310	77.00	3.00
A. awamori + SSP @ 30 kg P_2O_5/ha	2490	83.00	9.00
A. awamori+ RP @ 30 kg P_2O_5/ha	2260	75.33	1.33
P. strata + SSP @30 kg P_2O_5/ha	2350	78.33	4.33
P. strata + RP@ 30 kg P_2O_5/ha	2230	74.33	0.33
SSP @ 60 kg P_2O_5/ha	2400	40.00	3.00

Experiment conducted at Directorate of Soybean Research, Indore to study the effect of phosphate solubilizing micro-organisms inoculation on productivity and nutrient use efficiency of soybean. The application of 30 kg phosphorus through single super phosphate was most efficient (PFP and AE) than applied through rock phosphate (Table 26). The inoculation had very little effect on soybean yield, however, inoculation with *P. strata* was more efficient than *A. awamori* when soybean grown without phosphorus.

Table 27: Effect of co-inoculation of PGPR with *Bradyrhizobium japonicum* on nodulation and seed yield of soybean

Treatment	Nodules /plant	Nodule dry biomass/ plant (mg)	Shoot dry biomass/ plant (g)	Seed yield (kg/ha)	Per cent increase over control	Percent increase over B. j
Un inoculated (control)	8.67	141.11	4.01	1391	-	-
Bradyrhizobium japonicum	28.67	163.33	6.00	1633	17.40	-
Bj + *Pseudomonas fluorescens isolate-1*	31.33	48.39	6.50	1759	26.46	7.72
Bj + *Pseudomonas fluorescens isolate IV*	30.11	160.22	6.80	1731	24.44	6.00
Bj + *Entrobacter* Spp	9.11	52.78	4.14	1885	35.44	15.37
Bj + *Proteus vulgaris*	19.22	45.56	5.31	1742	25.23	6.67
Bj + *Bacillus sphaeriticus*	34.66	97.89	8.09	1785	28.32	9.30
Bj + *Klbiesiella planticola*	35.33	161.44	4.11	1673	20.27	8.57
unidentified bacterial isolate	31.00	141.00	7.09	1836	31.99	12.43
unidentified bacterial isolate	19.78	33.44	6.98	1591	14.38	-
CD	2.25	5.17	0.19	311	-	-

Source: Billore *et al.* (2006).

The results revealed that the seed inoculation with *B. japonicum* alone increased the nodule number and their dry biomass significantly over un-inoculated control (Table 27). Further significant enhancement was also recorded when seeds were inoculated with *B. japonicum* + PGPR i.e. *Pseudomonas fluorescens isolate-1*, *Bacillus sphaeriticus*, *Klbiesiella planticola* and unidentified bacterial isolate over *B. japonicum* alone. However, the co-inoculation of *B. japonicum* with PGPR with *Entrobacter* sp., *Proteus vulgaris* and unidentified bacterial isolate showed significantly lower nodule number and their dry biomass as compared to *B. japonicum* alone. The maximum nodule number and their dry biomass was associated with *B. japonicum* + *Klbiesiella planticola*, while the lowest was with *B. japonicum* + *Entrobacter* Spp. Plant dry matter accumulation at flowering also showed similar trend. A significant variation was noted in yield also. Seed inoculation with *B. japonicum* alone produced 17.4% higher yield as compared to un-inoculated control. Further improvement in seed yield was noted when seed inoculated with *B. japonicum* + PGPR, which was to the tune of 14.4 to 35.4% as compared to un-inoculated, control. When comparing the effect of *B. japonicum* alone and *B. japonicum.* + PGPR, the magnitude of response was to the tune of 6.0 to 15.4%. The highest seed yield was recorded with *B. japonicum* + *Entrobacter* Spp, which remained at par with all the treatment except un-inoculated control. Similar trend was also observed in yield attributes.

Water management

Nutrients move to plant roots in the soil water by various means, N by mass flow, P by diffusion and K by both mass flow and diffusion. The efficiency of added nutrients may be diminished where water supplied by irrigation or rainfall leaches soluble nutrients below the root zone, or facilitates other reactions that result loss of nutrients. The times and methods of application of both water and fertilizers should be carefully coordinated to ensure that nutrients are not lost from the system and that the position of nutrients in the soil facilitates rapid uptake by the crop root. It is not the total rainfall that is important for crop production and higher nutrient use efficiency, but its distribution during the cropping season. Drought stress is recognized as the most damaging abiotic stress for soybean production. Increased irrigation has been showing large yield increases of over 1000 kg/ha under irrigated *vs* non-irrigated conditions. Crop response to fertilizer can vary with amount of rainfall or precipitation or irrigation received during growing season. Adequate moisture content in the root zone is an important factor for efficient use of available nutrients. For most of the annual crops, soil moisture content should not fall below 70% of field capacity. Low soil moisture contents reduce P and K uptake by plant roots because P and K diffusion rates decreased as soil moisture content decreases. Fertilizer productivity is closely related to soil moisture availability. Fertilizer has the strongest relationship with water (95%) followed by seed (83%) and electricity (72%). Crop production in water stressed conditions depends on rainfall (Sandhu *et al.*, 1992). Thus enhancing soil water storage ability and water conservation in such regions may be one of the most crucial agricultural management practices (Mosavi *et al.*, 2009).

Soil moisture availability must be considered in determining the appropriate N recommendations for every crop and field situations. When too little water is a vailable (drought conditions), plant demand for nitrogen will be decreased. Significant interactions between moisture and nutrients have been recorded with various crops. Another physiological process sensitive to drought stress is nitrogen fixation. Decreased BNF starts when water potential of root nodules starts falling below -0.2 to -0.4 MPa. Because of the high protein content of seed, soybean has a greater demand for N compared with other crops. Soybean obtains N from fixation and directly from the soil. During seed filling, much of seed N demand is met by remobilization from the leaves. The contribution of nitrogen fixation to the plant's nitrogen supply varies inversely with soil nitrogen availability.

Table 28: Effect of irrigation scheduling on productivity and partial factor productivity of soybean over years in different agro-climatic regions of India (2009-2011)

Irrigation at	NPZ (SL 525)		CZ (JS 97 52)		SZ (RKS 18)	
	Yield (kg/ha)	PFP (kg/kg)	Yield (kg/ha)	PFP (kg/kg)	Yield (kg/ha)	PFP (kg/kg)
Seedling stage (15-20 DAS)	1099	8.45	1666	11.90	1784	11.89
Flower initiation	1021	7.85	1693	12.09	1872	12.48
Seed filling (20 days after flower initiation)	1091	8.39	1931	13.79	1848	12.32
Seedling stage + flower initiation	1199	9.22	1752	12.50	1996	13.30
Seedling stage+ seed filling	1566	12.05	1943	13.88	2023	13.49
Flower initiation + seed filling	1416	10.89	2017	14.41	2156	14.37
Seedling stage + flower initiation + seed filling	1647	12.70	2044	14.60	2175	14.50
Control (no irrigation)	752	5.78	1569	11.21	1646	10.97
CD (P=0.05)	139.09		97.68		75.16	

Source: Modified from Billore and Srivastava (2014)

Soil moisture and fertilizer had a positive interaction for achieving the higher FUE/NUE in soybean crop. Soybean being a rainfed crop in India and experienced moisture stress during the crop period. The moisture deficit at critical stage is found to be detrimental for the exploitation of full yield potential of soybean. The full yield potential of soybean could be achieved only when soil moisture should not be below the 70% of the field capacity. The three year results (Table 28) indicated that the supplemental irrigation (s) significantly enhanced the fertilizer PFP in all the zones. The PFP increased as the number of irrigations increased and the maximum with three irrigations at critical stages. The magnitude of response was higher in central and southern zones than north plain zone.

Crop rotation

Appropriate crop rotation is an effective way to alleviate weed and allelopathic problems (Fageria, 1992). Among the cropping systems, rotation gave higher yields than sole crop and intercropping systems and increased yield by 46% without fertilizer. Partial factor productivity of fertilizer was similarly higher in rotations than monoculture system. The crop diversification either rotations or intercropping systems is an insurance against total crop failure. Rotation of soybean and maize improved maize yield and FUE. Others have ascribed benefits of rotations to improved soil physical properties and to the ability of leguminous crops to increase P availability through secretion of enzymes or acids in the rhizosphere to solubilize P bounded to sesquioxides. Legume integration into the farming systems is an important component of ISFM because of their BNF potential and hence reduces farmers' costs for purchase of nitrogen fertilizers. Legumes have also been reported to improve the soil physical and chemical attributes as well as provide protein supplements for poor families. Crop rotation and use of cover crops and green manure crops are known to improve soil fertility and physical properties and to minimize pest and weed problems (Fageria *et al.*, 1997a and Delgado *et al.*, 1999). Carefully selected cropping systems result in increased fertilizer use efficiency as determined by crop equivalents or net returns (Gill and Ahlawat, 2006).

The synergizing effects of cropping system diversification have been found in weed management in the 3 and 4 years rotations. Weeds were suppressed as effectively in these systems as in the 2 year rotation, with declining soil seed banks and negligible weed biomass, yet herbicide inputs in the 3 and 4 years rotation were 6 to 10 times lower, and fresh water toxicity 200 times lower, than in the 2 year rotation. Improved efficiency and environmental sustainability of weed management in the 3 and 4 years rotations resulted from integrating multiple, complementary tactics in an ecological weed management frame work. Mounting evidence for unintended effects resulting from heavy reliance on herbicides highlights the need to re-think the role of herbicides in weed management. Non-target impacts of herbicides include reproductive abnormalities and mortality in vertebrates and potential for diminished non-crop nectar resources for key pollinator species. Herbicide overuse has also resulted in widespread, accelerating evolution of weed genotypes resistant to one or more modes of herbicide action (Tranel *et al.*, 2011). In the context of a cropping system with weed suppressive characteristics, small herbicide inputs may contribute to a diverse suite of tactics that cumulatively provide effective, reliable, and more durable weed management.

In irrigated or high-rainfall production regions, soybean–corn rotations have high NUE and can reduce the amount of residual N available for leaching when compared with continuous corn (Varvel, 1994). Unfortunately, rotations are not easily adopted by farmers who have become accustomed to mono-culture production systems, since a new crop often requires purchase of additional equipment and learning to integrate new cultural practices. Nitrogen use efficiency for wheat following legumes is greater than that for wheat following fallow or continuous wheat. More intensive dry-land cropping systems lead to increased water use efficiency and also better maintain soil quality. The advantage of soybean/corn rotation over continuous corn was consistent across years. Under continuous corn, there is trade-off when switching from conventional to conservation tillage, while efficiency increases, yields decrease.

Significantly higher uptake of N, P and K by *kharif* and *rabi* crops was recorded in integrated practice compared to inorganic nutrient management practice. The available N, P_2O_5, K_2O and S were significantly higher in legume based cropping systems during both the seasons of the study than non-legume system (Vidhyavathi *et al.*, 2011). The results from farmers field trials (3 years) revealed that the planting of soybean on ridge furrow system showed highest nutrient PFP and AE followed by broad bed furrow planting, and minimum tillage and lowest efficiencies with flat bed planting (Table 29). The soybean –chickpea system was more nutrient use efficient than soybean wheat cropping system in all the planting systems. The results clearly demonstrated that the *in-situ* moisture conservation practices namely ridge furrow, broad bed furrow planting and minimum tillage enhanced the NUE in soybean based cropping systems.

The micronutrient use efficiency in soybean –wheat cropping system was assessed for 3 years and results indicated that PFP and AE over estimate the micronutrients use efficiency. The integrated use of micronutrients and FYM substantially improved the NUE as compared to alone application of micronutrients (Table 30). Six years results from soybean –wheat cropping system indicated that the NUE of the system decreased as the levels of nutrients increases (Table 31). The use of single nutrients showed highest NUE as compared to combined application of NP or NPK or NPKZn. The AE of sulphur increased as the levels of sulphur increased up to 60 kg/ha and thereafter it declined (Table 32). The highest AE was being with 60 kg S/ha in soybean.

While in case of chickpea, the maximum AE was with 20 kg sulphur and thereafter it declined. PFP linearly declined as the levels of sulphur increases in both the crops and the highest PFP was associated with 20 kg S/ha. The maximum PE and ARE were observed with 60 kg S/ha in both the crops and thereafter it declined. The application of sulphur in both the crops showed highest AE, PFP, PE and ARE in both the crops (Girothia, 2013). Three years study on soybean based cropping systems revealed that the soybean – wheat cropping system was to be the most nutrient efficient system than soybean-linseed and soybean-mustard (Table 33). Soybean- linseed possesses higher PFP than soybean – mustard whiles the just reverse with AE. The higher levels of nutrients led to low nutrient use efficiency in all the soybean based cropping systems.

Table 29: Effect of in-situ moisture conservation techniques on productivity and nutrient use efficiency of soybean based cropping systems (2005-2008)

Treatment	Soybean-Wheat			Soybean-Chickpea		
	SEY (kg/ha)	PFP(kg/kg)	AE(kg/kg)	SEY (kg/ha)	PFP(kg/kg)	AE(kg/kg)
Ridge furrow planting						
RDF + FYM	6396	19.99	3.58	6206	34.48	6.66
RDF	5917	18.49	2.08	5592	31.06	3.25
Control	5252			5007		
Mean	5855	19.24	2.83	5603	32.77	4.96
Broad bed furrow planting						
RDF + FYM	6159	19.24	3.60	6075	33.75	5.25
RDF	5737	17.92	2.28	5611	31.17	2.67
Control	5008			5130		
Mean	5634	18.58	2.94	5605	32.46	3.96
Minimum tillage planting						
RDF + FYM	5677	17.74	1.24	5680	31.55	5.56
RDF	5325	16.64	0.14	5160	28.67	2.67
Control	4842			4679		
Mean	5281	17.19	0.69	5173	30.11	4.12
Flat bed planting						
RDF + FYM	5246	16.39	1.98	5033	27.96	3.82
RDF	4988	15.89	1.18	4792	26.62	2.48
Control	4612			4345		
Mean	4948	16.14	1.58	4723	27.29	3.15

Table 30: Effect of integrated micronutrient management on productivity and nutrient use efficiency of soybean – wheat cropping system (pooled data for 3 years)

Treatment	Seed yield (kg/ha)		System NUE (kg/kg)	
	Soybean	Wheat	PFP	AE
Zn @ 5 kg/ha	1583	3969	1110	33
Seed treatment with $NaMoO_4$ @ 4 g/kg seed	1609	3927	18453	500
Boron @ 0.5 kg/ha	1513	4034	11094	322
Zn @ 5 kg/ha + FYM @ 10 t/ha	1595	4050	1129	52
Boron @ 0.5 kg/ha + FYM @ 10 t/ha	1539	3991	11060	288
Seed treatment with $NaMoO_4$ @ 4 g/kg seed + FYM @ 10 t/ha	1513	4324	19456	1503
Zn @ 5 kg/ha + Lime @ 2.5 l/ha foliar spray at 25 – 30 DAS	1541	3890	1086	9
Zn @ 5 kg/ha + B@ 0.5 kg/ha as basal	1516	4031	1009	29
Control	1520	3866	-	-
CD	50	137		

Source: Modified from Billore and Joshi (2005).

Table 31: Effect of long term fertilizer application on seed yield of soybean - wheat cropping system (1993-98)

Treatment	Yield (kg/ha)		Soybean		Wheat		Total	
	Soybean	Wheat	PFP	AE	PFP	AE	PFP	AE
50%NPK	1 431	3 026	28.62	0.58	27.51	9.07	27.85	6.42
100%NPK	1 628	3 620	16.28	2.26	16.45	7.23	16.40	5.68
150%NPK	1 594	3 513	10.62	1.28	10.65	4.50	10.64	3.49
100%NPK +HW	1 707	3 490	17.07	3.05	15.86	6.65	16.24	5.52
100%NPK +Zn	1 654	3 534	15.03	2.29	15.36	6.55	15.72	5.32
100%NP	1 546	3 711	19.32	1.80	20.62	9.35	20.21	7.02
100%N	1 602	3 528	80.10	10.00	29.40	12.50	36.64	12.14
100%NPK + FYM	1 767	3 865	17.67	3.65	17.57	8.35	17.60	6.88
100%NPK-S	1 736	3 644	17.36	3.34	16.56	7.34	16.81	6.09
Control	1 402	2 028	-	-				
CD	65	283						

Source: Modified from Joshi and Billore (2004).

Table 32: Effect of sulphur levels and their application frequencies on agronomical efficiency and partial factor productivity of sulphur in soybean –chickpea cropping system (2008-09)

Treatment	Agronomic Efficiency (kg/ha)		Partial Factor Productivity (kg/ha)		Physiological Efficiency (kg/kg S uptake)		Apparent Sulphur Recovery (%)	
	Soybean	Chickpea	Soybean	Chickpea	Soybean	Chickpea	Soybean	Chickpea
S level (kg/ha)								
20	7.39	9.78	89.9	98.3	123.3	175.0	5.97	5.59
40	9.11	8.64	50.4	52.9	139.3	166.3	6.54	5.21
60	10.64	8.53	38.2	38.1	148.7	157.5	7.15	5.42
80	5.88	5.45	26.5	27.6	135.0	140.2	4.36	3.89
S- frequency								
Kharif only	8.5	3.81	51.5	49.9	132.5	134.5	6.72	2.97
Rabi only	3.61	7.74	46.6	53.9	147.7	161.2	2.41	4.72
Kharif + *Rabi*	12.66	12.75	55.7	58.9	142.8	163.1	8.88	7.40
Control	0.0	0.0	0.0	0.0	0.0	0.0	0.0	0.0

Source: Girothia (2013)

Table 33: Yield and economical parameters of soybean based cropping systems under different fertility levels (1994-1997)

Cropping system	Fertility level	Seed yield (kg/ha)		PFP (kg/kg)	AE (kg/kg)
Soybean-wheat		*kharif*	*Rabi*		
	0 %	1089	2410	-	-
	50 %	1158	3085	28.29	4.96
	100 %	1243	3803	16.82	5.16
	Mean	1164	3100	22.56	5.06
Soybean-linseed	0 %	1077	846	-	-
	50 %	1091	946	16.98	0.95
	100 %	1326	1053	9.91	1.90
	Mean	1164	948	13.45	1.43
Soybean-mustard	0 %	1108	340	-	-
	50 %	1264	477	13.39	2.25
	100 %	1366	600	7.56	1.99
	Mean	1243	472	10.48	2.12

Inter-cropping system

Legume/cereal intercrops are known to yield more because of legume N_2 fixation. However, in a P-deficient intercropping system, P can be mobilized by legume species because legume crops can acidify the rhizosphere through proton release in association with P deficiency and N fixation (Zhang *et al.*, 2010).

The seed yield of intercrop soybean was 8.7% less than sole soybean whereas the grain yield of intercrop sorghum was 9.5% more than that of sole sorghum. However, the productivity in terms of soybean equivalent yield (SEY) was relatively high in intercropping system. The increasing NPK dose from 0 to 100% significantly improved SEY in sole sorghum and soybean + sorghum intercropping system and the integrated use of organics and inorganics recorded significantly more SEY than inorganics , which indirectly indicated improvement in nutrient use efficiency in intercropping system than sole cropping (Ghosh *et al.*, 2004). Most studies on intercropping have focused on the legume-cereal intercropping, a productive and sustainable system, its resource utilization (water, light, nutrients), and its effect on N input from symbiotic nitrogen fixation into the cropping system and reduction of negative impacts on the environment. The 53% increase of grain yield in wheat intercropped with soybean was due to a 30 and 23% above- and below-ground interaction between the two species, respectively. The contributions of above-ground and below-ground interactions to the increase of N uptake by wheat were 23 and 19%, respectively, for wheat/ soybean intercropping. Similarly, the contribution of above-ground and below-ground interactions to the increase of P uptake was 26 and 28%, respectively, for wheat/soybean intercropping (Zhang *et al.*, 2001).

Rao *et al.* (1999) found that acquisition of P by the legume was markedly greater than that by the grass, regardless of the P form being inorganic or organic. In wheat/soybean intercropping systems, a significant yield increase of intercropped wheat over sole wheat was observed, which resulted from positive effects of the border row and inner rows of intercropped wheat. The border row effect was due to interspecific competition for nutrients as wheat had a higher competitive ability than soybean. There was also compensatory growth, or a recovery process, of subordinate species such as soybean, offsetting the impairment of early growth of the subordinate species. Finally, both dominant and subordinate species in intercropping obtain higher yields than that in corresponding sole wheat and soybean. The processes summarized as the 'competition-recovery production principle'. It was observed that interspecific facilitation, where maize improves iron nutrition in intercropped peanut, faba bean enhances N and P uptake by intercropped maize, and chickpea facilitates P uptake by associated wheat from phytate-P. However, iron deficiency chlorosis frequently occurs on calcareous soils. Interestingly, iron deficiency chlorosis of the crop is more severe in sole cropping systems than in the intercropping system on these soils.

Intercropping could reduce the nitrate accumulation in soil profiles. Nitrogen uptake by intercropping of wheat and maize was greater than that by corresponding sole cropping under same N supply. Furthermore, intercropping

Table 34: Integrated nutrient management in soybean based intercropping systems in different zones across India (2002-04)

Treatment	Central zone			Southern zone			North plain and North eastern zone		
	Yield (kg/ha)		PFP (kg/kg)	Yield (kg/ha)		PFP (kg/kg)	Yield (kg/ha)		PFP (kg/kg)
	Soy	P.pea		Soy	P.pea		Soy	Maize	
RDF	1278	1098	23.76	1358	746	21.04	1027	2500	35.27
75% RDF	1203	1031	30.45	1271	762	27.10	953	2267	42.93
50% RDF	1050	923	39.46	1159	706	37.30	891	1931	56.44
FYM + RDF	1394	1197	25.91	1436	823	22.59	1066	2610	36.76
FYM + 75% RDF	1377	1144	33.61	1373	764	28.49	900	2424	44.32
FYM + 50% RDF	1184	1058	44.84	1371	767	42.76	892	2062	59.08
FYM + Zn**	1135	987	424.40	1323	678	400.20	952	2162	622.80
FYM + Zn + RDF	1475	1315	26.57	1477	835	22.01	1137	2747	36.51
FYM + Zn + 75% RDF	1456	1217	33.41	1477	816	28.66	927	2665	44.90
FYM + Zn + 50% RDF	1261	1097	42.87	1350	743	38.05	891	2255	57.20
Zn + RDF	1328	1192	24.00	1360	734	19.94	1213	2628	36.58
Zn + 75% RDF	1301	1116	30.21	1307	737	25.55	1098	2491	44.87
Zn + 50% RDF	1129	1006	38.81	1241	694	35.18	977	2099	55.93
Sole soybean	1684	-	16.84	1858	-	18.58	2220	-	22.20
Sole pigeon pea	-	1802	18.02	-	1242	12.42	-	4277	13.80
CD (P=0.05)	35.0	42.8		66.8	58.4		140.9	233.7	

*FYM @ 5 t/ha, ** Zn @ 5 kg/ha

reduced the nitrate content in the soil profile as intercropping uses soil nutrients more efficiently than sole cropping.

Interspecific root interactions between intercropped legume and cereal maize played an important role in the yield advantage and N and P acquisition by the intercropping system (Li *et al.*, 1999). Possible mechanisms for interspecific facilitation are: (1) the phosphorus that is mobilized by bean may have been made available for maize; (2) nitrogen that was fixed by legume may have been transferred to cereal.

The INM resulted in higher yields for maize/soybean mixture compared to other inter-croppings (Muyayabantu *et al.*, 2013).The cassava nitrogen use efficiency increased with increase in applied nitrogen up to 60 kg N/ha and then decreased beyond this point with or without soybean intercropping. The greater NUE was recorded when cassava was intercropped with soybean as compared to sole cassava (Umeh *et al.*, 2012). A multi-location (10) and years (3) data accrued from soybean based intercropping experiment revealed that the intercropping of soybean + maize or pigeon pea were more nutrient use efficient than their respective sole crops (Table 35). The conjunctive use of nutrients and manure increased the nutrient use efficiency in all the zones. Soybean + maze intercropping was to be more efficient than soybean + pigeon pea intercropping.

Weed management

Competition among crops and weeds has been recognized since biblical times. Weeds compete with crops for water, light, nutrients and space. Because of this competition between weeds and crops, yields are reduced and consequently, NUE are also reduced. The weeds are more efficient to extract nutrients from soil than soybean crop.

Adequate mineral nutrition is essential for the growth and development of plants. When the essential elements are missing or when there is competition between plants for a particular element, the fixation of other elements can also be affected. The competition of competing plants by sources of nitrogen and other minerals in the soil depends on its specific ability to capture these sources (root architecture and absorption properties of root tissue). The high extraction capacity of soil nutrients by plants is an important factor in the delimitation of competitive parameters. In this sense, it was found that increasing plant density (increased competition) caused a decline in the absolute concentration of N, P and K in leaves, stems and vegetables in soybean. Ronchi *et al.* (2003) noted that *Bidens pilosa* accumulates 5.53, 11.19 and 5.32 times more nitrogen, phosphorus and potassium, respectively compared to the crop, with maximum accumulation of dry matter mass at the pre-flowering stage. For soybean, Pitelli *et al.* (1983) also found increased intake of phosphorus and potassium on dry matter of *Cyperus rotundus* in relation to the crop, which demonstrates the high potential of these species in the uptake of soil nutrients.

Soybean yield up to 80% is lost due to weed competition in many parts of the world. In the case of weeds, water and nutrients extraction reduce the availability of these resources for the target culture, which causes stress and ultimately reduces the growth of both and also the yield of the crop. Competition for water and, consequently, the effects of its stress are undoubtedly factors that also contribute to lowering the productivity of crops, as the occurrence of droughts become more and more frequent.

The weeds can reduce the soybean yield to the tune of 62.7 to 1296.4%, the magnitude of reduction depends on the density, type of weed and time of occurrence/presence (Table 35). The efficient management of weeds may double the fertilizer PFP in all the zones.

Table 35: Effect of weed management on fertilizer partial factor productivity in soybean

Treatment	North plain zone		North eastern zone		Central zone		Southern zone	
	Seed yield (kg/ha)	PFP (kg/kg)	Seed yield (kg/ha)	PFP (kg/kg)	Seed yield (kg/ha)	PFP (kg/kg)	Seed yield (kg/ha)	PFP (kg/kg)
Protected	1417	10.90	2081	11.56	1528	10.91	1655	11.03
Weedy Check	753	5.79	1232	6.84	675	4.82	1017	6.78
CD (P=0.05)	63.97		226		143.79		201	

Source: AICRPS- Annual report 2012-13.

Selected References

Adams F.1984. Soil Acidity and Liming. Agronomy 12. Am. Soc. Agron., Madison, WI.

Adesemoye AO, Torbert HA , Kloepper JW. 2008. Enhanced plantnutrient use efficiency with PGPR and AMF in an integratednutrient management system. *Canadian Journal of Microbiology* 54:876–886.

Bakker PAHM, Pieterse CMJ, van Loon LC. 2007. Induced systemicresistance by fluorescent *Pseudomonas* spp. *Phytopathology* 97:239–243.

Baligar VC , Ahlrichs J. 1998. Nature and distribution of acid soils in theworld. Pp. 1–11. *In*: Schaffert RE. Proc. of workshop: *Develop and Strategy for Collaborative Research and Dissemination of Technology*. PurdueUniversity,W. Lafayette, IN.

Baligar VC , Bennet OL.1986b. NPK-fertilizer efficiency- a situation analysis for the tropics. *Fertiliser Research* 10:147-164.

Baligar VC , Bennett OL. 1986a. Outlook on fertilizer use efficiency in thetropics. *Fertilizers Research*10:83–96.

Baligar VC and Duncan RR. 1990. Crops as Enhancers of Nutrient Use. Academic Press, San Diego, CA.

Baligar VC , Fageria NK. 1997. Nutrient use efficiency in acid soils: nutrient management and plant use efficiency pp. 75–93. In: Monitz AC,Furlani AMC, Fageria NK, Rosolem CA, and Cantarells H. (eds.),*Plant-Soil Interactions at Low pH: Sustainable Agriculture and ForestryProduction*. Brazilian Soil Science Society Compinas, Brazil.

Banerjee MR, Yesmin L , Vessey JK. 2006. Plant growth-promoting rhizobacteria as biofertilizers and biopesticide. *In:* Rai MK (ed). *Hand book of microbial biofertilizers*. Food Products Press, NewYork, pp 137–181.

Barber SA. 1995. Soil nutrient bioavailability: A mechanistic approach. 2nd ed. John Wiley and Sons, Inc., New York.

Barlog P, Grzebisz W. 2004. Effect of timing and nitrogen fertilizerapplication on winter oilseed rape. II.Nitrogen uptake dynamics and fertilizer efficiency. *Journal of Agronomy and Crop Science,* 190:314–323.

Bashan Y, Holguin G, de-Bashan LE.2004. Azospirillum-plantrelationships: physiological, molecular, agricultural and environmentaladvances (1997–2003). *Canadian Journal of Microbiology* 50:521–577.

Bennie AT. 1996. Growth and mechanical impedance. pp. 453– 470. *In:*Waisel Y, Eshel A and Kafkafi U. (eds.), *Plant roots—The Hidden Half.* Marcel Dekker, New York, NY.

Billore S D , Joshi O P. 2004. Production potential and economic feasibility of soybean based cropping systems. *Journal of Oilseeds Research*, 21(1): 55-7.

Billore SD , Joshi OP. 2005. Direct and residual effect of integrated micronutrient application in soybean –wheat cropping system. *Indian Journal of agriculture Science* 75(9):566-8.

Billore SD , Srivastava SK. 2014. Effect of seed rate and row spacing on productivity of soybean varieties under different agro-climatic zones of India. *Soybean Research* (In press).

Billore SD , Srivastava SK. 2013. Sustainability and stability of yield of soybean varieties under various planting time in different agro-climatic regions of India. *Soybean Research* 11(2):8-16.

Billore SD , Srivastava SK. 2014. Assessment of soybean varieties for nutrient tolerance. *Soybean Research* (in Press).

Billore SD , Srivastava SK. 2014. Soybean productivity and sustainability as affected by irrigation scheduling under different agro-ecological regions of India. *Soybean Research* (In press).

Billore SD, Ramesh A, Vyas AK , Joshi OP. 2009. Potassium use efficiencies and economic optimization as influenced by levels of potassium and soybean genotypes under staggered planting. *Indian Journal of agriculture Science* 79(7):510-14.

BiswasTD, Jain BL , Mandal SC. 1971. Cumulative effect of different levels of manuresonthe physical properties of soil. *Journal of Indian Society of Soil Science* 19:31-34.

Board JE, Harville BG , Saxton AM. 1990. Narrow-row seed-yield enhancement in determinate soybean. *Agronomy Journal* 82:64–68.

Board JE, Zhang W , Harville BG. 1996. Yield rankings for soybean cultivars grown in narrow and wide rows with late planting dates. *Agronomy Journal* 88:240–245.

Burke, IC, Yonker CM, Parton WJ, Cole CV, Flach K, Schimel. 1989. Texture, climate and cultivation effects on soil organic matter content in US grassland soils. *Soil Science Society of American Journal* 53:800-805.

Camp CR , Lund ZF. 1964. Effect of soil compaction on cotton roots. *Crops and Soils Magazine* 17(2): 13 15.

Cassman KG, Dobermann A, Walters DT, Yang H. 2003. Meeting cereal demand while protecting natural resources and improving environmental quality. *Annual Review of Environmental Resources* 28:315–358.

Chen XP, Cui ZL and Vitousek PM. 2011. Integrated soil–crop system management for food security. *Proc. Natl. Academy of Sciences, USA* 108:6399–6404.

Clark RB. 1970. Mineral deficiencies of corn. *Agrichem. Age.*13:4–8.

Clark RB. 1982. Plant response to mineral element toxicity and deficiency.pp. 71–73. *In:* Christiansen M N and Lewis CF (eds), *Breeding Plants forLess Favorable Environ*ments. John Wiley and Sons, New York.

Clark RB , Duncan RR. 1991. Improvement of plant mineral nutritionthrough breeding. *Field Crops Research* 27: 219–240.

Clarkson DT , Hanson JB. 1980. The mineral nutrition of higher plants. *Annual Review of Plant Physiology* 31:239-298.

Cooke GW. 1987. Maximizing fertilizer efficiency by overcoming constraints. *Journal of Plant Nutrition* 10:1357-1369.

Cope JT Jr. 1973. Use of a fertility index in soil test interpretation. *Communication sin soil science and Plant analysis* 4:137-146

Crasswell ET , Godwin DS. 1984. *The efficiency of nitrogen fertilizers applied to cereals in different climates.* Pp. 1-55. *In:* Tinker PB and Lauchli A (eds.). *Advances in plant nutrition.* 1. Praeger Scientific, New York.

Delgado JA, Sparks RT, Follett RF, Sharkoff JL, Riggenbach RR.1999. Use of winter cover crops to conserve soil and water quality in the San Luis Valley of South Central Colorado. pp 125–142.R. Lal (ed.) *SoilQuality and Soil Erosion.* CRC Press, Boca Raton, FL.

Diacono M , Montemurro F. 2010. Long-term effects of organic amendmentson soil fertility. *Agronomy for Sustainable Development* 30:401–422.

Dixit AK, Tomar DS, Saxena A, Kaushik SK. *2011.*Assessment of Potassium Nutrition in Soybean for Higher Sustainable Yield in Medium Black Soils of Central India. IPI, *e-ifc* No. 26, March.

Dudal R.1976. Inventory of the major soils of the world with special reference to mineral stress hazards. Pp. 3-14. In- Wright MJ. (ed.). *Plant adaptation to mineral stress in problem soils.* Cornell University, Ithaca, New York.

Edmeades DC. 2003. The long-term effects of manures and fertilizerson soil productivity and quality: a review. *Nutrient Cycling in Agroecosystems* 66:165–180.

Egli DB, Yu ZW. 1991. Crop growth rate and seeds per unit area in soybean. *Crop Science* 31:439–442.

Egli DB. 1997. Cultivar maturity and response of soybean to shade stress during seed filling. *Field Crops Res.* 52:1-8.

EgliDB. 2010. Soybean reproductive sink size and short-term reductions in photosynthesis during flowering and pod set. *Crop Science* 50: 1971-1977.

Epstein E. 1972. *Mineral Nutrition of Plants: Principles and Perspective.* Wiley Publisher, New York.

Fageria NK , Baligar VC. 1997a. Phosphorous—use efficiency by corngenotypes. *Journal of Plant Nutr.* 20: 1267–1277.

Fageria NK and Baligar VC. 1997b. Integrated plant nutrient managementfor sustainable crop production—An over. *International Journal of Tropical Agriculture* 15:7–18.

Fageria NK, Morais OP, Baligar VC , Wright RJ. 1988a. Response of rice cultivars to phosphorus supply on an Oxisol. *Fertilizer Research* 16:195-206.

Fageria NK, Wright RJ, Baligar VC. 1988b. Rice cultivar evaluation for phosphorus use efficiency. *Plant and Soil* 111:105-109.

Fageria NK,Baligar VC , Edwards DG. 1990. Soil-Plant nutrient relationshipsat low pH stress. pp. 475–507. *In:* Baligar VC and Duncan RR (eds.).*Crops as Enhancers of Nutrient Use.* Academic Press. San Diego, CA.

Fageria NK. 1992. *Maximizing crop yields*. Marcel Dekker, New York.

Fageria NK , Baligar VC. 1994. Screening crop genotypes for mineralstresses. pp. 152–159. *In:* Maranville JW, Baligar VC, DuncanRR and Yohe JM (eds.), *Adaptation of plants to soil stress*. Univ. Nebraska,INTSORMIL-USAID, Lincoln, NE.

Fixen PE. 2006. Turning challenges into opportunities. *In:* Proceedings of the Fluid Forum, Fluids: Balancing Fertility and Economics. Fluid Fertilizer Foundation, February 12-14, Scottsdale, Arizona.

Foulkes MJ, Reynolds MP , Sylvester-Bradley R. 2009. Genetic improvement of grain crops: yield potential. *In:Crop Physiology: Applications for Genetic Improvement and Agronomy.* Sadras VO and Calderini DF (Eds.), pp. 355-385. Elsevier, Burlington, MA.

Frink CR, Waggoner PE , Ausubel JH. 1999. Nitrogen fertilizer:retrospect and prospect. *Proceedings of National Academy of Science* 96:1175–1180.

Gerloff GC, Gabelman WH. 1983. Genetic basis of inorganic plant nutrition.pp 453– 480. *In*: Lauchli A and Bieleski RL (eds.), *Inorganic Plant Nutrition*. Encyclopedia and Plant Physiology New Series, Volume 15B.Springer Verlag, New York.

Ghosh PK, Ramesh P, Bandyopadhyay KK, Tripathi AK, Hati KM, Misra AK, Acharya CL. 2004.Comparative effectiveness of cattle manure, poultry manure, phosphocompost and fertilizer-NPK on three cropping systems in vertisols of semi-arid tropics. I. Crop yields and system performance. *Bioresource Technology* 95:77–83.

Gill MS, Ahlawat IPS. 2006. Crop diversification –its role towards sustainability and profitability. *Indian Journal of Fertilisers* 2(9): 125–150.

Giovannetti M, Avio L, Fortuna P, Pellegrino E, Sbrana C, Strani P. 2006. At the root of the wood wide web. Self recognition andnonself incompatibility in mycorrhizal networks. *Plant Signal Behaviour* 1:1–5.

Girothia OP. 2013. Direct and residual effect of sulphur levels on yield, quality and S-use efficiency of soybean – Gram cropping system. Ph.D. Thesis submitted to Mahatma Gandhi Chitrakoot Gramodaya Vishwavidyalaya, Chitrakoot, Satna.

Graham RD. 1984. *Breeding for nutritional characteristics in cereals*. Pp. 57-102. *In*- Tinker PB and Lauchli A. (eds.). *Advances in plant nutrition*, 1. Praeger Scientific, New York.

Gyaneshwar P, Kumar GN, Parekh LJ, Poole PS. 2002. Role of soil microorganisms in improving P nutrition of plants. *Plant and Soil* 245:83–93.

Hashemi JM. 2001. Sowing date on the developmental stages and some agronomic and physiological characteristics of five soybean cultivars grown in the second planting. *Crop Science Journal* 3(4): 49-59.

Hati KM, Mandal KG, Misra AK. 2000.Effect of nutrient management on water use and yield of rainfed soybean grown on cracking claysoil. *Environmental Ecology* 18(1): 202-206.

Hati KM, Swarup A, Dwivedi AK, Misra AK, Bandyopahhyay KK. 2007. Changes in soil physical properties and organic carbon status at the topsoil horizon of a vertisol of central India after 28 years of continuous cropping, fertilization and manuring. *Agriculture Ecosystem and Environment.* 119:127–134.

Herridge DF, Roughley RJ, Brockwell J. 1984. Effect of rhizobia and soil nitrate on the establishment and functioning of the soybean symbiosis in the field. *Australian Journal of Agriculture Research* 35:149-161.

Hungaria M, Franchim J, Campo R , Graham P. 2005.The importance of nitrogen fixation to soybean cropping in South America. In: Werner D and Newton W (Eds.). *Nitrogen fixation agriculture, forestory, ecology and the environment,* Springer, Netherlands, pp. 25-42.

Hungaria M, Franchim J, Campo R, Crispino CC, Moraes JZ, Stbaldell RNR, Mendes IC , Arihara J. 2006. Nitrogen nutrition of soybean in Brazil: Contributions of biological N2 fixation and N fertilizer to grain yield. *Canadian Journal of Plant Science* 86:927-939.

Ismail I, Blevins RL, Frye WW. 1994. Long-term no tillage effects on soilproperties and continuous corn yields. *Soil Science Society of American Journal* 58:193–198.

Israel DW, Rufty TW Jr. 1988. Influence of phosphorus nutrition on phosphorus and nitrogen utilization efficiencies and associated physiological responses in soybean. *Crop Science* 28:954-960.

Jagadeeswaran R, Murugappan V , Govindasamy M 2005. Effect of slow release NPK fertilizer sourceson the nutrient use efficiency in turmeric. *World Journal of Agriculture Science*. 1 (1): 65-69.

Jarecki MK, Parkin TB, Chan ASK, Hatfield JL , Jones R. 2008.Greenhouse gas emissions from two soils receiving nitrogenfertilizer and swine manure slurry. *Journal of Environmental Quality* 37:1432–1438.

Jiang H , Egli DB. 1995. Soybean seed number and crop growth rate during flowering. *Agronomy Journal* 87:264-267.

Joshi OP , Billore SD. 2004. Fertilizer management in soybean –wheat cropping system. *Indain Journal of Agricultural Science* 74(8): 430-2.

Kafkafi U , Bernstein N. 1996. Root growth under salinity stress. pp. 435–451. In Y.Waisel, A. Eshel, and U. Kafkafi (eds.), *Plant Roots-The Hidden Half*. Marcel Dekker Inc. New York.

Kapland DI , Estes GO. 1985. Organic matter relationships to soil nutrient status and aluminum toxicities in alfalfa. *Agronomy Journal J.*77:735-738.

Kazemi SH, Ghaloshi S, Ghanbari A , Kianoosh GH.2005. Effects of planting date and seed inoculated with bacteria on yield and yield components of two soybean cultivars. *Journal of Agriculture and Natural Resources Sciences* Twelfth year.

Kessel Cvan , Hartley C. 2000. Agricultural man- agemen to f grain legumes: has it led to an increase in nitrogen fixation? *Field Crops Research* 65:165–181.

Ladha JK, Kundu DK, Coppenolle Angelo- van MG, Peoples MB, Carangal VR , Dart PJ. 1996. Legume productivity and soil nitrogen dynamicsin lowland rice-based cropping systems. *Soil Science Society of American Journal* 60:183–192.

Ladha JK, Pathak H, Krupnik IJ , Kessel C Van. 2005. Efficiency of fertilizer nitrogen in cereal production: retrospects and prospects. *Advances in Agronomy* 87:86-156.

Li L, Yang SC, Li XL, Zhang FS, Christie P. 1999. Interspecific complementary and competitive interaction between intercropped maize and faba bean. *Plant and Soil* 212:105–114.

Liu A, Hamel C, Hamilton RI, Ma BL , Smith DL. 2000. Acquisitionof Cu, Zn, Mn and Fe by mycorrhizal maize grown in soil at different P and micronutrient levels. *Mycorrhiza* 9:331–336.

Mahboubi AA, Lal R , Faussey NR.1993. Twenty-eight years of tillage effectson two soils in Ohio. *Soil Science Society of American Journal* 57:506 –512.

Mahmood S, Finlay RD, Erland S , Wallander H. 2001. Solubilisationand colonisation of wood ash by ectomycorrhizal fungi isolatedfrom a wood ash fertilized spruce forest. *FEMS Microbiology Ecology* 35:151–161.

Makokha S, Kimani S, Mwangi W, Verkuijl H , Musembi F. 2001. Determinants of fertilizer and manure use for maize production in Kiambu District, Kenya. D.F.: International Maize and Wheat Improvement Center (CIMMYT) and Kenya Agricultural Research Institute (KARI).

Mandal KG, Misra AK , Hati KM. 2000. Effectof combination of NPK and FYM on growth, yieldand agronomic efficiency of soybean in Vertisol. *Environmental Ecology*, 18(1):207-209.

Mengal K , Kirkby EA. 1982. *Principles of plant nutrition*, 3rded.Intl. Potash Institute, Berne, Switzerland.

Mosavi SB, Jafarzadeh AA, Neishabori MR , Ostan SH. 2009. Application of rye green manure in wheat rotation system alters soil water content and chemical characteristics under dryland condition in Maragheh. *Pakistan Journal of Biological Science*12:178-182.

Moser M , Wandter KH. 1983. Ecophysiology of mycorrhizal symbioses. Pp. 391-421. *In:* Lange OL, Nobel PS, Osmand CB and Ziegler M. (eds.). *Physiological plant ecology*, vol.3. *Response to the chemical and biological environment*. Springer Verlag, New Yok.

Muyayabantu GM, Kadiata BD , Nkongolo K K. 2013.Assessing the effects of integrated soil fertility management on biological efficiency and economic advantages of intercropped maize and soybean in DR Congo. *American Journal of Experimental Agriculture* 3(3): 520-541.

Peoples MB, Freney JR , Mosier AR. 1995 Minimizing gaseous lossesof nitrogen. pp. 565–602. *In:* Bacon PE (ed.), *Nitrogen fertilization in theenvironment*. Marcel Dekker, Inc. NY.

Pitelli RA, Durigan JC , Benedetti NJ. 1983. Interspecific and intraspecific competition of *Glycine max* and *Cyperusrotundus* in greenhouse conditions. *Planta Daninha* 6:129-137.

Prasad, R. 1996. Nitrogen use efficiency. In: *Nitrogen Research and Crop Production*. Tandon, H.L.S. (ed.).Fertilizer Development and Consultation Organization, New Delhi, pp. 104–115.

Prasad R, Power JF. 1995. Nitrification inhibitors for agriculture, health andenvironment. *Advances in Agronomy* 54:233–28.

Purcell LC , Specht JE. 2004. Physiological traits for ameliorating drought stress. *In:Soybeans: Improvement, production, and uses (3rd ed.),* Boerma HR and Specht JE (Eds.), Am. Soc. Agron. Madison, WI.

Purcell LC, Serajj R, Sinclair TR , De A. 2004. Soybean N2 fixation estimates, uride concentration and yield response to drought. *Crop Science* 44:484-492.

Radhika, Hemalatha S, Maragatham S, Praveena Kathrine S. 2013. Placement of Nutrients in Soil: A Review. *Research and Reviews: Journal of Agriculture and Allied Science* 2(2):12-19.

Rao IM, Borrero V, Ricaurte J , Garcia R. 1999. Adaptive attributes of tropical forage species to acid soils. V. Differences in phosphorus acquisition from less available in organic and organic sources of phosphate. *Plant Nutrition* 22:1175–1196.

Ronchi CP, Terra AA, Silva AA , Ferreira LR. 2003. Nutrient contents of coffee plants under weed interference, *Planta Daninha* 21(2):217-227.

Salvagiotti F, Cassman KG, Specht JE, Walters DT, Weiss A and Dobermann A. 2008. Nitrogen uptake, fixation and response to fertilizer N. in soybeans: A review. *Field Crops Research* 3(1):1-19.

Samra JS, Swarup A. 2005. Impact of long-term intensive cropping on soil potassium and sustainability of crop production. www.ipipotash.org/. pp. 167-183.

Sandhu KS, Benbi DK, Prihar SS , Saggar S. 1992. Dryland wheat yield dependence on rainfall, applied N and mulching in preceding maize. *Nutrient Cycling in Agroecosystems* 32:229-237.

Sattelmacher B, Horst WJ , Becker HC. 1994. Factors that contribute togeneticvariation for nutrient efficiency of crop plants. *Journal of Plant Nutrition and Soil Science* 157: 215–224.

Shafigh MMH, Rashid M , Nassiri M. 2006. The effect of cattle cotton on yield and yield components of different soybean plant density and different planting dates. *Iranian Journal of Agricultural Research* 4(1): 71-81.

Sharpley AN, Weld JL, Beegle DB, Kleiman PJA, Gburek WJ, Moore PA Jr, Mullins G. 2003.Development of phosphorus indices for nutrient management planning strategies in the United States. *Journal of Soil and Water Conservation* 58:137–152.

Shen Jianbo, Li Chunjian, Mi Guohua, Li Long, Yuan Lixing, Jiang Rongfeng , Zhang Fusuo. 2012. Maximizing root/rhizosphere efficiency to improve crop productivity and nutrient use efficiency in intensive agriculture of China. *Journal of Experimental Botany* 342:2-12.

Singh Muneshwar, Tripathi AK, Kundu S, Takkar PN , Singh M. 1999. Nitrogen requirement ofsoybean - wheat cropping system and biological N fixation asinfluenced by integrated use of fertilizer Nandfarmyard manure in Typic Haplustert. *Indian Journal of Agriculture Science* 69(5): 379-381.

Singh RB. 2002. The state of food and agriculture in Asia and the Pacific: challenges and opportunities. UN-FAO and IFA, Rome.

Smith S E, Roson AD , Abbott LK. 1993. The involvement of mycorrhizasin assessment of genetically dependent efficiency of nutrient uptakeand use. pp. 221–231. *In:* Randall P J, Delhaize E, Richards RA and Munns R (eds.), *Genetic aspects of plant mineral nutrition.* Kluwer Academic Pubisher. The Netherlands.

Sousa DMG, Lobato E, Ritchey KD , Rein TA. 1992. Response ofannual crops and leucaena to gypsum in the Cerrado. *In:* 2nd Seminar on the use of gypsum in agriculture. 2:277–306. IBRAFOS, Brazil.

Specht JE, Hume DJ , Kumudini SV. 1999. Soybean yield potential—agenetic and physiological perspective.*Crop Science* 39:1560–1570.

Steinshamn H, Thuen E, Bleken MA, Brenoe UT, Ekerholt G , Yri C. 2004. Utilization of nitrogen and phosphorus in anorganic dairy farming system in Norway. *Agriculture Ecosystems and Environment* 104:509–522.

Swarup A 1998. Emerging soil fertility managementissues for sustainable crop production in irrigatedsystem. *In:* Swarup A, Reddy DD and PrasadRN (eds.), *Long-term soil fertility management through integrated plant nutrient supply*. pp. 54-68. Indian Institute of Soil Science, Bhopal.

Tawaraya K, Naito M, Wagatsuma T. 2006. Solubilization of insoluble inorganic phosphate by hyphal exudates of arbuscularmycorrhizal fungi. *Journal of Plant Nutrition* 29:657–665.

Thurlow DL, Hiltbolt AE. 1985. Dinitrogen fixation by soybeanin albama. *Agronomy Journal* 77:432-436.

Tranel PJ, Riggins CW, Bell MS, Hager AG. 2011. Herbicide resistances in *Amaranthus tuberculatus*: A call for new options. *Journal of Agriculture and Food Chemistry* 59:5808–5812.

Umeh, SI, Eze SC, Eze EI, Ameh GI. 2012. Nitrogen fertilization and use efficiency in an intercrop system of cassava and soybean. *Journal of Biotechnology* 11(41):9753-9757.

Varvel GE.1994. Monoculture and rotation system effects on precipitation use efficiency in corn. *Agronomy Journal* 86:204–208.

Vidhyavathi, Dasog GS, Babalad HB, Hebsur NS, Gali SK, Patil SG, Alagawadi AR. 2011. Influence of nutrient management practices on crop response and economics in differentcropping systems in a vertisol. *Karnataka Journal of Agricultural Science* 24 (4):455-460.

Von Uexkull HR. 1986. Efficient fertilizer use in acid upland soils of the HumidTropics. FAO and *Plant Nutrition Bull.*10. FAO Rome Italy.

Vyas AK, Billore SD, Chauhan, GS , Pandya N. 2008. Agro-economic analysis of soybean – based cropping systems. *Indian Journal of Fertilizer* 4(5):41-51.

Vyas AK, Billore SD, Ramesh A, Joshi OP, Gupta GK, Sharma AN, Pachlania NK, Pandya N, Imas P. 2008. Role of potassium in balanced fertilization of soybean-wheat cropping system. *In*: Proc. of Regional seminar on “Recent advances in potassium nutrition management for soybean based cropping systems” (eds. Vyas AK and Imas Patricia) pp.49-53.

Wani SP, Rupela OP , Lee KK. 1995. Sustainable agriculture in the semi-arid tropics through biological nitrogen-fixation ingrain legumes. *Plant and Soil* 174: 29-49.

Weil RR, Beneditto PW, Sikora LJ and Bandel VA. 1988. Influence of tillage practices on phosphorus distribution and forms in three Ultisols. *Agronomy Journal* 80:503-509.

Wilcox JR. 2001. Sixty years of improvement in publicly developed elite soybean lines. *Crop Science* 41:1711-1716.

Wilcox JR. 2004.World distributionand tradeof soybean.*In:*Boerma HR and Specht JE (eds), *Soybeans:improvement, productionand uses.* ASA,CSSA,ASSA, pp.1–13.

Yadav RL. 2003. Assessing on-farm efficiency and economics of fertilizer N, P and K in ricewheat systems of India. *Field Crops Research* 18: 39-51.

Zhang FS, Li L , Sun JH. 2001. Contribution of above and belowground interactions to intercropping. *In:* Plant Nutrition – Food Security and Sustainability ofAgro-ecosystems. Ed. Horst *et al.* pp. 979–980. KluwerAcademic Publishers, Dordrecht, The Netherlands.

Zhang FS, Shen JB, Zhang JL, Zuo YM, Li L , Chen XP. 2010. Rhizosphere processes and management for improving nutrient use efficiency and crop productivity: implications for China. *In*: DL Sparks, ed, *Advances in Agronomy* 107: 1–32. Academic Press, San Diego.

Zhang H, Wang B, Xu M , Fan T. 2009a. Crop yield and soil responsesto long-term fertilization on a red soil in southern China. *Pedosphere* 19:199–207.

Zhao B, Li XY, Li XP, Shi XJ, Huang SM, Wang BR, Zhu P,Yang XY, Liu H, Chen Y, Poulton P, Powlson D, Todd A, Payne R. 2010. Long-term fertilizer experiment network in China: cropyields and soil nutrient trends. *Agronomy Journal* 102:216–230.

23

Enhancing Nutrient Use Efficiency, pp. 343-361
Editors: K. Ramesh, A.K. Biswas, B.L. Lakaria, S. Srivastava and A.K. Patra

Sugarcane Based Cropping Systems

Nitin Gudadhe[1], J.D. Thanki[1], Kulsekaran Ramesh[2], P.S. Bodake[3] and S.R. Imade[1]

[1]Navsari Agricultural University, Navsari – 39645, India
[2]ICAR-Indian Institute of Soil Science, Bhopal – 462038, India
[3]Zonal Agricultural Research Station, Igatpuri, M.P.K.V., Rahuri, India

Introduction

Sugarcane is an important cash crop of India. Since 1930-31 the area of sugarcane in the country has increased by about 4.3 times and production by about 9.8 times because of the rise in productivity of sugarcane by about 2.28 times, however, during the last decade there is no perceptible increase in productivity of sugarcane in the country. The domestic demand of sugar is around 22-23 million tonnes annually whereas the production of sugar in India is about 26 million tonnes annually. Data on nitrogen use efficiency of sugarcane from 1998-99 to 2013-14 in India is depicted in Fig. 1 and 2 show a decreasing trend (Anonymous, 2015). However, the escalating price of fertilizers in the market has become a cause of concern to sustain the productivity of the crop.

Of late, it is seen that yields of many crops are plateauing due to imbalanced fertilization and a decrease in soil organic matter. Over-application of fertilizers will not only affect the profitability of cane operations, but also the loss of applied nutrients such as nitrogen through surface runoff and leaching, will impact adversely on aquatic environments and groundwater quality. Nitrogen applied to sugarcane is eroded (up to 92%) from plant and soil system is the prime reason for poor nitrogen fertilizer efficiency. Also eutrophication of the aquatic ecosystem and groundwater pollution is seen due to excessive application of phosphorus and nitrogen. The future food production has to be much more efficient as we have the cases of nitrogen and phosphorus use efficiency 20-50 and 10-25 percent respectively in the current era. (Shaviv, 2000; Chinnamuthu and Boopathi, 2009). During nitrification, fertilizer N is lost due to volatilization of ammonia and also through erosion, leaching, denitrification and runoff, and the form and method of fertilizer application to the agroecosystem decides the relative importance of these processes (Chen, 2008). Loss of N through denitrification and volatilization is ultimately contributing to emission of greenhouse gas.

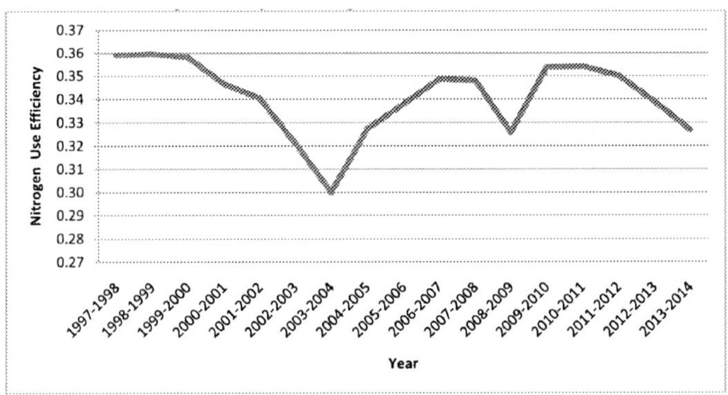

Fig. 1: Fertilizer N use efficiency (tonnes cane/kg N) for sugarcane production in India, 1997–2014.

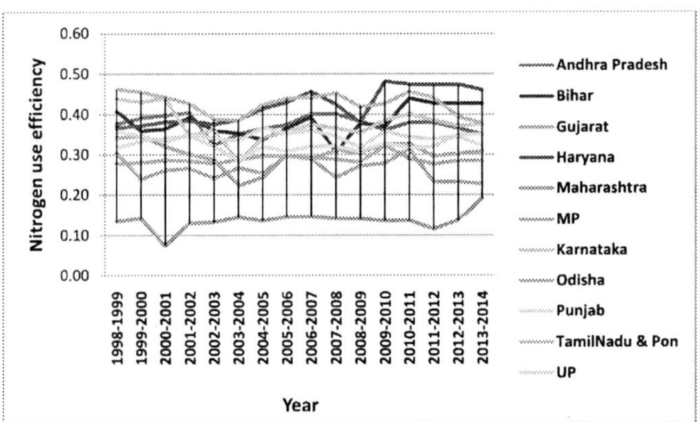

Fig. 2: Fertilizer N use efficiency (tonnes cane/kg N) for sugarcane production in different states, 1998–2014.

Agronomists and field managers must understand the basic nutritional need of sugarcane to have optimum balance between crop nutrient requirement, nutrient supplying capacity of soil and management of fertilizer in terms of timing and placement of fertilizer application, without altering original establishment of environment and soil fertility (Meyer, 2011).

In this chapter, promising agronomic measures are outlined to enhance nutrient use efficiency of sugarcane based cropping systems.

Nutrient requirement of sugarcane

Water, mineral matter of a wide range of elements and organic materials are essentially responsible for the composition of sugarcane. Carbon, hydrogen and oxygen are the three structural elements which comprise mostly (95%) the fresh biomass of the sugarcane crop and it is coming mainly from water and air. For the reproductive cycle and good growth of sugarcane 13 elements are said to be necessary and which are contributing as remaining 5% mineral matter composition of sugarcane. N, P and K contributes as a primary macronutrient and calcium, magnesium and sulphur contributes as a secondary macronutrient while zinc, copper, iron, manganese, boron, chlorine and molybdenum contribute as a micronutrient. In the recent past, silicon has gained attention and it is the 14th important element in sugarcane mineral matter. Silicon is largely accumulated in sugarcane hence it is considered as a beneficial element through its functional role, although it is not an essential element for sugarcane plants vegetative and reproductive growth (Kingston, 1999; Meyer and Keeping, 1999; Savant *et al.*, 1999; Berthelsen *et al.*, 2001).

Nutrient removal by sugarcane

The macronutrients N and potassium are rapidly depleted by the heavy feeder sugarcane crop. Wide range variations are seen in nutrient use efficiency of sugarcane because of varying climatic conditions, cultivars availability, fertilizer management practices and soil fertility. Uptake of nutrients through sugarcane are more under high temperature conditions with irrigation supply, however sugarcane uptake is lower under rainfed conditions with cool climate (Meyer, 2011). Sugarcane can remove 208 kg N, 53 kg P_2O_5, 280 kg K_2O, 30 kg S, 3.4 kg Fe, 1.2 kg Mn and 0.6 kg Cu from soil if it is yielding 100 t/ha. Hence to attend the sustainable productivity of sugarcane the soil has to be first nourished with sufficient nutrients (Anonymous, 2013).

Strategies for enhancing NUE in sugarcane

Screening of nutrient efficient cultivars

Varietal selection and cane breeding are the tools through which potentially high N use efficient sugarcane varieties can be developed. Location wise sugarcane varieties have varied acknowledgement to applied N. Hence, efficient cultivars for nutrients should always be selected for sugarcane planting. If the cultivars has the ability to incorporate and use the applied nutrients for the production of high economic biomass of sugarcane then the cultivar can be called as nutritionally efficient (Blair, 1993). Hence the plantation of nutritionally less efficient cultivars with high application of nutrients may increase the cost of cultivation and ultimately lead to profitability losses. Sundara (1994) has studied varietal differences in P absorption and utilization to select P efficient varieties which can perform well under conditions of low soil phosphorus availability and at low P application rates.

Schumann *et al.* (1998) found significant varietal differences in internal N use efficiency (g sucrose/g accumulated N) in the ratoon cane at the first N increment in seven commercial South African varieties (NCo376, N12, N14, N16, N19, N24, N25). Later, classification into three categories depending on the sucrose yield to N accumulation ratio was made by Schumann (2000) viz., (i) efficient N use responders, (ii) inefficient non-responders and (iii) inefficient responders. Looking to the potential of the specific N efficient cultivar, recommendation of N fertilizer should be made, rather than multiple nutrient application and agronomic management recommendations on single famous location specific variety. Later, Kist (2015) has evaluated eight sugarcane cultivars in Brazil for N, P, K, Ca and Mg content. He found N, P, Ca and Mg elements had best utilization efficiency by the cultivar RB835486 in two cycles and turned out to be high yielding cultivar among all eight cultivars tested.

Laser levelling of the field

Laser levelling is a process of leveling the land surface from its average elevation using laser-equipped drag buckets. Precision land levelling with laser leveller is the basic requirement for uniform crop stand. Uniform application of irrigation water gets affected due to unevenness in soil surface which ultimately affects the total plant population. This inefficiency of uniform distribution of moisture/ water and transmission losses can be corrected throughlaser leveling. Laser land levelling (LLL) improves the uniformity of sugarcane maturity and 3-5% cultivable area can also be increased. LLL not only improves nutrient use efficiency but also increases crop nutrient uptake. Jat *et al.* (2009) presented the on farm result of 71experiments conducted in western Uttar Pradesh through the intervention of LLL in rice-wheat cropping system which has improved the N use efficiency due to improvement in nutrient-water interactions. Sustainable sugarcane productivity and economic use of depleting water resources can be maintained through economically viable and ecofriendly options (Shrivastava *et al.*, 2011).

Planting pattern

Improper planting method and geometry is one of the reason for lower sugarcane yield among all agronomic practices. Appropriate production technology in relation to planting method and optimum row spacing of sugarcane need to be developed.

To overcome the problem of lower productivity, high density planting of sugarcane is one of the practices being used. Increased yield due to high plant population; efficient use of land, inputs, sunlight; lower weed infestation due to thick canopy; higher operational efficiencies like harvesting and trickle irrigation; lower harvesting losses and reduced soil compaction are its major advantages. Over single row plant, the dual row strip planting has its own benefits like 1. Close and thick crop canopy 2. Minimize weed competition and result in high yield 3. Better quality over single row planting. Hence even without use of any extra inputs the row spacing manipulation can be done provided recommended NPK are supplied to the crop. Dual row planting has an encouraging overall performance and it can be recommended for highest sugarcane productivity. If cane planting is shifted from March to September, the overall cane yield can be increased up to 30-32% (Sajjad *et al.*, 2014).

To accommodate more number of millable canes/unit area Indian Instittute of Sugarcane Research, Lucknow has developed another method of managing cane population where ring-pit method of planting was developed, in which nutrients are efficiently utilized as fertilizer is applied in the pits in the vicinity of the cane roots. In ring pit method agronomic efficiency and partial factor productivity were significantly greater over conventional flat method of planting. Ring pit method and conventional flat method of planting received 356 and 441 kg/ha of N respectively. Sugarcane crop requires only 2.34 kg N in ring pit method of planting over 8.48 kg in conventional flat method of planting to produce one tonne of cane yield (Yadav, 2004).

Use of organic amendments

A system need to be evolved in which there will be assured supply of organic manures with addition of chemical fertilizers to maintain high level sugarcane productivity on a sustainable long term basis. In many soils nutrient related stress is increasing and becoming a chronic problem of cane growing areas. With addition of biochar or boiler ash and use of urea + gypsum mixture at lowest rate @ 223 + 522 kg respectively was found equivalent to 100 kg N and 120 kg S by Nurhidayati and Basit (2015). Hence to optimize the productivity of sugarcane crop it is suggested to add organic source such as biochar and boiler ash as an organic soil amendment. Singh *et al.* (2007) tried the combination of various organic manures with and without *Gluconacetobacter diazotrophicus* (Gd) and they concluded that vermicompost + Gdhas registered highest cane yield (78.6 t ha^{-1}), however yield of ratoon was highest fromsulphitation press mud cake (SPMC) + Gd was 80.8 and 74.9 t ha^{-1}, respectively during first and second ratoon. SPMC + Gd recorded highest benefit cost ratio of 1.28, 2.36 and 2.03 respectively for plant and two ratoon crops with positive nutrient balance.

Farm waste can be used through recycling or use of FYM can be done to increase organic carbon pool of soil for intensive cultivation and production in the tropical climates of Indian sub continent. If green manure is raised as an intercrop in sugarcane, it can save up to 25% N requirement through chemical fertilizer. Biofertilizers like *Acetobacter*, *Azotobacter* and *Azospirillum* are N fixing bioagents and inoculation of same can improve soil residual N content and it can also bring 20-25% saving of N fertilizer. Phosphorus solubilisers like *Bacillus polymyxa*, *Bacillus megatherium*, *Aspergillus awamori*, *Pseudomonas striata etc.*, can improve phosphorus use efficiency. In sugarcane and sugarcane based cropping systems the productivity and soil fertility can be maintained sustainably through precise combination organic, inorganic and biofertilizers as a potential tool (Ramesh *et al.*, 2004).

Balanced nutrition

Nowadays due to non-application of other essential micronutrients in crops fertilizer schedule, farmers are experiencing declining response to major nutrients like N and phosphorus. The cane productivity and sugar yield can be enhanced through judicious and balanced use of all macro and micro nutrients. In a long term experiment of balanced nutrition by Singh *et al.* (2008) observed that application of N_{200} P_{100} $K_{150}S_{60}$ Mg_{30} kg ha^{-1} obtained highest cane yield (111.7 to 112.8 t ha^{-1}) and it was 80-83% higher than 200 kg N ha^{-1} alone. Significant yield reduction was observed by omission of P, K, S and Mg, shows the importance of these nutrients for replenishment and for reaching highest yield target. The range of increasing N use efficiency by application of P, K, S and Mg was 364-557 kg cane kg^{-1} nutrient application. Efficiency of individual nutrient was increased by addition of P_2O_5, K_2O, S and Mg was respectively in the range of 1,652 to 2,532, 692 to 906, 1615 to 1857 and 3687 to 3713 kg cane kg^{-1} application of these nutrients. Multi location trials in India have shown the significance of balanced use of fertilizer for increasing crop productivity and increasing N use efficiency (Ghosh *et al.*, 2015).

Use of AM and PGPR

Interest of farmers and researchers towards the use of Arbascular Mycorrhizae (AM) has been rejuvenated because of reduction in phosphorus reserves and increasing prices of phosphate fertilizers, AM can efficiently supply available phosphorus to plants in sugarcane production systems as uptake of P has direct relation for increasing the sugarcane juice quality. Uptake of essential nutrients, especially of phosphorus can be improved by colonisation of cane roots with fungi AM, this can also increase quality of soil. Without sacrifice of sugarcane yield and soil fertility sustainability in the medium P soils the addition P fertilizer

@ 75% + AM of 12.5 kg ha^{-1}can reduce P requirement up to 25% (Surendran and Vani, 2013).

Inoculation of *Gluconacetobacter diazotrophicus* with other diazotrophs or AM can enhance cane yield significantly as it has a N fixing ability and popularly known as plant growth promoting bacteria (PGPR) which is associated with sugarcane roots, shoots and leaves in high quantity (Cavalcante and Dobereiner, 1988). Uptake of N and nutrient use efficiency was increased when indigenous *G. diazotrophicus* isolates were inoculated in sugarcane by Suman *et al.* (2005).

In eight sugarcane cultivars inoculation of beneficial microbes with three levels of N were studied by Suman *et al.* (2008). Partial factor productivity (PFP) was more in all varieties at N dose 75 N ha^{-1} over 150 N ha^{-1} in terms of kg cane yield per kg application of N. Highest PFP of 68% was recorded on an average at 75 kg N ha^{-1} as compared to 150 kg N ha^{-1}. This study has shown that when N was applied at 75 kg ha^{-1} the population of *G. diazotrophicus* was increased but when the dose was increased up to 150 kg N ha^{-1}, there was no change in the population of these microbes. Hence it is concluded that in Indian subtropical climate the sugarcane varieties had more *G. diazotrophicus* population by application of N at 75 kg ha^{-1} over 150 kg N ha^{-1} and 0 kg N ha^{-1}. Association of microbes and plants indicated a positive role at low N application and which has registered high PFP and recovery efficiency. Sugarcane has shown positive correlation with N uptake, PFP, cane yield and other quality parameters at 75 kg N ha^{-1}. Hence Agronomic efficiency (AE) of sugarcane shown higher cane yield per kg N fertilizer application, as soon as sugarcane is established at lower dose of N application. When organic sources are integrated with PGPR it gives positive response towards cane yield. When combination of vermicompost + *Gluconacetobacter diazotrophicus* used in sugarcane it has given highest cane yield (78.6 t ha^{-1}), however in ratoon crop during first and second cycle recorded the yield of 80.8 and 74.9 t ha^{-1}, respectively by application of sulphitation press mud cake + *Gluconacetobacter diazotrophicus* (Singh *et al.*, 2007).

Nutrient recycling through cropping systems

Rotation of cereals with pulses can absorb more nitrates which are released by decaying roots and root nodules of legume crops, however incorporation of french bean, sunhemp *etc*. Kind of legumes as the intercrops have enhanced the N use efficiency (Shankaraiah *et al.,* 1999). Greengram intercropping has shown highest N use efficiency (198.2 kg cane kg^{-1} N) over intercropping of *Sesbania* (154.7 kg cane kg^{-1} N) in the cereal crops (Singh *et al.*, 2007).

Site specific nutrient management

a. Right place and method

It is always critical to determine the right placement of nutrient fertilizers for high nutrient use efficiency. Placement of fertilizer at 12 cm distance from the planting row registered significantly higher cane yield by Ahmed (2007). Mandal and Thakur during 2010 has developed a sub soiler cum differential rate fertilizer placement machine. This machine had choice to place the fertilizer upto 50 cm deep in the soil by main winged tine and deliver nutrients up to 25cm depth by leading tines and helped to place fertilizer at various depths in soil profile in a single pass. Hence these methods can increase fertilizer use efficiency and save fertilizer cost up to 20%.

Proper placement of fertilizer has additional advantage of lowering fixation of P fertilizers and reduction of volatilization losses besides increasing cane yield. Placement of fertilizers can be done at 8–10 cm deep furrows on both sides of rows using this kind implements and further fertilizers can be placed in the furrow and then covering can be done. When urea was buried in soil the recovery proportion of N was 33% over broadcasting of urea (18%) (Chapman *et al.*, 1994). Kwong *et al.* (1999) concluded that N requirement in sugarcane crop can be reduced up to 30% by using drip irrigation method in the prevailing soil and climatic condition of Mauritius by using N^{15} labeled fertilizer. As compared to conventional method of fertilizer application the fertigation in sugarcane can reduce the requirement of N up to 25-50% hence this can maintain the productivity and environmental security (Thorburn *et al.*, 2000). Results of application of 75 and 100% application of recommended fertilizer dose has shown on par yield which indicates a fertilizer saving up to 25%. As the dose of fertilizer level and number of splits were increased the quality parameters viz., brix, pol, purity and CCS percent were improved (Raskar and Bhoi, 2001). The abiotic stress situations like limited water supply and water logging, a well-known technique of N supply to sugarcane crop are foliar application of urea. Sub surface application of water and urea can either save N loss or it can increase the crop yield for the same inputs of N fertilizer was shown by Thorburn *et al.* (2003) through simulation studies. Drip fertigation of N and K at an interval of six days from 30 to 240 days after planting has increased the productivity of cane by 24.34% and save water up to 46.52% as compared to surface method of irrigation with recommended fertilizer dose under medium balck soils of Belgaum (Karnataka). Also fertigation has not affected the quality parameters of sugarcane like brix, pol and commercial cane sugar percent (Rajanna and Patil, 2003).

b. Right time

N use efficiency is known to increase by split doses of N during crops growing season, than application of single dose of fertilizer and large application before the planting of sugarcane (Cassman *et al.*, 2002). Maximum number of benefits can be assumed by right time of fertilizer application. The seed cane had favourable influence on its quality with low sucrose content and high glucose content by extending the duration of N application up to 135 days (Lakshmi *et al.*, 2006). Higher accumulation of reducing sugars can be noticed due to delay in N application which is affecting tiller production and ultimately cane maturity get prolonged. This may result in poor juice quality. Too high dose of N with delayed application may lead to reduction in quality of cane juice.

The broad guidelines for different nutrients are summarized below:

Nitrogen

- N should be applied in split doses, depending on soil and physiological stages of crop growth. In light textured soil and late varieties more N splits are recommended.
- Banded application of N should be followed at planting of crop; immediate irrigation should be done just after top dressing splits to avoid volatilization losses.

Phosphorus

- 1/3 of the phosphate fertilizers should be mixed with 2 two parts of moist and decomposed FYM for >12 hours before soil application.
- Apply the P through banding method at the side of set.
- Dissolve P fertilizer in water and apply it with first irrigation which can improve efficiency by 20-30%.

Potash

- Band application of K at sowing.
- At light textured soils go for two split applications.

Micronutrients

- B and Zn should be banded with macro nutrients, or broadcast them after mixing with 5 times of their volume of well tilled soil.

c. Controlled fertilizer/right fertilizer

If N is lost through various known sources it is causing economic losses to farmers but also making its impact equally on environment and human health. During nitrification the N from fertilizers can be lost by volatilization of ammonia and also further by runoff, leaching, erosion and denitrification and intensity of these losses vary widely according to method of fertilizer application and agroecosystem situations. Use of controlled release of fertilizers and application of nitrification and urease inhibitors are the new technologies through which N losses can be curtailed (Chen *et al.*, 2008).

25% saving of recommended dose NPK is possible if 75% of NPK the recommended doses of fertilizer is applied to crop with two equal split doses (50% at ratooning and 50% at 135 days after ratooning) through briquettes and resulted in high cane yield, CCS yield and B:C ratio. Briquette application has significantly improved all growth characters and cane yield than no application of fertilizers in the form of briquettes (Deshmukh *et al.*, 2014).

Crowbar method of fertilizer application

In Maharashtra, sugarcane production is mainly concentrated on black cotton soils with a productivity of 88 t/ha^{-1}, higher than the national productivity due to high fertilizer use efficiency. A technology has been developed by Central Sugarcane Research Station Padegaon, Satara, Maharashtra (MPKV, Rahuri) since placement of fertilizer plays the key role in its efficiency.

Table 1: Standard recommended doses of fertilizers for all cropping seasons of sugarcane

Sr. No	Seasons	Nitrogen kg/ha^{-1}	Phosphorous kg/ha^{-1}	Potassium kg/ha^{-1}
1.	SURU	300	140	140
2.	Pre-seasonal	400	170	170
3.	Adsali	500	200	200
4.	Ratoon	300	140	140

At field capacity, the total quantity of fertilizer is applied into two equal splits within 15 days after sowing or irrigation with the implement as shown in figure (1). The hole should be dug at 10-15 cm apart from the plant and 10-15 cm deep on one side of ridge only. The distance between two holes should be kept 30 cm only.

Table 2: The fertilizer doses (kg/ha) of seasons to be applied with crowbar method

Sr. No	Time of fertilizer application	Adsali			Pre-seasonal			Suru/Ratoon		
		N	P	K	N	P	K	N	P	K
1	1st Split application	250	100	100	200	85	85	150	70	70
2	2nd Split at 135 days after sowing	250	100	100	200	85	85	150	70	70
	Total	500	200	200	400	170	170	300	140	140

Advantages of crowbar technology

1. The deep placement near the roots facilitatesnutrient availability to crops.
2. Deep placement reduces the losses through volatilization and erosion.
3. Reduction in weed competition and cost of weeding reduced by 50-75%.
4. Slow release of fertilizer.
5. Due to uniform application of fertilizers in field the growth is uniform and improves the yield by 10-15%.
6. The one side application of fertilizers helps in increasing the beneficial microorganism from other side.
7. Improves the fertilizer efficiency by 25-30% due to continuous and uniform availability of all macronutrients leads to higher cane and sugar yield.

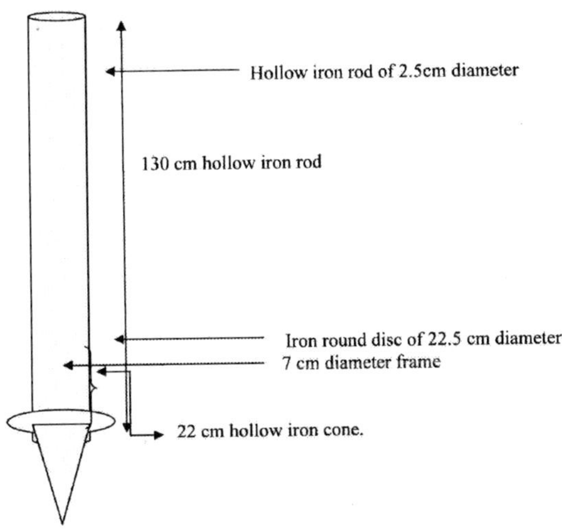

Fig. 3. Crowbar

Precision farming

Precision farming starts with yield maps and analysis of the yield variability. In the sugar industry, few accurate yield maps are available but other tools used in precision farming are increasingly common. This has led to the situation where the sugar industry perception of the term precision farming is very different to other agricultural industries that use yield maps as the starting point for precision agriculture. Precision farming is currently being investigated by the sugar industry with many industry participants seeing this minimalist adoption as an end in itself. Integrating these technologies offer much as drivers of cultural and industry reform. Precision agriculture within the industry are perceived to mean different things including GPS technology, controlled traffic farming, zero/minimum tillage farming, yield monitoring, variable rate fertilizing, harvesting feedback, crop modeling *etc*. Prammanee *et al.* (2005) have developed, Cane Fert 1.0, a computer programme to accurately predict the time for the application of major nutrient fertilizers by using soil properties of the cultivated area throughout Thailand. This programme consists of sugarcane growth simulation by using CANEGRO 3.5 to determine the effect of water, N and economic analyses of the recommended chemical fertilizer. The validations trials have confirmed the recommendations produced by Cane Fert 1.0. Canegro-N model is based on the model DSSAT and it may be useful for improving our ideas of N dynamics in sugarcane production system (Van Der Laan *et al.*, 2011). Simulated results showed its potential for curtailing fertilizer N requirement and its losses through soil inorganic N levels monitoring and adjustment of fertilizer application accordingly. Also during this period of active crop growth considerable amount of inorganic form of N could be available in soil due to mineralization. Accounting for this N enables fertilizer N application to be delayed to sometime after planting or commencement of ratoon growth, thereby significantly reducing the risky period during which applied N may be leached.

APSIM-Sugar is a computer model developed to simulate crop growth and estimate the extent of various management practices on losses of N and N use efficiency. Models can be useful to assist with decision making for prescriptive and corrective management of N fertilizers also they can be useful for adjustments between environmental and economic considerations (Wood *et al.*, 2010). Jose *et al.* (2014) explored alternative environment protection methods like precision agriculture methods over traditional agricultural practices for higher crop productivity at the same time. A GreenSeeker® hand-held optical active sensor was used to obtain normalized difference vegetative index (NDVI) readings which would need to delay in-season N fertilization by one month in order to integrate yield potential into an N management scheme without the risk of yield reduction (Lofton, 2012).

Fertilizer application based on tissue analysis/soil test

If one knows the peak time of uptake of N and its preferred form one can make comparatively better use of N fertilizer in sugarcane production system which can enhance its efficiency in plants. Generally the peak time of nutrient requirement in sugarcane occurs during initial stages within 90 days of fertilizer application before its full development. N status of sugarcane crop can be estimated by using plant sap analysis as an alternate method. Sap extract is the liquid outcome from the cell tissues of plants and plant nutrients can be quantified from it and the results of which are helpful for successful management of fertilizers. N fertilizer application to crops shows its immediate results in the plant sap and analysis of this sap may be useful for decision making in sugarcane for right time of fertilizer application and also to study the behaviour of N in sugarcane (Joris *et al.*, 2014).

In cane growers fertilizer application programme the analysis of leaf for nutrients has great scope to play. Clements (1980) developed crop logging concept in which critical nutrient concentration and tissue standardizing index has been used to guide fertilizer application in sugarcane comprehensively. N content at 3-6 leaf blade and the same leaf sheath for P, K and moisture content are used to estimate crop logging and these indices can be useful to fix nutrients critical levels in plant tissue and recommendations for fertilizer application and irrigation water application on economic basis can be made for improving cane yield and crop ripening. Diagnosis and Recommendation Integrated System (DRIS) was tested by Meyer (1981) for calculating fertilizer recommendations for sugarcane. Ratios of tissue nutrient concentration are useful to estimate DRIS indices, which helps to diagnose various nutrient deficiencies and imbalances. Schroeder *et al.* (2005) given an combined technology approach in which "Six Easy Step approached" is used in Australia, which is an integration of understanding soils and soil related processes, regular soil testing, adoption of soil specific nutrient guidelines, leaf analysis and good record keeping.

Microirrigation

Microirrigation has became an excellent mode for supplying water and fertilizers to crops. Ravikumar (2011) made an attempt to evaluate the reactive transport of urea in the root zone of sugarcane crop under drip irrigation and to quantify the fluxes of urea, ammonium and nitrate into the crop roots, volatilization fluxes and deep drainage using a numerical model. He has tested HYDRUS-2D model, based on the recommended total quantity of urea for sugarcane under drip irrigation was tested. Urea was applied in split doses with an interval of fortnight, depending on crop stage. In this event, the modeled crop uptake was observed to be 30% more than the crop nutrient demand. Further, a perfect fertigation schedule was chalked out and dose of urea was reduced by 30% while supplying enough N for crop uptake and use in all stages of sugarcane. HYDRUS model can be very applicable in altering the fertigation schedules using the growth curve nutrition strategy which can enhance nutrient use efficiency and reduce cost of production. Rain gun sprinkler irrigation is another micro irrigation technique which was found to be effective in sugarcane crop and saving irrigation water by 32%, saving fertilizers by 25%, increase in cane yield by 15% and 1.7 times more water use efficiency compared to traditional surface irrigation (Shinde, 2007).

Alla (2014) fertilized sugarcane ratoons using fertigation predominantly through the injection of dissolved urea during the entire irrigation schedule which has distributed N uniformly as compared to the injection during first or second half of the irrigation schedule. This has curtailed and saved tractor horse-power and operation cost and time. Deshmukh *et al.* (2010) concluded that by using drip irrigation with recommended fertilizer dose, it was possible to achieve more than two times higher water use efficiency and at the same time reduced fertilizer requirement. This was achieved in addition to 42.5 percent water saving using 30 percent less N and K fertilizer.

Fertilizer management in ratoon crop

Ratoon crops follow the plant crop or the preceding crop on the same soil. Due to the diminished soil physical conditions and poor root system development, assimilation of nutrients by the ratoon cane get affected. To have good yields the ratoons have to nourish with enough quantities of manures and fertilizers. Several experimental trials have showed the need for early nutrient application to ratoon sugarcane. For ratoon crop, nitrogenous fertilizers may be applied in two or three splits. Even in cases of split application, 1/3 to 1/2 dose of N should be applied immediately at the commencement of ratoon to have enough quantity of available N in soil to surpass the temporary immobilization of N due to microbial activity on the decomposing sugarcane stubbles. Total dose of P may be applied at first dose of N application at ratoon initiation (Verma, 2002). To produce one tonne of ratoon cane, more N is needed than in regular sugarcane planting. Regular cane planting have higher nutrient use efficiency and it decreases with each successive ratoons. Imbalances in root shoot ratio, altered root shoot development, inefficiency of stubble roots are the causes of reduced nutrient use efficiency in ratoons. Ratoon crop is reported to be responding to higher doses of N in all states of India. Ratoon crop requires 25-50% more N than normal cane planting. In Tamil Nadu after 5-7 days of ratoon initiation application of 25% extra N produced the highest cane and sugar yields (Mahendran *et al.*, 1995). Kumar *et al.* (2007) studied the effect of phosphorus and potassium doses on the productivity, uptake and efficiency of planted sugarcane and subsequent ratoon. Normal sugarcane planting responded significantly at full dose of 60 kg P and 40 kg K ha^{-1} and ratoon gave mean cane yield of 88.0 t ha^{-1}. Nutrient use efficiency was also highest with ratio of 60 kg P:40 kg K to both the crops, being 256 kg kg^{-1} of P_2O_5 and 384 kg kg^{-1} of K_2O applied as compared to different combinations of P_2O_5 : K_2O at sugarcane planting and ratoon.

Fertilizer management under moisture stress

Nutrient application during the time of moisture stress should help to developdeeper root system and sufficient tillering before the stress sets in. Therefore basal P fertilizer application along with little quantities of N and K, succeeded by an early N and K top dressing may be desirable which is further followed by another top dressing with last irrigation before the stress period. Just after the stress is over final manuring should be done. If six month moisture stress hasoccurred in sugarcane (Wiedenfeld, 2000), its response to application of N was not influenced and interaction between N and irrigation was found non-significant. However, significantly strong relation, influence and interaction were seen between irrigation level and N application when moisture stress was given for whole growing period of sugarcane (Wiedenfeld, 1995). Productivity of cane and sugar can be improved under drought conditions by application of potassium. Under drought conditions of sandy loam soils the application of K has reported 30.8 and 28.2% increase in cane and sugar yields by Naidu *et al.* (1983). If urea and muriate of potash applied each at 2.5% at 15–20 days interval was helpful to keep more vigorous shoots till the moisture condition becomes favourable for sugarcane (Gopalasundaram *et al.*, 2012). Externally applied silicon can reduce the shoot:root ratio by enhancing root growth. It can maintain higher photosynthetic rate and stomatal conductance over no application of silicon under stress conditions (Hattori *et al.*, 2005).

Conclusion

Good agronomic measures will improve sugarcane yields as well as NUE. Planting N-efficient sugarcane genotypes is a very attractive strategy for reducing the costs of crop production, enhance crop yields with keeping a healthy environment. Microirrigation has became an optimal means for providing water and nutrients to crops, drip irrigation with models like HYDRUS combined with fertigation has been found beneficial for high efficiency of fertilizer use. Improper planting method and geometry reduce sugarcane productivity which can be arrested by using dual row high density planting with ring pit method of planting. Sustainable sugarcane yield is possible by maintaining balanced use of macro and micro nutrients with use of suitable organic sources. Laser land leveling is a good resource conservation technology which can save irrigation, fertilizer and land at a time in sugarcane. Cropping system approaches wherein leguminous crops are raised along with sugarcane are found to enhance the nutrient use efficiency. Generally it is necessary to supply adequate quantities of manures and fertilizers to ratoon cane, which requires 25-50% more N but it should be applied in split dose where as full P and K should be applied at ratoon initiation to have more nutrient use efficiency. In anticipation of moisture stress in sugarcane early manuring of N and K induce deep root with adequate tillering followed by early N and K top dressing. Application of K under drought conditions may increase cane and sugar yield by 25-30% whereas silicon application during moisture stress can lower the shoot to root ratio, and can maintain higher photosynthetic rate and stomatal conductance.

Selected References

Ahmad, N. 2007. Fertilizer best management practices in Pakistan. Papers presented at the IFA International Workshop on Fertilizer Best Management Practices, Brussels, Belgium, 7-9 March, 2007.

Alla, D. 2014. Introduction of fertigation in sugarcane production for optimization of water and fertilizers use. *Agricultural Sciences* 5:945-957.

Anonymous, 2013.Status paper on sugarcane, Government of India, Ministry of Agriculture, 8th Floor, Kendriya Bhavan, Aliganj, Lucknow, U. P.

Anonymous, 2015.Indian Sugar, Indian Sugar Mills Association, Vol. no. LXVI, September, 2015.

Baldani, IJ, Reis VM, Baldani VLD, Dobereiner J. 2002. A brief story of nitrogen fixation in sugarcane- reasons for success in Brazil. *Functional Plant Biology* 29:417–423.

Berthelsen S, Hurney AP, Kingston G, Rudd A, Garside AL, Noble AD. 2001.Plant cane response to silicate products in the Mossman, Innisfail and Bundaberg districts. *Proceedings of the Australian Society* of *Sugar Cane Technology* 23: 297-303.

Blair G. 1993. Nutrient efficiency what do we really mean? In: Randall, P. J. *et al.* (ed.) Genetic Aspects of Plant Mineral Nutrition, Dordrecht: Kluwer Academic p. 205-213.

Cassman KG, DobermannA, Walter SDT. 2002.Agroecosystems, nitrogen use efficiency and nitrogen management. *Journal of the Human Environment* 31(2): 132-140.

Cavalcante VA, Dobereiner J. 1988. A new acid-tolerant nitrogen fixing bacterium associated with sugarcane. *Plant Soil* 108: 23–31.

Chapman LS, Haysom MBC, Saffigna PG. 1994. The recovery of N^{15} from labelled urea fertilizer in crop components of sugarcane and in soil profiles. *Australian Journal of Agricultural Research* 45(7):1577-1585.

Chen DH, Suter A, Islam R, EdisFreney JR, Walker CN. 2008. Prospects of improving efficiency of fertilizer nitrogen in Australian agriculture: a review of enhanced efficiency fertilizers *Australian Journal of Soil Research* 46:289–301.

Chinnamuthu CR, Boopathi PM. 2009. Nanotechnology and Agroecosystem. *Madras Agricultural Journal* 96:17-31.

Clements HF. 1980. Sugarcane crop logging and crop control: Principles and practices. The University Press of Hawaii, Honolulu p.520.

Deshmukh AS, Shinde PP, Katake SS, Phonde DB, Mali VS, Imas P. 2010. Effect of various levels of potash application through drip irrigation on yield and quality of sugarcane. International Potash Institute e-ifc No. 24 p. 1-6.

Deshmukh SU, Potdar DS, Pawar SM. 2014. Nutrient management through NPK briquettes for sugarcane ratoon. *International Journal of Current Research* 6(8):8214-8216.

Ghosh BN, Singh RJ, Mishra PK. 2015. Soil and input management options for increasing nutrient use efficiency. *Nutrient Use Efficiency: from Basics to Advances* 17-27.

Gopalasundaram P, Bhaskaran A, Rakkiyappan P. 2012. Integrated nutrient management in sugarcane. *Sugar Technology* 14(1).3–20.

Hattori T, Inanaga S, Hideki A, Ping A, Shigenori M, Miroslava L, Lux, A. 2005. Application of silicon enhanced drought tolerance in Sorghum bicolor, *Physiologia Plantarum* 123:459-466.

Jat ML, Gathala MK, Ladha JK, Saharawat YS, Jat AS, Vipin K, Sharma SK, Gupta R. 2009. Evaluation of precision land levelling and double zero-till systems in the rice-wheat rotation: Water use, productivity, profitability and soil physical properties. *Soil and Tillage Research* 105:112-121.

Joris HAW, Souza TR, Montezano ZF, Vargas VP,Cantarella H. 2014. Evaluating Nitrogen behavior in sugarcane after fertilization using leaf and sap extract analyzes. *American Journal of Plant Sciences* 5:2655-2664.

Jose AMD, Jose LLD, Evandro RA, Roberto N, Jorge LM. 2014. Precision agriculture for sugarcane management: a strategy applied for Brazilian conditions. *Acta Scientiarum* 36(1): 111-117.

Kingston G. 1999. A role for silicon, nitrogen and reduced bulk density in yield response to sugar mill ash and filter mud/ash mixtures. *Proceedings of theAustralian Society* of *Sugar Cane Technology* 21:14-121.

Kist V, Gustavo S, Paulo, Mafra de Almeida Costa, Mauro, Wagner de Oliveira, Márcio H PB. 2015. Nutrient use efficiency in sugarcane cultivars. *Cientifica Joboticable* 43(2): 117–125.

Kumar S, Rana NS, Chandra R, Sandeep K. 2007. Effect of phosphorus and potassium doses and their application schedule on yield, juice quality and nutrient use efficiency of sugarcane-ratoon crop sequence. *Journal of the Indian Society of Soil Science* 55(4):122-142.

Kwong KF, NgK, Paul JP, Deville J. 1999. Drip-fertigation–A means for reducing fertilizer nitrogen to sugarcane. *Experimental Agriculture* 35(1):31-37.

Lakshmi MB, Chitkala DT, Raju DVN. 2006. Nitrogen management in sugarcane seed crop *Sugar Technology* 8(1):91-94.

Lofton J. 2012. Improving nitrogen management in sugarcane production of the mid South using remote sensing technology. A Ph. D. dissertation submitted to the Graduate Faculty of the Louisiana State University and Agricultural and Mechanical College in The School of Plant, Environmental & Soil Sciences, U.S.A.

Mahendran S., Karamathullah J, Porpavai S, Ayyamperumal A. 1995. Effect of planting systems and ratoon management on the yield and quality of ratoon cane. *Bhartiya Sugar* 22(1): 123-127.

Mandal S, ThakurTC. 2010. Design and development of subsoiler-cum-differential rate fertilizer applicator. *Agricultural Engineering International: the CIGR Ejournal.* Manuscript No. 1394, Vol.XII.1-14.

Meyer J. 2011. Good management practices manual for the cane sugar industries (Final) PGBI Sugar & Bio-Energy (Pty) Ltd. PGBI House, 8 Wolseley Street, Woodmead East, 2191, Johannesburg, South Africa.

Meyer J, Keeping M.1999. Past, present and future silicon research in the South African sugar industry.Proceedings International Conference on Silicon, Miami. In Press.

Meyer JH. 1981. An evaluation of DRIS based on leaf analysis for sugarcane in South Africa. *Proceedings of The South African Sugar Technologists Association* 55:169-176.

Naidu KM, Srinivasan TR, Michaelraj S. 1983. Varietal behavior and management for improvement of sugarcane yield and quality under moisture stress conditions. *SISSTA Sugar Journal* 9(2):47-52.

Nurhidayati N, Basit A. 2015. Improvement of nitrogen use efficiency derived from ammonium sulfate substitute fertilizer in sugarcane cultivation through the addition of organic amendment. *International Journal of Plant & Soil Science* 6(6): 341-349.

Prammanee P, Jintrawet A, Lairungreung C, Kongton S. 2005. Canefert 1.0: A chemical fertilizer recommendation software for sugarcane production in Thailand .*Proceedings of International Society of Sugarcane Technology* 25:198-203.

Rajanna MP, Patil VC. 2003. Effect of fertigation on yield and quality of sugarcane.*Indian Sugar* 52:1007-1011.

Ramesh T, Chinnusamy C, Jayanthi, C. 2004. Bioorganic nutrient management in sugarcane production-a review. *Agricultural Reviews* 25(3): 201-210.

Raskar BS, Bhoi PG. 2001. Productivity of sugarcane as influenced by planting techniques and sources of fertigation scheduling of irrigation using a casdose zone flow and transport model.*Indian Sugarcane* 53: 685-692.

Ravikumar V, Vijayakumar G, Simunek J, Chellamuthu S, Santhi R, Appavu K. 2011. Evaluation of fertigation scheduling for sugarcane using a vadose zone flow and transport model.*Agricultural Water Management* 98:1431-1440.

Sajjad M, Bari A, Nawaz M, Iqbal S. 2014. Effect of planting pattern and nutrient management on yield of spring planted sugarcane. *Sarhad Journal of Agriculture* 30(1):67-71.

Savant NK, Korndörfer GH, Datnoff LE, Snyder GH. 1999. Silicon nutrition and sugarcane production - A review. *Journal of Plant Nutrition* 22(12):1853-1903.

Schroeder B, Wood AW, Moody PW, Bell MJ, Garside AL. 2005.Nitrogen fertilizer guidelines in perspective. *Proceedings of the Australian Society of Sugar Cane Technology* 27: 291-304.

Schumann AW. 2000. Prospects for improving nitrogen fertilizer use efficiency with a new soil test and ammonia volatilisation model. *Proceedings of the South African Sugar Technologists Association* 74:70–78.

Schumann AW, Meyer JH, Nair S.1998. Evidence for different nitrogen use efflclencles of selected sugarcane varieties. *Proceedings of the South African Sugar Technologists Association* 78: 77–80.

Shankaraiah C, Vishwantha SR, Nagaraju MS, Rudraswamy P. 1999. Soybeanvarietal interaction under sugarcane plus soybean intercropping system. *MysoreJournal of Agricultural Sciences* 33(4): 306-311.

Shaviv A. 2000. Advances in controlled release of fertilizers.*Advances in Agronomy* 71:1-49.

Shinde PP. 2007. Increasing water and fertilizer use efficiency through rain gun sprinkler irrigation in sugar cane agriculture, USCID International Conference on Irrigation and Drainage, *The Role of Irrigation and Drainage in a Sustainable Future* 1291-1294. Fourth Sacramento, California — October 3-6, 2007.

Shrivastava AK, Srivastava AK, Solomon S. 2011. Sustaining sugarcane productivity under depleting water resources. *Current Science* 101(6):748-754.

Singh AK, Mehni L, Singh KP,Suman, A. 2007. Enhancing nitrogen use efficiency of spring planted sugarcane through *insitu* residue management of intercropped vegetable cowpea at varying levels of fertilizer nitrogen. *Journal of Vegetable Science* 34(2):163-166.

Singh KP, Suman A, Singh PN,Menhi L. 2007. Yield and soil nutrient balance of a sugarcane plant–ratoon system with conventional and organic nutrient management in sub-tropical India.*Nutrient Cycling in Agroecosystems* 79:209–219.

Singh VK, Shukla AK, Gill MS, Sharma SK, Tiwari KN. 2008.Improving sugarcane productivity through balanced nutrition with potassium, sulphur, and magnesium.*Better Crops*-India. A Publication of the International Plant Nutrition Institute (IPNI) 2(1): 12-14.

Suman A, Gaur A, Shrivastava AK, Yadav RL. 2005. Improving sugarcane growth and nutrient uptake by inoculating *Gluconacetobacter diazotrophicus*. *Plant Growth Regulation* 47:155–162.

Suman A, Shrivastava AK, Gaur A, Singh P, Singh J, Yadav RL. 2008. Nitrogen use efficiency of sugarcane in relation to its BNF potential and population of endophytic diazotrophs at different N levels. *Plant Growth Regulation* 54:1–11.

Sundara, B. 1994.Phosphorus efficiency of sugarcane varieties in a tropical alfisol. *Fertilizer Research* 39:83-85.

Surendran U, Vani D. 2013.Influence of arbuscularmycorrhizal fungi in sugarcane productivity under semiarid tropical agro ecosystem in India. *International Journal of Plant Production* 7(2): 269-278.

Thorburn PJ, Dart IK, Baillie CP. 2000. Nitrogen balances in trickle irrigated sugarcane. Paper presented at the 6th International Micro irrigation Congress, Cape Town, South Africa, October, 2000.

Thorburn PJ, Dart IK, Biggs IM, Baillie CP, Smith MA, Keating BA. 2003. The fate of nitrogen applied to sugarcane by trickle irrigation. *Irrigation Science* 22:201-209.

Van LM, Miles N, Annandale JG, Preez Du CC. 2011. Identification of opportunities for improved nitrogen management in sugarcane cropping systems using the newly developed Canegro-N model.*Nutrient Cycling in Agroecosystems* 90:87-96.

Verma RS. 2002. Sugarcane Ratoon Management. International Book Distributing Co. Charbagh, Lucknow.

Wiedenfeld RP. 1995. Effects of irrigation and N Fertilizer application on sugarcane yield and quality. *Field Crops Research* 43:101-108.

Wiedenfeld RP. 2000. Water stress during different sugarcane growth periods on yield and response to N fertilization.*Agricultural Water Management* 43(2):173–182.

Wood AW, Schroeder BL, Dwyer R. 2010. Opportunities for improving the efficiency of nitrogen fertilizer in the Australian sugar industry. *Proceedings of theAustralian Society of Sugar Cane Technology* 32: 221-233.

Yadav RL. 2004. Enhancing efficiency of fertilizer N use in sugarcane by ring pit method of planting. *Sugar Technology* 6(3):169-171.

24

Enhancing Nutrient Use Efficiency, pp. 363-378
Editors: K. Ramesh, A.K. Biswas, B.L. Lakaria, S. Srivastava and A.K. Patra

Cotton Based Cropping Systems

D. Blaise

ICAR-Central Institute for Cotton Research, Nagpur – 441108, India

Introduction

Cotton belongs to the genus *Gossypium*, family *Malvaceae*. Of the 50 species, only four are cultivated, namely, *G. arboreum*, *G. herbaceum*, *G. hirsutum* and *G. barbadense*. The *G. arboreum* and *G. herbaceum* are popularly known as the *desi* cottons. The *G. hirsutum* and *G. barbadense* evolved in the new world and are known as American or Upland (*G. hirsutum*) and Egyptian or Pima (*G. barbadense*) cottons. India is the only country in the world where all the four cultivated species of cotton are grown on a commercial scale besides hybrids (both inter and intra specific) in three distinct agro-ecological regions and wide range of soil types.

Historically, the Asiatic *desi* cottons were cultivated with minimal external inputs. The biggest change in cotton cultivation was witnessed after the introduction of Bt cotton hybrids in March 2002. The area planted with Bt cotton hybrids increased from a meager 50,000 ha in 2003 to more than 11.2 m ha in 2011-12. At present it occupies more than 95% of the total cotton acreage (Blaise *et al.*, 2014).

In three distinct agro-ecological regions and wide range of soil types, the acreage of cotton and the wide-ranging soil types are presented in Table 1. The north zone comprising of Punjab, Haryana and Rajasthan has major soil orders as Entisols, Inceptisols and Aridisols on which cotton is grown. The productivity in the region is high due to irrigation facilities. The central zone comprises of Gujarat, Madhya Pradesh and Maharashtra that is predominantly rainfed and has nearly 68% of the cotton area. Maharashtra occupies nearly one-third of the area under cotton in the country but less than 10% of the area is irrigated. Vertisols and associated soils are the major soil types in the region. In the south zone, cotton is grown in Karnataka, Andhra Pradesh and Tamil Nadu with about 60% of the area rain-dependent. Vertisols, Alfisols and Entisols are the major soil orders in this region.

Table 1: Major cotton growing areas in the country (2012-13)

Zone	States	Major soil type	Acreage (m ha)
North	Punjab, Haryana, Rajasthan	Inceptisols, Entisols, Aridisols	1.57
Central	Gujarat, Maharashtra, Madhya Pradesh	Vertisols	7.10
South	Tamil Nadu, Andhra Pradesh, Karnataka	Vertisols and mixed red and black soils	2.77
Others	Odisha, Northeast	Red and black soils	1.69

Although, area under cotton has increased in the last few years, no further increase is likely due to other competing factors for land. Therefore, a vertical growth -increase in crop productivity is the only option left. Productivity levels have, however, plateaued indicating signs of fatigue. One major reason is the appearance of multiple nutrient deficiencies (Brar *et al.*, 2008). Interestingly, wide gap exists between the yields achieved on the farmers' fields compared to those on-station experiments and genetic potential. This suggests that there is ample scope to meet future production targets by reducing yield gaps. Adopting good agronomic practices not only ensures achieving yield targets but also high profitability and returns because of an improved input efficiency. Furthermore, cotton farming has become less profitable due to enhanced cost of cultivation (high input costs). Thus, improving the efficiency of inputs such as fertilizers can make cotton farming more profitable.

Why the concerns for nutrient use efficiency?

Among the inputs, fertilizers continue to play a major role in increasing cotton productivity. However, low nutrient use efficiency is a major concern because the efficiency of fertilizer N is around 50%, while the efficiency of P is less than 20% and micronutrients less than 5% (NAAS, 2006).

Nutrient use efficiency depends on several agronomic practices such as tillage, crop variety, timely planting, plant population, optimal irrigation, balanced fertilizer, weed control, pest and disease management. In this chapter, agronomic practices to improve the nutrient use efficiency in cotton based cropping systems are discussed.

Tillage

Tillage is basically done to provide a well-prepared seedbed in order to enhance rate of seed germination and seedling growth. Tillage for seedbed preparation should modify the soil to allow desired depth of seed placement, effective control of weeds, improve water infiltration and reduces erosion. A good soil management lays the foundation for a better crop development.

Cotton is a deep rooted crop and intensive tillage with a mouldboard plough were recommended. This is still in vogue with clean cultivation as a practice in the entire cotton belt of India. Furthermore, several interculture operations are done during the crop growing season. Frequently disturbing or tilling the land causes rapid mineralization of the soil organic carbon. The soil also becomes more prone to erosion and ultimately leads to soil degradation. These could be the reasons for the decline in factor productivity. Interest in conservation tillage practices gained ground the world over. In long-term tillage experiments conducted at CICR, Nagpur, significant yield increases were evident in the reduced tillage practices. Because of these yield improvements, expectedly the partial factor productivity (PFP) of the applied fertilizer nutrients is increased (Table 2). This was mainly due to a better soil hydro-thermal regime and improved weed control (Blaise, 2006).

Table 2: Effect of tillage methods on seed cotton yield and factor productivity (kg seed cotton/kg NPK) in *Bt* transgenic cotton at Nagpur

Tillage method	Yield (kg/ha)	PFP_{NPK}
Non-Bt Cotton		
Conventional till	1281	13.1
Reduced till + 2 interculture + 2 hand weeding	1387	14.2
Reduced till + two hand weeding)	1417	14.5
Bt transgenic hybrid		
Conventional till	1526	9.8
Reduced till+ 2 interculture + 2 hand weeding	1874	12.0
Reduced till+ 2 hand weeding	2054	13.2

Source: Blaise and Ravindran (2003); Blaise (2011)

From the data presented in Table 2, it was also observed that non Bt cotton utilize the nutrients more efficiently than the Bt hybrids. A total of 98 kg NPK was applied to the non *bt* cotton variety on the other hand, Bt hybrid received 156 kg NPK per hectare.

In the rainfed areas, soil moisture conservation mechanisms need to be made an integral part of the farming systems. It is being followed in some areas, such as opening of furrows in alternate rows in the month of August. This facilitates *in situ* rainwater conservation. Adoption of practices such as green manuring, residue recycling and other locally developed moisture conserving practices will help to improve the productivity and consequently the fertilizer use efficiency. Singh *et al.* (2004) validated the technology on farmers fields through a participatory approach and found that the ridge and furrow method of planting a better option than the flat bed method of planting (Table 3). Because of improvements in the soil moisture status, the available and applied nutrients are utilized efficiently. However, most often sowing on ridge and furrow may not be possible due to the non-availability of machinery. Under such situations, the furrows can be opened up during the inter-cultivation operations.

Table 3: Effect of soil moisture conserving practices on seed cotton yield and partial factor productivity (kg seed cotton/kg NPK)

Sowing method	Yield (kg/ha)	PFP_{NPK}
Flat bed	756	5.2
Ridge and furrow	848	5.8

Source: Singh *et al.* (2004)

In recent years, new concepts of using polythene mulch have come up. In the Bt cotton hybrids, which is spaced at a wide row spacing on drip irrigation, large areas are covered with polythene mulching by resourceful farmers. Nalayini *et al.* (2011) demonstrated an increase in seed cotton yield with the polythene mulch compared to the non mulched plots, both in the non-*Bt* as well as Bt hybrids. Yield increases were greater with the *Bt* hybrids (Table 4). PFP was calculated from the available yield and fertilizer input data and is presented in Table 4. PFP of the applied NPK (90-19-37 kg/ha), was greater for the mulched plots. Soil moisture in the mulched plots was 24.9% compared to 19.8% in the plots without mulch. Higher soil moisture would result in greater availability of nutrients and consequently better utilization.

Table 4: Effect of polythene mulching on PFP (kg seed cotton/kg NPK applied) in non *Bt* and *Bt* hybrids

Mulching	Seed cotton yield (kg/ha)	PFP_{NPK}
Non Bt hybrids		
Polythene	3175	21.7
Without mulch	2579	17.7
Bt hybrids		
Polythene	5280	36.2
Without mulch	3376	23.1

Source: Nalayini *et al.* (2011)

Crop variety/hybrid

It is alarming that when the Bt cotton hybrids reached a level of saturation in the country, yield levels plateaued off (Blaise *et al.*, 2014). One major reason is the multiplicity of hybrids available on the market that caused a mismatch of the hybrid and the soil type on which these are cultivated (Kranthi, 2012). Suitable cultivars, recommended for an agro-eco region need to be grown for realizing not only high yields but also improved fertilizer use efficiency. The differences in the cultivar could be either (i) better nutrient utilization or its acquisition. A better nutrient utilization results in apportioning of dry matter to the economic yield. Genotypes with efficient nutrient acquisition tend to perform better than the inefficient ones under nutrient stress conditions. Venugopalan and Pundarikakshudu (1998) observed that the *desi* cotton varieties had higher N and P utilization efficiency than the American upland cottons (Table 5).

Table 5: Use efficiency (kg seed cotton/kg N or P uptake) bydesi and American cotton

Cotton species	N utilization efficiency	P utilization efficiency
G. arboreum (*desi*)	18.7	90.3
G. hirsutum (American)	14.9	81.8

Source: Venugopalan and Pundarikakshudu (1998)

Similarly, for *Bt* hybrids, too, it is essential to grow the ones that are recommended for the region. Taking up hybrids that are not suitable for the area, not only results in lower seed cotton yield but use efficiency of inputs is low (Table 6) which is ultimately a waste of fertilizer nutrients.

Table 6: PFP of applied fertilizer NPK (kg seed cotton/kg NPK applied) to Bt cotton hybrids

Bt hybrids	Seed cotton yield (kg/ha)	PFP_{NPK}
MECH-184	1718	11.8
MECH-162	1717	11.8
MECH-12	1143	7.8

Source: Singh *et al.* (2003)

Timely planting

The crop takes full advantage of rainfall with timely planting. In addition, date of planting can be adjusted to help avoid or escape the pest. A delay in sowing brings about a significant reduction in yield and a consequent reduction in the FUE. Thus, for a better use of the inputs, timely planting is essential.

Venugopalan *et al.* (2012) reported lower nutrient use efficiency, uptake and utilization efficiency were under delayed sowing compared to the normal planting time of the *Bt* cotton hybrids on rainfed Vertisols (Table 7). Some hybrids are more suited for late sown conditions and may be cultivated under such situations.

Table 7: Nutrient use efficiency parameters under normal and delayed sowing

Sowing time	MECH-184	RCH-2	NCS 145
PFP			
Normal	12.6	15.9	11.7
Delayed	10.7	9.2	8.6
Nutrient uptake efficiency			
Normal	73.2	89.8	76.7
Delayed	70.0	64.6	71.6
Nutrient utilization efficiency			
Normal	17.3	17.7	14.7
Delayed	15.3	13.8	11.2

Source: Venugopalan *et al.* (2012)

Also in the irrigated cotton-wheat zone, timely planting was found advantageous. This is because; early planting ensures sufficient thermal units for crop maturity. Singh *et al.* (2013) in a study conducted at Faridkot, Punjab showed the yields were greater in the 24 April sown crop than the 5 May sown plots. Consequently, the use efficiency was greater in the early sown than the late sown crop (Table 8). This clearly indicates that such non-monetary inputs also contribute to improved use efficiency.

Table 8: Effect of date of sowing on seed cotton yield and PFP_{NPK} at Faridkot, Punjab

Date of sowing	Seed cotton yield (kg/ha)	PFP_{NPK}
24 April	3521	15.6
05 May	3391	15.0

Source: Singh *et al.* (2013)

Bt seed is a costly input and crop is sown only after receipt of rains. In some irrigated regions, cotton planting is delayed due to a delay in release of canal water. In such situations, transplanting is a viable option. Salakinkop (2011) suggested these transplants were far superior to the cotton planted after the release of canal water (Table 9). This is important because huge amounts of fertilizers are applied in such systems. In the irrigated tract of Raichur, Karnataka; nearly 300 kg NPK are added per hectare. Timely planting may not be possible in the irrigated tracts. Adopting transplanting is a technique to overcome this problem. From this data, the PFP calculated suggests that additional 2 kg seed cotton is produced for a given input with the transplanting method compared to the normal sowing. Thus, for the 300 kg fertilizer dose, a loss of 600 kg seed cotton occurred. Put in other words, the normal sown crop would require more than 50 kg additional fertilizer to achieve the similar level of yield achieved with the transplanting technique.

Table 9: Effect of methods of planting cotton on PFP_{NPK}

Method of planting	Seed cotton yield (t/ha)	PFP_{NPK}
Direct sowing	3.27	10.9
Transplanting	3.83	12.8

Source: Salakinkop (2011)

Plant population

Lack of optimum plant stand is a major factor that affects crop yields both in the rainfed as well as irrigated cotton. Sub-optimal or greater than the recommended plant stands will affect the crop productivity and nutrient use efficiency. This became more evident with the Bt hybrids. Conventional hybrids were sown at wider row spacing and in a square planting pattern for a better interculture. However, the Bt cotton hybrids were found to perform better at closer spacing (Bhalerao and Gaikwad, 2010). The optimal plant stand not only enhances yield, but also the input use efficiency (Table 10).

Table 10: Effect of plant stand on PFP_{NPK} (kg seed cotton/kg NPK applied)

Plant spacing (cm)	Seed cotton yield (kg/ha)	PFP_{NPK}
90 x 90	639	6.4
90 x 60	942	9.4
90 x 45	1031	10.3

Source: Bhalerao and Gaikwad (2010)

Similarly, for the varieties of both the American as well as *desi* cottons, closer spacings were found to be superior to the wide row spaced crop. Venugopalan *et al.* (2013), showed that varieties when grown at a plant population as high as 166,000 per hectare, the yields are the highest for some cultivars. Thus, maintaining an optimal plant stands based on the varietal characteristic is an important feature. Bushy types are not amenable for closer row spacing, while the erect and compact types are suitable for the ultra narrow row spacing.

Venugoaplan and Blaise (2001) pointed out that the plant population density interacts further with the inputs such as fertilizer N. Fertilizer N requirement varies with each of the spacing (Table 11). The response curves obtained using the data points out that yield is a function of N uptake while the uptake is a function of N applied (Venugopalan and Blaise, 2001). The study also brought out that the N recovery was the highest at closer plant spacing than the low plant stands. This was probably because of the close row and plant spacing resulted in greater soil exploration. In general, yield and nutrient use efficiency patterns were similar for the 45 x 30 and 45 x 45 cm spacing and significantly greater than the 60 x 45 cm spacing.

Table 11: Effect of N and population density on N-use efficiency (kg seed cotton/kg N)

Spacing (cm)	N_0	N_{50}	N_{75}	N_{100}
45 x 30 (74,100)	1004	1389	1527	1455
45 x 45 (49,400)	1005	1405	1507	1495
60 x 45 (37,050)	940	1320	1258	1185
N use efficiency				
45 x 30 (74,100)	-	7.7	7.0	4.5
45 x 45 (49,400)	-	8.0	6.7	4.9
60 x 45 (37,050)	-	7.6	4.2	2.5

Source: Venugopalan and Blaise (2001)

Weed management

Weeds compete for resources – water, nutrients and in some cases solar radiation. Several studies have pointed out significant yield advantages with integrated weed management practices. One case is the use of herbicides which has fast caught up with the farmers due to labour shortages. Further, availability of the herbicides on the market and its practical feasibility has resulted in widespread adoption of this technology. In the previous section, improved seed cotton yields and factor productivity in the reduced tillage systems observed was mainly due to a better weed control in the reduced tillage systems because of the use of herbicides. Effective weed control apart from improving crop yields, improves fertilizer use efficiency.

Crop rotation

Modern agriculture led to monocropping systems which were easier to manage as well as was highly remunerative. For instance, historically cotton was rotated with grain sorghum in rainfed central India. However, low prices for sorghum resulted in a shift to continuous cotton cultivation in rainfed central India. Some areas also witnessed a change in cotton area being replaced with soybean. Experimental studies clearly indicated a 2-year rotation of cotton-soybean was more productive with high nutrient use efficiency as compared to cotton alone (Blaise *et al*., 2008).

Similar is the case in the high intensive cropping systems of north India. Introduction of high yielding short duration wheat cultivars resulted in intensive cotton-wheat cropping a shift from the earlier systems relying on inclusion of pulses in a rotation. Legumes are known to fix substantial amounts of N and incorporating a legume in the cropping systems can reduce the N requirement of the cropping system. Besides, there are other benefits such as a break in pest and disease incidence.

Fertilizers and nutrient use efficiency

Two major strategies involved are: 1) product strategy and 2) fertilizer management strategy. In the former, a modification of the product is done to enhance the efficiency of the fertilizer and consequently the nutrient use efficiency. Coating of urea with materials such as sulphur (sulphur coated urea), polymers and neem cake are examples. Nitrification (*ex.* N serve, DCD) and urease inhibitors (*ex.* PPD) are also added to improve the use efficiency of N fertilizers. These materials reduce N losses through de-nitrification and enhance use efficiency. Seshadri (1985) reported significantly higher use efficiency with neem coated urea compared to prilled urea.

In the fertilizer management strategy, emphasis is on improving efficiency of fertilization process; *i.e.* after the fertilizer is added as an input. Nutrient use efficiency depends very much on rate, time and method of application and sources (Prasad, 2007).

Use efficiency vs. rate of application

Use efficiency declines with an increase in rate of nutrient addition. This is mainly because yield follows the law of diminishing returns with increasing addition of a given factor when all other factors are in adequate amounts and the pattern gets reflected in use efficiency. The best management strategy is to apply inputs at the optimal doses. Excess of this would be undesirable as it would make the plant move more towards the vegetative growth from a reproductive phase. More leaf N concentrations result in the plant becoming susceptible to sucking pest infestation.

Among the fertilizer nutrients, there is an uncertainty for N requirement for obtaining optimal cotton yields under different environmental conditions. This is due to the indeterminate growth habit of cotton. Response vary up to 90 kg N/ha in central India under rainfed conditions and as high as 150 kg/ha under irrigated conditions (Table 12).

Table 12: Response of Bt cotton hybrids to nutrient application

Location	Soil	Nutrient	Optimum dose (kg/ha)	PFP
Sirivilliputur, Tamil Nadu (I)	Sandy clay loam	N	100	15.2
Coimbatore, Tamil Nadu (I)	Red loam	N	90	24.5
Faridkot, Punjab (I)	Sandy Loam	N	187	16.9
Bapatla, Andhra Pradesh (R)	Clay	N	180	17.0
Warangal, Andhra Pradesh (I)	Sandy Loam	N	150	19.5
Khedbrahma (I)	Clay	N	240	10.2
Guntur, Andhra Pradesh (R)	Clay	N	120	23.9
Warangal, Andhra Pradesh (R)	Clay	N	120	28.0
Parbhani, Maharashtra (R)	Clay	N	100	21.1
Warangal, Andhra Pradesh (I)	Sandy Loam	P	26	109.7
Khedbrahma (I)	Clay	P	8.7	276.8
Warangal, Andhra Pradesh (I)	Sandy Loam	K	50	58.4
Khedbrahma (I)	Clay	K	72.7	72.7
Punjab (I)	Sandy loam	K	60	50.0
Bathinda, Punjab (I)	Sandy loam	Zn	10	382.0

Source: Venugopalan *et al.* (2011)

An alteration in the N supply could impact protein synthesis and metabolism. This is critical in transgenic crops since anything that affects protein synthesis could potentially alter cry toxin expression. Pettigrew and Adamczyk (2006) reported 14% greater Cry1Ac concentration in treatments receiving additional N. Hallikeri *et al.* (2011), too, reported additional N and split application of fertilizer enhanced Cry1Ac toxin content in Bt transgenic cotton hybrids.

Split application and timing of application

Among the various fertilizer management strategies, this is one strategy that has caught the imagination of the farmers. It is well understood that instead of applying entire N during early stages of the crop growth, splitting its use when the crop needs the most results in enhanced efficiency. Fertilizer application time and method of application are flexible and need to be adjusted to local climatic conditions and cropping systems (Blaise *et al.*, 2014).

Placement

Urea the most commonly used nitrogenous fertilizer is predominantly broadcasted. When fertilizer is broadcast and incorporated the entire topsoil, containing the major portion of roots, would be nutrient rich. But this also leads to more amounts of fertilizer-P getting fixed. Placement of fertilizer-N would reduce N losses due to NH_3 volatilization and probably enhance N use efficiency. With placement, immobilization is also reduced. P fertilizers when placed are more available to the crop because of reduced P fixation.

Balanced fertilization

Among the good agronomic practices, adopting balanced fertilization is a major approach to improve crop productivity. Blaise *et al.* (2006) observed in long-term studies that yield level stability is achieved only in systems receiving a balanced nutrition supply (Table 13). Furthermore, AE and PFP in the balanced fertilizer plots were better compared to the plots wherein one or more of the fertilizer nutrients were omitted. Similarly, with the *Bt* hybrids, too, application of the recommended doses of fertilizer in a balanced manner was the best option (Table 14).

Table 13: Mean yield response (kg ha^{-1}), yield stability and PFP(kg seed cotton/kg N applied) as affected by fertilizer application

Treatment	Mean yield response	Slope	PFP
N	634.7	0.63* (<1)	10.6
NP	920.5	0.92	15.3
NPK	950.2	1.09	15.8
NPK + FYM	1100.8	1.47* (>1)	18.4

Source: Blaise *et al.* (2006)

Table 14: Effect of balanced (NPK) fertilizers on the PFP (kg grain/kg NPK applied)

Fertilizer dose	Seed cotton yield (kg/ha)	PFP_{NPK}
75% RDF (37.5-18.75-18.75)	798	10.6
RDF (50-25-25)	873	8.7
125% RDF (62.5-31.25-31.25)	910	7.3

Source: Bhalerao and Gaikwad (2010)

Among fertilizer nutrients, potassium application has increased in the recent years. Earlier, use of potash fertilizers was negligible, although the SAU's recommended its application for the American cotton varieties and hybrids. This was because most of the cotton growing soils had medium to high available K and response to K application was either low or inconsistent. This changed with the large scale cultivation of the *Bt* cotton hybrids. Cotton grown in irrigated as well as under rainfed conditions responded positively to K application. Brar *et al.* (2008) reported the irrigated *Bt* cotton hybrids (RCH-134 & RCH-317) in Punjab responded up to 41.6 kg K/ha on soils testing low to medium exchangeable K. No response was observed on soils with high available K content. Increase in seed cotton yields of *Bt* cotton hybrids to an extent of 13-20% was observed in the rainfed and irrigated conditions in Karnataka (Biradar *et al.*, 2011). On the rainfed Vertisols of central India, response to K application was noticed with the Bt transgenic hybrids on the farmer field trials having low exchangeable K content. These soils also had low non exchangeable

K content (Blaise, 2012). More recently, *Bt* hybrid cotton (Mallika) grown on Vertisols in Adilabad, Andhra Pradesh, was found to respond to K as high as 90 kg/ha (Das *et al.,* 2013). This is because K requirement is high during boll formation stage and the uptake is limited during this phase, particularly under rainfed conditions. Such responses were not observed in the past with the old varieties and hybrids (Mannikar, 1993). Due to such significant responses, farmers have begun to apply potash fertilizers to their *Bt* cotton crop.

Presently, blanket recommendation of fertilizers is done for a region/variety. Better understanding of residual nutrient status would benefit fertilizer management and improve efficiency. This led to the concept of soil-test based fertilizer application to realize a given yield target. Further, concept of site-specific nutrient management (SSNM) evolved with nutrient additions done on real-time basis taking into account peak plant nutrient demand, indigenous soil supply and potential yield levels. At CICR, Nagpur, it was observed that seed cotton yield in the SSNM was equal to or better than the existing recommended practice. In the SSNM, the N dose was reduced by 15 kg/ha.

In order to achieve potential crop yields, higher levels of fertilizer nutrients are needed (Vidyavathi *et al.*, 2013). For a 3000 kg yield target, 400 kg NPK needed per hectare whereas the present recommended dose is 300 kg/ha. With an application of 402 kg NPK/ha, yield of 2976 kg/ha was obtained. PFP for the two systems was 7.4 and 6.7 kg seed cotton per kg NPK applied, respectively.

Balanced fertilization is not a new concept and is not an exclusive story of application of NPK. According to Liebig's law of minimum, any nutrient that becomes limiting limits the efficiency of other elements applied, even in excess of their requirement. Nutrient use efficiency improved when Zn and B were applied along with NPK to cotton (Table 15).

Table 15: Effect of Zn and B application on the N, P and K use efficiency of cotton

	Apparent recovery (%)	Physiological efficiency (kg seed cotton/kg nutrient removed)	Agronomic efficiency (kg seed cotton/kg nutrient applied)
		N	
RDF	27.6	19.1	5.3
RDF +Zn + B	52.3	16.1	8.4
		P	
RDF	18.6	134.0	25.0
RDF +Zn + B	32.9	121.1	39.9
		K	
RDF	74.8	17.1	12.8
RDF +Zn + B	115.9	49.9	20.5

Source: Blaise *et al.* (2008)

Irrigation

Only 40% of the cotton area is irrigated and in these ecosystems, nutrients are major growth and yield limiting factors. Application of fertilizers along with drip irrigation has gained popularity especially in the rainfed central India where water is scarce. Drip method of fertigation has been found to not only improve yields and use efficiency, but the amount of fertilizers required for optimal crop yields is lowered (Table 16).

Table 16: Effect of irrigation method and fertilizer doses on seed cotton yield and PFP (kg seed cotton/kg NPK)

Method of irrigation	Seed cotton yield (t/ha)	PFP_{NPK}
Surface (120-26-50)	2.77	14.1
Drip with 75% RDF	3.91	26.6
Drip with 100% RDF	4.21	21.5
Drip with 125% RDF	4.53	18.5

Source: Pawar *et al.* (2013)

At Bhatinda, Punjab, on Ustochreptic Camborthid soil type, Thind *et al.* (2012) reported that the drip method of irrigation resulted in a saving of 50% irrigation water compared to the check basin method. Further, 25% saving of fertilizer was also observed as the yields of the drip irrigated plots with 75% of fertilizer was equivalent to the check basin with 100% RDF and ultimately the nutrient use efficiency.

Results of the combined application of water soluble fertilizers with drip irrigation systems were found to significantly improve seed cotton yields than the use of drip irrigation alone (Pawar *et al.*, 2013). Further, water use efficiency, water productivity and the nutrient use efficiency also improved compared to the traditional flatbed irrigation systems.

Foliar fertilization

Pre-plant and side-dress method of fertilizer applications have been recommended so far, because it is convenient. This meets the nutrient requirement during the vegetative stage. But in long duration crops such as cotton the nutrient requirement, especially N and K, are the greatest when the crop is in the fruit formation stage. At this time root activity is reduced and concomitantly is the nutrient uptake (McMichael, 1990). The situation is worse under rainfed conditions, where the crop also experiences moisture stress coinciding with the fruit formation stage. In such situations, nutrient demand can be supplemented through foliar application and improve the productivity and use efficiency.

In general, crop demand for the nutrients is the highest during the boll development stage more so in the *Bt* hybrids compared to the traditional varieties and hybrids. Further, accelerated uptake begins earlier than the conventional hybrids. A high proportion of nutrients taken up by the crop are eventually translocated to the bolls. As the bolls fill, there is an enormous drain on the plants nutrient reserves and the plants' ability to take up nutrients may not meet the demand (Rochester *et al.*, 2006).

Organics and biofertilizers

Besides application of fertilizers in recommended doses, application of microbial consortia helps in reducing the nutrient requirement. Further, these microorganisms make the nutrients more available to the crop plants. Nalayini *et al.* (2010) demonstrated this clearly with the cotton varieties grown on deep Vertisols at Coimbatore. AE was better with the application of microbial inoculants (Table 17).

Table17: Agronomic efficiency (AE) of cotton at different levels of fertilizer with and without inoculation with microbials

Fertilizer level	AE_{NPK} (kg seed cotton/kg NPK)	
	Control	+ Microbials
0	-	-
50% RDF	7.3	9.4
75% RDF	7.6	13.5
100% RDF	8.4	7.1

Source: Nalayini *et al.* (2010)

As discussed previously, application of organic manures along with fertilizers in an integrated manner helps to build not only soil fertility but also sustains crop productivity. This has greater implications in rainfed ecosystems wherein soil organic matter helps to improve soil water retention. Earlier sections clearly indicated the benefits of soil moisture conservation on improvements in nutrient use efficiency.

Adopting sound agronomic practices would enable to reduce existing yield gaps, improve use efficiency and profitability of the production systems and also seem to be environmentally friendly.

Selected References

Bhalerao PD, Gaikwad G.S. 2010. Productivity and profitability of Bt cotton (*Gossypium hirsutum*) under various plant geometry and fertilizer levels. *Indian Journal Agronomy* 55: 60-63.

Biradar DP, Aladakatti YR, Basavanneppa MA, Shivamurthy D, Satyanarayana T. 2011. Assessing the contribution of nutrients to maximize transgenic cotton yields in Vertisols of northern Karnataka. *Better Crops – South Asia* 5(1): 22-25.

Blaise D. 2006. Effect of tillage systems on weed control, yield and fibre quality of upland (*Gossypiumhirsutum* L.) and Asiatic tree cotton (*G. arboreum* L.). *Soil and Tillage Research* 91: 207-216.

Blaise D. 2011. Tillage and green manure effects on Bt transgenic cotton (*Gossypium hirsutum* l.) hybrid grown on rainfed Vertisols of central India. *Soil and Tillage Research* 114: 86-96.

Blaise D. 2012. Fertilizer-K recommendation for cotton grown on Vertisols: is there a need for revision? In: *IPI-FAI-IPNI Roundtable on Refinement of K recommendations in Vertisols*, 20 March 2012, New Delhi. http://www.ipipotash.org/udocs/presentation_dr_blaise.pdf

Blaise D, Ravindran CD. 2003. Influence of tillage and residue management on growth and yield of cotton grown on a Vertisol over 5 years in a semi-arid region of India. *Soil Tillage Research* 70: 163-173.

Blaise D, Ravindran CD, Singh JV. 2006. Trends and stability analyses to interpret results of long-term effects of application of fertilizers and manure to rainfed cotton. *Journal of Agronomy and Crop Science* 192: 319-330.

Blaise D, Venugopalan MV, Raju A. R. 2014. Introduction of Bt cotton hbrids in India: Did it change the Agronomy: *Indian Journal of Agronomy* 59: 1-20.

Brar MS, Gill MS, Sekhon KS, Sidhu BS, Sharma P, Singh A. 2008. Effect of soil and foliar application of nutrients on yield and nutrient concentration in Bt cotton. *PAU Journal of Research* 45: 125-131.

Hallikeri SS, Halemani HL, Patil BC, Nanadagavi RA. 2011. Influence of nitrogen management on expression of cry protein in Bt-cotton (*Gossypium hirsutum*). *Indian Journal of Agronomy* 56:62-67.

Kranthi KR. 2012. Bt cotton Q&A Questions and Answers. *Indian Society for Cotton Improvement*, Mumbai. p. 70.

Mannikar ND. 1993. Fertilizer management in cotton.In: *Fertilizer management of commercial crops* (H.L.S.Tandon ed.), FDCO Publication, New Delhi, p. 26-46.

McMichael B.L. 1990. Root-shoot relationships in cotton. pp. 232-251. In: J.E. Box and L.C. Hammond (eds.). Rhizosphere Dynamics. Westview Press: Boulder, Colo.

NAAS. 2006. Low and declining crop responses to fertilizers. Policy paper N. 35, National Academy of Agricultural Sciences, New Delhi.

Nalayini P, Raj SP, Sankaranarayana K. 2011. Growth and yield performance of cotton (*Gossypium hirsutum*) expressing *Bacillus thuringiensis* var: Kurstaki as influenced by polyethylene mulching and planting techniques. *Indian Journal of Agricultural Sciences* 81: 55-59.

Nalayini P, Sankaranarayana K, Anandham R. 2010. Bioinoculants for enhancing the productivity and nutrient uptake of winter irrigated cotton (*Gossypium hirsutum*) under graded levels of nitrogen and phosphatic fertilizers. *Indian Journal of Agronomy* 55: 64-67.

Pawar DD, Dingre SK, Bhakre BD, Surve US. 2013. Nutrient and water use by Bt cotton (*Gossypium hirsutum*) under drip fertigation. *Indian Journal of Agronomy*58: 237-242.

Pettigrew WT, Adamczyk Jr JJ. 2006. Nitrogen fertility and planting date effects on lint yield and Cry1Ac (Bt) endotoxin production. *Agronomy Journal* 98: 691-697.

Prasad R. 2007. Strategy for increasing fertilizer use efficiency. *Indian Journal of Fertilisers* 3(1): 53-62.

Rochester IJ, Constable GA, Dowling C. 2006. Understanding Bollgard II nutrition. *Australian Cottongrower* 26(7): 14-16.

Salakinkop SR. 2011. Enhancing the productivity of irrigated *Bt*cotton (*Gossypium hirsutum*) bytransplanting technique and planting geometry *Indian Journal of Agricultural Sciences* 81: 150–153.

Seshadri V. 1985. Efficiency of applied nitrogen on varalaxmi hybrid cotton as influenced by different agro techniques. *Journal of Agronomy* 30 : 305-309.

Singh K, Singh H, Singh K, Rathore P. 2013. Effect of transplanting and seedling age on growth, yield attributes and seed cotton yield of Bt cotton (*Gossypium hirsutum*). *Indian Journal of Agricultural Sciences* 85: 508-513.

Singh J, Blaise D, Rao MRK, Mayee CD, Deshmukh MS. 2003. Assessment of agronomic efficiency of Bt cotton in rainfed Vertisols. *Journal of Indian Society of Cotton Improvement* 28:185-190.

Singh J, Blaise D, Deshmukh MS, Rao KV, Patil BC, Dhawan AS. 2004. Impact of integrated nutrient management and moisture conservation on rainfed cotton (*Gossypiumhirsutum*) under farmers' conditions. *Indian Journal of Agricultural Sciences* 74: 649-653.

Venugopalan MV, Blaise D. 2001. Effect of planting density and nitrogen levels on the productivity and nitrogen use efficiency of rainfed upland cotton (*Gossypium hirsutum*). *Indian Journal of Agronomy* 46:346-353.

Venugopalan MV, Pundarikakshudu R. 1998. Long term effects of nutrient management and cropping system on cotton yield and soil fertility in rainfed vertisols. *Nutrient Cycling in Agroecosystems* 55:159-164.

Venugopalan MV, Blaise D, Yadav MS, Deshmukh R. 2011. Fertilizer Response and Nutrient Management Strategies for Cotton. *Indian Journal of Fertilisers*, 7 (4) 82-94.

Venugopalan MV, Deshmukh R, Hebbar KB, Tandulkar NR. 2012. Productivity and nitrogen-use efficiency yardsticks in conventional and *Bt* cotton hybrids on rainfed Vertisols. *Indian Journal of Agricultural Sciences* 82: 641-644.

Venugopalan MV, Kranthi KR, Blaise D, Lakde S, Sankaranarayana K. 2013. High density planting system in cotton – The Brazil experience and Indian initiatives. *Cotton Research Journal, ISCI* 5(2): 172-185.

Vidyavathi GY, Ravi MV, Yadahalli GS, Upperi SN, Latha HS. 2013. Response of Bt cotton to different methods of fertilizer application under irrigated situation in north-eastern dry zone of Karnataka. In: Global Cotton Production Technologies *vis-a-vis* Climate Change, Cotton Research Development Association, Hisar, Haryana, pp. 197-201.

25

Enhancing Nutrient Use Efficiency, pp. 379-396
Editors: K. Ramesh, A.K. Biswas, B.L. Lakaria, S. Srivastava and A.K. Patra

Tobacoo and Oilseed Based Cropping Systems

A.K. Vishwakarma

ICAR-Indian Institute of Soil Science, Nabi Bagh, Bhopal – 462 038, India

Introduction

In India, most of the agricultural soils are deficient in one or more of the essential plant nutrients required for raising healthy crops and harvesting potential yield. Acidity, alkalinity, salinity, anthropogenic processes, nature of farming, and erosion are the main factors that can lead to soil degradation either alone or in combination of more than one factors. Addition of fertilizers and amendments are essential for ensuring a proper nutrient supply system for these soils and harvesting satisfactory yields. Besides, another reason for lower yields is low nutrient use efficiencies of native and applied nutrients. The general estimates of overall efficiency of applied fertilizer have been reported to be about or less than 50% for N, 10-15% for P, and about 40% for K. Fertilizer N recovery efficiencies from researcher managed experiments for major grain crops range from 46% to 65%, compared to on-farm N recovery efficiencies of 20% to 40%. A recent review of worldwide data on N use efficiency for cereal crops from researcher-managed experimental plots reported that single-year fertilizer N recovery efficiencies averaged 65% for corn, 57% for wheat, and 46% for rice (Ladha *et al.*, 2005).

Fertilizer use efficiency can be optimized by best management practices that apply nutrients at the right rate, time, and place. The highest nutrient use efficiency always occurs at the lower parts of the yield response curve, where fertilizer inputs are lowest, but effectiveness of fertilizers in increasing crop yields and optimizing farmer profitability should not be sacrificed for the sake of efficiency alone. There must be a balance between optimal nutrient use efficiency and optimal crop productivity. Plants that are efficient in absorption and utilization of nutrients greatly enhance the efficiency of applied fertilizers, reducing cost of inputs, and preventing losses of nutrients to ecosystems. In the face of the rapid rise in prices of synthetic fertilizers, emphasis in the agronomy research and development programme is focused on maximizing the efficiency of fertilizer use by crops while at the same time minimizing the nutrient losses to the environment. In this context, alternative sources of nutrients to mineral fertilizers and a more accurate prediction of the soil supply of the major nutrients are being advocated. There is need for developing techniques and practices that not only enhance nutrient absorption, transport, utilization, and mobilization in plant cultivars but also improve crop yields and improve nutrient use efficiency.

Nutrient use efficiency

Nutrient use efficiency (NUE)may be defined as yield per unit input of added nutrients. Nutrient use efficiency can be expressed in several ways. Mosier *et al.* (2004) described four agronomic indices commonly used to describe nutrient use efficiency: partial factor productivity (PFP, kg crop yield per kg nutrient applied); agronomic efficiency (AE, kg crop yield increase per kg nutrient applied); apparent recovery efficiency (RE, kg nutrient taken up per kg nutrient applied); and physiological efficiency (PE, kg yield increase per kg nutrient taken up). Crop removal efficiency (removal of nutrient in harvested crop as % of nutrient applied) is also commonly used to explain nutrient efficiency. Improvement of NUE is an essential pre-requisite for expansion of crop production into marginal lands with low nutrient availability. The nutrients most commonly limiting for plant growth are N, P, K and S. NUE not only depends on the ability of the plant to efficiently take up the nutrient from the soil, soil physical, chemical and biological properties governing nutrient availability in available form to the plants for a longer duration but also on transport, storage, mobilization, usage within the plant, and even on the environment. Based on the nutrient supplying power of soils as determined by soil test values, fertilizer recommendations are made. The aim is to get maximum economic yield with minimum inputs.

Agronomic interventions for enhanceing FUE

The fertilizer nutrients through fertilizers should be applied to the crop in conjunction with best management practices (BMPs) for achieving optimum nutrient efficiency. The following are some common agronomic techniques to improve the Fertilizer use efficiency (FUE).

1. Using best fertilizer source

In selecting the fertilizer care should be taken to select such type of fertilizer which will have minimum interaction with soil so as to ensure minimum immobilization of nutrients contained in the fertilizers. Ammonium and potassium ions in fertilizer may be immobilized by strong adsorption by 2:1 type clay minerals. High soil pH enhances this type of fixation. Identification of best source of fertilizer is pre-requisite for better crop production. Source of fertilizer depends on crop and variety, climatic and soil condition, availability of fertilizer, etc.

- Nitrogen: Ammonical or Nitrate
- Phosphorus: Water soluble or Citrate soluble
- Potassium: Muriate of potash
- Sulphur: Sulphate or Elemental S
- Multi-nutrient fertilizers: MAP, DAP, SSP, Nitrophosphates
- Multi-nutrient mixtures: Several combinations of NPK
- Fortified fertilizers: Neem-coated urea, Zincated urea, Boronated SSP, NPKS mix.
- Liquid formulations.

2. Using adequate rate and diagnostic techniques

The fertilizer recommendation must be in adequate quantity so as to meet the demand of crop at any point of growth. The fertilizer supply is made by diagnosing its requirement by any of the following method.

- State recommended generalized fertilizer dose or blanket recommendation
- Soil-test based fertilizer recommendations
- Soil-test crop response based recommendation
- Plant analysis for diagnosing nutrient deficiencies
- Chlorophyll meter and Leaf colour charts, etc.

3. Balanced fertilization

Balanced fertilization includes adequate supply of all essential nutrients, proper method of application, right time of application and nutrient interrelationships. A farmer participatory survey conducted in U.P. revealed that growers generally apply >200 kg N/ha and 45 to 60 kg P_2O_5/ha. However, use of K, secondary nutrients, and micronutrients is altogether missing. Farmers are experiencing declining responses to N and P due to omission of other essential nutrients in their fertilizer schedules. Adoption of balanced and judicious use of all needed nutrients can help improve crop productivity (Yadav *et al.*, 1993). In a recent review based on 241 site-years of experiments in China, India, and North America, balanced fertilization with N, P, and K resulted in first-year recoveries of 54% compared to recoveries of only 21% where N was applied alone (Fixen *et al.*, 2005).

a. **Adequate supply of all essential nutrients:** Due to more concentration and application of primary nutrients (NPK), soils develope deficiency symptoms for secondary and micro-nutrients. Hence, ignored elements must be added with the NPK (may be in minor quantity) to get higher yields in crops. Experimental results show that micro-nutrient application in soil or two foliar sprays increase the yield of crops up to 20%. Most crops are location and season specific - depending on cultivar, management practices, climate, etc., so it is critical that realistic yield goals are established and the nutrient are applied to meet the target yield. Over- or under-application will result in reduced nutrient use efficiency or losses in yield and crop quality.

b. **Proper method:** N and K can be applied as broadcasting and band placement. Water soluble P fertilizers are preferred to apply as band placement in neutral & alkaline soils. Citrate soluble P fertilizers are applied as broadcast method in acidic soils. Sulphate forms of S fertilizers are applied as broadcasting or band placement, whereas, elemental S and pyrite are applied as broadcasting method. Micronutrients are applied in minor quantity as foliar sprays and water soluble fertilizers are applied in fertigation.

c. **Right time:** Rate of application and timing are also important because balancing crop demand with nutrient supply improves nutrient uptake and increases crop response. Greater synchrony between crop demand and nutrient supply is necessary to improve nutrient use efficiency, especially for N. Split applications of N during the growing season, rather than a single, large application prior to planting, are known to be effective in increasing N use efficiency (Cassman *et al.*, 2002). Tissue testing, chlorophyll meters and leaf color charts have been highly successful in guiding split N applications in rice and maize production in Asia. Another approach to synchronize release of N from fertilizers with crop need is the use of N stabilizers and controlled release fertilizers. Nitrogen stabilizers (e.g., nitrapyrin, DCD [dicyandiamide], NBPT [n-butyl-thiophosphorictriamide]) inhibit nitrification or urease activity, thereby slowing the conversion of the fertilizer to nitrate (Havlin *et al.*, 2005).

d. **Right place:** Application method has always been critical in ensuring that fertilizer nutrients are being used efficiently. Determining the right placement is as important as determining the right application rate.

Numerous placements methods are available, but most generally involve are surface or sub-surface applications before or after planting. Prior to planting, nutrients can be broadcast (*i.e.* applied uniformly on the soil surface and may or may not be incorporated), applied as a band on the surface, or applied as a subsurface band, usually 5 to 20 cm deep. Applied at planting, nutrients can be banded with the seed, below the seed, or below and to the side of the seed. After planting, application is usually restricted to N and placement can be as a top dress or a subsurface side dress. In general, nutrient recovery efficiency tends to be higher with banded applications because less contact with the soil lessens the opportunity for nutrient loss due to leaching or fixation reactions. Placement decisions depend on the crop and soil conditions, which interact to influence nutrient uptake and availability.

e. **Nutrient interrelationships:** Antagonistic nature of fertilizers is to be considered while applying into the soil. Some of the fertilizer application in excess quantities, results in loss of yield and quality of crops. Ex. Application of excessive phosphorus @120 kg P ha^{-1} created an imbalance and reduced the seed and oil yields in soybean compared to 80 kg P ha^{-1}.

4. Integrated nutrient management

Organic manures, crop residues, green manures, bio-fertilizers etc. are to be blended in right manner along with inorganic fertilizers to meet the crop demand. All the possible and available organic sources are to be utilized efficiently to reduce the usage of inorganic fertilizers.

5. Utilization of residual nutrients

Some of the strategies to utilize the crop residues in efficient manner are:

- Knowledge on climatic conditions & carry-over effects of residues
- Blending rightly on cereal-legume rotations
- Mixing shallow-deep rooted crop rotations

6. Reducing losses of applied nutrients

a. Leaching losses: Nitrate fertilizers are easily lost in leaching. The extent of leaching is more in sandy soil than in clayey soils. The loss is more in bare soil than cropped soil. The losses can be reduced to an extent by suitable method and timing of application. Losses in leaching also occur when ammonical, calcium cyanamide and urea fertilizers are applied to soils. In extreme acid soils or sandy acidic soils, ammonium fertilizers are lost because ammonium ions cannot easily replace the aluminium ions in exchange sites. In alkali soils ammonium ions are subjected to volatilization loss. During summer losses from the ammonical, calcium cyanamide and urea are more because they are rapidly oxidized by nitrifying organisms. It has been proved that when urea mixed with neem seed crush is applied to paddy soil, the efficiency of urea is more. The activity of nitrifying organism is reduced and thereby the leaching loss is minimized. Some chemical compounds that inhibit nitrifying organisms are shown in table 1.

Table.1. Chemical compounds when applied with nitrogen fertilizers inhabit the nitrifying organisms and reduces the leaching loss.

Name	Rate of application
N- Serve (2 – chloro – 6 – trichloromethyl pyridine)	0.15 – 0.5 Kg/ha
AM (2 – Amino-4-chlore-6- methyl pyridine)	0.3 – 0.4% (of applied fertilizer)
Thiourea	1.5 – 2.5% (of applied fertilizer)

Applied potassic fertilizers are easily lost in drainage water in sandy soils and acid soils. However in clayey soils there is no appreciable loss. The loss can be minimized by adjusting the time of application to synchronize with maximum plant uptake period and also applying the fertilizer in 2 or more split doses. There are some slow release potassic fertilizers which are not subjected to leaching losses easily. *E.g.* potash frits, potassium meta phosphate and fused potassium phosphate.

b. Gaseous losses: The nitrogen compounds present in the fertilizer are lost as gases under certain soil condition. The following kinds of gaseous losses are noted.

1. Loss as ammonia under high pH conditions i.e. under alkaline conditions.
2. Loss as N_2, N_2O, NO due to denitrification.
3. Loss as N_2, N_2O, or NO under nitrification of ammonium fertilizers.

The above losses are determined by soil pH, fresh organic matter, moisture, temperature and type of micro- organisms present in soil. Losses in the form of ammonia under high pH conditions can be controlled by proper placement of urea. As far as possible ammonium fertilizer should be avoided. If there is no alternative to ammonium fertilizers, the fertilizer should be placed at least 4-6 inches below the surface. Wherever loss of N by denitrification processes is observed, urea should be used instead of ammonium and nitrate fertilizer.

c. Immobilization of fertilizer nutrients: Nutrient elements may be immobilized or fixed or converted into unavailable forms by one or more of the following three means

1. Chemical immobilization
2. Physicochemical immobilization
3. Microbiological immobilization

While selecting the fertilizer care should be taken to select those fertilizers which will have minimum interaction with soil and the time and mode of application should be such as to ensure minimum immobilization of nutrients contained in the fertilizers. Ammonium and potassium ions in fertilizer may be immobilized by strong adsorption by 2:1 type clay minerals. High soil pH enhances this type of fixation. In acid soils, the efficiency of water soluble phosphorus is very low. The water soluble phosphorus is immediately converted into insoluble phosphorus compounds. In such soils insoluble phosphatic fertilizer like rock phosphates should be utilized. Further a thorough mixing of the phosphorus with soil increases the efficiency of the fertilizers. In calcareous soils applied phosphatic fertilizers invariably converted into tricalcium phosphate – a compound from which phosphorus is not easily available. Under such conditions water soluble phosphorus are relatively more efficient than water insoluble P like rock phosphate. Microbiological fixation of fertilizer N may be a serious problem when un-decomposed organic materials of high C/N ratio are present in the soil. This type of immobilization is of short duration only. It can be overcome by application of larger quantities of water soluble N fertilizers or by allowing enough time for complete decomposition of unrecompensed organic matter.

d. **Interaction between different fertilizers:** It is a common practice to mix fertilizers containing different nutrient carriers, just prior to application. The efficiency of the following fertilizers will be lowered if mixed with the fertilizer or amendment noted against them. Ammonium sulphate - Basic slag Ammonium sulphate - Calcium carbonate, Super phosphate - Basic slag Ammonium phosphate - Basic slag, Ammonium phosphate - Calcium carbonate Super phosphate - Calcium carbonate, Urea can be mixed with all fertilizers.

e. Compaction and fertilizers efficiency: Soil compaction brings about the soil particles closer resulting in the decreased bulk density or apparent density. The combined effect of the changes in different physical properties of soils due to compaction is the poor response for nitrogen and phosphorus fertilizers. Under mechanized farming, soil compaction is most common observation.

f. Soil temperature – fertilizer response: Soil temperature is one of the important environmental factors affecting plant growth and fertilizers response of crops. To an extent the soil temperature is manageable by common management practices like tillage, mulching and irrigation. Soil temperature affects the fertilizers efficiency by changing solubility of fertilizers, concentration of solubilized fertilizer cation exchange and also the ability of the plants to absorb and use nutrients. Variation in temperature causes differential response to crop uptake, of NH_4^+ and NO_3^-N fertilizers. Below 13°C the uptake of N from NH_4^+ or NO_3^- fertilizers is almost nil. Maximum uptake of N by plants from NH_4^+ or NO_3^- fertilizers is observed in the range of 19° - 24°C. The efficiency of phosphatic fertilizers increases significantly with the increase in soil temperature from 10-35°C.

g. Soil moisture: One of the most important aspects in agriculture is the lack of moisture or presence of sufficient moisture in the soil. Efficient water management is complementary to efficient fertilizer management. Maximum efficiency of fertilizers can be obtained only in the presence of adequate soil moisture and vice versa. Excessive moisture leads to leaching loss of added fertilizers whereas lack of moisture results in poor availability of the added fertilizer and high osmotic pressure of soil solution due to concentration of fertilizers soils.

h. Plant characteristics: Different crops remove varied amount of plant nutrients from soils. There is also appreciable variation within varieties of same plants, between dicots and monocots etc. Since the roots are the principal organs through which plants take up nutrients, the rooting pattern and habit have an important bearing on the nutrient removal. Plants which develop a vigorous deep-root system during their early stages of growth require a large quantity of fertilizer as basal dressing. The fertilizer needs of deep rooted crops are generally lower than shallow rooted crops.

i. Fertilizer characteristics: The mobility of the fertilizer nutrients in the fertilizer, the type of fertilizer and the time and mode of application decide the efficiency of a fertilizer. The nitrogenous fertilizers are highly mobile and subjected to both downward and sideward mobility. Phosphorus is highly immobile. Potassium is also mobile but compared to nitrogen its mobility is lower. To get maximum efficiency N and K fertilizer should be applied in frequent split doses and phosphorus as basal dressing or near the root zone.

7. Use of liquid formulations

In order to utilize the land to its optimum potential and maximize crop production and conserving soil, liquid Fertilizers can be very helpful. These products can improve crops' nutrient use efficiency and reduce the amount of applied fertilizer needed. As a result, it is possible to get top crop performance while still employing sustainable agricultural practices. Here are four sustainable farming practices that can help increase nutrient use efficiency and boost crop yields by using liquid formulations.

a. Improve nutrient utilization: Rising fertilizer prices, environmental concerns and stagnant crop prices have forced the growers to look for ways to increase the nutrient use efficiency of their crops. For example, eNhance™ is a nitrogen additive that works within the plant, nutritionally fortifying it so that it can use nitrogen more effectively. This formulation helps reduce the needed rate of urea ammonium nitrate (UAN) solutions, often with a yield response comparable to full rate UAN applications. Liberate Ca™ aids with nutrient mobilization throughout the plant and improves nutrient availability in conservation tillage environments. It also helps with availability of nutrients in the root zone.

b. Maximize crop response to nutrient applications: Liquid formulations have high levels of absorption and effectiveness within the plant, which maximizes crop response to them. For example NResponse™, a premium nitrogen product that quickly assimilates into the crop for fast-acting results. Rate of application and timing are also important because balancing crop demand with nutrient supply improves nutrient uptake and increases crop response.

c. Utilize fertilizers with a low salt index: Fertilizers with a high salt index can damage plant tissues, making nutrient absorption more difficult. Liquid Fertilizers' products have low salt indexes, which makes it safer to use them for foliar applications and apply them near the active root zone.

d. Reduce amount of applied fertilizer: Volume is rarely the determining factor when it comes to eliciting a positive plant response; rather what matters is the type of product used and the plant's ability to utilize it. Over application of nutrients is wasteful and unnecessary when smaller amounts of product can produce the same or better results. Balanced formulations that include micronutrients make liquid products highly usable and efficient. There is less risk to the environment when more of the applied nutrients end up in food and fiber and not in our streams and lakes.

8. Inter cultivation

Cultivation practices taken up after sowing of crop is called inter-cultivation. It is otherwise called as after operation. There are three important after cultivation processes *viz.*, Thinning and gap filling, weeding and hoeing and earthing up.

1. Thinning and gap filling

The objective of thinning and gap filling process is to maintain optimum plant population. Thinning is the removal of excess plants leaving healthy seedlings. Gap filling is done to fill the gaps by sowing of seeds or transplanting of seedlings in gap where early sown seed had not germinated. It is a simultaneous process. Normally, these are practiced a week after sowing to a maximum of 15 days.

In dry land agriculture, gap filling is done first. Seeds are dibbled after 7 days of sowing. Thinning is done after gap filling; in order to avoid drought. It is a management strategy to remove a portion of plant population to mitigate stress is referred to as mid-season correction.

2. Weeding and hoeing

Weeding is removal of unwanted plants. Weeding and hoeing is a simultaneous operation. Hoeing is disturbing the top soil by small hand tools and helps in aerating the soil. This reduces the competition and more uptake of nutrients by the crop and hence improves nutrient use efficiency.

3. Earthing up

It is a dislocation of soil from one side of a ridge and to be placed nearer the cropped side. It is carried out in wide spaced and deep rooted crops. It is done around 6-8 weeks after sowing / planting in sugarcane, tapioca, banana, etc.

4. Other inter cultivation practices

Topping

Removal of terminal buds. It is done to stimulate auxiliary growth. Practiced in cotton and tobacco crops to encourage auxiliary growth and prevent apical dominance.

Propping

Provision of support to the crop is called propping. Practiced in sugarcane commonly. It is done to prevent lodging of the crop. Cane stalks from adjacent rows are brought together and tied with their own trash and old leaves.

De-trashing: Removing of older leaves from the sugarcane crop.

De-suckering

Removal of auxiliary buds and branches which are considered non-essential for crop production and which removes plant nutrients considerably are called suckers. *Ex.* Tobacco.

Intercropping in sugarcane for better NUE

Sugarcane + Potato

- Seed rate: Sugarcane- 60 q/ha, Potato-25 q/ha

- 1:2 row ratio, sugarcane planted at 90 cm and two rows of potato are accommodated at 30 cm spacing
- Weed control through Simazine @ 1 kg a.i./ha as pre-emergence followed by hoeing and earthing up at 30 and 50 DAP respectively
- Apply N:P:K fertilizers for Sugarcane @ 150:60:60, for Potato @ 120px:80:100.

System yields: Potato-272 q/ha and Sugarcane-90.6 t/ha with a profit margin of Rs.1, 06,736 /ha

Sugarcane + Rajmash

- Seed rate: Sugarcane-60 q/ha, Rajmash- 80 kg/ha
- 1:2 row ratio, sugarcane planted at 90 cm accommodating two rows of rajmash at 30 cm spacing
- Apply N:P:K fertilizers for sugarcane @ 150:60:60, for Rajmash @ 80:40:30.
- Control weeds through Pendimethalin as pre-emergence @ 2 kg a.i./ha followed by 2 to 3 hoeing after harvest of rajmash

System yields: Sugarcane-86.8 t/ha, Rajmash grain-17.5 q/ha with profit margin of Rs.89, 884/ha

Sugarcane + Mustard

- Seed rate: Sugarcane- 60 q/ha, Mustard: 5 kg /ha
- 1:2 row ratio, sugarcane planted at 90 cm and two rows of mustard accommodated at 30 cm spacing
- Apply N:P:K fertilizers for Sugarcane @ 150:60:60, for Mustard @ 30:20:0.
- Control weeds through Pendimethalin @ 2 kg a.i./ha as preemergence followed by two hoeing at 30 and 60 days after harvest of mustard

System yield: Sugarcane-86.8 t/ha, Mustard-17.5 q/ha with a profit margin of Rs.89, 884/ha

Sugarcane + Wheat

- Seed rate: Sugarcane: 60 q/ha, Wheat: 75 kg /ha
- 1:3 row ratio, sugarcane planted at 90 cm and three rows of wheat accommodated at 20 cm spacing through IISR Planter cum Seeder under FIRB system

- Apply N:P:K fertilizers for Sugarcane @ 150:60:60, for Wheat @ 90:45:45.
- Control weeds through Pendimethalin @ 2 kg a.i./ha as pre-emergence followed by two hoeing at 30 and 60 days after harvest of wheat

System yield: Sugarcane-74.5 t/ha, Wheat-39.4 q/ha with a profit margin of Rs.56329 /ha.

Sugarcane ratoon management

- At initiation dismantling of ridges, stubble shaving and off-barring is recommended for good ratoon yield.
- Gap filling with slip setts / pre-germinated setts / polybag raised settlings is a must if gaps exceed 15% of normal crop stand. More than 45 cm distance between subsequent clumps is taken as gap.
- Paired row system of planting (120px:30) reduces gaps and optimizes plant population in subsequent ratoon. Thus it produces higher yields compared to sole planting at 90 cm.
- Trash mulching (10 cm thick) in alternate rows for conserving soil moisture, minimizing weed infestation and maintaining soil organic carbon.
- Application of potassium (80 kg K_2O/ha) with irrigation water in standing plant cane one month prior to harvesting improves bud sprouting, number of millable canes and yield of succeeding ratoon crop.

Nutrient Management in tobacco

- The inorganic constituents of the tobacco plant range from 15 to 25% of its dry matter, the distribution being about 50% in the leaf, 30% in the stem and 20% in the roots.
- The mineral content varies widely, depending on soil and climatic conditions, in different types and varieties of tobacco.
- The major nutrients essential for growth the development of tobacco plant are nitrogen, phosphorus, potassium, calcium, magnesium and sulphur.
- The minor elements equally essential but in micro-quantities are boron, manganese, iron, zinc, molybdenum and copper.
- Chlorine is also recognized as an essential micro nutrient, at least for certain plants including tobacco.

When the supply of these elements becomes limited, the plants exhibit characteristic symptoms of deficiency.

Nitrogen

- Nitrogen plays a key role in the production of tobacco. Nitrogen requirement for different types of tobacco varies widely from 20 kg N/ha for FCV tobacco on black clayey soils to 180 kg N/ha for bidi-tobacco.

Types of tobacco	Nitrogen in kg/ha
Flue-cured Virginia in heavy soils of A P	40-45
Flue-cured Virginia of light soils of A.P.	40-50 (S.Light soils) 60-70 (N.Light soils) ??
Natu tobacco of AP	44
Bidi in Karnataka	45 (As Ammonium sulphate)
Bidi in Gujarat	90 (As Ammonium sulphate)
Cigars and cheroot in Tamil Nadu	50 (As Ammonium sulphate)
Chewing in Bihar	56 (As Ammonium sulphate)
Hookah and chewing tobacco in U .P .	150
Wrapper tobacco in W.B	125

Doses based on type of tobacco

Method of application

- For majority types of tobacco nitrogen is applied as basal i.e one week before planting.
- But for FCV tobacco grown in light soils of AP, Cigar & cherrot tobacco in Tamil Nadu and for chewing tobacco in Bihar nitrogen is applied in two equal splits. One at two weeks before planting and another at 3-4 weeks after transplanting (FCV in Light soils of AP) or 6-7 weeks after transplanting.

Phosphorus

- The requirement of phosphorus by flue-cured tobacco is generally low. It ranges from 20 to 40 kg/ha. Only 10% of the applied phosphorus is recovered by the tobacco crop. Better utilization of phosphorus takes place when applied along with nitrogen and particularly when the fertilizers are applied as a band at a depth of 15 cm.

Dosage

- 30 kg/ha for FCV tobacco grown in heavy soils of Andhra Pradesh
- 60-80 kg/ha for FCV tobacco grown in light soils of AP

Potassium

- Potassium is an essential element for the normal and healthy growth of tobacco plant. The quality parameters that are related to potassium are physical appearance and burning quality. Among the different sources of potassium, potassium sulphate or potassium nitrate is preferred but not potassium chloride.

Dosage

- 50 kg/ha for FCV tobacco grown in heavy soils of Andhra Pradesh
- 80-100 kg/ha for FCV tobacco grown in light soils of AP
- 150 kg/ha for Hookah and chewing tobacco in west Bengal
- 112-125 kg/ha in Wrapper tobacco in West Bengal.

Calcium

- The actual requirement of Calcium by tobacco plant appears to be somewhat less than that for K under average conditions; the content of Ca in the leaf exceeds that of K because of its more abundant supply.

Dose

- Normally Ca need not be applied to the crop. However, it is being supplied with P in superphosphate.
- In acidic soils liming is found to be beneficial.

Magnesium

- A normal tobacco plant contains MgO to an extent of 0.8% of dry matter, 1.10% in the leaves and 0.35% in the stalk. Magnesium deficiency, called 'sand drown'. Magnesium-deficient leaf when cured, produces chlorotic area which are relatively dark and of uneven colour. In sandy loams of Karnataka and East and West Godavari Districts of Andhra Pradesh 12-15 kg of MgO in the form of dolomite is recommended to safeguard against Mg-deficiency which may sometimes occur under repeated irrigation or heavy rainfall.

Sulphur

- Sulphur deficient plants are pale from top to bottom and lower leaves do not burn up. Sulphur deficiency can be partly corrected with side dress application of fertilizers such as potassium sulphate, magnesium sulphate, and ammonium sulphate.

Micronutrients

- Micronutrients play a significant role in influencing the yield and quality of tobacco.
- Micronutrients act as a co-factors in several enzyme systems (like catalases, peroxidases, polyphenol oxydases etc.) that bring about certain vital transformations during curing effecting the final quality of the leaf.
- It has been established by several investigators that the micronutrients when present in toxic or deficient quantities change the quality of the tobacco leaf. Gray tobacco is attributed to the toxic leaves of Fe, Mn and Zn.
- Copper deficiency causes metabolic changes in leaf at maturity, resulting in low total sugars and high total and protein nitrogen which leads to poor quality.
- Boron deficiency is considered injurious to leaf quality as it hinders the translocation on carbohydrates.
- Micronutrient status of tobacco growing areas in India revealed that the soils in general are having available micronutrients higher than the critical limits except Zinc in southern light soils and boron in northern light soils.
- Also in some areas of northern light soils, Zn content of soil is marginally lower than the critical limits. In traditional black soils it is higher than the critical limits.
- In spite of low levels of micronutrients in some areas, the leaf analysis showed sufficient quantities of micronutrients.

Chloride

- Chloride is an essential micro-nutrient for tobacco and plays an important role in influencing the leaf quality and leaf burn. One of the principle effects of chloride in growing leaf is to increase water content and turgor, which in turn tends to produce a larger, smoother and thinner leaf. When present in small quantities, it improves the yield and certain quality factors like color, moisture content and keeping quality. Larger amount of chloride produces cured leaves of muddy and uneven color with excessive hygroscopicity and poor burn and such leaf is commonly known as saline leaf. The main sources of chloride supply are soil, fertilizers and irrigation water. Soils containing >100 ppm of chlorides are not suitable for tobacco cultivation. Irrigation water containing >40 ppm are not suitable for irrigation.

Practical approaches to increase efficiency of fertilizers

1. High yielding varieties should be grown wherever possible because high yielding varieties of crops give higher yields than local varieties without fertilization as well as a higher per unit response to fertilizer even at the lower rate of application.
2. Sowing of crop at optimum time because deviation from normal planting or sowing time suited for a particular crop variety in a particular locality will adversely affect the use efficiency of fertilizers.
3. Sowing at optimum spacing and maintenance of optimum plant population is essential to get maximum benefit from the applied fertilizers.
4. Ensure effective organic matter recycling, in order to maintain fertility and productivity of soil. Response for a nutrient is generally higher in soils supplied with adequate amount of organic matter.
5. Ensure adequate supply of nitrogenous fertilizers to soils where organic manures having high C: N ratio is added to compensate biological locking up of N in microorganisms.
6. Include a legume either in rotational sequence or as an intercrop. Legumes besides fixing atmospheric N, transform non-available native phosphorus and precipitated or fixed fertilizer P into available forms.
7. There should not be excess water in the soil particularly at the time of fertilizer application. Fertilizer application should be made after draining the excess water. Excessive irrigation should be avoided as it results in the loss of N and K fertilizers. Further, there is also gaseous loss of N under waterlogged condition.
8. Crop response to phosphatic fertilizers is generally more in dry seasons.
9. Balanced fertilization should be practiced based on the soil test values. Efficiency of a straight fertilizer containing nutrient) depends on the sufficiency of other nutrient in the soil.
10. The phosphates in general are more efficient when the entire dose applied as basal dressing, potash entire quantity as basal dressing or part as basal and rest in split doses depending on the soil texture and the nitrogen in 2-3 (or 4) split doses.
11. Water soluble phosphatic fertilizers should be placed 4-6 cm below the soil and 4-6 cm away from the seeds to ensure maximum availability to plants. Insoluble fertilizers should be thoroughly mixed in soil.

12. Sometimes it is better to cure the urea by mixing the urea with 5-10 parts of soil thoroughly and keeping the mixture for overnight. This enhances the conversion rate of urea into ammonia carbonate.
13. Whenever N loss in drainage water is suspected due to efficiency of the N fertilizer can be increased by mixing the fertilizer with crushed neem seed (5:1 parts).
14. Foliar application of fertilizers should be resorted to under certain soil conditions and climatic conditions.
15. Zinc deficiency is becoming widespread day by day, application of zinc sulphate at the rate of 10-25 kg of zinc sulphate as basal dressing, not only corrects the zinc deficiency but also enhances the efficiency of the other applied fertilizers.
16. Adverse soil condition should be corrected by using appropriate amendments, to get maximum benefit from fertilizer.
17. Ensure weed, pest and disease free environment to enhance the efficiency of applied fertilizers.

Conclusion

Improving nutrient efficiency is a worthy goal and fundamental challenge for increasing crop production and ensuring environmental security. The opportunities are there and tools are available to accomplish the task of improving the efficiency of applied nutrients. Agronomic practices play an important role in determining crop productivity and in turn influence the nutrient use efficiency significantly. While applying techniques and practices for improving nutrient use efficiency we must be cautious that improvements in efficiency do not come at the expense of the farmers' economic viability or the environment. Judicious application of fertilizer BMPs, right rate, right time, and right place targeting both high yields and nutrient efficiency will benefit farmers, society, and the environment alike.

Selected References

Cassman KG, Dobermann A,Walters DT. 2002. Agroecosystems, nitrogen use efficiency, and nitrogen management. *Ambio*. 31: 132-140.

Fixen PE. 2005. Understanding and improving nutrient use efficiency as an application of information technology. In: *Proceedings of the Symposium on Information Technology in Soil Fertility and Fertilizer Management, a satellite symposium at the XV International Plant Nutrient Colloquium*, Sep. 14-16, 2005.Beijing, China.

http://www.iisr.nic.in/research/technologies.htm

http://www.ctri.org.in/pages/agronomy.pdf

Ladha JK, Pathak H, Krupnik TJ, Six J, van Kessel C. 2005. Efficiency of fertilizer nitrogen in cereal production: retrospects and prospects. *Advances Agronomy* 87: 85-156.

Mosier AR, SyersJK,FreneyJR. 2004. Agriculture and the Nitrogen Cycle. Assessing the Impacts of Fertilizer Use on Food Production and the Environment.Scope-65. Island Press, London.

Prasad R. 1996. Management of fertilizer nitrogen for higher recovery. In: *Nitrogen Research and Crop Production*. (Ed. H.L.S.Tandon), FDCO, New Delhi, India. pp. 104-115.

Reddy SR. 2004. Principles of Agronomy. pp.197-249 Kalyani Publishers

Yadav RL. 1993. Agronomy of sugarcane: Principle and Practice. Ist Eds. International Book Distributing Co. Lucknow, India.

26

Enhancing Nutrient Use Efficiency, pp. 397-410
Editors: K. Ramesh, A.K. Biswas, B.L. Lakaria, S. Srivastava and A.K. Patra

Rapeseed-Mustard Based Cropping Systems

S.S. Rathore, Kapila Shekhawat, B.K. Kandpal and O.P. Premi

ICAR-Directorate of Rapeseed-Mustard Research, Sewar, Bharatpur Rajasthan–321303, India

Introduction

Poor nutrient use efficiency result in loss of nearly 0.8 million tonnes of nitrogen, 1.8 million tonnes of phosphorus and 26.3 million tonnes of potassium annually. Soil health is further deteriorated by imbalanced application of nutrients and excessive mining of micronutrients, leading to deficiency of macro- and micronutrients in the soils (ICAR Vision 2030). Due to the imbalanced use of plant nutrients, mining of nutrients is considered as the main cause for decline in crop yield and crop response ratio. About 8–10 Mt of NPK is mined from soil annually in India. Soils are also being depleted of secondary and micronutrients. About 42% of the soils are deficient in sulphur, 48.5% deficient in zinc and 33% deficient in boron (Gupta *et al.*, 2007). Nutrient use efficiency has been defined in many ways in diverse contexts (Clark, 1990; Blair, 1993). Most definitions mean, efficiency is the ability of a system to convert inputs into preferred outputs, or to minimize inputs waste. On genetic and physiological basis plants have profound effects on their abilities to absorb and utilize nutrients under various environmental and ecological conditions. When considering nutrient efficiency, supply or availability or amount of a mineral nutrient is the input, and plant growth, physiological activity or yield are typical outputs. "Efficiency" is the relationship of output to input. This is expressed as a simple ratio, such as kg yield per kg fertilizer, or g of plant dry weight per mg of nutrient, but as the amount of input and output vary, the ratio between them is rarely fixed, so efficiency is most comprehensively described by the entire relationship of output as a function of input. Agricultural productivity could only be enhanced through increasing the resource use efficiency by improving soil fertility. Rapeseed-mustard area, production and productivity in India during 2012-13 was 5.92 Mha, 7.82 mt and 1145 kg/ha (Agricultural Statistics at a Glance, 2013). Enhancement of NUE by plants could lessen fertilizer input costs, reduces the nutrient losses, and boost the crop productivity. There is scope to increase the mustard productivity up to 2000.0 kg/ha, this can only be achieved by enhancing input use efficiency and among inputs the fertilizers nutrient sources has great role to play.

Rapeseed-mustard production zones in India

Zone I

This zone comprises Jammu & Kashmir (Khudwani) and Himachal Pradesh (Kangra, Dhaula Kuan, Bajaura).

Zone II

Zone II includes Jammu & Kashmir (Chatha), Punjab (Ludhiana, Bathinda), Haryana (Hisar, Bawal, Kaul), Rajasthan (Navgaon, Sriganganagar, Bikaner) and Delhi (TERI, New Delhi).

Zone III

States of Uttar Pradesh (Kanpur, Faizabad, Varanasi), Uttaranchal (Pantnagar), Madhya Pradesh (Morena, Indore) and Rajasthan (Kota, Bharatpur)

Zone IV

Zone IV spreads in Rajasthan (Jobner, Sumerpur, Mondor) ,Gujarat (Junagadh, S.K. Nagar) and Maharashtra (Jalgaon, Nagpur).

Zone V

This zone comprises Chhattisgarh (Raipur, Jagdalpur, Bihar (Dholi), Jharkhand Kanke (Ranchi), West Bengal (Berhampore), Orissa (Bhubaneswar), Assam (Shillong) and N.E.Hill (Imphal).

Delineation of crop ecological zones can be used in formulating the strategies for enhancing NUE. For example, the entire geographical area under any crop can be classified into four basic groups based on different combinations of two segregating criteria , *viz* contribution to total area and productivity. Out of these, our efforts must be concentrated on two groups, *viz*, High area –Low productivity and Low area- High Productivity zones. This will give better marginal efficiency in the efforts to increase production and productivity of oilseed crops. An indicative state-wise crop zoning of three major oilseed crops is given in table 1.

Table 1: Indicative state-wise crop zoning based on area and productivity

Crop/ Zone	Low Area Low Productivity	Low Area High Productivity	High Area Low Productivity	High Area High Productivity
Rapeseed Mustard	Jammu & Kashmir Himachal Pradesh	Gujarat Haryana	West Bengal Uttar Pradesh	Rajasthan

The rapeseed-mustard group broadly includes Indian mustard, yellow sarson, brown sarson, raya and toria crops. Indian mustard (*Brassica juncea* L.) (*Czernj. & Cosson*) is predominantly cultivated in Rajasthan, UP, Haryana, Madhya Pradesh and Gujarat. It is also grown under some non-traditional areas of South India including Karnataka, Tamil Nadu and Andhra Pradesh. The crop can be raised well under both irrigated and rainfed conditions.

Definition of nutrient efficiency

It is difficult to explain nutrient use efficiency due the fact that there is no sole or generally accepted definition of Nutrient use efficiency. Different definitions are used in literature to describe the agronomic and physiological range of Nutrient efficiency referring to external and internal Nutrient-status. However, the evaluation of NUE is useful to differentiate plant species, genotypes, and cultivars for their ability to absorb and utilize nutrients for maximum yields. The Baligar *et al.*, 2001 reported that NUE is based on (a) uptake efficiency (acquire from soil, influx rate into roots, influx kinetics, radial transport in roots are based on root parameters per weight or length and uptake is also related to the amounts of the particular nutrient applied or present in soil), (b) incorporation efficiency (transports to shoot and leaves are based on shoot parameters) and (c) utilization efficiency (based on remobilization, whole plant *i.e.* root and shoot parameters).

Most definitions of Nutrient use efficiency refer to the external nutrient-supply in terms of the agronomic meaning benchmarking seed yield as the essential objective. Nutrient-efficient cultivars are resulting in high seed yield under conditions of limited nutrient-supply (Graham, 1984).

Enhancement of NUE in oilseed *Brassica*

Nutrient use efficiency of Oilseeds *Brassica* is greatly influenced by the rate, source and method of fertilizer application. The rate of particular nutrient major or minor depends upon the initial soil status, climate, topography, cropping system in practice and crop. In plants, N management can be divided into two main phases , first phase, i.e. the vegetative phase, during which, sink organs (roots, young leaves) evolved for the assimilation of inorganic N via the nitrate assimilatory pathway (Hirel and Lea, 2001). In oilseed rape, the requirement for N per yield unit is higher than in cereal crops (Hocking and Strapper, 2001). Oilseed rape has a high capacity to take up nitrate from the soil (Laine *et al.*, 1993), and thus to accumulate large quantities of N that is stored in vegetative parts at the beginning of flowering. However, in oilseed rape, yield is half that of wheat, due to the production of oil, which is costly in carbohydrate production. Since oilseed rape N content in the seed is not much higher (3% in oilseed rape and 2% in wheat on average), an important part of the N stored in the vegetative organs is not used. Besides, a large quantity of N is lost in early falling leaves (Malagoli *et al.*, 2005).

Improved agro-techniques for enhancement of nutrient use efficiency

Improved agro-techniques for enhancement of nutrient use efficiency also enhance nutrient use efficiency of Oilseed *Brassica*. Advanced agronomic management improves nutrient use efficiency of Oilseed *Brassica* (Table 2).

Table 2: Integrated N-management strategies affecting nutrient use efficiency-efficiency in plant production

S N	Techniques	Aim
1	Cropping system	Increased uptake and utilization of soil and fertilizer nutrient by cultivation of nutrient specific -efficient crops, reduction of fallow frequency and rotation of shallow/deep rooting crops. Uptake of soil-nutrient and mineralized plant residue-nutrient therebyreducing nutrient losses by leaching and increasing supply to succeeding crops.
2	Oilseed *Brassica* cultivar	Increased uptake and utilization of soil and fertilizer nutrient bycultivation of nutrient-efficient cultivars
3	Irrigation and management practices	Increased uptake and utilization of soil and fertilizer nutrient by well-grown crops. Timing, intensity, and depth of soil cultivation control the soil mineral-nutrient release.
4	Precise forecast of fertilizer nutrient-requirement (*e.g.* soil and plant nutrient-tests, sensor controlled fertilization, modelling soil nutrient-supply)	Increased uptake and utilization of soil and fertilizer nutrient by considering available soil mineral nutrient at the beginning of the growing season and nutrient-mineralization during the growing season.

(*Contd.*)

5	Form of fertilizer (*e.g.* mineral fertilizer *vs*. organic manure, urea *vs*. ammonium vs. nitrate, use ofurease and nitrification inhibitors)	Avoidance of nutrient-losses caused by specific nutrient forms/nutrient-transformations in the soil, increased physiological efficiency of nutrient by considering plant specific preferences of certain forms of N (NH_4^+ *vs*. NO_3)
6	Timing of nutrient application	Reduction of nutrient losses (as in case of N, NO_3, N_2) at the beginning of the growing season,increased physiological efficiency by specific stimulation of harvestable organs
7	Technique of fertilizer nutrient application (*e.g.* surface *vs*. incorporation vs.broadcast vs. banded)	Reduction of losses (NH_3^+), improved spatial availability of nutrient, reduction of N-immobilization
8	Management of crop residues	Control of N-mineralization during fallow and Immobilization of soil mineral-nutrients

Conservation tillage in the form of zero tillage is also more productive and enhanced nitrogen use efficiency with 80 kg N/ha (AICRP-RM, 2007). Increase in the nitrogen level up to 60 kg N/ha consistently and significantly increased the number of primary branches, number of seeds per siliquae and 1000 seed weight (Sharma *et al.,* 2007), however, increasing the nitrogen level up to 90 kg/ha increased the number of secondary branches per plant, number of siliquae per plant, seed and straw yield with maximum cost benefit ratio of 3.03 (Sah *et al.*, 2006). Split application of total nitrogen in three equal doses one-each as basal, second after first irrigation and remaining one-third after second irrigation resulted in maximum increase in yield attributes and yield of *Brassica juncea* compared to application of total nitrogen in two split doses (Reager *et al.,* 2006). Top-dressing of N fertilizers should be done immediately after first irrigation. Delaying of first irrigation, results in yield reduction of mustard crop. In case of nitrogen applied with pre-sowing irrigation single application of nitrogen was at par with split application (Sidhu and Sandhu, 1995).

The dry matter/plant significantly enhanced with the application of phosphorus up to 60 kg/ha. P application up to 40 kg /ha increased the plant height, branches per plant and leaf chlorophyll content. The uptake of NPK and sulphur by both seed and stover increased significantly with successive increase in nitrogen levels up to 120 kg N/ha, sulphur levels up to 60 kg S/ha and P_2O_5 level up to 60 kg P_2O_5/ha. Seed yield and yield attributes increased while oil content decreased with increasing level of nitrogen up to 120 kg/ha. Different levels of phosphorus increased seed yield, maximum being at 80 kg P/ha due to higher number of secondary branches/plant and consequently siliquae/plant. Oil content also increased with increase in levels of N, P_2O_5 and S. Activities of all nitrogen assimilating enzymes named; nitrate reductase, nitrite reductase, glutamine synthetase and glutamate synthetase were found to be maximum at 100 kg N/ ha.

Sulphur fertilization and nutrient use efficiency

Among the oilseed crops, rapeseed-mustard has the highest requirement of sulphur. Sulphur promotes oil synthesis. It is an important constituent of seed protein, amino acid, enzymes, glucosinolate and is needed for chlorophyll formation (Holmes, 1980). Sulphur increased the yield of mustard by 12 to 48% under irrigated, and by 17 to 124% under rainfed conditions (Aulakh and Pasricha, 1988). In terms of agronomic efficiency, each kilogram of sulphur increases the yield of mustard by 7.7 kg (Katyal *et al.,* 1997).

Oil content in Canola-4 and Hyola-401 is 3% higher than the hybrid 'PGSH-51' due to the effect of various doses of nitrogen and sulphur, while the oleic acid content in these hybrids are double than 'PGSH-51'. 'PGSH-51' had erucic acid ranging from 23.2 and 29.4%. At higher sulphur level there is 2-3% reduction in erucic acid content. However, lower level of nitrogen reduced erucic acid content by 3% with a concomitant increase in oleic acid (Table 3). Higher doses of sulphur along with low doses of nitrogen affect the chain elongation enzyme system thereby leading to reduction in erucic synthesis.

A significant increase in yield was observed with increase in sulphur levels up to 40 kg S/ha in mustard based cropping system. At Bawal, the highest seed yield of mustard was recorded in green gram-mustard cropping sequence while the lowest (2686 kg/ha) in pearl millet-mustard sequence. In rice-mustard sequence, the optimum seed yield of mustard was obtained at 40 kg S/ha at Behrampore and for blackgram- mustard at Dholi. Each successive increase in S level increased seed yield up to 20 kg S/ha at Dholi and Ludhiana, 40 kg S/ha at SK Nagar and 60 kg S/ha at Berhampur and Morena conditions (AICRP-RM, 2008).

Role of micronutrients in enhancing nutrient use efficiency of oilseed Brassica

Mustard, in general is very sensitive to micronutrient deficiency, especially zinc and boron. The increase in seed yield was 8.5% at 12.5 kg $ZnSO_4$/ha. The harvest index (HI) was significantly affected by Zn application, although seed yield showed diminishing return with additional $ZnSO_4$ doses (Table 4).

Table 3: Effect of N and S levels (kg/ha) application on Fatty acid composition and glucosinolate content in *Brassica juncea* cv. Varuna at Ludhiana.

N (kg/ha)	S (kg/ha)	Glucosinolate content (μ moles/g in defatted meal)	Palmitic acid	Stearic acid	Oleic acid	Linoleic acid	Linolenic acid	Eicosenoic acid	Erucic acid
75	0	64	2.61	1.17	11.78	14.99	6.48	50.91	11.80
75	20	72	2.88	1.31	10.15	14.53	5.14	52.75	12.28
100	0	52	2.58	1.58	13.16	15.31	7.01	49.55	10.57
100	20	42	2.91	1.65	11.94	15.06	6.13	49.63	12.18
125	0	52	3.01	1.33	12.19	16.17	5.91	47.71	12.26
125	20	42	4.42	1.31	16.12	16.55	6.57	44.77	9.55

Source: AICRP-RM, 2007

Table 4. Effect of Zn on yield and yield attributes of Indian mustard

$ZnSO_4$ (kg/ha) levels	Seed yield (kg/ha)	Secondary branches/ plant	Oil content (%)	Oil yield (kg/ha)	Protein (%)	Protein yield (kg/ha)	Harvest index (%)
0	1161	6.5	40.2	465.6	22.1	255.2	21.6
12.5	1260	8.1	39.9	501.1	22.5	281.9	22.4
25.0	1336	9.6	39.9	532.4	22.6	301.6	22.9
50	1414	12.4	39.9	570.0	22.5	318.6	22.2
CD at 5%	33	0.7	NS	22.8	NS	18.8	0.8

Source: AICRP-RM, 2000

The response of various ideotype to the applied micronutrients varies considerably. The response of Indian mustard varieties, *viz.* 'Pusa Bold' and 'Vardan' to applied zinc was found higher (AICRP-RM, 2000) as compared to Varuna, RH- 30 and Aravali (Fig. 1).

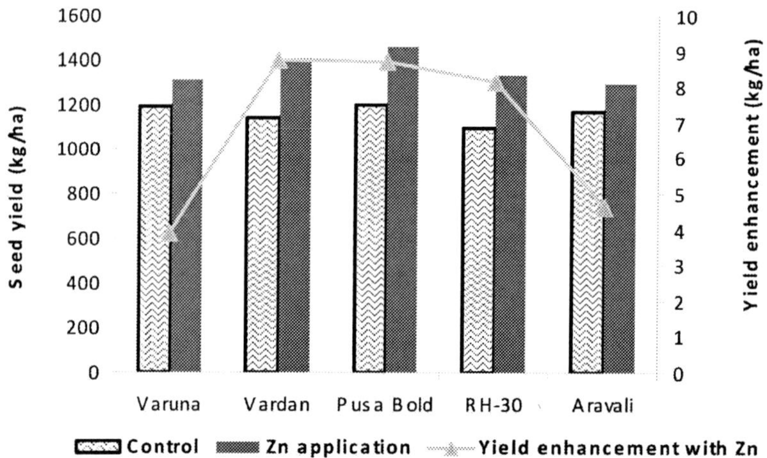

Fig. 1: Influence of zinc application on seed yield of different cultivars of mustard.

The concentration of Zn at flowering, pod formation stage, concentration and uptake of Zn in straw and grain at maturity and uptake of Zn in grain and straw at maturity of Indian mustard increased significantly with increase in Zn levels (Gupta and Kaushik, 2006). Similarly, the seed yield increased significantly (16-47%) with the application of boron. The average response to boron application ranged from 21-31%. The yield increase was due to 27% and 10% increase respectively in seeds/siliqua and 1000-seed weight, indicating the importance role in seed formation (AICRP-RM, 2005).

Integrated approaches for enhancing nutrient use efficiency of oilseed Brassica

It is important to exploit the potential of organic manures, composts, crop residues, agricultural wastes, bio-fertilizers and their synergistic effect with chemical fertilizers for increasing balanced nutrient supply and their use efficiency for increasing productivity, sustainability of agriculture and improving soil health and environmental safety. Balanced fertilization at right time by proper method increases nutrient use efficiency in mustard. Experiments have been conducted at different AICRP centres with the integrated use of organic manure, green manure, crop residue and bio-fertilizers along with inorganic fertilizers. INM not only reduces the demand of inorganic fertilizers but also increases the efficiency of applied nutrients due to their favourable effect on physical, chemical and biological properties of soil. The introduction of leguminous crops in the rotational and intercropping sequence and use of bacterial and algal cultures play an important role in increasing the nutrient use efficiency (Prasad *et al.*, 1992).

INM improves the nutrient uptake by mustard and hence enhances the use efficiency of various nutrients from the soil. The incorporation of 25% nitrogen through FYM+ 75% by chemical fertilizer + 100% sulphur significantly enhanced the uptake use efficiency and of nitrogen and sulphur in both seed and stover of crop followed by 100% NS and 50% N through FYM + 50% by chemical fertilizer+ 100% S (Bhat *et.al.*, 2005). The highest mustard-equivalent yield, based on market price of the crops (24.88q/ha), net monetary returns (Rs. 15,537/ha), B: C ratio (2.07) and agronomic efficiency (16.1) were recorded with the application of 100% recommended N in the rainy season through FYM and 100% recommended NP in the winter season through inorganic fertilizers (Kumpawat, 2004). Agronomic efficiency is the response in terms of increase in mustard seed yield per unit use of nitrogen.

Organic sources of nutrients

Bulky organic manures are applied to improve overall soil health and reduce evaporation losses of soil moisture. Depending upon the availability of raw material and land use conditions various organic sources viz. clusterbean (green manure), Sesbania (green manure), mustard straw@ 3t/ha and vermicompost (2.5-7.5 t/ha) have been evaluated at Bharatpur. Green manure with sesbania gave significantly higher mustard seed yield at Bharatpur and Bawal. Sesbania green manuring has shown higher mustard yield and improved soil environment (AICRP-RM, 2006).

Many bio-stimulants also encourage higher production. At Hisar, foliar spray of bioforce (an organic formulation) 2 ml/l at the flowering and siliqua formation stage enhanced mustard seed yield (2059 kg/ha) (AICRP-RM, 2007).

Growth promoter and bio-fertilizer as a component of INM

Biofertilizers are inoculants or preparation containing micro-organims that apply nutrients especially N and P. Two types of N-fixing micro-organisms *viz.*, free living (*Azotobacter*) and associative symbiosis (*Azospirillum*) and two P supplying micro-organisms *viz.* phosphate solubilizing bacteria and vesicular arbuscular mychorriza (VAM) were extensively tested at various AICRP-RM centers. Inoculation of mustard seeds with efficient strains of *Azotobacter* and *Azospirillum* enhanced the seed yield up to 389 and 305 kg respectively with 40 Kg N/ha. The total NPK uptake was also higher with *Azotobacter* inoculation. The combined application of 10 t FYM+ 90:45:45 NPK kg/ha with *Azotobacter* inoculation gave the highest B: C ratio of 1.51. At lower N levels, without inoculation, the seed yield decline was more as compared to inoculated treatment. Growth promoter's formulations like bioforce and bio-power contains bio-amino acid, plant growth promoting terpenoid, siderophores and attenuated bacteria fortified with BGA helped to increase water and nutrient absorption from the soil. Similarly, bioforce contains natural free amino acid, phytohormones, macro and micro elements and plant growth promoting terpenoid activated the cell division and stimulates plant growth, development and photosynthate translocation. RDF (80:40:0) along with 25 kg Biopower/ha+ spray of Bioforce (one liter in 500 litres of water) at 50% flowering and pod filling stage gave significant higher yield of mustard over other combinations (AICRP-RM 2005). Premi *et al.* (2012) reported a synergistic effect of PSB and PSB + VAM and antagonistic effect of VAM was observed in agronomic P efficiency and apparent recovery efficiency at 17.6 kg P/ha basal application and at 15.4 kg P/ha basal and 2.2 kg P/ha foliar spray at 35 DAS.

Effect of INM on quality of mustard oil

At Kanpur, INM studies were evaluated in maize-mustard, bajra-mustard and fallow mustard sequence. In maize-mustard sequence, 100/75% of RDF + 2 t FYM gave highest seed yield and quality of the oil (Table 5).

Table 5: Effect of INM on quality of mustard (Kanti-RK 9807) under maize-mustard sequence

Treatment	Legends	Oil (%) content	Fatty Acid composition (%)					
			16:1 Palmtic acid	18:1 Oleic acid	18:2 Linoleic acid	18:3 Linolenic acid	20:1 Eicosenoic acid	22:1 Erucic acid
RDF (120-40-40)	T_1	40.4	2.8	18.4	10.1	10.6	4.3	52.7
T_1 + 10 t FYM /ha	T_2	40.9	2.8	16.3	13.3	10.4	4.1	52.2
T_2+ 40 kg S/ha	T_3	40.4	2.9	18.0	14.4	12.2	3.2	48.6
T_3 + Zn SO4 25 kg/ha	T_4	40.3	2.8	17.8	14.9	10.1	6.1	47.3
T_4 + B 1 kg/ha	T_5	40.7	2.7	23.0	16.2	9.0	5.2	43.3
T_1+ Crop residue (Maize)	T_6	40.1	2.7	20.0	14.3	9.2	4.4	48.6
75% RDF		40.4	2.6	17.8	15.1	7.9	6.3	49.7

Source: Modified from AICRP-RM, 2002

At Bharatpur and Jobner, 17.8 and 8.6 % increase in seed yield was recorded with 50% RDF + 50% N through FYM and vermin-compost. Sole organic treated plot recorded 29.9% lesser seed yield over RDF at Jobner (AICRP-RM, 2008). Amount of available phosphorus increased over initial value when organic manures and crop residues were incorporated. Organic carbon status builds up in organic source-incorporated plots. The application of 10 t FYM/ha in addition with recommended dose of fertilizer (RDF) improved soil physical condition by improving aggregation, increased saturated hydraulic conductivity and reducing bulk density and penetration resistance of the surface soil (Hati *et al.,* 2006). Nutrient use efficiency of primary, secondary and micro-nutrient can be enhanced through sound integrated nutrient management of oil seed *Brassica* especially rapeseed-mustard crops.

Increased NUE of oilseed *Brassica* spp. is vital to enhance the yield and quality of crops, reduce nutrient input cost and improve soil, water and air quality. Selection of nutrient efficient genotypes and incorporation of these in breeding programs will result in better NUE. Nonetheless, the poorly developed nutritional genetics of crop plants and its response to external environmental factors and the complexity of identifying nutrient efficiency traits by rapid, reliable techniques have contributed to a lack of progress and success in breeding plant cultivars with high NUE. The different crop cultivars differ in absorption and utilization of nutrients and such differences are attributed to morphological, physiological and biochemical processes in plants and their interaction with climatic, soil, fertilizer, biological and agronomic practices. Nutrient use efficiency of oilseed Brassica could be enhanced by cautious manipulation of plant, soil, fertilizer, biological, environmental factors and best management practices.

Conclusion

In present scenario of increasing demand for edible oil, it is imperative to enhance nutrient use efficiency for increasing crop productivity and reducing the nutrient waste, which is very high presently. The efficiency implies the capacity of a system to translate inputs into chosen outputs, or to minimize inputs waste. Enhancement of nutrient use efficiency (NUE) by plants could lessen fertilizer input costs, reduces the nutrient losses, and boost the crop productivity. There is scope to increase the mustard productivity up to 2000.0-2500.0 kg/ha, from present national average of 1145 kg/ha. This can only be achieved by enhancing input use efficiency and among inputs the fertilizers nutrient sources has great role to play. Nutrient use efficiency enhancement is prerequisite not only for primary nutrients but it necessitates high NUE for secondary and micro nutrient for oilseeds Brassica. Mustard, in general is very sensitive to micronutrient deficiency, specially zinc and boron. The response of various ideotype to the applied micronutrients varies considerably. The precise information of the bio-physiological mechanism for adaptation to nutrient stress will help in enhancing NUE at plant level. It is important to exploit the potential of organic manures, composts, crop residues, agricultural wastes, bio-fertilizers and their synergistic effect with chemical fertilizers for increasing balanced nutrient supply and their use efficiency for increasing productivity, sustainability of agriculture and improving soil health and environmental safety. INM improves the nutrient uptake by mustard and hence enhances the use efficiency of various nutrients from the soil.

Selected References

AICRP-RM, 2002. Annual Progress Report of National Research Centre on Rapeseed-mustard. 2001-02. pp.29-34.

AICRP-RM, 2005. Annual Progress Report of National Research Centre on Rapeseed-mustard. 2004-05. pp: 9-11.

AICRP-RM, 2007. Annual Progress Report of All India Coordinated Research Project on Rapeseed-Mustard, pp. A1-16.

AICRP-RM, 2007. Annual Progress Report of All India Coordinated Research Project on Rapeseed-Mustard, pp. A1-16.

AICRP-RM, 2008. Annual Progress Report of All India Coordinated Research Project on Rapeseed-Mustard, pp. A1-22.

Aulakh MS, Pasricha NS. 1988. S fertilization of oilseeds for yield and quality. TSI-FAI symposium. S in agriculture-S-1 1/3.

Baligar VCN, Fageria K , He ZL. 2001. Nutrient use efficiency in plants. *Communication in Soil Science and Plant Analysis*, 32(7&8): 921–950.

Baligar VC, FageriaNK, ElrashidiMA. 1998. Toxicity and nutrient constraints in root growth. *Horticultural Science* 36 :960 –965.

Bennett WF. 1993. Plant nutrient utilization and diagnostic plant symptoms. pp. 1–7. In: W.F. Bennett (eds.). Nutrient Deficiencies and Toxicities in Crop Plants. The American Phytopathological Society Press. St. Paul, MN.

Bhat MA, Singh R, Dash D. 2005. Effect of INM on uptake and use efficiency of N and S in Indian mustard on an inceptisol. *Crop Research* (Hisar) 30:23-25.

Blair G.1993. Nutrient efficiency-what do we really mean? In Randall PJ, Delhaize E, Richards RA, Munns R (eds.) Genetic aspects of plant nutrition. Kluwer Academic Publishers, the Netherlands, pp 204-213.

Hilbert DW. 1989. Optimization of Plant Root: Shoot Ratios and Internal Nitrogen Concentration. *Annals of Botany* 66 (1): 91-99.

Eticha D, Schenk MK. 2001. Phosphorus efficiency of cabbage varieties. In: Horst et al. (eds.) Plant nutrition-food security and sustainability of agro-ecosystems through basic and applied research. Kluwer Academic Publisher Dordrecht, the Netherlands, pp 542-543.

Flügge UI, Freisl M., Heldt, H.W. 1980. Balance between metabolite accumulation and transport in relation to photosynthesis by spinach chloroplasts. *Plant Physiology* 65:574-577.

Fohse D, Claassen, N, Jungk A. 1988. Phosphorus efficiency of plants. I. External and internal P requirement and P uptake efficiency of different plant species. *Plant and Soil* 110, 101-109.

Gahoonia TS, Nielsen NE. 1996. Variation in acquisition of soil phosphorus among wheat and barley genotypes. *Plant and Soil* 178: 223-230.

Gahoonia TS, Nielsen NE. 1998. Direct evidence on participation of root hairs in phosphorus (32P) uptake from soil. *Plant and Soil* 198:147-152.

Graham RD. 1984. Breeding for nutritional characteristics in cereals. pp. 57–102. In P. B. Tinker and A. Lauchli (eds.), Advances in Plant Nutrition Vol. 1 Praiger Publisher, New York, NY.

Graham RD., 1984. Breeding characteristics in cereals. In: Tinker, P.B., La¨uchli, A. (Eds.), Advances in Plant Nutrition, vol. 1. Praeger Verlag, New York, pp. 57–90.

Gunes A, Inal A, Aplaslan M, Cakmak I. 2006. Genotypic variation in phosphorus efficiency between wheat cultivars grown under greenhouse and field conditions. *Soil Science and Plant Nutrition* 52: 470-478.

Gupta SP, Singh MV, Dixit ML. 2007. Deficiency and management of micronutrients. *Indian Journal of Fertiliers* 3: 57–60.

HatiKM, Mishra AK, Mandal KG, Ghosh PK, Bandopadhyay KK. 2006 Irrigation and nutrient management effect on soil physical properties under soybean-mustard cropping system. *Agricultural Water Management* 85 (3): 279-286.

Hirel B, Lea PJ. 2001. Ammonium assimilation. In: Lea PJ, Morot-Gaudry JF, eds. Plant nitrogen. Berlin: Springer-Verlag, 79–99.

Hocking PJ, Strapper M. 2001. Effect of sowing time and nitrogen fertliser on canola and wheat, and nitrogen fertliser on Indian mustard. II. Nitrogen concentrations, N accumulation, and N use efficiency. *Australian Journal of Agricultural Research* 52: 635–644.

Holmes MRJ. 1980. Nutrition of the oilseed rape crops. Applied Science Publishers Ltd. Essex, England, pp 158. In TSI/FAI/IFA Symposium.

Horst WJ, Behrens T, Heuberger H, Kamh M, Reidenbach G, Wiesler F.2002. Genotypic differences in nitrogen use-efficiency in crop plants. In: Lynch, J.M., Schepers, J.S., U¨ nver, I. (Eds.), Innovative soil-plant systems for sustainable agricultural practices, OECD Workshop 2002, OECD Publications Paris, pp. 75–92.

Jungk A. 2001. Root hairs and acquisition of plant nutrients from soil. *Journal of Plant Nutrition and Soil Science* 164: 121-129.

Kamara AY, Kling JG, Menkir A, Ibikunle G. 2003. Agronomic performance of maize (Zea mays L.) breeding lines derived from a low nitrogen maize population. *Journal of Agricultural Science* 141: 221–230.

Katyal JC, Sharma KL, Srinivas K. 1997. Sulphur in Indian agriculture. pp. KS-2/1-2/12.

Kumpawat BS. 2004. Integrated nutrient management for maize-mustard cropping system. *Indian Journal of Agronomy* 49:4-7.

Laine′ P, Ourry A, Macduff JH, Boucaud J, Salette J. 1993. Kinetic parameters of nitrate uptake by different catch crop species: effect of low temperatures or previous nitrate starvation. *Physiologia Plantarum* 88: 85–92.

Lambers H, Shane MW, Cramer MD, Pearse SJ, Veneklaas EJ. 2006. Root structure and functioning for efficient acquisition of phosphorus: matching morphological and physiological traits. *Annals of Botany* 98: 693-713.

Lauer MJ, Blevins DG, Sierputowska-Gracz H. 1989a. 31P-nuclear magnetic resonance determination of phosphate compartmentation in leaves of reproductive soybeans (Glycine max L.) as affected by phosphate nutrition. *Plant Physiology* 89: 1331-1336.

Malagoli P, Laine P, Rossato L, Ourry A. 2005. Dynamics of nitrogen uptake and mobilization in field-grown winter oilseed rape (*Brassica napus*) from stem extension to harvest. *Annals of Botany* 95: 853–861.

Marschner H, Dell, B. 1994. Nutrient uptake in mycorrhizal symbiosis. *Plant and Soil*, 159: 89-102.

Parker JS, Cavell AC, Dolan L, Roberts K, Grierson CS. 2000. Genetic interactions during root hair morphogenesis in Arabidopsis. *Plant Cell* 12: 1961-1974.

Plaxton WC, Carswell MC. 1999. Metabolic aspects of the phosphate starvation response in plants. In: Lerner HR (ed.) Plant response to environmental stress: from phytohormones to genome reorganization. New York, NY, USA: Marcel-Dekker, pp 350-372.

Premi OP, Kandpal BK, Sandeep K, Rathore SS, Kapila S, Bhogal NS. 2012. Phosphorus Use Efficiency of Indian Mustard (Brassica juncea) under Semi Arid Conditions in Relation to Phosphorus Solubilizing and Mobilizing Microorganisms and P fertilization. *National Academy of Science Letters* 35(6):547–553.

Prasad R, Sharma SN, Singh S, Lakshaman R. 1992. Agronomic practices for increasing nutrient use efficiency and sustained crop production. Paper presented in National Seminar on resource management for sustainable production, Feb. 1992, New Delhi.

Raghothama KG. 1999. Phosphate acquisition. *Annual Review of Plant Physiology and Plant Molecular Biology* 50: 665-693.

Raghothama KG, Karthikeyan AS. 2005. Phosphate acquisition. *Plant and Soil* 274: 37-49.

Reager ML, SharmaSK, Yadav R.S. 2006. yield attributes, yield and nutrient uptake uptake of Indian mustard as influenced by N levels and its split application in arid western Rajasthan. *Indian Journal of Agronomy*. 51 (3):123-126

Richardson AE, Lynch JP, Ryan PR, Delhaize E, Smith FA, Smith SE, Harvey PR, Ryan MH, Veneklaas EJ, Lambers H, Oberson A, Culvenor RA, Simpson RJ. 2011. Plant and microbial strategies to improve the phosphorus efficiency of agriculture. *Plant and Soil* 349:121-156.

Sah D, Bohra JS, Shukla DN. 2006. Effect of N, P, S on growth attributes and nutrient uptake of mustard. *Crop Research* 31 (1): 234-236

Sharma R, Thakur KS, Chopra P. 2007. Response of N and spacing on production of Ethopian mustard under mid-hill conditions of Himachal Pradesh. *Research on Crops* 8(1):65-68.

Sidhu AS, Sandhu KS. 1995. Response of mustard to method of N application and timing of first irrigation. *Journal of Indian Society of Soil Science* 43 (3): 331-334.

Smith FW. 2001. Plant response to nutritional stresses. In: Hawkesford MJ, Buchner P (eds.) Molecular analysis of plant adaptation to the environment. Kluwer Academic Publishers. The Netherlands. pp 249-269.

Smith SE, Read DJ. 1997. Mycorrhizal symbiosis. Academic Press, San Diego, California.

Tinker PB, Nye PH. 2000. Solute movement in the rhizosphere. Oxford University Press, Inc.

Vance CP, Uhde-Stone C, Allan D. 2003. Phosphorus acquisition and use: critical adaptation by plants for securing non-renewable resources. *New Phytologist* 15: 423-447.

27

Enhancing Nutrient Use Efficiency, pp. 411-448
Editors: K. Ramesh, A.K. Biswas, B.L. Lakaria, S. Srivastava and A.K. Patra

Spices Based Cropping Systems

K. Kandiannan, R. Dinesh, V. Srinivasan, C.K. Thankamani and S. Hamza

ICAR-Indian Institute of Spices Research, Kozhikode – 673 012, Kerala

Introduction

Spice – definition and uses

SPICES are aromatic or pungent vegetable substances, used as whole, broken, or ground form, whose function in food is seasoning rather than nutrition and commonly used as a condiment. It may be a dried seed, fruit, root, rhizome, bark or vegetative substance or flower bud. American Spice Trade Association (ASTA) defines spices as "any dried plant product used primarily for seasoning purposes". Included are tropical aromatics (pepper, cinnamon, cloves, *etc.*), leafy herbs (basil, oregano, marjoram, *etc.*), spice seeds (sesame, poppy, mustard, *etc.*) and dehydrated vegetables (onions, garlic, *etc.*). Many spices have other uses like food preservation, medicine, religious rituals, cosmetics, and perfumery or as vegetables.

Spices have a profound influence on the course of human civilization as they permeate our lives from birth to death. In everyday life, spices succor, cure, relax, and excite us. In ancient times, they were as precious as gold; and as significant as medicines, preservatives, and perfumes. One of the first uses of spices was for religious purposes such as for incense, embalming, in sacrificial rites, and as charms, perfumes, or medicines. In the middle Ages, considered as luxury item and was accepted as currency. No wonder they were sought after in the manner same to gold and precious metals (Chomchalow, 1996).

According to International Organization for Standardization (ISO) there are 109 spices (31 families) are under cultivation. Although there are several classifications, based on the plant part used, they are classified as (I) rhizome and root spices, (II) bark spices, (III) leaf spices, (IV) flower spices, (V) fruit spices and (V) seed spices.

Spices production and export from India

Out of 109 spices, India cultivates more than 50 spices due to diverse agro-ecosystems present in the country. Spices export has registered substantial growth in the World Spice Trade. During the year 2012-13, a total of 7,26,613 tons of spices and spice products valued Rs.12112.76 crore (Table 1) (US$2212.13 Million) has been exported.

Table 1: Export of spices from India (2012-13)

Item	Quantity (Tonnes)	Value (Rs. in Lakhs)
Pepper	15,363	63810
Cardamom (S)	2,372	21215
Cardamom (L)	1,217	6255
Chilli	3,01,000	238061
Ginger	22,207	18725
Turmeric	88,513	55488
Coriander	35,902	20183
Cumin	85,602	115307
Celery	5,171	2977
Fennel	13,811	10466
Fenugreek	29,622	10488
Other Seeds (1)	18,442	11179
Garlic	22,872	6868
Tamarind	17,950	10753
Nutmeg & Mace	3,231	22592
Vanilla	55	683
Other Spices (2)	16,293	18773
Curry Powders/Paste	17,436	27516
Mint Products (3)	20,039	394050
Spice Oils &Oleoresins	9,515	155888
Total	7,26,613	1211276

(1) Include Mustard, Aniseed, Ajwan seed, Dill Seed, Poppy Seed *etc.*
(2) Include Asafoetida, Cassia, Saffron *etc.*
(3) Include Mint Oils, Menthol & Menthol Crystal.
Source: DGCI & S, Kolkata/Shipping Bills/Exporters return (Spices Board)

India is a traditional spice producer and supplier to the world spice market, in addition to its strong domestic consumption. At a annual growth rate of 10% in spices consumption, India's demand by 2050 will be around 16.6 mt. Estimated target of selected spices for 2050 are given Table 2. During 2012-13, we had 3.2 m ha under spices with a production of 5.9 mt

(Table3). Andhra Pradesh, Rajasthan, Gujarat, Karnataka, Madhya Pradesh, Tamil Nadu, Assam, West Bengal and Uttar Pradesh are the major producers (Table 4). Like many other crop production sectors, spices production is also facing several challenges and nutrient management is one of the key issues.

Table 2: Estimated production target for spices in India ('000 tons)

Area	Total demand by 2050	Extra to be produced over the present production	Productivity to meet the total demand by 2050 (t/ha)
Black pepper	239.25	197.25	0.94
Cardamom (S)	79.07	63.20	0.99
Ginger	2152.49	1380.30	9.35
Turmeric	2882.15	1819.70	10.03

Table 3: Area and production of important spices in India (2012-13)

Crop	Area (ha)	Production (tons)
Pepper	200279	40621
Ginger	155063	755617
Chillies	804792	1276301
Turmeric	218646	1166843
Garlic	242491	1228324
Cardamom	89006	15816
Coriander	557870	532947
Cumin	593980	394328
Fennel	99554	142949
F. Greek	93605	115929
Ajwain	35376	26778
Dill / Poppy / Celery	33470	32642
Cinnamon / Tejpat	2944	5035
Nutmeg	17485	12573
Clove	2387	1105
Tamarind	58428	202574
Saffron / Vanilla	7095	10742
Total	3212471	5951458

Table 4: Statewise area and production of spices (2012-13)

State	Area ('000 ha)	Production ('000 tons)
Andaman Nicobar	1.65	2.98
Andhra Pradesh	292.82	1129.31
Arunachal Pradesh	10.05	61.60
Assam	93.05	261.56
Bihar	13.01	12.54
Chhatishgarh	11.67	8.32
Goa	0.73	0.23
Gujarat	551.67	882.14
Haryana	12.80	61.69
Himachal Pradesh	4.77	19.26
Jammu & Kashmir	4.15	1.08
Karnataka	265.12	502.46
Kerala	254.55	112.80
Madhya Pradesh	299.91	461.17
Maharashtra	116.52	106.47
Manipur	10.47	24.14
Meghalaya	16.84	74.82
Mizoram	20.65	114.98
Nagaland	9.77	39.17
Odisha	123.92	187.50
Puducherry	0.09	0.12
Punjab	18.37	68.21
Rajasthan	730.51	871.64
Sikkim	24.38	54.41
Tamil Nadu	157.33	426.38
Tripura	5.68	18.04
Uttar Pradesh	58.29	201.97
Uttarakhand	6.60	38.77
West Bengal	97.12	207.70
Total	3212.47	5951.46

In this chapter, the best nutrient management practices for improving nutrient use efficiency of black pepper, cardamom, ginger and turmeric are summarised below:

Black Pepper

Black pepper (*Piper nigrum* L., Family - *Piperaceae*) , the 'King of Spices' or 'Black gold' is one of the oldest and highly valued spice.

Crop Ecology

Black pepper grows successfully between 20° North to 20° South of equator and from sea level up to 1500 m above MSL on wide variety of soils rich in organic matter with good water holding capacity and pH ranging from 5 to 6. Well-drained soil with near neutral pH, high organic matter and high base saturation enhances productivity. Soils of high yielding black pepper gardens are generally sandy to loam textured with near neutral pH, high in exchangeable bases, organic carbon and micronutrients, especially zinc. It is a plant of humid tropics, requiring 2000-3000 mm of rainfall, tropical temperature and high relative humidity with little variation in day length throughout the year. It does not like excessive heat and dryness.

Crop production

Black pepper is propagated both vegetatively (rooted cutting) as well as through seed. Advanced methods like 'Bamboo Method' (Bavappa & Gurusinghe, 1978; Sivaraman 1987) and serpentine method (Thankamani *et al.,* 2004) are also available. Pepper vine requires support for its establishment. Some of the common living supports are *Erythrina indica, E. lithosperma, Garuga pinnata, Gliricidia sepium* and *Grevillea robusta.*

Crop nutrition

Black pepper is a surface feeder and feeding roots are concentrated on top 50 - 60 cm layer. Owing to the heavy rains and unsustainable soil management practices, the soils have lost their fertility and, therefore, balanced manuring has become imperative. Some phosphate fertilizers, like rock phosphate can be directly applied to meet the requirement, whereas better use efficiency of applied nitrogen can be achieved by supplying it at the time when crop needs it. Potassium uptake by black pepper is more or less continuous throughout the different stages of growth. In order to minimize leaching loss, split application is generally recommended as black pepper is raised as a rainfed crop in heavy rainfall areas except in countries like Thailand where it is raised as an irrigated crop.

One third of the recommended dose of nutrients is applied during first year of planting, two third in the second year and full dose is given from the third year. In India, the recommended dose of fertilizers is applied in two splits, the first in May with the onset of monsoon and the second split in August-September. In Sarawak, Malaysia, the elite growers use 110 g per plant of either Nitrophoska or Nitrophoska special in the first year and 230 gm Nitrophoska per plant in the second year at bi-monthly intervals. From the third year the fertilizer dose is increased to 1.8-2.3 kg per plant in four applications mainly during July to January (Sadanandan, 2000). In Sarawak, the recommendation is 2043 g fertilizer mixture of 16 N: 6 P_2O_5: 18 K_2O: 4 Mg per yielding plant plus 454 g dolomite and 28 g trace elements and farmers are reported to apply 4 to 6 kg fertilizer per plant in

7-8 splits (George, 1982). Fertilizer is applied in ¾ circles around the vine or in alternating parallel bands on either side of the vine underneath the edge of the canopy. Over 90% of the feeder root activity was found within a 30 cm radius around the vine with the uniform vertical distribution down to a soil depth of 40 cm. Radiotracer studies on the field planted bush pepper showed that majority of the roots concentrated in the surface layer of the soil up to 20 cm depth and 30 cm lateral distance. In the deeper layers of 40 and 60 cm the root spread was within a lateral distance of 10 cm. About 60% of the roots were distributed in 20 cm depth and about 30% in 20-40 cm depth constituting about 70-87% roots within 40 cm depth and the remaining down below. Similarly 79-93% roots were distributed within 30 cm lateral distance (Thankamani *et al*., 2003). Foliar spray is effective for the micronutrient application since only very small quantities of these elements are required. Geetha & Sivaraman Nair (1990) reported that spraying 0.5% zinc sulphate reduced spike shedding by 48.4%.

Judicious use by combining organic manures and inorganic fertilizers in a proper manner is the most efficient and economical for pepper production. Addition of organic matter enhanced the pepper growth and biomass (Mustika, *et al*., 1994; Sivakumar & Wahid, 1994). And so organic farming can improve the pepper productivity (Zulkifly, 1996). In India, only 4 - 10 % of the farmers use fertilizers, as more than 95 % are cultivated in homesteads (Nair & Sreedharan, 1986) where multi species cropping systems is prevalent which helps in improvement of soil fertility (Bavappa *et al*., 1986). Studies carried out at IISR showed that application of NPK at 150:60:270 kg ha^{-1} with micronutrients Zn, B and Mo at 5:2:1 kg ha^{-1} recorded highest yield in Subhakara and Sreekara varieties of black pepper. ICAR-IISR developed crop specific foliar nutrient mixture suited for black pepper, as foliar sprays recommended @ 5 g/L of water at flowering and berry development stages which recorded 15-30% increase in yield (Anandaraj *et al*., 2014). The critical stages of nutrient requirement are (1) initiation of flower primordia and (2) flower emergence, berry formation and development (Raj, 1978). However, fertilizer application depends on soil moisture availability. Fertilizers are applied in two splits in India and 3 - 4 splits in Indonesia (Wahid & Sitepu, 1987). Fertilizer application is restricted to a lateral distance of 30 cm in full circle area around the vine (Geetha *et al*., 1993).

Integrated nutrient management with annual application of NPK at the rate of 100:40:140 kg ha^{-1} and organics like FYM at the rate of 5.0 kg $vine^{-1}$ has been reported to enhance the fertility of soils under black pepper. Application of organics like castor and cotton cakes @ 1-2 kg per plant or poultry waste @ 1-2 kg or cattle manure @ 3-5 kg, 200 to 300g NPK (12:12:17) mixture, 500g of lime and 300g thermo phosphate per plant per year in addition to foliar sprays of micro nutrients are the general practice followed by the growers (Sadanandan *et al.*, 2000). From a four year study conducted in 162 locations, Sadanandan & Hamza (2002) reported an increase in almost all the nutrients and yield due to the adoption of integrated nutrient management involving recycling of FYM.

A five year study on optimizing inorganic nutrient use in black pepper involving substitution of 50% of the recommended inorganic fertilizer dose with FYM on equivalent N basis and the remaining 50% of the dose either through inorganic N and P, inorganic N + Phosphobacterium (PB) + Vesicular Arbuscular Mycorrhiza (VAM) or inorganic P + Azospirillum revealed that inorganic nitrogen and phosphorus dose could be reduced by 50% in black pepper (Mathew and Nybe, 2004).

Cattle manure or chicken manure at the rate of 1-2 kg plant^{-1} for the first and second year and 3-4 kg plant^{-1} in subsequent years are also applied. Kanthaswamy *et al.* (1996) reported highest yield of black pepper due to application of Azospirillum and chemical fertilizers in addition to FYM. Nybe *et al.* (2004) suggested manuring @ 10 kg cattle manure/ compost/ green leaves with NPK @ 50:50:150 g vine^{-1} year^{-1} and liming @ 500g vine^{-1} in alternate years during April-May in plant basins at 50-75 cm radius for higher yield. Srinivasan *et al.* (2005) recommended application of coir pith compost at 1.25 t ha^{-1} integrated with half the recommended fertilizer dose and *Azospirillum* sp. @ 20 g vine^{-1} for higher yield and fertility build up in black pepper gardens.

Slow release fertilizers are useful to minimize fertilizer losses by leaching and will supply the nutrient for long duration. Studies of Yufdi (1991) indicated that avoiding liming can inhibit the growth of black pepper cultivars. Liming will eliminate adverse effects of exchangeable aluminium and iron by correcting the pH and provides better nutrient availability for growth and performance. Slow release formulations such as *Nimin* coated urea followed by cyclo-diurea and urea formaldehyde significantly increased availability of various fractions of N in the soil, tissue concentration of N and there by the utilization efficiency. Sadanandan & Hamza (1993) found that application of neem oil coated urea increased ammoniacal N, (NO_2 + NO_3)-N and total N in the soil and the yield by 280% compared to urea alone. Sadanandan and Hamza (2001) reported that Mussoorie phos at 80 kg P_2O_5 vine^{-1} year^{-1} was as efficient as super phosphate for var. Panniyur-1 and Karimunda with respect to soil P availability, yield response, relative agronomy efficiency (RAE) and economics. They also stated that ammonium polyphosphate at 50 kg P_2O_5 ha^{-1} was as efficient as diammonium phosphate with regard to nutrient availability, yield response and P uptake. Addition of 50% of recommended dose of K as sulphate of potash (SOP) for soils of high K status and 100% recommended K as SOP + 2% foliar spray of SOP for soil of low K status was recommended for higher yield and quality by Srinivasan *et al.* (2013a). Targeted yield equation for predicting nutrient requirements to get fixed yields of dry black pepper in soils with varying fertility levels was standardized based on field trials at IISR, Kozhikode. The economic optimum in terms of profitable response for money invested was found to be Rs. 1.60/standard for N, Rs. 2.40/standard for P and Rs. 5.40/standard for K (Srinivasan *et al.*, 2013b).

Efficient organic nutrient management

In recent years, organic agriculture has gained considerable importance and many Black pepper as a best component crop in agri-horti and silvi-horti systems, recycling of farm waste can be effectively done when grown as a component crop in the existing coffee, tea, coconut and arecanut plantations. Mixed cropping of black pepper with rubber, cocoa, orange *etc.* are common in other countries like Brazil. It is also commonly cultivated on live support shade trees which play a dominant role in biomass regeneration and utilization through nutrients recycling as a source of organic inputs. In black pepper and coffee production systems shade trees like *Michelia champaca* Linn., *Artocarpus integrifolia* and *Grevillea robusta A. Cunn*. recorded highest leaf fall ranging from 1855 to 4214 g m^2year^{-1} (Dinesh Kumar and Babitha, 2006). For certified organic production for at least 18-36 months the crop should be under organic management or the first yield of a newly planted plantation can be sold as organic, as the crop starts economic yielding from third year onwards. It is desirable that organic method of production is followed in the entire farm, but in the case of large extent of area, the transition can be done in a phased manner for which a conversion plan is to be prepared. Traditional varieties adapted to the local soil and climatic conditions are promoted. In order to avoid contamination of organically cultivated plots from neighboring non-organic farms, a suitable buffer zone is to be maintained. In sloppy lands adequate precaution should be taken to avoid the entry of run off water and drift from the neighboring farms.

For nutrient management under organic management system, a judicious application of a combination of organic manures such as FYM 5 kg, neem cake 1 kg and vermicompost @ 1 kg per vine per year can be made during May-June. Biofertilizers *viz.,* Phosphobacteria and *Azotobacter* are also applied @ 100g/vine mixed with farm yard manure in the second year of planting (Srinivasan *et al.*, 2017). Application of FYM (10 kg), burnt earth (10 kg) or wood ash (2 kg) and *Azospirillum* (50g) or Phosphobacteria (50g) per vine increased the yield by 37-173% over control. The studies at IISR showed that soil P, Ca, Mg, Zn and Cu availability increased significantly under organic system of black pepper cultivation with highest average fresh yield of 2.9 kg $vine^{-1}$ as compared to integrated and inorganically managed systems. The available soil N and K levels of organic sites were comparable with that of integrated nutrient management sites.

Several PGPRs (plant growth promoting rhizobacteria) isolated from the rhizosphere of black pepper were found growth productive, disease suppressive and are efficient P solubilizers. Leaf compost and vermicompost supplementation supported significantly highest population of free-living N-fixing bacteria and phosphate solublizing bacterial populations. Biofertilizers like Azospirllum, Azotobacter, phosphobacteria *etc* are also widely used in black pepper plantations. Bopaiah and Khader (1989) reported increased shoot and root weight of cuttings of pepper by dipping in a peat based culture slurry of *Azotobactor, Azospirillum* and *Glomus mosseae* (VAM). Kandiannan *et al.* (1994, 2000) reported enhanced growth of pepper in the nursery due to application of 10g *Azospirillum* into the growing medium and significant enhancement in plant height, leaf area, biomass and dry matter production and nutrient content of black pepper cuttings due to combined application of *Azospirillium*, phosphobacteria and VAM.

Application of *Pseudomonas fluorescens* and inclusion of *Trichoderma harzianum* in potting mixture for rooting black pepper cuttings were found to increase the biomass production by influencing nutrient uptake through P solubilization increased feeder root production and thereby enhancing the absorptive surface area (Anandaraj & Sarma, 2003) and phytohormone and siderophores production (Diby *et al.,* 2005) reported significant uptake of nitrogen (N) and phosphorus (P) and enhanced vigour of black pepper when inoculated with *Pseudomonas fluorescens* strains, due to higher root proliferation and nutrient mobilisation especially that of P in the rhizosphere. *P. fluorescens* strains enhanced P uptake by 122% over control (non bacterised plants) and N uptake by 65% over control resulting in increased root growth and biomass production.

Organic manures like guano, prawn, fish refuse and soybean cakes are extensively being used in quantities of about 85 g in every two months in Sarawak for growing black pepper. Potassium fortified sterilized animal meat and bone meal is also popular. Experiments conducted at various locations like Panniyur, Yercaud, Ambalavayal and Sirsi in India revealed that inoculation of Azospirillum (50g vine^{-1}) along with 10 kg FYM and 100% recommended inorganic N dose yielded 4.8 kg vine^{-1} recording 44% increased yield over control. Similarly phosphobacteria (50g vine^{-1}) inoculation with 10 kg FYM and 75% recommended inorganic P fertilizer dose recorded 185% increased yield (6.0 kg vine^{-1}) over control (2.1 kg vine^{-1}).

Cardamom

Cardamom *(Elettaria cardamomum* (Linn.) Maton) is the dried fruit of a perennial herb. The fruits are picked when they are fully matures, but not fullyripe. The seeds have a pleasant aroma and characteristic warm but slightly pungent taste. The spice is used for flavouring curries, cakes and bread and for other culinary purposes. Lesser cardamom, green cardamom, Malabar cardomom and small cardamom are synonyms and are true cardamom of commerce, and known as 'Queen of Spices'. It is one of the most valued spices of the world. India, Sri Lanka, Guatemala and Thailand are the major producers of cardamom in the world. India had the monoply of cardamom meeting 90-95% of the total world demand, but now a days, Guatemala and other countries emerged as strong contenders. Cardamoms occur wild in the evergreen monsoon forests of the Western Ghats in southern India and Sri Lanka. The geographical distribution of the crop in India extends northwards from Tirunelveli (Tamil Nadu) for over 1,000 kms to Sirsi (Karnataka) through the high ranges of Kerala.

Crop ecology

Cardamom is a crop of tropical humid climate and it grows upto 600 to 1500m MSL. Cardamom is very sensitive to temperature fluctuations. In India, optimum growth and development is observed in the warm and humid conditions at a temperature range of 10-35°C. In open area where there is no shade soil temperature may go up (36°C) coupled with strong wind increases evapo-transpiration and the soil dries up quickly. Moderate rainfall (2000 to 2500 mm/ annuum)with filtered natural shade is ideal for growing cardamom. In South India, 75-80% of the cardamom cultivation is under rainfed conditions. Cardamom generally grows well in forest loamy soils with good drainage. The cardamom growing soils of Karnataka are mostly clay loam. These soils are generally acidic in nature with pH from 4.2 to 6.8, high in organic matter and nitrogen, low to medium in available phosphorus and medium to high in available potassium. The crop is cultivated under natural shade trees and the soil naturally enriched by the addition of leaf litter. The soils of major cardamom growing areas in India comes under the order *Alfisols*, formed under alternate wet and dry conditions and the sub order *ustalfs* derived from *schists*, granite and gnesis and lateritic in nature.

Crop production

Cardamom propagated both vegetatively as well as through seed. Being a cross-pollinated crop, clones are ideal for generating true-to-type planting materials from high-yielding clumps. However, due to lack of adequate clonal planting materials, farmers still prefer seedlings. Cardamom is traditionally an understory crop in the natural forest. The shade of tall trees provides suitable environment for cardamom growth (Abraham, 1965). Cardamom is generally cultivated in forest area. Clearing the forest and thinning out overhead canopy in order to get one umbrella canopy to get filtered shade is initial works.

Spacing depends on variety and duration of crop. 'Mysore' and 'Vazhukka' types are vigorous and need wider spacing. 'Malabar' types need closer spacing as comparatively smaller plants. Cardamom seedlings planted at 2 × 1m in hill slopes and 2.1 × 2.1m in flat lands. The trench method of planting with a spacing of 2 × 1 m resulted in better growth and yield.

Crop nutrition

Cardamom is a surface feeder, therefore in the first year of planting, frequent weeding is necessary to avoid the root competition between the young seedlings and weeds. Cardamom is generally grown in the rich fertile soils of the forest eco-system. The nutrient analysis data obtained from cardamom-growing tracts in south India revealed that the limiting leaf nutrients were in the order Zn > K > P > Ca > Mg > Cu > Mo > Fe > Mn > N (Hamza *et al.*, 2009). A steady absorption and utilization of plant nutrients take place throughout the life cycle of cardamom and hence a regular application of nutrients should be followed for higher yields. Good plant growth, yield and capsule quality depends on an adequate supply of all the nutrients through the growth period. A deficiency of any one of them will affect the cardamom yields detrimentally. Deficiency symptoms of N, P, K, Ca and Mg have been observed and were found to be expressed, when their contents in leaf were 0.84, 0.32, 2.16, 0.65 and 0.11% respectively. Nutrient uptake studies in cardamom have indicated the uptake of N, P, K , Ca and Mg to be in the ratio of 6:12:3:0:8 respectively and for the production of one kg of cardamom capsules, 0.122 kg N, 0.014 kg P and 0.2 kg K are removed by the plant (Kulkarni *et al.*, 1971). Of the various nutrients, the total K uptake was more followed by N and P in both stages of crop suggesting that the crop requires more of K than N and P. Among the different plant parts, highest contribution to the total N uptake was by green leaves (44.69%) and tillers (29.6%) in one-year-old and prepotents respectively. The highest P and K uptake was by roots and grown up tillers in one-year-old plants, and green leaves and tillers in prepotent plants, respectively. Among secondary nutrients, the highest uptake of Ca was observed in shoots at both the stages, whereas, for Mg it was the leaves which showed highest values. Uptake of Ca by shoots at harvest stage was more than that of pre-bearing stage.

Efficient inorganic nutrient management

The present recommendation of nutrients for cardamom is NPK at 75:75:150 kg ha^{-1}. The fertilizers may be applied in two splits before and after the southwest monsoon, in a circular band 20 cm wide at 30-40 cm away the base of the clump and incorporated with the soil. For rapid multiplication of suckers, application of 32.5 g of N, 25 g of P_2O_5 and 50 g of K_2O plant was found desirable. A fertilizer dose of 37.5:37.5:75 kg of NPK ha^{-1} for one year old crop during September and 75:75:150 kg of NPK ha^{-1} from second year onwards in two - split doses (during May and September) were found to be optimum.

Studies indicated that a fertilizer dose of 45: 45: 45 kg N, P_2O_5 and K_2O ha^{-1} for Kerala, 67: 34: 100 kg N, P_2O_5 and K_2O ha^{-1} for Karnataka with half N in organic form and 45: 34: 45 kg N, P_2O_5 and K_2O ha^{-1} for Tamil Nadu were optimum. Considering the low nutrient requirement and the high status of N and K in cardamom-growing soils, a maintenance dose of 30:60:30 kg N, P_2O_5 and K_2O ha^{-1} was recommended (Zachariah, 1978). For a crop yielding 100 kg dry capsules ha^{-1} under rainfed conditions, the present fertilizer dose of 75:75:150 kg N, P_2O_5 and K_2O ha^{-1} was recommended. During soil application, it would be necessary to give one-third of this dose during the first year of planting both under rainfed and irrigated conditions. During the second year, the dose may be increased to one-half of the recommended dose (75:75:150 and 125:125: 250 kg N, P_2O_5 and K_2O ha^{-1} for rainfed and irrigated conditions, respectively) and the full dose may be applied from the third year of planting. Additional fertilizer dose of 0.65 kg N, P_2O_5 and 1.30 kg K_2O needs to be applied for every increase of 2.5 kg dry capsules over normal yield. When cardamom is cultivated under controlled artificial shade, a fertilizer dose of 100: 25: 100 kg NPK ha^{-1} is recommended (Korikanthimath *et al.*, 1998). However, for efficient utilization of applied fertilizers, time of application is very important.

Thimmarayappa *et al.* (2000) observed that plants treated with 100% inorganic fertilizer (IF), 75:75:100 kg NPK ha^{-1} produced the greatest number of bearing suckers per clump (9.96) and panicles per clump (23.15), and green (707 kg ha^{-1}) and dry (184 kg ha^{-1}) yields. Treatment with 100% IF also gave the highest total and net returns, and B: C ratio. Contrarily, Murugan *et al.* (2007) reported that increasing the levels of N, P and K increased the yields of cardamom up to 125:125:200 kg ha^{-1}. Application of fertilizer nutrients at the present level of recommendation (75:75:150 kg NPK ha^{-1} yr^{-1}) along with 0.5 kg neem cake per plant did not increase the yield significantly over the control.

Further, significant yield increase and savings in fertilizer requirements would be possible by way of soil-cum-foliar application. This is well supported by the findings of Srinivasan *et al.* (1998), who suggested that in Karnataka state

(India) , N, P_2O_5 and K_2O ha^{-1} at 37.5: 37.5: 75 kg ha^{-1} through soil application and 2.5 % urea + 0.75% single super phosphate + 1.0 per cent muriate of potash through foliar spray gave 64 kg yield increase ha^{-1} over no fertilizer application. The yield increase recorded in Tamil Nadu state (India) was 43 kg ha^{-1} when a fertilizer dose of 20: 40: 20 kg N, P_2O_5 and K_2O ha^{-1} through soil and 3% urea + 1% single super phosphate + 2% muriate of potash as foliar spray was given. Under irrigated conditions, the general fertilizer recommendation is 125:125:250 kg N, P_2O_5 and K_2O ha^{-1} applied in three splits.

Under high production technology, where crop is harvested from 18 months onwards, fertilizer recommendation for full-grown plantation could be adopted from the second year onwards. Fertilizers would be applied in smaller doses in four or more splits after every harvest or combining both soil and foliar application of fertilzers. Whenever, the plant growth is affected due to root damage (root grub/fusarium disease/soil compactness), foliar application of DAP (2 %) + MOP (2 %) could be adopted. The fertilizer schedule for cardamom adopted in major cardamom growing states (Krishnakumar and Potty, 2002) and the schedule recommended by ICRI based on plant age is given in Tables 5 and 6.

Among the N forms, ammoniacal forms of N was found to be the best for acidic soils to enhance nutrient use efficiency. Hence, complex fertilizers 20:20:15 or 10:26:26 or 18:18:18 or DAP are much useful. In case of P source, acid soluble form of P either as rock phosphate or bone meal is better for acid soils. A mixture containing N: P: S association in the form of ammonium phosphate sulphate enhances the efficiency of applied nutrients. Spray of water soluble form of P viz. polyfeed (13: 40: 13- NPK), mono ammonium phosphate (0: 52: 53 NPK) is helpful for profuse flowering and yield. Similarly, water soluble form of K viz., K nitrate (13: 0: 45 NPK), sulphate of potash (0:0:50:20- NPKS) polyfeed (13:40:13-NPK) and mono potassium phosphate (0:52:34 NPK) would help in getting bold capsules with high oil content.

Table 5: Fertilizer rates recommended for cardamom

Region	Soil application NPK, kg ha^{-1}	Soil-cum-foliar application	Time of application	
			Soil	Foliar
Karnataka	75: 75: 150	37.5: 37.5: 75 kg ha^{-1} Urea (2.5%). Single super phosphate (0.75%), Muriate of potash (1.0%)	May-June, August-September	September, November, January
Kerala	75: 75: 150	37.5: 37.5: 75 kg ha^{-1} Urea (2.5%). Single super phosphate (0.75%), Muriate of potash (1.0%)	May-June, August-September	September, November, January
Tamil Nadu	40: 80: 40	20: 40: 20 kg ha^{-1} Urea (3.0%) Single super phosphate (1.0%) Muriate of potash (2.0%)	September, November	June, August, November-December

Source: Krishnakumar and Potty (2002)

While lime application is not required as a routine practice in cardamom plantations (Nair 1988), liming is essential if pH of the soil is less than 5.5. Though the quantity of lime is to be arrived at by assessing the lime requirement of the soil, for practical purpose, application of agricultural lime is recommended at 1.0 kg plant^{-1} year^{-1} for soils with pH below 5.0. Lime is to be applied in one or two splits during May and September. Fertilizer shall be applied only after 15-20 days of lime application.

Table 6: Schedule for fertilizer application in cardamom based on plant age
Soil application

Age of plants	Rainfed (kg ha^{-1})	Irrigated (kg ha^{-1})
First year of planting	25 -25 -50 (Two split application)	25 -25 -50 (Two split application)
Second year of planting (Non stabilized yield)	40-40-80 (Two split application)	60-60-80 (Three split application)
Third year of planting (Stabilized yield)	75-75-150 (Two split application)	125-125-250 (Three split application)
Zinc (Zinc sulphate) shall be applied as foliar spray at 250 g 100 litres^{-1} twice a year.		

(*Source*: Spices Board, 2009)

Based on DRIS norms, Sadanandan *et al.* (2000) have reported that the optimum status of micronutrients for realizing high yield is 253 ppm Fe, 371 ppm Mn, 33 ppm Zn, 28 ppm Cu and 0.56 ppm Mo. Micronutrient survey conducted by

Indian Cardamom Research Institute, Myladumpara, Idukki, Kerala showed that zinc deficiency is widespread in cardamom soils and B deficiency is observed in certain areas. Application of Zn to the foliage is found to enhance not only the growth and yield but also the quality of the produce. Hence, it is recommended that Zn may be applied as a foliar spray as $ZnSO_4$ at 250 g 100 liters^{-1} of water during April-May and September-October. It should be applied alone and should not be mixed with any insecticide or fungicide or fertilizer since Zn may precipitate and become unavailable to the plants. Besides, Zn, application of B seems to be imperative in soils deficient in B. In such cases, soil application of B in the commercial grade borax at the rate of 7.5 kg ha^{-1} is recommended in B deficient areas. In an experiment involving B and Mo, application of B at 20 kg ha^{-1} and Mo at 0.25 kg ha^{-1} independently or in combination increased the capsule yield of cardamom by 20%. ICAR-IISR developed crop specific foliar nutrient mixture for cardamom as foliar sprays recommended @ 5 g/ L of water at flowering, capsule formation and development stages of the crop that recorded 10-15% increase in yield (Anandaraj *et al.*, 2014).

Efficient organic nutrient management and INM

Studies on exclusive organic nutrition in cardamom are very few. However, organic manures are considered essential for improving physical characteristics of soil, apart from their nutritional values, and they are indispensable for cardamom irrespective of whether fertilizers are applied or not (Krishnakumar and Potty, 2002). The limited literature available indicated that application of organic manures such as neem cake at 1-2 kg plant^{-1} or poultry manure/ FYM/compost/vermicompost at 5 kg plant^{-1} may be done once in a year during May-June. If necessary, application of Mussorie rock phosphate or bone meal is done based on soil analysis. Nevertheless, application of compost or FYM at kg plant^{-1} year^{-1} in May-June is recommended. Nearly 70% of cardamom roots are confined to shallow depth of 5-40 cm and 30-50 cm radius. Therefore, for the maximum efficiency of applied manures it would be necessary to apply at a radius within 50 cm and being a surface feeder, deep placement of manures are not advisable.

Studies on integrated use of organic and inorganic sources of nutrients in cardamom are many. Thimmarayappa *et al.* (2000) reported that integrated nutrient management with 25% organic manure + 75 % inorganic fertilizers and 50 % organic manure + 50 % inorganic fertilizers recorded yields at par with 75 % organic manure + 25 % inorganic fertilizers. Field studies in cardamom have shown that application of recommended NPK nutrients as organic fertilizer (50 % each as FYM and neem cake + 50 % P each as bone meal and rock phosphate + 50 % K as wood ash) was effective in increasing the yield and quality of cardamom (Sadanandan and Hamza, 2006).

Three species of vesicular arbuscular mycorrhizal (VAM) fungi (*Glomus macrocarpum*, *G. fasciculatum* and *Gigaspora coralloidea)* were recorded in the roots of cardamom plants. Per cent colonization in roots varied from 40 – 100 % in pre monsoon samples and 63-94 % in post monsoon samples. Among the various mycorrhizal fungi tested, seedlings inoculated with *Gigaspora margarita* and *Glomus monosporum* exhibited significantly greater growth with increased uptake of nutrients. Results revealed that application of 100% organic manure in the form of FYM enhanced capsule yield by 34.1% over control (Kumar *et al.*, 2009a). They also found that as the proportion of inorganic nutrient application increased, the response of yield also increased. Application of 100% inorganic recommended dose of fertilizer yielded 188.81 kg ha^{-1} followed by 75% inorganic RDF+25% FYM (144.14 kg ha^{-1}). Field experiments conducted for seven years (2000 to 2007) revealed that application of FYM at 5 or 10 kgplant^{-1} with or without *Azospirillum* did not influence the yield components as well as yield levels appreciably. However, application of FYM at 5 kg plant^{-1} + 50% recommended N + *Azospirillum* yielded 141.20 kg ha^{-1} similar to that of FYM at 5 kg plant^{-1} + 75% recommended N (146.34 kg ha^{-1}), thereby providing 25% saving in inorganic N (Kumar *et al.*, 2009b).

Ginger

Ginger (*Zingiber officinale* Rosc.) is a perennial herb and the economic part is underground stem (rhizome), which is used as a spice. It is native to tropical South-East Asia, introduced into the West Indies, African countries and other tropical belt of the world. The ginger rhizome is harvested between 6 and 12 months after planting and used as spice in the fresh condition, peeled and split dried or powdered forms. It is also useful for cold-induced diseases, nausea, asthma, cough, heart palpitation, syperia, and is home remedy in India. Ginger is cultivated in more than 25 countries. India contributes around 30.27% to the world production.

Ginger requires a tropical or subtropical climate, prefers a warm, humid climate and cannot withstand very low temperature. It comes up well up to an altitude of 1500 m above MSL, the optimum being 300-900 m. The base temperature requirement is 13°C with an optimum range of 19-28°C for ginger crop growth and performance. Seed rhizome sprouts better at a soil temperature of 25-26°C and for growth and development of the crop it is 27.5°C. In Kerala, an important ginger producing state in India, the crop is being cultivated in the air temperature range of 28-35°C. Ginger also requires partial shade for better rhizome yield. Frost is injurious to foliage and rhizomes of ginger. Hence, in areas where frost is common as that of hill regions of North/North eastern India, harvesting is done before the frost.

Ginger can be grown on a wide variety of soils such as sandy loams, clay loams, alluvial and lateritic soils. However, it is mainly grown in red and laterite soils of Kerala, Karnataka, Orissa, West Bengal, Maharashtra and Northeastern states. Deep soils with rich organic matter content and nutrient availability are more suitable for cultivation. However, it performs well on medium loams having enough quantity of humus. Deep slopes in hilly areas are not advisable for ginger cultivation as it leads to soil erosion during heavy rainfall. The optimum soil pH preferred for ginger is 5-7 and if the pH is more than 8, growth is retarded. Maximum rhizome yield can be achieved in sandy loam soil having moderately acidic (pH 5.7), high organic matter and available K and yield decreased with increasing clay content and decrease in pH (Sahu and Mitra, 1992). The soil with low bulk density (BD) promotes the growth of ginger and improved yield.

Crop production

Ginger is propagated through vegetative seed rhizome called seed 'piece' or 'sett' or 'knob' or 'cutting', the length and weight of pieces vary from place to place and variety to variety. In general, bigger size of seed material results in quicker emergence and better growth and yield. In India, seed rate varies from 1500 to 2500 kg per hectare depending on seed size and spacing. Seed treatment induces early germination and prevents seed-borne pathogens and pests. The planting time depends on the onset of the monsoon. In India, generally planting is done during March-June. Different spacings are (15-45×15-45 cm) adopted, however, planting on raised beds at a spacing of 20-25 × 20-25 cm and a depth of 4-10 cm with the viable bud facing upward is recommended. Mechnical or chemical (herbicide) weeding or combination of different methods would help in weed control in ginger. Earthing up is essential and it is combined with hand hoeing (weeding), and mulching/manuring. Irrigation is based on need and availability of water. When rhizomes is used for vegetable or for preparation of ginger preserve, candy, soft drinks, pickles and alcoholic beverages, harvesting should be done 4-5 months after planting, whereas, when it is used for dried ginger and preparation of value added products like ginger oil, oleoresin, dehydrated and bleached ginger, harvesting should be done between 8-10 months.

Crop nutrition

Ginger is a nutrient exhaustive crop and application of organic manures and fertilizers are absolutely essential. Ginger rhizomes are mainly N and K exhausting, intermediary in P and Mg removal and the least in Ca removal. Haag *et al.* (1990) reported an accumulation of macronutrients in the decreasing order *viz.,* N, K, Ca, Mg, S and P and micronutrients in the order of Fe, Mn, Zn,

B and Cu. However, nutrient uptake varies considerably with soil type, climatic conditions, nutrient levels in the soil, variety adopted *etc*. Govender *et al.* (2009) found that the soil quality influenced the elemental distribution within the ginger rhizome and the plant has the inherent ability to control the amount of elements absorbed. The ginger 'flesh' tends to accumulate Mn and Mg. A synergistic relationship between Cr and Mn and an antagonistic relationship between Fe and Cu and Fe and Cr were observed.

The development of ginger with respect to the development of aerial tissues can be classified into three distinct growth phases *viz.,* active growth [90-120 days after planting (DAP)], slow vegetative growth (120-180 DAP) and senescence (180 DAP) in which the rhizome development continues till harvest. The uptake of N, P and K in leaf and pseudostem increases up to 180 day and then decreases, whereas that of rhizome uptake steadily increases till the harvest (Johnson, 1978). Xu *et al.* (2004) observed that ginger shoots and leaves were growth centers at the seedling stage, and 80.7 % of the carbon (C) assimilation was transferred to these parts. Later, the distribution rate for shoots and leaves decreased gradually with the growth, whereas the distribution rate for the rhizome increased. Up to the vigorous growth stage of rhizome, C assimilation is mainly transported from leaves into the rhizome, thus the rhizome is the growth center. The absorption and utilization of N were the same as C assimilates. About 48.41 % of the N absorbed from fertilizer applied at seedling stage was distributed to the shoots and leaves. While 65.43 % of the N derived from fertilizer applied at vigorous growth of rhizome was distributed into rhizomes, only 32.04 % distributed into shoots and leaves. The results indicated that the rate of fertilizer-N utilization increased with delay of application. The exchange of C and N nutrition between the above ground parts and underground seed of ginger and their transportation and distribution during ginger growth ensures that the seed-ginger will not shrivel.

An average dry yield of 4.0 t ha^{-1} dry ginger rhizomes removes 70 kg N, 17 kg P_2O_5, 117 kg K_2O, 8.6 kg Ca, 9.1 kg Mg, 1.8 kg Fe, 500 g Mn, 130 g Zn and 40 g Cu ha^{-1}. For healthy growth of ginger, very low external Ca is required. A heavy ginger crop removes 35-50 kg P ha^{-1}. The leaves of healthy plants contained 1.1 to 1.3% of Ca and concentrations as low as 2 ppm is sufficient to achieve 90 % of maximum yield. In order to calculate nutrient requirement of the crop, Johnson (1978) recommended 5th pair of leaf in a period between 90 to 120 DAP for foliar diagnosis of N, P and K.

Efficient organic nutrient management and INM

The requirement of organic matter may be met from various sources such as green/organic manures and mulches. This was very much evidenced by good performance of ginger crop shown in high fertility conditions of Wayanad, Kerala, India supplied with 10 t of organic manure and 15 t of green leaf mulch per hectare, without any fertilizer application. Farm Yard Manure (FYM), poultry manure, green leaf, compost, press mud, oil cakes and biofertilizers are used as sources of organics. The quantity of organics applied may vary with availability of material and generally it varies between 5-30 t ha^{-1}. Organic manures are mostly applied as basal doses and in certain places it is also applied after the emergence of crop as mulch. However, farmers of Maharashtra apply heavy doses of FYM, about 40 to 50 t ha^{-1}. In Kerala the recommended dose of organic manure is 25-30 t ha^{-1} of FYM and 30 t ha^{-1} of green leaf mulch applied in three splits. Application of FYM up to 48 t ha^{-1} resulted in highest yield and benefit: cost ratio under Kerala conditions.

Rhizome yield increases with the increase in the level of FYM. Application of FYM at 6 t ha^{-1} was found to be superior for getting high yield (3.2 t ha^{-1}) under Madhya Pradesh conditions. Application of organic cakes increased nutrient availability, improved physical condition of the soil, increased the yield and oleoresin production. Maximum availability of exchangeable-Ca in the soil was recorded by FYM (10 t ha^{-1}; Sadanandan and Iyer, 1986). Application of coconut cake (0.3% w/w) neem cake at 2 t ha^{-1} (Sadanandan and Iyer, 1986), neem cake at 2.5 t ha^{-1} or groundnut cake at 1 t ha^{-1} (Sadanandan and Hamza, 1998b) improved yield and quality parameters and reduced soft/rhizome rot incidences in ginger. In addition to neem cake, phosphobacteria and P as rock phosphate increased the availability of P, Ca, Mg, Zn and Mn. Incorporation of inorganic N with neem cake (50% N), poultry manure (25% N), and groundnut cake (50% N) were found to be best for getting higher yield of up to 29 t ha^{-1} of ginger. Also, application of neem cake at 2t ha^{-1} along with NPK at 75:50:50 kg ha^{-1} increased yield of ginger and reduced the rhizome rot incidence under Kerala conditions. Among other organic sources, application of coir compost (Terracare) at 2.5 t ha^{-1} increased the yield by 37.5% over control. Gradual availability of nutrients through decomposition of organics throughout the growth phase may be the probable cause for better growth and development of plant and ultimate yield when inorganics were substituted with organics at different levels.

The trials on different management systems on ginger at Indian Institute of Spices Research (IISR), Calicut, India showed that higher soil nutrient build up with highest organic carbon content (2.33%) in organic system on par with integrated system and among the different systems organic management systems recorded the highest concentration of soil P, Ca, Mg, Zn and Cu availability. In ginger (Table 7), oil content did not vary significantly with the sources of nutreints. However, higher starch and oleoresin content and reduced fibre content were observed in organic system compared to inorganic and integrated system. The highest yield and oleoresin was obtained with application of 10 t of FYM ha^{-1} + 1.25t ha^{-1} of coir compost + 20 kg ha^{-1} of *Azospirillum* with higher nutrient uptake.

Table 7: Effect of different management systems on quality of ginger

Management systems	Oil	Oleoresin	Starch	Fibre
	%			
Organic	1.20	3.96	70.07	1.69
Inorganic	1.20	3.15	62.21	1.90
Integrated	1.25	3.36	55.82	1.90
CD 5%	NS	0.23	9.89	0.077

Mulches add organic matter, check soil erosion, conserve moisture, reduce soil temperature, improve soil physical properties and suppress weed growth. Mulching can alter the physical and chemical properties of the soil and increase the availability of P and K. In general, 10 to 30 t ha^{-1} mulch is applied twice or thrice, one at planting, second at 45th day and third at 90th day after planting. Some of the commonly used mulch materials are green and dry forest leaves, Coconut leaves, banana leaves, dry *sal* leaves, *shisam* residues like sugarcane trash, paddy, wheat, finger millet, little millet and barley straws and also weeds and other vegetation are used as mulch. FYM and compost are also used as mulch. If these materials are not available in sufficient quantity, live mulches like sunhemp, greengram, horsegram, blackgram, niger, common sesbania, cluster bean, french bean, soybean, cowpea, daincha and red gram can be grown as intercrop and used for *in situ* mulching after 45-60 days of planting. Application of dried coconut leaves after removing the petiole or paddy straw (2-3 kg/bed) as mulch in ginger is recommended for effective weed control (up to 85%) and higher rhizome yield (10-36%) (Thankamani *et al.* 2016). Fodder cowpea and daincha are suitable green manure crops for a self sustainable system to provide mulch material of 89% and 61%, respectively of the total requirement in the first sowing itself.

Studies at Kerala Agricultural University (KAU), Ambalavayal, Kerala, India showed that organic amendments such as coconut oil cake, ground nut cake, sesamum cake and cashew shell when applied to the soil to give N requirement of 50 kg ha^{-1} were found to reduce the soft rot incidence of ginger caused by *Pythium aphanidermatum*. Lime application is practiced in acidic soils to correct the pH and nutrient imbalance. Under Kerala conditions where the soils are lateritic with high P fixing capacity, available P increased with the application of rock phosphate incubated with FYM and among the P sources, Rajphos was found to be superior. The apparent phosphate recovery, agronomic efficiency of the applied P and percentage yield response was higher for Gafsaphos, followed by Rajphos incubated with FYM in ginger.

Both the nature and level of the organic manures markedly influence the growth and yield attributes of ginger. Roy and Hore (2007) reported that the maximum values for plant height, tiller number, leaf number, clump weight and length and plot yield (13.76 kg 3 m^{-2}) and projected yield (29.72 t ha^{-1}) were recorded when 1/4 N was incorporated through poultry manure and 3/4 N as urea. Plants raised under groundnut cake (1/2 N) + urea (1/2 N) had the highest number of primary fingers compared to the lowest number under mustard cake (1/4 N) + urea (3/4 N). Similarly, a field experiment on ginger (cv. Nagaland local) conducted on an Alfisol in Medziphema, Nagaland, India, revealed that the highest number of tillers per plant (10.1), highest fresh yield (294.1 q ha^{-1}), highest oil (0.47 ml 100 g^{-1}) and oleoresin (5.57 ml 100 g^{-1}) were observed with the application of pig manure and it was significantly superior over FYM and poultry manure. Singh and Singh (2007) also observed the highest uptake of N (136.7 kg ha^{-1}) and K (73.9 kg ha^{-1}) with the application of pig manure.

Efficient inorganic nutrients management

A. Major nutrients

Response of fertilizer varies with variety, soil type and climate. A fertilizer dose of 36-225: 20-115: 48-200 N, P_2O_5, K_2Okg ha^{-1} has been adopted in different states in India. A fertilizer dose of 200:229:199 in Australia, 66:82:66 in West Indies and 105:241:126 in Nigeria have been recommended. For ginger production, straight fertilizers, and fertilizer mixtures containing NPK 8:8:16 are also used.

The recommendation of N, which is an extremely important element for ginger production varies from 36 to 225 kg ha^{-1} in India. Mohanty *et al.* (1993) recorded maximum yield with the application of 125 kg N ha^{-1} in Orissa, India, while Randhawa and Nandpuri (1969) with 100 kg N ha^{-1} in Himachal Pradesh, India. Sadanandan and Sasidharan (1979) obtained the highest yield with 50 kg N ha^{-1} in the main crop, where as Tarle *et al.* (2007) recorded highest yield and yield attributing characters with local variety at 150 kg ha^{-1} N application, in Maharashtra, India.

A heavy crop removes 35 to 50 kg ha^{-1} of P and it is mostly applied as a basal dose at the time of planting. Application of P at 20 and 40 kg ha^{-1} increased the yield by 21.5 and 11.5 %, respectively. Plant height, number of leaves per clump, number of tillers per clump, leaf length, fresh and dry yield, crude protein and fibre, and volatile oil content were highest when ginger was sown in flat beds followed by earthing up and application of 50 kg P ha^{-1}. Under Kerala conditions, in an Ustic Humitropept, Bray available P increased with the application of rock phosphate mixed with FYM and among the P sources, Raj phos was superior. Rhizome P concentration was higher when superphosphate and FYM incubated rock phosphates were applied. Significantly higher dry yield in ginger was registered with Rajphos and Gafsaphos application with FYM, and it was significantly higher than other sources of P applied. Not much variation was observed in oleoresin content of ginger due to application of different sources of P. Among all the sources, apparent phosphate recovery, agronomic efficiency of the applied P and percentage yield response was higher for Gafsa phos, followed by Raj phos incubated with FYM in ginger.

K is one of the most important limiting factors for ginger production, since the crop removes a large amount of soil K (up to 500 kg ha^{-1}). The main practices to obtain high rhizome yield with optimal nutrient use efficiency include fertilizer application based on soil testing, top dressing K fertilizer at growth stages with peak demand, and applying enough K to balance the appropriate N and P application rates. In the lower range of replaceable K (0.3 me %), a positive field response was observed with application of up to 325 kg K_2O ha^{-1}, while above 0.3 me % no response was recorded with rates above 60 kg K_2O ha^{-1}. Even though, K increased the yield, a higher dose of K_2O reduced the height of plants and yield. Besides sole application of K, an obvious response of ginger growth and quality was observed due to the combined application of N and K. They observed that a suitable rate and ratio of K and N combined application markedly promoted ginger growth, increased rhizome yield, improved nutrition qualities and enhanced K recovery efficiency. Besides, the maximum rhizomatous yield was attained by the application of 450 kg ha^{-1} each of N and K.

For healthy growth of ginger, very low external Ca is required since relatively large amounts of Ca are added to the soil through super phosphate and organic manures. The upper leaves of Ca deficient plants contained 0.05 to 0.07% Ca while corresponding leaves of healthy plants contained 1.1 to 1.3%. Ca concentration as low as 2 ppm appears sufficient to achieve 90% of maximum yield.

In Kerala, India three N and K combinations *viz*., 150 kg N + 50 kg K, 150 kg N + 100 kg K and 75 kg N + 150 kg K ha^{-1} were found to be superior for yield. However, the optimum dose was found to be 144 kg N and 109 kg ha^{-1} K. Rhizome yield increased due to the application of 70 kg N and 140 kg K ha^{-1} but was unaffected by P and they found greater association of K content in the leaf with higher yield. Under low shade, 150 per cent (112.5: 75: 75 kg NPK ha^{-1}) of the recommended dose (75:50:50 kg ha^{-1}) enhanced the yield (3415 kg ha^{-1}). Gowda *et al.* (1999) obtained higher yield (15.62-22.53 t ha^{-1}) with the combined applications of 150 kg N + 75 kg P ha^{-1}; 150 kg N + 50 kg K ha^{-1}; and 25 kg P + 50 kg K ha^{-1}. However, the yield of 'Rio-de-Janiero' could be increased by the application of 150 Kg N, 75 kg P_2O_5 and 50 kg K_2O ha^{-1} in Bangalore, India. Targeted yield equation for predicting nutrient requirements to get fixed rhizome yields in soils with varying fertility levels was standardized in ginger at IISR and the economic optimum in terms of profitable response for money invested was found to be Rs. 3.75/ bed (3 m2) for N, Rs. 1.30/ bed for P and Rs. 0.60/bed for K (Srinivasan *et al.*, 2014).

Hepperly *et al.* (2004) used slow-release fertilizer mixture containing equal parts by volume of Nutricote® 13-13-13 240-day release Nutricote 16-40-0 70-day release Nutricote 1-0-38 240-day release Nutricote 12-0-0 240-day release. Six ounces of this mixture to the 6 gallons of growing medium in each bag at planting time provided most of the basic nutrient needs for the ginger plant during the growing season, when grown under green house conditions with an average production of 15 lb per bag. Studies by Haque *et al.* (2007) revealed that the effects of combined application of N and K was found more pronounced than the single effect of N and K and also the effect of N was more distinct than K. The combined effect of N and K up to 180 and 160 kg ha^{-1} respectively had significantly increased the yield and other yield contributing characters like plant height, number of leaves, finger numbers, weight and rhizome yield. The highest mean yield (26.95 t ha^{-1}) was also recorded by N and K application at 180 and 160 kg ha^{-1} respectively. Paliyal *et al.* (2008) obtained highest rhizome yield of 16 t ha^{-1} due to balanced nutrient application of FYM at 20 t ha^{-1} and 100-75-40-5-1 kg ha^{-1} of N-K-S-Zn-B fertilizers. Application S, Zn and B along with NPK increased the uptake of N, P, K, Ca and S. Soil availability of organic carbon, available N, P, S, Ca, S, Cu, Mn, Fe and Zn increased after the harvest of the crop. The fertilizer requirement of ginger in different states of India was studied and the recommendations are provided in Table 8.

Table 8: Manures and fertilizers recommended for ginger in various locations of India

State	Recommendation
Kerala	FYM 30 t ha^{-1}; NPK 70:50:50 kg ha^{-1}. Full dose of P and 50% K may be applied as basal dose. Half the quantity of N applied at 60 DAS. The remaining quantity of N and K applied at 60 DAS.
Karnataka	FYM/compost 25 t ha^{-1}; NPK 100:50:50 kg ha^{-1}. Apply the entire dose of P and K at planting. Half quantity of N to be applied at 30-40 DAS and other half at 60-70 DAS.
Odisha	FYM 25 t ha^{-1}; NPK 125:100:100 kg ha^{-1}. Full P and half K applied as basal dose in furrows before planting and N and K in 2 splits at 45 and 90 DAS.
Himachal Pradesh	FYM 20-30 t ha^{-1}; CAN @ 400 kg ha^{-1}, NPK 100:50:60 kg ha^{-1}. Apply P and K at the time of planting and N in 3 equal splits, first at the time of planting and subsequent 2 doses at monthly interval. K_2O also can be applied in two splits, half at sowing and other half at rhizome initiation.
Bihar	FYM 20-30 t ha^{-1}; NPK @ 60:60:120 kg ha^{-1}
Andhra Pradesh	FYM 20-30 t ha^{-1}; NPK @ 75:50:50 kg ha^{-1}
Chattisgarh	FYM 20-30 t ha^{-1}; NPK @ 150:125:125 kg ha^{-1}
Sikkim	Manures 40-50 t ha^{-1}; Few farmers apply fertilizers*
Meghalaya	FYM 10 t ha^{-1}; NPK @ 60:90:60 kg ha^{-1}**

Source: Ramana *et al.* (2003)

Micronutrients

Soil and plant analysis of samples across the country revealed 49% mean deficiency of zinc (Zn) with acid soils of Meghalaya having the highest deficiency rate of 57% (Srinivasan *et al.,* 2009). Roy *et al.* (1992) obtained maximum yield with a combined spraying of Zn (0.3%) + Fe (0.2%) + B (0.2%), twice at 45 and 75 days after planting. Foliar application of Zn at 0.25% (twice-May-June and August-September) increased the rhizome yield (16.2 kg/3m^2 bed), as compared to soil application (14.4 kg). Studies conducted at Solan, India on the effect of micronutrients namely, zinc (20 kg ha^{-1}), B (10 kg ha^{-1}), Mo (1 kg ha^{-1}), Mg (10 kg ha^{-1}) and Jagromin (chelated form of micronutrients) along with recommended dose of NPK (100:50:50 kg ha^{-1}) showed significant increase in the yield with NPK along with two sprays of Jagromin (0.7%), once at rhizome initiation and again one month after the first spray (Ramana *et al.*, 2003). In an Ustic Humitropept soil of Kerala, highest rhizome yield (13.78 kg bed^{-1} of 3 m^2) was obtained in plots where Zn (zinc sulfate) at 5 kg ha^{-1} was applied. The optimum fertilizer rate for obtaining maximum rhizome yield was determined as 6 kg Zn ha^{-1}. The maximum limit of soil DTPA-extractable zinc for obtaining higher rhizome yield was 3.4 mg kg^{-1} (Srinivasan *et al.*, 2004). An appropriate critical limit of 2.1- 3.74 mg Zn kg^{-1} for soil and 27.0-53.8 mg

Zn kg^{-1} for leaf was fixed for getting maximum ginger yield in Ustic Humitropept of Kerala (Srinivasan *et al.*, 2009). ICAR-IISR developed crop and location specific foliar nutrient mixtures suited for ginger for varying soil pH conditions (Anandaraj *et al.*, 2014). Ginger specific micronutrient foliar sprays are recommended @ 5 g/ L of water at 60 & 90 days after planting the crop and recorded 10-20% increase in yield.

Turmeric

Turmeric (*Curcuma domestica* Val. (syn. *C. longa* Koenig *non* L.) is an important spice, a native of India and South East Asia. is a 'wonder spice', is traditionally used in Asian countries. India is the largest producer, consumer and exporter of Turmeric. Other producers in Asia include Bangladesh, Pakistan, Sri Lanka, Taiwan, China, Burma (Myanmar), and Indonesia. In addition to its use as a spice, it has other uses that are prominent in the life of the people of southern Asia, and connected with birth, marriage, death and in agriculture.

Turmeric is cultivated in the tropical and subtropical humid climate. It is successfully cultivated from subtropical to wet tropical zone at an altitude of 1200 m above MSL, with an optimum range of 450-900 m. In India it is widely grown in warm to hot, per-humid to humid eco sub regions. The crop is rainfed where rainfall is bimodal and with irrigation in plains where rainfall is uni-modal and low. It tolerates an annual rainfall of 640 to 4290 mm. Moderate rainfall of 1500 mm at sowing, fairly heavy and well distributed rain during growing period and dry weather about one month before harvest are much suitable. Kandiannan *et al.* (2002) reported that turmeric yield had positive and significant relation ($r = 0.6024$) with the rainfall received during second month after planting. The temperature range of 18.2–27.4°C is optimum.

Turmeric is grown in a wide range of soil types of varying fertility grouped under inceptisols, entisoils, vertisols, alfisols and ultisols. Soils that are having high organic carbon, base saturation, major and secondary nutrients are suitable for turmeric cultivation. Well drained, deep loamy to clay loam soils with good organic matter status are well suited for this crop and very coarse and heavy soils are unsuitable for rhizome development. The crop can thrive well in the soil pH range of 4.3 to 7.5. Sahu and Mitra (1992) noted the variations in the performance of turmeric varieties with respect to soil types.

Crop nutrition

Nutrient requirement of turmeric is shown in the Table 9 below. According to Kumar *et al.* (2003) approximately 9% of the turmeric growing area in Tamil Nadu is severely limited by mineral nutrition and about 20% of samples were identified as having deficiencies. They worked out optimum levels for 12 nutrients (N, P, K, Ca, Mg, Na, S, B, Zn, Cu, Fe and Mn) in the leaves of turmeric using Diagnosis and Recommendation Integrated System (DRIS)/Modified Diagnosis and Recommendation Integrated System (MDRIS) and Compositional Nutrient Diagnosis (CND) approaches. The reference norms for optimum concentration in leaves of turmeric in Erode district of Tamil Nadu ranged as 1.22-2.75% for N, 0.36-1.27% for P, 3.66-6.6% for K, 0.18-0.33% for Ca, 0.61-1.25% for Mg, 0.16-0.31% for Na, 0.13-0.29% for S, 14.3-26.3 mg kg^{-1} for B, 41.1-93.2 mg kg^{-1} for Zn, 15.2-40.3 mg kg^{-1} for Cu, 143-1568 mg kg^{-1} for Fe and 66-219 mg kg^{-1} for Mn.

Table 9: Removal of nutrients by turmeric at harvest (kg ha^{-1})

Location/soil type	Nutrients					Reference
	N	P_2O_5	K_2O	Ca	Mg	
Kasaragod-Laterite	124	30	236	73	84	Nagarajan and Pillai (1979)
Vellanikkara-Laterite	72-115	14-17	141-233	—	—	Rethinam *et al.* (1994)
Bhavanisagar-Sandy loam	166	37	285	—	—	Rethinam *et al.* (1994)
Coimbatore-Clayey loam	187	39	327	—	—	Rethinam *et al.* (1994)
Calicut – Laterite	86	31	194			Sadanandan and Hamza (1998a)

Dixit and Srivastava (2000) also observed decrease in plant growth, fresh weight, rhizome size, photosynthetic rate and chlorophyll content and significant increase in curcumin content in all the genotypes under Fe deficiency. N, Mn and Zn contents of rhizomes had a positive significant correlation with curcumin content, whereas P, Ca, Mg, S, Cu and Fe contents had no correlations (Kumar *et al.*, 2000a). Total Zn concentration in the third leaf was considered for delineation of deficiency that varied from 10 to 51.3 mg kg^{-1}. Zinc concentration was significantly and positively correlated with leaf weight and rhizome yield and negatively with Fe: Zn and P: Zn concentration ratios. P: Zn ratio in the absence of added Zn proved as a good indicator of efficiency of response to added Zn. Leaf P: Zn ratio had a positive correlation and soil P: Zn ratio had negative correlation with rhizome yield.

Efficient organic nutrient management and INM

Many experimental evidences show the beneficial effects of organic matter alone or in combination with inorganic fertilizers in turmeric. Compost or FYM at 40 t ha^{-1} is applied by broadcasting and ploughing at the time of preparation of land or as basal dressing by spreading over the beds to cover the seed after planting. The fresh yield increase was over 37%. Kerala Agricultural University (KAU) recommends 40 t ha^{-1}compost or cattle manure as basal dressing, for Western Ghats and Coastal Plains of Kerala; whereas in Orissa 15 kg has been applied in 5 m^2 bed area in three splits along with chemical fertilizers. Organic manures like groundnut cake, vermicompost and neem cake can also be applied. In such cases the dosage of FYM can be reduced. Application of FYM increased the organic carbon, available P content and effective cation exchange capacity of the soil there by improving the fertility which in turn significantly improved plant height, number and weight of mother, primary and secondary rhizomes and fresh rhizome yield. The N, P and K contents in leaves and rhizome were also improved with farmyard manure application. Application of FYM at 50 t ha^{-1} increased the fresh rhizome yield up to 11.72 t ha^{-1} in warm per humid conditions of North Eastern India.

Farmyard manure + 90 kg N ha^{-1} is optimum for turmeric production in acidic alfisols of Meghalaya, with considerable N use efficiency (34.3%) and nutrient build up in the soil. Gill *et al.* (1999) observed significant increase in rhizome yield and curcumin content with increased FYM (up to 60 t ha^{-1}) and wheat straw mulch. In turmeric, application of cowdung (50 t ha^{-1}) followed by 90, 60 and 90 kg N, P_2O_5, K_2O ha^{-1} yielded the highest with maximum profit in Mizoram, NE India. In Wayanad (Kerala, India), application of 100 kg N per ha along with FYM (15 t ha^{-1}) and green leaf mulch (50 t) produced maximum yield. Among other organic manures, groundnut cake (1.1 t ha^{-1}) significantly increased the dry yield and the highest curcumin production (Table 10), on par to neem cake application (2.5 t ha^{-1}; Sadanandan and Hamza, 1998a). Contrarily, no significant differences in dry yield and curcumin content of turmeric rhizomes was observed with the combination of chicken manure and inorganic fertilizer under Chinese conditions. Fields that have received neem cake as a dressing were also found to be free from Mimegralla fly (Reddy and Reddy, 2000).

Turmeric shows good response to the application of organics and biofertilizers. Integrated application of coir compost (2.5 t ha^{-1}) combined with FYM, *Azospirillum* and half the recommended NPK significantly increased yield and quality of turmeric (Srinivasan *et al.*, 2000b). Application of coir pith compost, *Azospirillum*, phosphobacteria and VAM induced maximum IAA oxidase and peroxidase activity in turmeric. Similarly, application of FYM (5 t ha^{-1}) with 50% inorganic N and *Azospirillum* (5 kg ha^{-1}) produced higher yield of mother, primary and secondary rhizomes with a cost benefit ratio of 1:2.28. Up to 16% increase in yield by the application of 25 kg *Azospirillum* with 50% of the recommended dose of inorganic N and 5 t FYM ha^{-1} over the recommended dose of fertilizer application was observed by Selvarajan and Chezhiyan (2001). Jena *et al.* (1999) noted that the rhizome yield and nutrient uptake by turmeric were significantly higher in both individual and combined inoculation of *Azotobacter* and *Azospirillum* and the percentage increase in yield due to inoculation integrated with fertilizer N ranged from 15.2 to 30.5% with enhanced N-use efficiency. The soil was left with a positive N-balance in the integrated treatments indicating a build up of soil fertility. Among the quality characters, curcumin content was more with neem cake 1.25 t ha^{-1} + FYM 12.5 t ha^{-1} + RFD, whereas, essential oils and oleoresins were more in vermicompost 1.0 t ha^{-1} alone over RFD.

Besides, organic manures and lime, application of biofertilizers has been found to benefit soil quality as well as yield of turmeric. Velmurugan *et al.* (2008) obtained the highest curing percentage and cured rhizome yield (21.26 and 7080.17 kg ha^{-1}, respectively) as well as the highest curcumin, oleoresin and essential oil contents (4.577, 9.477 and 3.817% respectively) with the application of FYM + *Azospirillum lipoferum* + phosphate-solubilizing bacterium (*Bacillus megaterium*) + VAM (vermiculite-based inoculum *Glomus fasciculatum, G. mosseae* and *Gigaspora* sp.). An earlier study by (Padmapriya *et al.*, 2007) revealed that shade with application of 100% recommended dose of NPK+ 50% FYM (15 t ha^{-1}) + coir compost (10 t ha^{-1}) + *Azospirillum* (10 kg ha^{-1}) + phosphobacteria (10 kg ha^{-1}) + 3% panchagavya showed increased total chlorophyll content, total phenol content and registered the highest turmeric yield per plot.

The profound effect of zinc solubilizing bacteria (ZSB) on the increase of dry matter yield of turmeric was found to be 14.0, 14.3 and 18.1% for $ZnSO_4$, Zn enriched FYM and Zn enriched coir pith along with ZSB, respectively than in treatments without the organism. All the genotypes of turmeric are found to have effective mycorrhizal association by extra-, intra- and intercellular hyphae of VAM fungi in their roots. Among cultivars, Suguna followed by Prabha and Sugandham, were observed to be the most heavily colonized with mycorrhizal fungi with mycelium, arbuscules, vesicles and spores of *Glomus*, *Gigaspora* and *Sclerocystis* with domination of *Glomus* population. The mycorrhizal inoculation is advantageous in improving plant growth and plants inoculated with different species of VAM fungi recorded a significant increase in growth compared to uninoculated plants . The highest growth rate was found in plants inoculated with *Glomus fasciculatum*. Mohan *et al.* (2004) recorded a linear response of turmeric growth, yield and quality of turmeric with inoculation of *Azospirillum* in combination with N levels compared to that of *Azotobacter* in Karnataka state, India. Poinkar *et al.* (2006) reported that application of NPK 120:60:60 kg ha^{-1} followed by FYM (10 t ha^{-1}) + Azotobacter + Phosphate

Solubilizing Bacteria (PSB) (250 g per 10 kg of seed) increased plant height, number of leaves, size and surface area of leaves, girth of pseudo-stem, number of tillers per plant and fresh yield per ha significantly under Maharashtra (Eastern Plateau, hot dry subhumid) conditions and C : B ratio was the highest (2.17) in treatment FYM + Azotobactor + PSB.

Efficient inorganic nutrient management

Major nutrients

Being an exhaustive crop, turmeric responds well to judicious application of fertilizers according to the agroecological situations. Yield increases to the tune of 81 to 282% have been observed due to fertilizer application compared to non-fertilized control. Location specific fertilizer recommendations have been made based on agro climatic situations (Table 11). Among the major nutrients, the response to N varies from location to location depending on soil type, variety and other factors. Application of varying levels of N, P and K significantly increased the growth attributes like number of tillers per plant, number of leaves per plant, plant height, leaf area and total dry matter production. Main growth characters like number of leaves per plant, number of tillers per plant, leaf area index, weight of mother rhizomes, weight of finger rhizomes and total curing percentage except plant height, had positive and significant correlation with rhizome yield . The optimum levels of N, P and K recommended for turmeric at different locations and agro-ecological situations of India are given in Tables 11 and 12.

Table 10: Yield and quality of turmeric as affected by organic and inorganic fertilizers

Treatment	Dry yield(kg ha^{-1})	Curcumin(kg ha^{-1})
Check	2374	169
FYM	2596	250
Neem cake	2602	287
Cotton cake	2640	284
Brassica cake	2784	243
Groundnut cake	2669	277
Gingily cake	2768	249
NPK fertilizer*	2480	268
CD at 5%	—	32.2

* NPK @ 60, 50, 120 kg ha^{-1}.

Source: Sadanandan and Hamza (1998a)

Table 11. Optimum levels of N, P and K (kg ha^{-1}) recommended for turmeric in different locations

Optimum N P K	Recommended Location	Reference
150:125:250 kg ha^{-1}	Central Karnataka Plateau, Hot Semiarid eco region	Venkatesha *et al.* (1998)
120: 60:120 kg ha^{-1}	Hill zone of Karnataka	Sheshagiri and Uthaiah (1994)
60: 50: 120 kg ha^{-1}	Western Ghats and Coastal Plains of Kerala, Hot humid eco region	Sadanandan and Hamza (1996)
130: 90: 70 kg ha^{-1}	Subdued Eastern Himalayas (Arunachal Pradesh), warm to hot perhumid region	Dubey and Yadav (2001)
200: 60: 200 kg ha^{-1}	Alluvial plains of West Bengal	Medda and Hore (2003)
75: 60: 150 kg ha^{-1}	Middle Gangetic plain, hot moist semi dry eco region (Allahabad)	Thomas *et al.* (2002)
	100: 50: 50 kg ha^{-1}	
	Western Himalayas, hot moist subhumid transition eco region (Shimla hills)	Randhawa *et al.* (1973)
200: 100: 100 kg ha^{-1}	Eastern Plateau, hot dry subhumid eco region (Maharashtra)	Yamgar *et al.* (2001)
100: 100: 100 kg ha^{-1}	Middle Gangetic plains of Uttar Pradesh	Upadhyay and Misra (1999)

Table 12. Fertilizer recommendations for turmeric in different agro-ecological situations

Location	YM (t ha^{-1})	Recommended dose N	P_2O_5 (kg ha^{-1})	K_2O
Andhra Pradesh - Deccan Plateau, hot semi arid eco region	25	300	125	200
Assam - Plains, hot subhumid to humid eco region	20	30	30	60
Bihar - Eastern plateau, hot subhumid eco region	NA*	150	50	100
Kerala - Western Ghats and Coastal Plains, hot humid eco region	40	30	30	60
Maharashtra - Eastern Plateau, hot dry subhumid eco region	NA	120	60	60
Tamil Nadu - Upland and Plains, hot, moist semiarid eco region	25	120	60	60

NA* - Not available; *Source*: Rethinam *et al.* (1994)

Micronutrients

Up to 24% increase in rhizome yield with the application of $FeSO_4$ at 30 kg ha^{-1} was observed in Fe deficient soils of Tamil Nadu. Application of $ZnSO_4$ at 15 kg ha^{-1} increased the rhizome yield by 15%. Combined application of 50 kg ha^{-1} each of $FeSO_4$ and $ZnSO_4$ increased yield up to 21.4 t ha^{-1}. The optimum dosage of Zn, B and Mo has been formulated as 5, 2 & 1 kg ha^{-1} with 20 and 15 tonnes of FYM and green mulching for sustainable yield in turmeric along with recommended fertilizer (Sadanandan and Hamza 1996). Application of 120 kg N and 120 kg K_2O ha^{-1}, together with the trace elements B (2 kg ha^{-1}) and Zn (10 kg ha^{-1}) gave the maximum economic yield in turmeric intercropped under partial shade of coconut. Foliar application at 0.25% $ZnSO_4$ twice has given higher rhizome yield on par to soil application of 7.5 kg Zn ha^{-1}. Studies have also revealed constant increase in the uptake of all the major nutrients from the early stage of the crop to harvest in FYM + Zn solubilizing bacteria (ZSB) and soil and foliar application of Zn and Fe. FYM + ZSB recorded 21.6% higher rhizome yield than the FYM alone application. Turmeric cultivars (Krishna, Suvarna, Salem and Waigaon) registered the greatest plant height, number of leaves per plant, leaf area index, total dry matter per plant, highest average and total yield of fresh and dry rhizomes at recommended fertilizer rates in Akola, Maharashtra (RFR, 120 kg N + 60 kg P_2O_5 + 60 kg K_2O) + 30 kg Zn ha^{-1} (Jadhao *et al.*, 2005).

The treatments with Zn either alone or in combination with Fe as soil application recorded lower Nutrient Imbalance Index (NII) that were under balanced range where as the foliar spray treatments recorded higher NII values exhibiting severe imbalances among the nutrients in the turmeric leaves. They also recorded significant positive correlation (0.92**) between yield and zinc indices as compared to the foliar concentration of Zn indicating the importance of nutrient indexing. Vishwakarma *et al.* (2006) found that Cu at 16 kg ha^{-1} and B at 20 kg ha^{-1} increased the growth and yield of turmeric under shaded conditions under mango orchards. Jirali *et al.* (2007) reported that foliar application of $FeSO_4$ at 0.50% (at 60 and 120 DAS) was very effective, followed by $MnSO_4$ at 0.40% and $ZnSO_4$ at 0.50% and these treatments significantly increased the important biophysical parameters, *i.e.* photosynthetic rate, stomatal conductance and transpiration rate and biochemical parameters, *i.e.* nitrate reductase activity and total chlorophyll content. These treatments were also effective in increasing both the fresh and dry rhizome yield. The application of these micronutrients significantly increased the curcumin content (Velmurugan *et al.*, 2007). Halder *et al.* (2007) found that the highest rhizome yield was recorded with the combination of Zn and B at the rate of 4.5 and 3.0 kg ha^{-1}, respectively. ICAR-IISR developed crop and location specific foliar nutrient mixtures suited for turmeric for varying soil pH conditions. Turmeric specific micronutrient foliar sprays are recommended @ 5 g/ L of water at 60 & 90 days after planting the crop and recorded 10-20% increase in yield (Anandaraj *et al.*, 2014).

Conclusion

Spices export have registered substantial growth during the last five years, registering acompound annual average growth rate of 23% in value and 11% in volume and Indiacommands a formidable position in the World Spice Trade. Spices shares around 13 % of the area and 2.0% in production of the total horticultural crops of India. Balanced nutrient management is very important for sustained production of spices. In addition to major and secondary nutrients, deficiency of micronutrients especially Zn, B and Mo are seen. It is therefore necessary to develop sound soil fertility management program that encompasses nutrient recommendations based on sound soil tests. Sound crop management programs that are location specific should form the basis for manure and fertilizer management that optimizes economic return while protecting water quality and the environment. Overall, it is apparent that in majority of the soils the reasons for low productivity of spices are acidic pH, high clay content, low CEC, low levels of organic carbon, K, Ca, Mg and Zn. Inorganic fertilizer schedules for different agro-ecological locations exits for these crops. Similarly organic nutrition schedules have also been devised for these crops. However, among the different nutrient management strategies, integrated nutrient management appears to be the best bet. Use of organic manures and biofertilizers in place of chemical fertilizers is energy efficient to the extent of saving half to two-third that is presently used in fertilizers inputs. As a best component crop in agri-horti and silvi-horti systems, recycling of farm waste can be effectively done when these spices are grown as a component crops with other plantations.

A major factor in the humid tropics is soil erosion due to intensive tillage and this can be efficiently counteracted by minimum tillage practices. The problem is that such an approach to sustainability of soil fertility may work well for one crop and not for a second. Moreover, social and economic circumstances still prevail in the choice of viable options for soil management. In the long term, however, and particularly for continuous cropping systems, available methods of soil fertility conservation should be adopted. It is imperative that we develop radical scientific solutions capable of contributing to new nutrient management systems, which are intensive but nevertheless sustainable.

Selected References

Anandaraj M, Sarma YR. 2003. "The potential of PGPRs in disease management of spice crops." In: Reddy M S, Anandaraj M, Eapen S J, Sarma Y R & Kloepper J W (eds.), *Proceedings of the VI International Workshop on Plant Growth Promoting Rhizobacteria*, IISR, Calicut, Kerala, India, 27-39.

Anandaraj M, Dinesh R, Srinivasan V, Hamza S, Bini, YK. 2014. Feeding Spice crops for quality produce. *Indian Horticulture* 59 (6):22-24.

Bavappa KVA, Gurusinghe P, De S .1978. Rapid multiplication of black pepper for commercial planting. *Journal of Plantation Crops* 6 : 92-95.

Bavappa KV, Kailasam C, Khader KBA, Biddappa CC, Khan HH, Bai KKV, Ramadasan A, Sundararaj P, Bopiah, BM, George V. Thomas, Misra LP, Balasimha D, Bhat NT, Shama Bhat K. 1986. Coconut and arecanut based high density multi species cropping system. *J. Plantation Crops* 14 : 74-87.

Bopaiah BM, Khader KB. 1989. Effect of biofertilizers on growth of black pepper. *Indian Journal of Agricultural Science* 59: 682-683.

Chomchalow N. 1996. Spice Production in Asia - An Overview. Unpublished paper presented at the IBC's Asia Spice Markets '96 Conference, Singapore, 27-28, May 1996 (www.journal.au.edu/au_techno/2001/oct2001/article6.pdf) online accessed on 08-3-2007.

Diby P, Anandaraj M, Kumar A , Sarma YR .2005. Antagonistic mechanisms of fluorescent pseudomonads against *Phytophthora capsici* in black pepper (*Piper nigrum* L.). *Journal of Spices Aromatic Crops* 14: 94-10.

Dinesh Kumar M , Babitha J. 2006. Rates of leaf fall and nutrient cycling of shade trees in coffee (*Coffea arabica* L.) cardamom (*Elettaria cardamomum* Maton) and black pepper (*Piper nigrum* L.) production systems of Mudigere, Karnataka. *Journal of Spices and Aromatic Crops* 15: 108-114.

Dixit D, Srivastava NK. 2000. Effect of iron deficiency stress on physiological and biochemical changes in turmeric (*Curcuma longa*) genotypes. *Journal of Medicinal and Aromatic Plant Science* 22: 652-658.

Dubey AK, Yadav DS. 2001. Response of turmeric (*Curcuma longa* L.) to NPK under foothill conditions of Arunachal Pradesh. *Indian Journal of Hill Farming* 14: 144-146.

Geetha CK, Sivaraman Nair PC.1990. Effect of plant growth regulators and zinc on spike shedding and quality of pepper. *South Indian Horticulture*14: 10-12.

Geetha CK, Aravindakshan M, Wahid PA. 1993. Influence of method of fertilizer application on nutrient absorption by black pepper vines. *South Indian Hort.* 41: 95-100.

George CK.1982. Pepper cultivation in Malaysia. *Indian Cocoa Arecanut Spices J.* 5: 75-76.

Gill BS, Randhawa RS, Randhawa GS, Singh J. 1999 Response of turmeric (*Curcuma longa*) to nitrogen in relation to application of farm yard manure and straw mulch. *Journal of Spices and Aromatic Crops* 8: 211-214.

Govender A, Kindness A, Jonnalagadda SB. 2009. Impact of soil quality on elemental uptake by *Zingiber officinale* (ginger rhizome). *International Journal of Environmental Analytical Chemistry* 89: 367-382.

Gowda KK, Melanta KR, Prasad TRG 1999. Influence of NPK on the yield of ginger (*Zingiber officinale* Rosc.). *Journal of Plantation Crops* 27: 67-69.

Haag HP, Saito S, Dechen AR, Carmello QAC. 1990. Anais da Escola Superior de Agriculture *Luiz de Queiroz*47: 435-457.

Hamza S, Srinivasan V, Dinesh R. 2009. Nutrient diagnosis of cardamom (*Elettaria cardamomum*) gardens in South India. *Indian Journal of Agricultural Sciences* 79: 429-432.

Haque MM, Rahman, KMM, Ahmed M, Masud MM, Sarker MMR. 2007 Effect of nitrogen and potassium on the yield and quality of turmeric in hill slope. *International Journal of Sustainable Crop Production* 2: 10-14.

Hepperly P, Francis Z, Russell K, Claire A, Mark M, Bernard K, Kert H, Dwight S. 2004 Producing bacterial wilt–free ginger in greenhouse culture, *Soil and Crop Management (SCM-8)*. University of Hawaii, College of Tropical Agriculture and Human Resources, Cooperative Extension Service, 1-6.

Jirali DI, Hiremath SM, Chetti MB, Patil SA. 2007. Biophysical, biochemical parameters yield and quality attributes as affected by micronutrients in turmeric. *Plant Archives* 7: 827-830.

Jadhao BJ, Gonge VS, Panchbhai DM, Mohariya A, Hussain, IR. 2005 Performance of turmeric varieties (*Curcuma longa* L.) under varying levels of zinc and iron. *International Journal of Agricultural Science* 1: 94-98.

Jena MK, Das PK, Pattanaik AK. 1999 Integrated effect of microbial inoculants and fertilizer nitrogen on N-use efficiency and rhizome yield of turmeric (*Curcuma longa* L.). *Orissa Journal of Horticulture* 27: 10-16.

Johnson PT. 1978 Foliar Diagnosis, yield and quality of ginger in relation to N, P and K. *M.Sc. (Ag.) Thesis*, KAU, Kerala, India.

Kandiannan K, Chandaragiri K K, Sankaran N, Balasubramanian TN, Kailasam C. 2002 Crop-weather model for turmeric yield forecasting for Coimbatore district, Tamil Nadu, India *Agricultural and Forest Meteorology.* 112: 133–137.

Kandiannan K, Sivaraman K, Thankamani C K 1994 Growth regulators in black pepper production. *Indian Cocoa Arecanut Spices Journal* 17: 119-123.

Kandiannan K, Sivaraman K, Anandaraj M, Krishnamurthy KS. 2000. Growth and nutrient content of black pepper (*Piper nigrum* L.) cuttings as influenced by inoculation with biofertilizers. *Journal of Spices and Aromatic Crops* 9: 145-147.

Kanthaswamy V, Pillai A O A, Natarajan S , Thamburaj S. 1996. Studies on nutrient requirement of black pepper. *South Indian Horticulture* 44: 3-4.

Korikanthimath VS, Hegde R, Hosmani MM. 1998 Influence of yield and yield parameters of cardamom grown under controlled shade. In: *Spices and Aromatic Plants*, Indian Society for Spices, IISR, Calicut, pp. 179-180.

Kulkarni DS, Kulkarni SV, Suryanarayana RB, d Pattanshetty HV. 1971 Nutrient uptake by cardamom (*Elettaria cardamomum* Maton). *Proceedings of the International Symposium on soil fertility evaluation*, New Delhi, Vol. 1, pp. 293-296.

Kumar MD, Devaraju KM, Madaiah D. 2009b. Effect of integrated nutrient supply on yield and uptake of cardamom (*Elettaria cardamomum* L. Maton.). *Journal of Plantation Crops* 37: 129-133.

Kumar MD, Devaraju KM, Madaiah D, Shivakumar KV. 2009a. Effect of integrated nutrient management on yield and nutrient content by cardamom (*Elettaria cardamomum* L. Maton.). *Karnataka Journal of Agricultural Sciences* 22: 1016-1019.

Kumar PSS, Geetha SA, Savithri P, Jagadeeswaran R, Mahendran PP. 2003. Diagnosis of nutrient imbalances and derivation of new RPZI (Reference Population Zero Index) values using DRIS/MDRIS and CND approaches in leaves of turmeric (*Curcuma longa* var.). *Journal of Applied Horticulture* 5: 7–10.

Mathew J , Nybe EV. 2004. Optimizing inorganic nutrient use in black pepper. *Proceedings ICAR National Symposium on Input Use Efficiency in Agriculture- Issues and Strategies*, Trichur, Kerala, India.pp. 103.

Medda PS, Hore JK. 2003. Effects of N and K on the growth and yield of turmeric in alluvial plains of West Bengal.*Indian Journal of Horticulture* 60: 84-88.

Mohan E, Melanta KR, Guruprasad TR, Herle PS, Gowda NAJ, Naik CM. 2004. Effects of graded levels of nitrogen and biofertilizers on growth, yield and quality in turmeric (*Curcuma domestica* Val.) cv. D K Local. *Environment and Ecology* 22: 715-719.

Mohanty DC, Naik BS, Panda BS. 1990. Ginger research in Orissa with reference to its varietal and cultural improvement. *Indian Cocoa, Arecanut and Spices Journal*14: 61-65.

Murugan M, Backiyarani S, Josephrajkumar A, Hiremath MB, Shetty PK. 2007. Yield of small cardamom (*Elettaria cardamomum* M) variety PV1 as influenced by levels of nutrients and neem cake under rain fed condition in southern western Ghats, India. *Caspian Journal of Environmental Sciences* 5: 19-25.

Mustika I, Rachmat A, Sudrajat D. 1994. Influence of organic matters on the growth of black pepper and antagonistic micro-organisms. Int. *Internationals Pepper News Bulletin*18 (2) : 19-24.

Nagarajan M, Pillai NG. 1979. A note on nutrient removal by ginger and turmeric rhizomes. *Madras Agricultural Journal* 66: 56-59.

Nair MA, Sreedharan C. 1986. Agroforestry farming systems in the homesteads of Kerala, Southern India. *Agroforestry Systems* 4: 339-363.

Nair CK. 1988. Phosphatic fertilizers for small cardamom.*Proceedings of the seminar on the use of rock phosphate in West Coast soils.* University of Agricultural Sciences, Bangalore and PPCL, p. 79.

Nybe EV, Peter KV, Mini Raj N. 2004. Integrated cropping in coconut. *Coconut Journal,* 36: 4-9.

Padmapriya S, Chezhiyan N, Sathiyamurthy VA. 2007. Effect of shade and integrated nutrient management on biochemical constituents of turmeric (*Curcuma longa* L.). *Journal of Horticultural Sciences* 2: 123-129.

Paliyal SS, Kanwar K, Sharma R, Bedi MK. 2008. Productivity and soil fertility under different nutrient applications in ginger-french bean cropping system in Western Himalayas. *Journal of Farming Systems Research and Development* 14: 20-

Raj HG. 1978. A comparison of the system of cultivation of black pepper - *Piper nigrum* L. in Malaysia and Indonesia. In: Silver Jubilee Souvenir- Pepper Research Station, Panniyur. pp.65-74 -Kerala Agricultural University, Trichur.

Ramana KV, Shiva KN, Johny AK. 2003 Production of quality planting materials of ginger. In: *National consultative meeting for improvement in productivity and utilization of ginger,* Aizawl, Mizoram, pp. 37-45.

Randhawa KS, Nandpuri KS. 1969. Response of ginger (*Zingiber officinale* Roscoe) to nitrogen, phosphate and potash fertilizers. *Journal of Research Punjab Agricultural University* 6: 782-785.

Randhawa KS, Nandpuri KS, Bajwa MS. 1973. Response of turmeric (*Curcuma longa*) to NPK fertilization. *Journal of Research Punjab Agricultural University* 10: 45-48.

Reddy MRS, Reddy PVRM. 2000. Preliminary observations on neem-cake, against rhizome fly of turmeric (*Curcuma longa*). *Insect Environment* 6: 62.

Rethinam P, Sivaraman K, Sushama PK. 1994. Nutrition of turmeric. In: Chadha, K.L. and Rethinam, P. (eds.) *Advances in Horticulture*, Vol 9 – Plantation and Spice Crops Part 1. Malhotra Publishing House, New Delhi, pp. 477-489.

Roy A, Chatterjee R, Hassan A, Mitra SK. 1992. Effect of Zn, Fe, and B on growth, yield and nutrient content in leaf of ginger. *Indian Cocoa Arecanut Spices Journal*15: 99-101.

Roy SS, Hore JK. 2007. Influence of organic manures on growth and yield of ginger. *Journal of Plantation Crops* 35: 52-55.

Sadanandan AK, Hamza S. 2001. "Studies on heavy metal residues due to continuous application of rock phosphate for sustainable black pepper production." In: Gupta A K (ed.), *Proceedings of the national symposium on combating pollution accumulation in ecosystem for sustainable agriculture*, Allahabad Agricultural Institute, Allahabad, India, pp. 59-65.

Sadanandan AK, Hamza S. 2002. "Long term effect of inorganic fertilizers on black pepper in laterite soil." In: Rethinam *et al.* (eds.), *Plantation Crops Research and Development in the New Millennium*, Coconut Development Board, Kerala, pp. 336-341.

Sadanandan AK. 2000. "Agronomy and nutrition of black pepper."In: Ravindran P N (ed.). *Black Pepper*. Academic Publishers, New Delhi, India, pp. 163-223.

Sadanandan AK, Hamza S, Bhargava BS, Raghupathi HB. 2000. "Diagnosis and Recommendation Integrated System (DRIS) norms for black pepper (*Piper nigrum* L.), growing soils of South India".In: Muraleedharan N & Raj Kumar P (eds.), *Recent Advances in Plantation Crops Research,* Allied Publishers Limited, New Delhi, India, pp. 203-205.

Sadanandan AK, Hamza S. 1993. Comparative efficiency of slow release N fertilizers on transformation of nitrogen and yield response of black pepper in an oxisol. *Journal of Plantation Crops* 21 (Suppl.): 58-66.

Sadanandan AK, Hamza S. 1996. Response of four turmeric (*Curcuma longa* L) varieties to nutrients in an oxisol on yield and curcumin content. *Journal of Plantation Crops* 24 (Suppl.): 120-125.

Sadanandan AK, Hamza S. 1998a. Effect of organic manures on nutrient uptake, yield and quality of turmeric (*Curcuma longa* L.). In: Mathew, N.M. and Kuruvilla Jacob, C. (eds.) *Developments in plantation crops research.* Proceedings of plantation crops symposium XII (PLACROSYM XII), Indian Rubber Research Institute, Kottayam, pp. 175-181.

Sadanandan AK, Hamza S. 1998b. Effect of organic farming on nutrient uptake, yield and quality of ginger (*Zingiber officinale* R.). In: Sadanandan, A.K., Krishnamurthy, K.S., Kandiannan, K. and Korikanthimath, V.S. *Water and nutrient management for sustainable production and quality of spices*, Madikeri, Karnataka, pp. 89-94.

Sadanandan AK, Hamza S. 2006. Effect of organic farming on soil quality, nutrient uptake, yield and quality of Indian spice, In: *Abstracts of 18th World Congress of Soil Science*, Philadelphia, USA.

Sadanandan N, Iyer R. 1986. Effect of organic amendments of rhizome rot of ginger. *Indian Cocoa Arecanut Spices Journal* 9: 94-95.

Sadanandan N, Sasidharan VK. 1979. A note on the performance of ginger under graded doses of nitrogen. *Agricultural Research Journal of Kerala* 17: 103-104.

Sahu SK, Mitra GN. 1992. Influence of physicochemical properties of soil on yield of ginger and turmeric. *Fert. News* 37 (10): 59-63.

Selvarajan M, Chezhiyan N. 2001. Studies on the influence of Azospirillum and different levels of nitrogen on growth and yield of turmeric (*Curcuma longa* L.). *South Indian Horticulture* 49: 140–141.

Sheshagiri KS, Uthaiah BC. 1994. Effect of nitrogen, phosphorus and potassium levels on growth and yield of turmeric (*Curcuma longa*) in the hill zone of Karnataka. *Journal of Spices and Aromatic Crops* 3: 28-32.

Sivakumar C , Wahid P A. 1994. Effect of application of organic materials on growth and foliar nutrient contents of black pepper (*Piper nigrum* L.). *Journal of Spices and Aromatic Crops* 3 : 135-141.

Sivaraman K. 1987. Rapid multiplication of quality planting material in black pepper. *Cocoa Arecanut Spices Journal* 11 : 115-118.

Srinivasan V, Hamza S, Sadanandan AK 2005. Evaluation of composted coir pith with chemical and biofertilizers on nutrient availability, yield and quality of black pepper (*Piper nigrum* L.). *Journal of Spices and Aromatic Crops,* 14: 15-20.

Srinivasan V, Kandiannan K, Hamza S. 2013a. Efficiency of sulphate of potash (SOP) as an alternate source of potassium for black pepper (*Piper nigrum* L.). *Journal of Spices and Aromatic Crops*, 22(2): 120-126.

Srinivasan V, Dinesh R, Ankegowda SJ, Hamza S, Thankamani CK, Krishnamurthy KS. 2013b. Soil test based site specific fertilizer recommendation and its economic optimum for targeted yield of black pepper. In: Souvenir & Abstracts – SYMSAC VII: Post-Harvest Processing of Spices and Fruit Crops, Sasikumar *et al.* (Eds), ISS, ICAR-IISR, Kozhikode, Kerala. pp. 202.

Srinivasan K, Krishnakumar V, Potty SN. 1998. Evaluation of fertilizer application methods on growth and yield of small cardamom (*Elettaria cardamomum* Maton).In: Muralitharan, N. and Rajkumar, R. (eds.) *Recent Advances in Plantation Crops Research*, Allied Publishers, New Delhi, pp. 199-202.

Srinivasan V, Hamza S, Dinesh R. 2009. Critical limits of zinc in soil and plant for increased productivity of ginger (*Zingiber officinale* Rosc.)*Journal of Indian Society of Soil Science* 57: 191-195.

Srinivasan V, Thankamani CK, Hamza S, John Zachariah T, Dinesh R, Praveena R. 2017. Organic Spices Production (Black pepper, Ginger and Turmeric). ICAR-Indian Institute of Spices Research, Kozhikode, Kerala, India. 45p.

Srinivasan V, Sadanandan AK, Hamza S. 2000. Efficiency of rock phosphate sources on ginger and turmeric in an *Ustic Humitropept. Journal of the Indian Society of Soil Science* 48: 532-536.

Tarle SG, Jadhao BJ, Panchabhai DM, Nandre DR, Khewale AP. 2007. Effect of nitrogen levels on growth and yield of ginger varieties. *Plant Archives* 7: 305-306.

Thankamani CK, Ashokan PK, Kamalam N V. 2003. Root distribution pattern of bush – black pepper (*Piper nigrum* L) employing radiotracer technique. *J. Nucl. Agric. Biol.,* 32: 23-28.

Thankamani CK, Kandiannan K, Hamza S, Saji KV. 2017. Effect of mulches on weed suppression and yield of ginger (*Zingiber officinale* Roscoe). Scientia Horticulturae, 207: 125-130.

Thankamani CK, Mathew PA, Kandiannan K. 2004. Production of healthy black pepper rooted cuttings. *Indian J. Arecanut, Spices and Medicinal Plants*, 6(4): 135-136.

Thimmarayappa M, Shivashankar KT, Shanthaveerabhadraiah SM. 2000. Effect of organic manure and inorganic fertilizers on growth, yield attributes and yield of cardamom (*Elettaria cardamomum* Maton). *Journal of Spices and Aromatic Crops* 9: 57-59.

Thomas A, Swati B, Singh DB. 2002. Influence of different levels of nitrogen and potassium on growth and yield of turmeric (*Curcuma longa* L.). *Journal of Spices and Aromatic Crops* 11: 74-77.

Velmurugan M, Chezhiyan N, Jawaharlal M. 2008. Influence of organic manures and inorganic fertilizers on cured rhizome yield and quality of turmeric (*Curcuma longa* L.) cv. BSR-2. *International Journal of Agricultural Sciences* 4: 142-145.

Velmurugan M, Chezhiyan N, Jawaharlal M, Anand M. 2007. Micronutrient studies in turmeric.*The Asian Journal of Horticulture* 2: 291-293.

Venkatesha J, Khan MM, Farooqi AA, Sadanandan AK. 1998. Effect of major nutrients (NPK) on growth, yield and quality of turmeric (*Curcuma domestica* Val.) cultivars. In: Sadanandan, A. K., Krishnamurthy, K.S., Kandiannan, K. and Korkanthimath, V.S. (eds) *Proceedings on water and nutrient management for sustainable production and quality of spices*. Indian Society of Spices, Calicut, India, pp. 52-58.

Vishwakarma SK, Kumar A, Prakash S. 2006. The effect of micro-nutrient on the growth and yield of turmeric under different shade conditions in mango orchard. *International Journal of Agricultural Sciences* 2: 241-243.

Wahid, P. and Sitepu, D. 1987. Current status and future prospects of pepper development in Indonesia. Food and Agricultural Organisation, Regional office for Asia and Pacific, Bangkok.

Xu K, Guo YY, Wang XF. 2004. Transportation and distribution of carbon and nitrogen nutrition in ginger, *Acta horticulturae* 629: 347-353.

Yamgar VT, Kathmale DK, Belhekar PS, Patil RC, Patil P.S. 2001. Effect of different levels of nitrogen, phosphorus and potassium and split application of N on growth and yield of turmeric (*Curcuma longa*). *Indian Journal of Agronomy* 46: 372-374.

Yufdi MP. 1991. Effect of liming on the growth of different black pepper varieties in the red and yellow podzolic soil. *Industrial Crops Research Journal* 17: 31-36.

Zachariah PK. 1978 "Fertilizer management for cardamom". In: *Proceedings of the First Annual Symposium on Plantations Crops,* Kottayam, pp.141-156.

Zulkifly E 1996 Prospects of implementing organic farming on pepper. *International Pepper News Bulletin* 20 (2):24-30.

Colour Figures

Chapter 2: Soil Testing: Basic Tool for Enhancing the Nutrient Use Efficiency

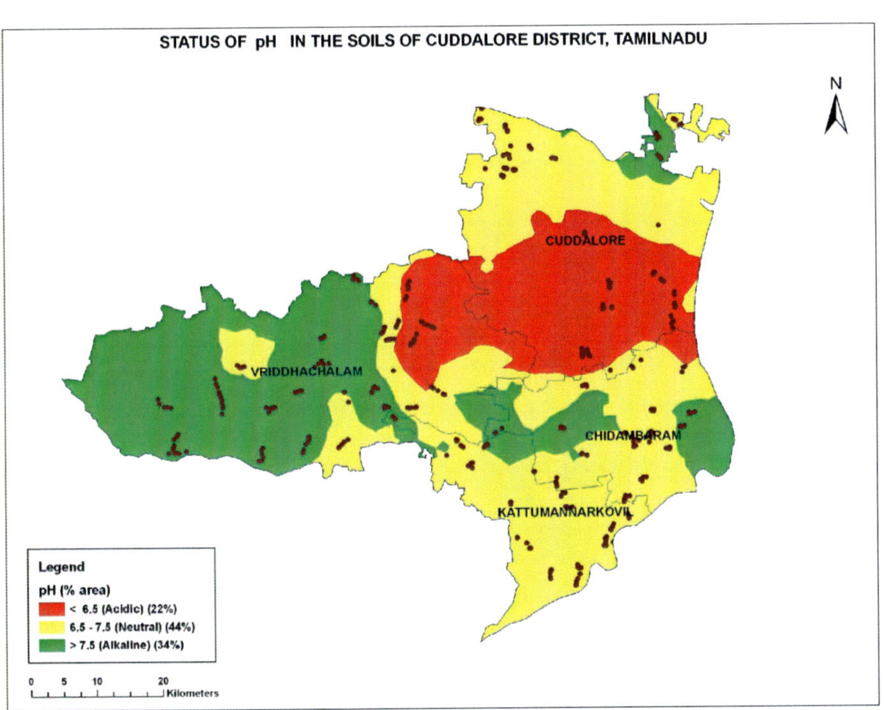

Fig 1: GIS/GPS based soil pH map of Cuddalore district, Tamil Nadu

Chapter 3: Management of Soil Physical Environment for Higher Nutrient Use Efficiency

Fig. 3. Paddy straw mulching in Mustard and Grass mulching in Tomato for higher nutrient use efficiency

Fig. 4. Land configuration in the form of RSB for increasing cropping intensity, nutrient and water use efficiency

Chapter 7: Handheld Devices for Judging Nutrients in Crops and Their Role in Enhancing Nutrient Use Efficiency

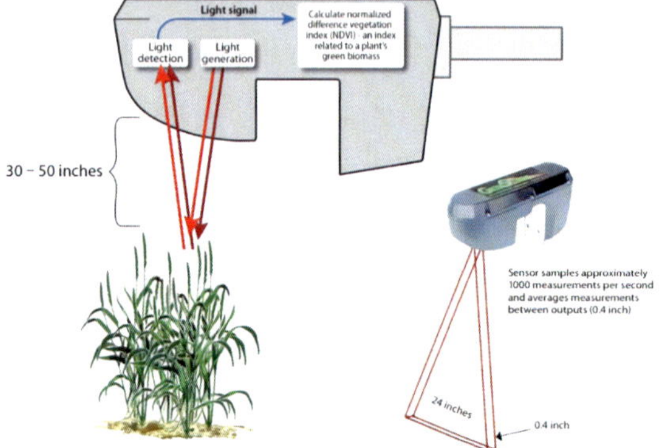

Fig. 1. A diagram representation of a Green Seeker optical sensor system and outputs measurement.

Fig. 4. Use of LCC for N estimation in rice

Fig. 5. Four panel and six panel leaf color chart

Fig. 6. Measurement of reflectance from individual leaves using leaf clip

Chapter 9: STCR-Based Fertilizer Recommendation and Improving Nutrient Response Ratio for Major Crops in India

Fig 2. Internet enabled soil test based fertilizer application software

Chapter 14: Enhancing Nutrient Use Efficiency through Conservation Agriculture

Fig 1. Residue retention under soybean-wheat system (left) and maize-gram system (right)

Fig 4. Crop stand under conventional tilled without residue (left) and RT with residue retained (right) under maize-gram.

About the Editors

Dr. K. Ramesh was born on 28th Jun 1974 in Trichirappalli, Trichy Distt. of Tamil Nadu. He studied B. Sc. (Agri.) during 1992 to 1996 from Pandit Jawaharlal Nehru College of Agriculture, Karaikal (UT of Pondicherry) under Tamil Nadu Agricultural University (Coimbatore), and M. Sc. and Ph. D. in Agronomy in 1999 and 2002, respectively, from AC&RI, Madurai and AC&RI., Coimbatore under TNAU, Coimbatore. He served as Assistant Professor (Agronomy) in an Agricultural college in Tamil Nadu during 2001-2004 and has introduced innovative teaching methods in agronomy. Later joined as Scientist Gr.IV(2) (Agronomy) in 2004 at Council of Scientific and Industrial Research and started his career at CSIR - Institute of Himalayan Bioresource Technology, Palampur, Himachal Pradesh (2004-2008). Thereafter he moved to Indian Council of Agricultural Research as a Senior Scientist at ICAR-Indian Institute of Soil Science, Bhopal. Presently, he is working as Principal Scientist at ICAR-IISS Bhopal, Madhya Pradesh, India, since September 2014.

During his scientific career at various institutions, he has worked on the management of cropping systems for enhancing income of the farmers as well as soil health. He has developed promising *Sesbania roastrata* based cropping systems for western Himalaya as well as strategies sustaining rice yields under double cropped wet lands of western region of Tamil Nadu. As a cropping system agronomist at CSIR-IHBT, Palampur, Himachal Pradesh, he has developed promising food crop+ natural sweetener (*Sesbania rebaudiana*) based cropping systems for Dhauladhar range of Kangra district of Palampur (HP). He was also involved in sustainable management of hilly areas with the introduction of Lavender (a medicinal aromatic perennial) in Himachal Pradesh and Arunachal Pradesh. He has developed molecule specific farming technology for *Tagetes minuta* in western Himalayas. Presently at ICAR-IISS, Bhopal, he has been working on zeolites for enhancing nutrient use efficiency, nanoparticle delivery in plant systems for improving nutrient use efficiency of agricultural crops and developing organic farming practices for agri-horticultural crops. Besides, he has been associated with projects on evaluation of new fertilizer products.

Dr K. Ramesh has contributed more than 90 publications, including 42 peer reviewed research papers, several book chapters, 20 popular articles, 25 conference proceedings and 4 training manuals which has been widely cited throughout the scientific community. He is a recipient of NAAS Young scientist award (1997-98), besides awards for bachelor and master's degree programs. He has served as Councilor of Indian Society of Agronomy for Madhya Pradesh state. He has served as expert of several important committees.

Dr. A.K. Biswas was born in village Bagna in North 24 Parganas district of West Bengal on May 5, 1963. He obtained his B.Sc. (Ag.) Hons. degree in 1984 from B.C.K.V., Mohanpur, W.B. He received his M. Sc. (Gold Medalist) and Ph. D. Degrees from I.A.R.I., New Delhi and is presently working as Head of the Division of Soil Chemistry & Fertility at the Indian Institute of Soil Science, Bhopal; Madhya Pradesh.

The major areas of work of Dr. A.K. Biswas have been concerning the development of farmers' resource-based integrated plant nutrient supply system in soybean-wheat system, farmers' participatory technology development of integrated nutrient management coupled with soil moisture conservation for rainfed pulse-based cropping systems, environmental impact and risk assessment of recycling of distillery and sewage effluents in agriculture, determination of sink capacity of mineralogically and physico-chemically variant soils for some heavy metal pollutants and phyto-remedial efficiency of some flowering plants to decontaminate Cd, Pb and Cr contaminated soils, delineation of nitrate contamination of groundwater in heavily fertilized and intensively cultivated districts of the country, evaluation of the effect of climate change on soil carbon stock in different agro-ecoregions of the country, development of soil organic C and N turnover model in light of C saturation and stabilization theory, development of methodology to estimate relative soil quality index to assess soil health under different agro-ecoregions of the country, and development of nano-rock phosphate for commercial utilization of low-grade indigenous rockphosphates of the country and modified urea products like oleoresin coated urea, zeolite-impregnated and biochar coated urea. He is part of the team who developed Mridaparikshak - A Soil Testing Minilab for which patent has been applied.

Dr. Biswas has contributed more than 200 publications, including 90 research papers, 20 review papers, 20 popular articles, 3 edited books, 40 book chapters and several other publications which has been widely cited throughout the scientific community.

Dr Biswas has been elected to the prestigious "Fellowship of National Academy of Agricultural Sciences (NAAS)" in 2016. He is also the recipient of the "Jawahar Lal Nehru Award" from ICAR for outstanding doctoral research in the field of Soil Science in 1992 and "The ISSS- Dr. J.S.P. Yadav Memorial Awards for Excellence in Soil Science- 2012" from the Indian Society of Soil Science, New Delhi. He has delivered lecture on "Maintaining soil health for evergreen revolution in the Agriculture and Forestry section of centenary Celebration of Indian Science Congress in Kolkata and got several best paper/ poster awards to his credit. He has guided several M.Sc. and Ph. D. students. He is a member of the Editorial Board of the Journal of the Indian Society of Soil Science since 2011, and a scientific reviewer of the Bioresource Technology, Journal of Hazardous Materials, Journal of the Indian Society of Soil Science, Agropedology, Legume Research, Annals of Soil and Plant Research and Journal of the Indian Society of Pulse Research and Development since 2003.

Dr. Brij Lal Lakaria was born in 1968 at Kanaid, Sunder Nagar, in Mandi district of Himachal Pradesh. He completed his B.Sc. (Agri.) and M. Sc. in Soil Science from Chaudhary Sarwan Kumar Himachal Pradesh Krishi Vishvavidyalaya, Palampur (formerly HPKV Palampur) during 1991 and 1993 respectively. He did his Ph. D. in Soil Science and Agricultural Chemistry in 1997 from Indian Agricultural Research Institute, New Delhi. He joined Agricultural Research Service in 1997 as Scientist at Central Soil and Water Conservation Research and Training Institute, Dehradun, Research Centre Datia (M.P.) and worked on soil natural resource management during 1997-2006. He joined Indian Institute of Soil Science, Bhopal in 2007 as Senior Scientist and was selected as Principal Scientist in 2011.

During his scientific career at various intuitions, he has worked on various aspects of Natural Resource Conservation, USLE parameters, erodibility indices, screening of efficient grasses for soil and water conservation, Assessment of soil erosion status and soil loss tolerance limits for the Madhya Pradesh and Chhatisgarh states, Development of Model Watershed etc. At Indian Institute of Soil Science, Dr. Brij Lal Lakaria, has focused his research on soil carbon sequestration, carbon pool dynamics under long term IPNS modules, organic farming practices for sustainable crop management. Presently he is conducting research on biochar for carbon sequestration, nutrient use efficiency and crop production. Besides, he has evaluated new nutrient/fertilizer formulations companies *vis-a-vis* nutrient use efficiency and crop performance.

Dr. Brij Lal Lakaria has contributed more than 80 publications including 45 peer reviewed papers in Journals of International and National repute and several book chapters, popular articles, conference proceedings and 5 training manuals/ technical bulletins. He had been a meritorious student throughout his academic career and was a recipient of college merit scholarships (B.Sc. and M.Sc.) and ICAR Senior Research Fellowship (Ph.D.). He is also a recipient of ISSS–Dr. J.S.P. Yadav Memorial Award for Excellence in Soil Science . He is life member of Indian Society of Soil Science, New Delhi and Indian Association of Soil and Water Conservationists, Dehradun, Uttarakhand.

Dr. Sanjay Srivastava was born on 1967 in Varanasi Dt. of Uttar Pradesh, has completed B.Sc. (Agri.) from Institute of Agricultural Sciences, Banaras Hindu University, Varanasi in 1988, and M.Sc. and Ph. D. in Soil Science & Agricultural Chemistry in 1990 and 1995, respectively, from Indian Agricultural Research Institute, New Delhi. Besides, he also did executive Post Graduate Diploma in Management from Indian Institute of Management, Indore. He served as Scientist (1996-2003), Senior Scientist (2003-2009), and Principal Scientist (2009 to till date) at ICAR-Indian Institute of Soil Science, Bhopal.

During his scientific career, he has worked on the soil test crop response, farmers'participatory integrated nutrient management, dynamics of potassium under long-term cropping, and soil fertility mapping for major nutrients. He has led the team who developed *Mridaparikshak* mini-lab for the assessment of soil health parameters. Presently, he has been working on developing low cost techniques of soil health assessment under resource limited conditions, and techniques to solubilize phosphorus from rock phosphate.

Dr Sanjay Srivastava has contributed more than 50 publications, including 30 peer reviewed research papers, edited two books, authored one bulletin (published by International Potash Institute) several book chapters, popular articles, conference proceedings and training manuals which has been widely cited. He has served as Councilor of Indian Society of Soil Science and Clay Minerals Society of India. He has served as expert of several important committees.

Dr. Ashok K. Patra is Director of the ICAR-Indian Institute of Soil Science located at Bhopal, Madhya Pradesh.

He did his early education at Araldihi High School and Dubrajpur Uttarayan Vidyayatan, Bankura, West Bengal. He studied B.Sc. (Agri.) during 1979-1983 from Banaras Hindu University (Varanasi), and M. Sc. and Ph. D. in the discipline of Soil Science & Agricultural Chemistry in 1985 and 1989, respectively, from Indian Agricultural Research Institute, New Delhi. He joined Agricultural Research Service (ICAR) in 1989 and started his career at the Indian Grassland & Fodder Research Institute (IGFRI), Jhansi as a Scientist/Scientist Sr. Scale (1990-1998). Then he moved to CIFE, Mumbai as a Senior Scientist (1998-1999), and to IARI, New Delhi as Senior Scientist (1999-2006) and Principal Scientist (2006-2014).

Under an ICAR-ICRISAT Collaborative programme, he was a postdoctoral scientist (1991-1993) at ICRISAT, Hyderabad, and under an Indo-UK Collaborative programme a Visiting Study Fellow (1996) at the Institute of Grassland and Environmental Research (IGER), Devon, UK. He was a recipient of the prestigious INRA Fellowship (2001-2003) of the French Research Ministry to work on molecular soil ecology in N cycling at the CNRS-Claude Bernard Université Lyon, France and made a significant contribution to unravel the complex processes of nitrogen cycling – its ecology, biodiversity and management in agro-ecosystems. For pursuing the frontier soil science research, DBT (GOI) awarded him the 'DBT Overseas Asssociateship'- 2008 for which he visited University of Notre Dame, USA during 2008-2009.

In addition to research, he was actively involved in teaching and guiding of postgraduate students of soil science at IARI, New Delhi (1999-2014). His research interest includes nitrogen cycling, ecology and microbial biodiversity, carbon sequestration and nanoparticles. His research work has been published in different leading professional journals and highly quoted. He contributed more than 220 publications, which includes refereed journals, reviews, books/book chapters, proceedings of seminars/conferences, etc. He is a recipient of several national and international awards/honours, namely British Council TCT Award 1996; DBT Overseas Associateship Award 2008; FAI Dhiru Morarji Memorial Award, 2011; Bharat Jyoti Award, 2012; Rajiv Gandhi Excellence Award, 2012; ISSS Dr. G.S. Sekhon Memorial Lecture Award of ISSS, 2012; IARI Hooker Award, 2013; Bioved Agri-innovation Award – 2015. He served as an Editor, Range Management and Agroforestry and Journal of the Indian Society of Soil Science. Currently he is Associate Editor, European Journal of Soil Science.

He was Councillor, Indian Society of Soil Science, New Delhi (2005-2006); Secretary, Delhi Chapter of Indian Society of Soil Science (ISSS) 2005-2007; Member, Nature's Reader Panel 2009; President (Delhi Chapter), ISSS 2012-2014; Joint Secretary, ISSS 2013-2014, Vice President, Indian Society of Soil Science (2016-2017) and President, Agriculture & Forestry Sciences Section, Indian Science Congress Association (2016-2017). He is a life member of several professional societies and acted as Member/Chairman of several committees.

Dr Patra is a Fellow of the National Academy of Agricultural Sciences, Range Management Society of India and Indian Society of Soil Science.

Other books by IISS, Bhopal, India

Advances in Nutrient Dynamics in Soil-Pant System for Improving Nutrient Use Efficiency
edited by R. Elanchezhian, A.K. Biswas K. Ramesh and A.K. Patra

ISBN: 9789385516962
Price: INR 3600.00
Weight: 784gm
Pages: 404
Binding: Hardcover
Available from: 2017

This book comprises 31 chapters on advances in soil-plant systems for improving nutrient use efficiency with four major themes viz. 1) Introduction and fundamentals of Soil plant atmosphere continuum and nutrient use efficiency; 2) Soil physical, chemical, biological and agronomic management for improving NUE; 3) Plant physiological, genetic & molecular biological basis for improving nutrient uptake & use efficiency and 4) Climate change aspects related to soil and plant systems for improving NUE. Besides the book also include few chapters on analytical techniques and instrumentation for the study of nutrient use efficiency with respect to physico-chemical and biological parameters.

Conservation Agriculture for Carbon Sequestration and Sustaining Soil Health
edited by J. Somasundaram, R.S. Chaudhary A., Subba Rao, K.M. Hati, N.K. Sinha and N. Vassanda Coumar

ISBN: 9789383305322
Price: INR 4550.00
Weight: 1kg 235gm; **Pages:** 528;
Binding: Hardcover; **Available from:** 2014

This book comprises 41 s dealing various issues, prospects and importance of conservation agriculture practices followed across different regions with special emphasis on rainfed regions. We hope this book on conservation agriculture will be highly useful to researchers, scientists, students, farmers and land managers for efficient and sustainable management of natural resources.

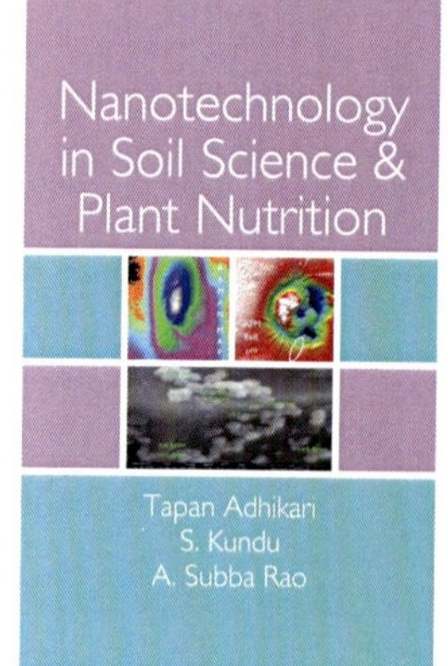

Nanotechnology in Soil Science and Plant Nutrition

edited by Adhikari, Tapan, S. Kundu and A. Subba Rao

ISBN: 9789381450789
Price: INR 2750.00
Weight: 810gm
Pages: 320
Binding: Hardcover
Available from: 2013

The book has 21 chapters addressing fundamentals and applied aspects of nanotechnology in soil science and plant nutrition research and written by explorers of a new frontier. The interpretation of subject matter in each is comprehensive, simple and lucid with relevant supporting data.

This book would offer a platform for basic, fundamental and advanced learning for students. It would also be useful and informative to researchers from SAUs and ICAR institutes.

Climate Change and Natural Resources Management

edited by Lenka, S., N.K.Lenka, S. Kundu and A. Subba Rao

ISBN: 9789381450673
Price: INR 2750.00
Weight: 820gm
Pages: 380
Binding: Hardcover
Available from: 2013

This book addresses the important issues of food security and sustainability of natural resources of India in the context of the projected climate change. Agroecosystems being the sites of intense interaction between human beings and natural world, global climate change is likely to affect the resource base, the crop productivity, input use efficiency and overall the profitability of agricultural production systems to a great extent. However, the adverse effects of climate change can be alleviated through mitigation and adaptation strategies which carry importance due to the increasing population and food demand in India. Thus, this compilation covers possible sources and sinks of greenhouse gases in Indian context including the potentials of soil carbon sequestration, crop pest and soil management and scientific livestock management as mitigation and adaptation options. The likelihood of carbon credits and trading through best management practices can help Indian farmers earning carbon credits in future. The book is useful for researchers, farm managers, policy makers and also students engaged in climate change related studies.

Murugan M, Backiyarani S, Josephrajkumar A, Hiremath MB, Shetty PK. 2007. Yield of small cardamom (*Elettaria cardamomum* M) variety PV1 as influenced by levels of nutrients and neem cake under rain fed condition in southern western Ghats, India. *Caspian Journal of Environmental Sciences* 5: 19-25.

Mustika I, Rachmat A, Sudrajat D. 1994. Influence of organic matters on the growth of black pepper and antagonistic micro-organisms. Int. *Internationals Pepper News Bulletin* 18 (2) : 19-24.

Nagarajan M, Pillai NG. 1979. A note on nutrient removal by ginger and turmeric rhizomes. *Madras Agricultural Journal* 66: 56-59.

Nair MA, Sreedharan C. 1986. Agroforestry farming systems in the homesteads of Kerala, Southern India. *Agroforestry Systems* 4: 339-363.

Nair CK. 1988. Phosphatic fertilizers for small cardamom.*Proceedings of the seminar on the use of rock phosphate in West Coast soils.* University of Agricultural Sciences, Bangalore and PPCL, p. 79.

Nybe EV, Peter KV, Mini Raj N. 2004. Integrated cropping in coconut. *Coconut Journal,* 36: 4-9.

Padmapriya S, Chezhiyan N, Sathiyamurthy VA. 2007. Effect of shade and integrated nutrient management on biochemical constituents of turmeric (*Curcuma longa* L.). *Journal of Horticultural Sciences* 2: 123-129.

Paliyal SS, Kanwar K, Sharma R, Bedi MK. 2008. Productivity and soil fertility under different nutrient applications in ginger-french bean cropping system in Western Himalayas. *Journal of Farming Systems Research and Development* 14: 20-

Raj HG. 1978. A comparison of the system of cultivation of black pepper - *Piper nigrum* L. in Malaysia and Indonesia. In: Silver Jubilee Souvenir- Pepper Research Station, Panniyur. pp.65-74 -Kerala Agricultural University, Trichur.

Ramana KV, Shiva KN, Johny AK. 2003 Production of quality planting materials of ginger. In: *National consultative meeting for improvement in productivity and utilization of ginger,* Aizawl, Mizoram, pp. 37-45.

Randhawa KS, Nandpuri KS. 1969. Response of ginger (*Zingiber officinale* Roscoe) to nitrogen, phosphate and potash fertilizers. *Journal of Research Punjab Agricultural University* 6: 782-785.

Randhawa KS, Nandpuri KS, Bajwa MS. 1973. Response of turmeric (*Curcuma longa*) to NPK fertilization. *Journal of Research Punjab Agricultural University* 10: 45-48.

Reddy MRS, Reddy PVRM. 2000. Preliminary observations on neem-cake, against rhizome fly of turmeric (*Curcuma longa*). *Insect Environment* 6: 62.

Rethinam P, Sivaraman K, Sushama PK. 1994. Nutrition of turmeric. In: Chadha, K.L. and Rethinam, P. (eds.) *Advances in Horticulture,* Vol 9 – Plantation and Spice Crops Part 1. Malhotra Publishing House, New Delhi, pp. 477-489.

Roy A, Chatterjee R, Hassan A, Mitra SK. 1992. Effect of Zn, Fe, and B on growth, yield and nutrient content in leaf of ginger. *Indian Cocoa Arecanut Spices Journal* 15: 99-101.

Roy SS, Hore JK. 2007. Influence of organic manures on growth and yield of ginger. *Journal of Plantation Crops* 35: 52-55.

Sadanandan AK, Hamza S. 2001. “Studies on heavy metal residues due to continuous application of rock phosphate for sustainable black pepper production.” In: Gupta A K (ed.), *Proceedings of the national symposium on combating pollution accumulation in ecosystem for sustainable agriculture*, Allahabad Agricultural Institute, Allahabad, India, pp. 59-65.

Sadanandan AK, Hamza S. 2002. “Long term effect of inorganic fertilizers on black pepper in laterite soil.” In: Rethinam *et al.* (eds.), *Plantation Crops Research and Development in the New Millennium*, Coconut Development Board, Kerala, pp. 336-341.

Sadanandan AK. 2000. "Agronomy and nutrition of black pepper."In: Ravindran P N (ed.). *Black Pepper*. Academic Publishers, New Delhi, India, pp. 163-223.

Sadanandan AK, Hamza S, Bhargava BS, Raghupathi HB. 2000. "Diagnosis and Recommendation Integrated System (DRIS) norms for black pepper (*Piper nigrum* L.), growing soils of South India".In: Muraleedharan N & Raj Kumar P (eds.), *Recent Advances in Plantation Crops Research,* Allied Publishers Limited, New Delhi, India, pp. 203-205.

Sadanandan AK, Hamza S. 1993. Comparative efficiency of slow release N fertilizers on transformation of nitrogen and yield response of black pepper in an oxisol. *Journal of Plantation Crops* 21 (Suppl.): 58-66.

Sadanandan AK, Hamza S. 1996. Response of four turmeric (*Curcuma longa* L) varieties to nutrients in an oxisol on yield and curcumin content. *Journal of Plantation Crops* 24 (Suppl.): 120-125.

Sadanandan AK, Hamza S. 1998a. Effect of organic manures on nutrient uptake, yield and quality of turmeric (*Curcuma longa* L.). In: Mathew, N.M. and Kuruvilla Jacob, C. (eds.) *Developments in plantation crops research.* Proceedings of plantation crops symposium XII (PLACROSYM XII), Indian Rubber Research Institute, Kottayam, pp. 175-181.

Sadanandan AK, Hamza S. 1998b. Effect of organic farming on nutrient uptake, yield and quality *of ginger* (*Zingiber officinale* R.). In: Sadanandan, A.K., Krishnamurthy, K.S., Kandiannan, K. and Korikanthimath, V.S. *Water and nutrient management for sustainable production and quality of spices*, Madikeri, Karnataka, pp. 89-94.

Sadanandan AK, Hamza S. 2006. Effect of organic farming on soil quality, nutrient uptake, yield and quality of Indian spice, In: *Abstracts of 18th World Congress of Soil Science*, Philadelphia, USA.

Sadanandan N, Iyer R. 1986. Effect of organic amendments of rhizome rot of ginger. *Indian Cocoa Arecanut Spices Journal* 9: 94-95.

Sadanandan N, Sasidharan VK. 1979. A note on the performance of ginger under graded doses of nitrogen. *Agricultural Research Journal of Kerala* 17: 103-104.

Sahu SK, Mitra GN. 1992. Influence of physicochemical properties of soil on yield of ginger and turmeric. *Fert. News* 37 (10): 59-63.

Selvarajan M, Chezhiyan N. 2001. Studies on the influence of Azospirillum and different levels of nitrogen on growth and yield of turmeric (*Curcuma longa* L.). *South Indian Horticulture* 49: 140–141.

Sheshagiri KS, Uthaiah BC. 1994. Effect of nitrogen, phosphorus and potassium levels on growth and yield of turmeric (*Curcuma longa*) in the hill zone of Karnataka. *Journal of Spices and Aromatic Crops* 3: 28-32.

Sivakumar C , Wahid P A. 1994. Effect of application of organic materials on growth and foliar nutrient contents of black pepper (*Piper nigrum* L.). *Journal of Spices and Aromatic Crops* 3 : 135-141.

Sivaraman K. 1987. Rapid multiplication of quality planting material in black pepper. *Cocoa Arecanut Spices Journal* 11 : 115-118.

Srinivasan V, Hamza S, Sadanandan AK 2005. Evaluation of composted coir pith with chemical and biofertilizers on nutrient availability, yield and quality of black pepper (*Piper nigrum* L.). *Journal of Spices and Aromatic Crops,* 14: 15-20.

Srinivasan V, Kandiannan K, Hamza S. 2013a. Efficiency of sulphate of potash (SOP) as an alternate source of potassium for black pepper (*Piper nigrum* L.). *Journal of Spices and Aromatic Crops*, 22(2): 120-126.

Srinivasan V, Dinesh R, Ankegowda SJ, Hamza S, Thankamani CK, Krishnamurthy KS. 2013b. Soil test based site specific fertilizer recommendation and its economic optimum for targeted yield of black pepper. In: Souvenir & Abstracts – SYMSAC VII: Post-Harvest Processing of Spices and Fruit Crops, Sasikumar *et al.* (Eds), ISS, ICAR-IISR, Kozhikode, Kerala. pp. 202.

Srinivasan K, Krishnakumar V, Potty SN. 1998. Evaluation of fertilizer application methods on growth and yield of small cardamom (*Elettaria cardamomum* Maton).In: Muralitharan, N. and Rajkumar, R. (eds.) *Recent Advances in Plantation Crops Research*, Allied Publishers, New Delhi, pp. 199-202.

Srinivasan V, Hamza S, Dinesh R. 2009. Critical limits of zinc in soil and plant for increased productivity of ginger (*Zingiber officinale* Rosc.) *Journal of Indian Society of Soil Science* 57: 191-195.

Srinivasan V, Thankamani CK, Hamza S, John Zachariah T, Dinesh R, Praveena R. 2017. Organic Spices Production (Black pepper, Ginger and Turmeric). ICAR-Indian Institute of Spices Research, Kozhikode, Kerala, India. 45p.

Srinivasan V, Sadanandan AK, Hamza S. 2000. Efficiency of rock phosphate sources on ginger and turmeric in an *Ustic Humitropept*. *Journal of the Indian Society of Soil Science* 48: 532-536.

Tarle SG, Jadhao BJ, Panchabhai DM, Nandre DR, Khewale AP. 2007. Effect of nitrogen levels on growth and yield of ginger varieties. *Plant Archives* 7: 305-306.

Thankamani CK, Ashokan PK, Kamalam N V. 2003. Root distribution pattern of bush – black pepper (*Piper nigrum* L) employing radiotracer technique. *J. Nucl. Agric. Biol.,* 32: 23-28.

Thankamani CK, Kandiannan K, Hamza S, Saji KV. 2017. Effect of mulches on weed suppression and yield of ginger (*Zingiber officinale* Roscoe). Scientia Horticulturae, 207: 125-130.

Thankamani CK, Mathew PA, Kandiannan K. 2004. Production of healthy black pepper rooted cuttings. *Indian J. Arecanut, Spices and Medicinal Plants*, 6(4): 135-136.

Thimmarayappa M, Shivashankar KT, Shanthaveerabhadraiah SM. 2000. Effect of organic manure and inorganic fertilizers on growth, yield attributes and yield of cardamom (*Elettaria cardamomum* Maton). *Journal of Spices and Aromatic Crops* 9: 57-59.

Thomas A, Swati B, Singh DB. 2002. Influence of different levels of nitrogen and potassium on growth and yield of turmeric (*Curcuma longa* L.). *Journal of Spices and Aromatic Crops* 11: 74-77.

Velmurugan M, Chezhiyan N, Jawaharlal M. 2008. Influence of organic manures and inorganic fertilizers on cured rhizome yield and quality of turmeric (*Curcuma longa* L.) cv. BSR-2. *International Journal of Agricultural Sciences* 4: 142-145.

Velmurugan M, Chezhiyan N, Jawaharlal M, Anand M. 2007. Micronutrient studies in turmeric.*The Asian Journal of Horticulture* 2: 291-293.

Venkatesha J, Khan MM, Farooqi AA, Sadanandan AK. 1998. Effect of major nutrients (NPK) on growth, yield and quality of turmeric (*Curcuma domestica* Val.) cultivars. In: Sadanandan, A. K., Krishnamurthy, K.S., Kandiannan, K. and Korkanthimath, V.S. (eds) *Proceedings on water and nutrient management for sustainable production and quality of spices*. Indian Society of Spices, Calicut, India, pp. 52-58.

Vishwakarma SK, Kumar A, Prakash S. 2006. The effect of micro-nutrient on the growth and yield of turmeric under different shade conditions in mango orchard. *International Journal of Agricultural Sciences* 2: 241-243.

Wahid, P. and Sitepu, D. 1987. Current status and future prospects of pepper development in Indonesia. Food and Agricultural Organisation, Regional office for Asia and Pacific, Bangkok.

Xu K, Guo YY, Wang XF. 2004. Transportation and distribution of carbon and nitrogen nutrition in ginger, *Acta horticulturae* 629: 347-353.

Yamgar VT, Kathmale DK, Belhekar PS, Patil RC, Patil P.S. 2001. Effect of different levels of nitrogen, phosphorus and potassium and split application of N on growth and yield of turmeric (*Curcuma longa*). *Indian Journal of Agronomy* 46: 372-374.

Yufdi MP. 1991. Effect of liming on the growth of different black pepper varieties in the red and yellow podzolic soil. *Industrial Crops Research Journal* 17: 31-36.

Zachariah PK. 1978 "Fertilizer management for cardamom". In: *Proceedings of the First Annual Symposium on Plantations Crops,* Kottayam, pp.141-156.

Zulkifly E 1996 Prospects of implementing organic farming on pepper. *International Pepper News Bulletin* 20 (2):24-30.